Processing Effects on Safety and Quality of Foods

Contemporary Food Engineering

Series Editor

Professor Da-Wen Sun, Director

Food Refrigeration & Computerized Food Technology
National University of Ireland, Dublin
(University College Dublin)
Dublin, Ireland
http://www.ucd.ie/sun/

Contemporary Food
Engineering Series
Da-Wen Sun, Series Editor

Processing Effects on Safety and Quality of Foods

Edited by
Enrique Ortega-Rivas

CRC Press
Taylor & Francis Group
Boca Raton London New York

CRC Press is an imprint of the
Taylor & Francis Group, an **informa** business

CRC Press
Taylor & Francis Group
6000 Broken Sound Parkway NW, Suite 300
Boca Raton, FL 33487-2742

First issued in paperback 2019

© 2010 by Taylor & Francis Group, LLC
CRC Press is an imprint of Taylor & Francis Group, an Informa business

No claim to original U.S. Government works

ISBN-13: 978-1-4200-6112-3 (hbk)
ISBN-13: 978-0-367-38511-8 (pbk)

Library of Congress Cataloging-in-Publication Data

Processing effects on safety and quality of foods / editor, Enrique Ortega-Rivas.
    p. cm. -- (Contemporary food engineering)
    Includes bibliographical references and index.
    ISBN 978-1-4200-6112-3 (alk. paper)
    1. Processed foods--Safety measures. 2. Food industry and trade--Quality control. 3. Food--Quality. I. Ortega-Rivas, Enrique. II. Title. III. Series.

    TP373.5.P76 2010
    664'.02--dc22
                                            2009003960

Visit the Taylor & Francis Web site at
http://www.taylorandfrancis.com

and the CRC Press Web site at
http://www.crcpress.com

*For Silvia, Cristina,
and Samantha*

Contents

PART I   Introduction: Processing, Safety, and Quality

PART II   Evaluating the Safety and Quality of Foods

PART III Effects of Specific Technologies on Safety and Quality

PART IV Integrative Approach: Safety and Quality in Food Plant Design

Series Preface

CONTEMPORARY FOOD ENGINEERING

Food engineering is the multidisciplinary field of applied physical sciences combined with the knowledge of product properties. Food engineers provide the technological knowledge transfer essential to the cost-effective production and commercialization of food products and services. In particular, food engineers develop and design processes and equipment in order to convert raw agricultural materials and ingredients into safe, convenient, and nutritious consumer food products. However, food engineering topics are continuously undergoing changes to meet diverse consumer demands, and the subject is being rapidly developed to reflect market needs.

In the development of food engineering, one of the many challenges is to employ modern tools and knowledge, such as computational materials science and nanotechnology, to develop new products and processes. Simultaneously, improving food quality, safety, and security remain critical issues in food engineering study. New packaging materials and techniques are being developed to provide more protection to foods, and novel preservation technologies are emerging to enhance food security and defense. Additionally, process control and automation regularly appear among the top priorities identified in food engineering. Advanced monitoring and control systems are developed to facilitate automation and flexible food manufacturing. Furthermore, energy saving and minimization of environmental problems continue to be an important food engineering issue, and significant progress is being made in waste management, the efficient utilization of energy, and the reduction of effluents and emissions in food production.

The Contemporary Food Engineering series, consisting of edited books, attempts to address some of the recent developments in food engineering. Advances in classical unit operations in engineering applied to food manufacturing are covered as well as such topics as progress in the transport and storage of liquid and solid foods; heating, chilling, and freezing of foods; mass transfer in foods; chemical and biochemical aspects of food engineering and the use of kinetic analysis; dehydration, thermal processing, nonthermal processing, extrusion, liquid food concentration, membrane processes, and applications of membranes in food processing; shelf life, electronic indicators in inventory management, and sustainable technologies in food processing; and packaging, cleaning, and sanitation. The books are aimed at professional food scientists, academics researching food engineering problems, and graduate level students.

The books' editors are leading engineers and scientists from many parts of the world. All the editors were asked to present their books to address the market need and pinpoint the cutting-edge technologies in food engineering.

Furthermore, all contributions are written by internationally renowned experts who have both academic and professional credentials. All authors have attempted to provide critical, comprehensive, and readily accessible information on the art and science of a relevant topic in each chapter, with reference lists for further information. Therefore, each book can serve as an essential reference source to students and researchers in universities and research institutions.

Da-Wen Sun, Series Editor

Preface

Food process engineering comprises a series of unit operations traditionally employed in the food industry and related fields. One major component of these operations involves heat application to provide foods free from pathogenic microorganisms, that may pose a threat to communities' health. Thermal processes have proved not only to control microbial populations, but can also affect the biochemical compositions of many foods, resulting in losses of both sensory and nutritional qualities. The last three decades have witnessed the advent and adaptation of many operations, processes, and techniques aimed at producing microbiologically safe foods, but with minimum alteration to sensory and nutritive properties. All these alternative techniques have been the subjects of detailed assessments of their capabilities in producing safe and quality processed foods.

Most of the above-mentioned operations that do not rely on heat to preserve foods have received a great deal of attention by researchers worldwide. The first step in validating any alternative food preservation technique is by comparison with approved thermal methods in terms of microbial inactivation. Once this objective is achieved, the next phase involves nutritive and sensory characterization, in order to establish whether alternative technologies are capable of maintaining most of the fresh attributes, safety, and storage stability of processed foods. A satisfactory evaluation of novel preservation technologies depends on the reliable estimation of many aspects, including microbiological, physicochemical, nutritive, and sensory characteristics. Each of these aspects involves and includes the interaction of many disciplines, such as chemistry and biochemistry, basic microbiology, materials science, and so on. Evaluation of processing effects on the safety and quality of preserved food materials is required to determine the viability of novel techniques as feasible alternatives to traditional processes in order to meet consumer demands.

Many books published in the general field of food processing and preservation technologies have been written within the context of describing engineering aspects of the design and operation of equipment and machinery conforming to specific techniques. Apparently, there are no texts in which all the different aspects involved in evaluating processing effects in food quality comprise a single volume. The student, professional, or practicing engineer dealing with processing effects in food production and distribution has to consult books on food processing, food properties, food microbiology, food chemistry, and sensory evaluation, to name just a few. A significant part of the research effort aimed at characterizing and validating alternative processing techniques makes use of deep knowledge of physical and chemical properties, food structure, processing variables, microbial kinetics, and biochemical reactions. A single volume covering all these topics would be an important addition to the literature. This book intends to fill such a gap by providing a compendium of methods and procedures used to quantify the safety and quality of preserved foods, and to underlie concepts used in such quantification in the proper design of food processing plants.

Enrique Ortega-Rivas

Series Editor

Professor Da-Wen Sun was born in Southern China and is a world authority on food engineering research and education. His main research activities include cooling, drying, and refrigeration processes and systems; quality and safety of food products; bioprocess simulation and optimization; and computer vision technology. His innovative studies on vacuum cooling of cooked meats, pizza quality inspection by computer vision, and edible films for shelf-life extension of fruits and vegetables have been widely reported in national and international media. Results of his work have been published in over 180 peer-reviewed journal papers and more than 200 conference papers.

Professor Sun received first class BSc honors and an MSc in mechanical engineering, and a PhD in chemical engineering in China before working in various universities in Europe. He became the first Chinese national to be permanently employed in an Irish university when he was appointed college lecturer at the National University of Ireland, Dublin (University College Dublin), Ireland, in 1995, and was then continuously promoted in the shortest possible time to senior lecturer, associate professor, and full professor. Dr. Sun is now professor of Food and Biosystems Engineering and director of the Food Refrigeration and Computerized Food Technology Research Group at the University College Dublin.

As a leading educator in food engineering, Dr. Sun has contributed significantly to the field of food engineering. He has trained many PhD students, who have made their own contributions to the industry and academia. He has also, on a regular basis, given lectures on advances in food engineering in academic institutions internationally and delivered keynote speeches at international conferences. As a recognized authority in food engineering, he has been conferred adjunct/visiting/consulting professorships from 10 top universities in China including Zhejiang University, Shanghai Jiaotong University, Harbin Institute of Technology, China Agricultural University, South China University of Technology, and Jiangnan University. In recognition of his significant contribution to food engineering worldwide and for his outstanding leadership in the field, the International Commission of Agricultural Engineering (CIGR) awarded him the CIGR Merit Award in 2000 and again in 2006, and the Institution of Mechanical Engineers based in the United Kingdom named him Food Engineer of the Year 2004; in 2008 he was awarded the CIGR Recognition Award in recognition of his distinguished achievements as the top one percent of agricultural engineering scientists around the world.

Dr. Sun is a fellow of the Institution of Agricultural Engineers and a fellow of Engineers Ireland. He has also received numerous awards for teaching and research excellence, including the President's Research Fellowship, and has received the President's Research Award from the University College Dublin on two occasions. He is a member of the CIGR executive board and honorary vice president of the CIGR; editor-in-chief of *Food and Bioprocess Technology—An International Journal* (Springer); series editor of *Contemporary Food Engineering* (CRC Press/ Taylor & Francis); former editor of the *Journal of Food Engineering* (Elsevier); and a editorial board member for the *Journal of Food Engineering* (Elsevier), the *Journal of Food Process Engineering* (Blackwell), *Sensing and Instrumentation for Food Quality and Safety* (Springer), and the *Czech Journal of Food Sciences*. He is also a chartered engineer.

Editor

Dr. Enrique Ortega-Rivas is currently a professor at the University of Chihuahua, México, and has held a visiting scientist appointment at Food Science Australia and a visiting lectureship at Monash University, Australia. He was also a Fulbright scholar, acting as adjunct associate professor at Washington State University. He has been recognized as "National Researcher," the highest recognition that the Mexican government confers on academics for their research efforts and endeavors. He has taught food process engineering in Chihuahua, Washington State, and Monash Universities, and has also taught unit operations involving particulate materials, as well as heat and mass transfer operations.

His research interests include food engineering, particle technology, and solid–fluid separation techniques. His research efforts have focused on biological applications of hydrocyclones, food powder processing, and employment of pulsed electric fields to pasteurize fluid foods.

Dr. Ortega-Rivas has published numerous papers in international indexed journals, apart from chapters in books and contributions in encyclopedias. He has coauthored a book, *Food Powders: Physical Properties, Processing, and Functionality* (Springer), and is a member of the editorial boards of *Food and Bioprocess Technology: An International Journal* and *Food Engineering Reviews* (both published by Springer). He also reviews manuscripts for different international journals.

Contributors

José Miguel Aguilera
Department of Chemical and
    Bioprocess Engineering
Pontificia Universidad
    Católica de Chile
Santiago, Chile

Stella M. Alzamora
Departamento de Industrias
Universidad de Buenos Aires
Buenos Aires, Argentina

Malek Amiali
Department of Bioresource
    Engineering
McGill University
Montreal, Quebec, Canada

Noelia Betoret
Institute of Food Engineering
    for Development
Polytechnic University of Valencia
Valencia, Spain

Katherine L. Cason
Department of Food Science and
    Human Nutrition
Clemson University
Clemson, South Carolina

Cielo D. Char
Departamento de Industrias
Universidad de Buenos Aires
Buenos Aires, Argentina

Xiao Dong Chen
Department of Chemical Engineering
Monash University
Clayton, Victoria, Australia

Liana Drummond
Food Refrigeration and Computerized
    Food Technology
National University of Ireland
Dublin, Ireland

Pedro Fito
Institute of Food Engineering
    for Development
Polytechnic University of Valencia
Valencia, Spain

Pedro José Fito
Institute of Food Engineering
    for Development
Polytechnic University of Valencia
Valencia, Spain

Sandra N. Guerrero
Departamento de Industrias
Universidad de Buenos Aires
Buenos Aires, Argentina

Guillermo Hough
Instituto Superior Experimental de
    Tecnología Alimentaria
Comisión de Investigaciones
    Científicas de la Provincia
    Buenos Aires
Buenos Aires, Argentina

Russell S.J. Keast
School of Exercise and
    Nutrition Sciences
Deakin University
Burnwood, Victoria, Australia

Maria Concepción Pérez Lamela
Departamento de Química
    Analítica y Alimentaria
Universidad de Vigo
Vigo, Spain

Marc Le Maguer
Innovation and Risk
    Management–France
Paris, France

Aurelio López-Malo
Departamento de Ingeniería
    Química y de Alimentos
Universidad de las Américas–Puebla
Puebla, Mexico

Luis Mayor
Institute of Food Engineering
    for Development
Polytechnic University of Valencia
Valencia, Spain

Juliana Morales-Castro
Graduate Program in Food Science
    and Biotechnology
Durango Technological Institute
Durango, Mexico

Michael O. Ngadi
Department of Bioresource
    Engineering
McGill University
Montreal, Quebec, Canada

Sergio Nieto-Montenegro
Hispanic Workforce Management
    LLC/Alimentos y Nutrición
El Paso, Ciudad Juarez
Chihuahua, Mexico

L. Araceli Ochoa-Martínez
Graduate Program in Food Science
    and Biotechnology
Durango Technological Institute
Durango, Mexico

Enrique Ortega-Rivas
Graduate Program in Food Science
    and Technology
Autonomous University of Chihuahua
Chihuahua, Mexico

Laura Otero
Departamento de Ingeniería,
    Instituto del Frío
Consejo Superior de Investigaciones
    Científicas
Madrid, Spain

Enrique Palou
Departamento de Ingeniería
    Química y de Alimentos
Universidad de las Américas–Puebla
Puebla, Mexico

Franco Pedreschi
Departamento de Ingeniería Química y
    Bioprocesos
Pontificia Universidad Católica de Chile
Santiago, Chile

Silvia Raffellini
Departamento de Tecnología
Universidad Nacional de Luján
Buenos Aires, Argentina

Marc Regier
Fachbereich V-Lebensmitteltechnologie
    Lebensmittelverfahrenstechnik
Technische Fachhochschule Berlin
Berlin, Germany

Matthias Rother
Lebensmittelverfahrenstechnik
Universität Karlsruhe
Karlsruhe, Germany

Jocelyn Sagarnaga-Lopez
Graduate Program in Food Science
    and Technology
Autonomous University of Chihuahua
Chihuahua, Mexico

Erika Salas-Muñoz
Graduate Program in Food Science
    and Technology
Autonomous University of Chihuahua
Chihuahua, Mexico

Marleny D. Aranda Saldaña
Department of Agricultural, Food,
and Nutritional Sciences
University of Alberta
Edmonton, Alberta, Canada

Pedro D. Sanz
Departamento de Ingeniería
Instituto del Frío
Consejo Superior de Investigaciones
Científicas
Madrid, Spain

Heike P. Schuchmann
Lebensmittelverfahrenstechnik
Universität Karlsruhe
Karlsruhe, Germany

Lucía Seguí
Institute of Food Engineering
for Development
Polytechnic University of Valencia
Valencia, Spain

Eileen M. Stewart
Agriculture, Food, and Environmental
Sciences Division
Agri-Food and Biosciences Institute
Queen's University Belfast
Belfast, Ireland

Hugo O. Suarez-Martinez
Graduate Program in Food Science
and Technology
Autonomous University of Chihuahua
Chihuahua, Mexico

Da-Wen Sun
Food Refrigeration and Computerized
Food Technology
National University of Ireland
Dublin, Ireland

Romeo T. Toledo
Department of Food Science
and Technology
The University of Georgia
Athens, Georgia

J. Antonio Torres
Department of Food Science
and Technology
Oregon State University
Corvallis, Oregon

Joy G. Waite
Department of Food Science
and Technology
The Ohio State University
Columbus, Ohio

Ahmed E. Yousef
Department of Food Science
and Technology
The Ohio State University
Columbus, Ohio

LiJuan Yu
Department of Bioresource
Engineering
McGill University
Montreal, Quebec, Canada

Part I

Introduction: Processing, Safety, and Quality

1 Food Preservation Technologies

Enrique Ortega-Rivas

CONTENTS

1.1　INTRODUCTION

The food processing industry is one of the largest manufacturing industries worldwide. Undoubtedly, it possesses a global strategic importance, so its growth is critically based on future research directions guided by an integrated interdisciplinary approach to problems in food process engineering. In this sense, food process engineers all around the world are faced with a common problem of compromising between severity of the processes, mainly preservation techniques, and the quality of the final products. Normally, severity of processing and overall quality are inversely proportional, so finding such a compromise represents a difficult task. The food industry has relied on chemical engineering principles, using them in the modeling and problem solving of routine food processing operations. Applying chemical engineering principles directly in food processing operations does not represent, generally, the best option since raw materials used in the chemical industry and the food industry differ considerably in properties. The chemical industry generally employs inert raw materials whose composition is definable and their changes in processing are fairly predictable based on the kinetics of chemical reactions. Raw materials in the food industry are, on the other hand, very complex in composition, so they are not as easily characterized as inert materials, while their changes on processing are hard to describe by chemical reaction kinetics. There is, therefore, an urgent need to define whether the food industry would be capable of coping with challenges in growth and development using the somewhat traditional approach of adapting chemical engineering principles, in order to provide nutritive, safe, and premium quality food products to an increasingly demanding population.

1.2   DISCIPLINES OF FOOD ENGINEERING AND FOOD SCIENCE

There has been a persister argument on whether food engineering can be considered an independent engineering discipline, or should be taken as part of chemical engineering. While chemical engineering principles have been traditionally employed in the design and control of food processing operations in a direct manner, there is an upsurge in activities aimed at adapting such principles to food processes in order to find better practical results. Theoretical principles of chemical engineering focus on the writing of mass and energy balance equations for individual pieces of equipment to scale up them to carry out a given unit operation, or determining and controlling the performance of a given unit. The application of these principles has been suitable to the chemical processing industry, but has not been found suitable when dealing with processes in the food industry. As a consequence, the many benefits achieved by the chemical industry by applying the above-mentioned principles cannot be equally advantageous for the food industry.

The principal factor discouraging the widespread use of chemical engineering principles for food processing systems is the complex composition of food materials, which makes their properties poorly understood as compared with well-defined chemical systems. Essentially, food materials consist of complicated matrices comprising a wide variety of chemical compounds including, often, diverse macromolecules. Chemical reactions involving different materials are governed by the laws of kinetics and thermodynamics. For most inert materials, such as nonmetallic minerals, the reactants and products are in well-defined thermodynamic states, that is, the reactants can be transformed into the products and the other way round. In contrast, biological substances such as many food materials cannot be precisely defined in a thermodynamic sense, as much as the various stages that a food material undergoes as it is processed cannot be described as transitions between thermodynamic states. This is due to the fact that the changes in food materials induced by processing are inevitably irreversible. For instance, dough can be baked to make bread, but neither the dough nor the bread can be thermodynamically characterized with any degree of precision. The bread, of course, cannot be reversed back to dough. Furthermore, process control strategies extensively applied in chemical processing cannot be easily applied to food processing. For example, a simple feedback control system where the feed parameters are controlled to yield a product with desired properties would be extremely difficult to apply to a food system. It may not be possible to establish desired properties in food processing using measurable physical parameters, because the food industry often relies on sensory panel assessments to define a number of quality attributes. Also, if the desired product properties are quantifiable through physical variables, appropriate online sensors are not normally available for food processes. To complicate matters even further, food materials act as suitable substrates for microbial growth, so kinetics of microorganism reproduction (as well as eradication, of course) have to be taken into account in processing of foodstuffs. Finally, understanding the effects of process alterations on the food product is grossly inadequate to implement any form of control strategy.

The established discipline dealing with the application of fundamental principles of physics, chemistry, and mathematics, along with transport phenomena criteria, on transformation of raw materials into finished products is, undoubtedly, chemical engineering. When these mentioned raw materials are biological in origin and used for human consumption, as stated above, chemical engineering principles fail to produce satisfactory practical results in the operation of food processing plants. It is, therefore, necessary to deeply understand the effects of different degrees of alterations by processing of food materials. The detailed study of biological materials used as foodstuffs corresponds to another established discipline known generally as "food science." Both, chemical engineering and food science can be considered mature disciplines, going by the number of programs of study in higher education institutions and well-established professional organizations supporting them. It may be thus fair to consider food engineering or food process engineering as an applied engineering discipline that may equally fit within the various chemical engineering topics or as one of the major components of the food science discipline (along, mainly, with food microbiology and food chemistry/biochemistry). In any case, food engineering as such has become an established engineering discipline, when considered part of chemical engineering, or a mature applied science within the context of food science. The relevance of food engineering lies in the fact that, comparing it with either chemical engineering or food science, it is also supported by programs of study, professional associations, periodical publications, as well as classical texts and reference books.

1.3 CHEMICAL ENGINEERING PRINCIPLES INVOLVED IN FOOD PROCESSING

As stated before, a number of chemical engineering principles can be applied to processing of food materials, bearing in mind theoretical aspects of food science. Chemical engineering interest in the food industry is rapidly increasing because the quantity of processed foods is increasing worldwide. The demand for a wider variety of foods will also increase, while the complexity of the products from the food industry may include "designed" foods, functional foods, custom-oriented foods, and so on. The consumer nowadays has become cultivated in diverse topics, and food is not an exception. Consumers are mainly concerned with the safety, nutritional quality, and sensory aspects of processed foods. They are also demanding functional foods which, apart from the nutrition aspect, may contribute in raising standards of life quality. In this sense, another increasing concern has to do with possible long-term effects of preservatives and additives used by food processors, as well as potential side effects of novel and unconventional manners of processing foodstuffs. It has, therefore, become imperative for food process engineers to become knowledgeable in the different technologies of food processing along the years. This knowledge will contribute in developing expertise in different food processes in order to modify traditional techniques, or suggest new directions in technology of food processing as a way of meeting consumer demands continuously on the increase.

The vested interest in food processing by the discipline of chemical engineering may be related to some deep understanding of the main physicochemical aspects of food materials through food science principles. Such knowledge has been benefited by advances in instrumental methods of chemical analyses, sophistication in technology, widespread of information, and interaction of scientific disciplines. Current instrumental and analytical techniques have revealed structures and the intimate configuration of food materials that needed to be somewhat "guessed" in the recent past. Simulations are possible in order to predict physical properties of foods under various conditions, such as temperature and moisture levels, in terms of mass fraction of water, protein content, or fat levels. The traditional principles of chemical engineering known collectively as transport phenomena, coupled with simulation and analytical methods to predict food properties, have allowed for more appropriate application of such principles in numerous processes of the food industry.

1.4 RELEVANCE OF CHARACTERIZATION OF PROCESSING EFFECTS

In the food industry, many processes have relied on typical unit operations based on heat transfer and mass transfer phenomena. The ultimate goal in the processing of food materials has been traditionally related to the need for controlling microbial populations that may become a threat to public safety. This is the reason why heat transfer, normally coupled with mass transfer, still represents the main principle on which unit operations used for preservation of foods are based. The severity of thermal processing, normally expressed as temperature and holding time, is directly proportional to safety, but indirectly proportional to quality as a whole (Figure 1.1). Principles of chemical kinetics dictate that, in general, reaction velocity duplicates roughly with increasing temperature by 10°C. A number of biochemical reactions triggered by temperature increase in food processing will cause irreversible effects in nutritional and sensory aspects of the foods being processed. The first upsurge of alternative technologies aimed at maintaining food quality of processed foods,

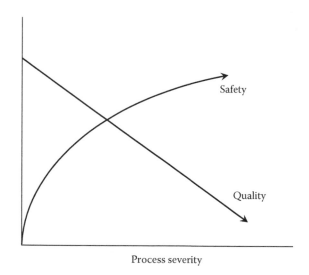

Safety

Quality

Process severity

FIGURE 1.1   Effects of increase of process severity on safety and quality.

while rendering them safe for human consumption, was based on manipulation of the temperature–time relationship. A second stage, focused on seeking a solution for the safety–quality dilemma in food processing, has been represented by a number of proposed technologies relying on an antimicrobial factor other than heat. Satisfactory evaluation of novel preservation technologies depends on reliable estimation of many aspects, including microbiological, physicochemical, nutritive, and sensory characteristics. The evaluation of processing effects on the safety and quality of preserved food materials is a task needed to determine the viability of novel techniques as feasible alternatives to traditional processes, in order to meet consumer demands.

1.5   CLASSIFICATION OF FOOD PROCESSING OPERATIONS

There are about 20 unit operations commonly employed in the chemical industry (Green and Maloney 1999, McCabe et al. 2001), while in the food industry that number could be twofold, at least (Brennan et al. 1990, Fellows 2000, Zeuthen and Bogh-Sorensen 2003). Due to the complexities of food processes discussed above, a standardized classification of unit operations for the food industry would be difficult to establish. A proposed list of unit operations as normally utilized in the food industry is given in Table 1.1. These operations may be broadly divided into the following four categories: preliminary processes, conversion operations, preservation techniques, and separation techniques. Several of the food operations listed could appear in more than one category, so the classification is not exhaustive. While some of the listed unit operations are common to either chemical or food processes (e.g., distillation or extraction), differing only in the raw materials used, some others can be considered specific to the food industry, such as puffing. This mentioned specificity lies mainly in the fact that, for some food-oriented unit operations, chemical engineering principles and concepts, or even idealized models, have not even been proposed. A number of unit operations well identified in food processing would be all those based on heat treatment. Pasteurization, sterilization, or evaporation can be considered unit operations of wide application in different industries and, as such, they rely on chemical engineering fundamentals.

When trying to substitute heat in food processing, at least partially, the so-called emerging technologies or nonthermal processes would constitute some of those unit operations in the food industry that are still at the stage of developing theoretical fundamentals. All these types of technologies have appeared in the food technology arena in the last few decades and they have shaped and modified the classifications or categorizations previously described. Many of these emerging technologies would be better described as nonpredominant, nonconventional or, simply, ambient temperature food processing technologies. While some ambient temperature food processing technologies are really traditional operations, such as mixing or comminution, some others are definitely novel as applied to food materials. Among these, some specific technologies like ultrahigh pressure and high-voltage pulsed electric fields have been extensively studied in recent times, and are considered potential methods for industrialization and wide commercialization. It is not foreseen, of course, that they may displace thermal processing technologies totally. It is more likely that they compete or complement traditional food preservation technologies, in order to

TABLE 1.1

Plausible Classification of Food Processing Operations

Unit Operation	Example
Preliminary Processes	
Classification	Sorting, grading
Depuration	Cleaning, peeling, deskinning, dehusking
Preparation	Coring, filleting, trimming, stemming
Physical Conversions	
Size reduction	Cutting, slicing, dicing, shredding, milling, grinding, pulverizing
Size enlargement	Agglomeration, granulation, instantiation, coagulation, flocculation
Shaping and reshaping	Sheeting, tabletting, flaking, puffing
Mixing and emulsification	Blending, dispersing, aerating, kneading, homogenization
Mass transfer operations	Extraction, leaching, rehydration
Others	Encapsulation, coating
Chemical Conversions	
Biochemical	Fermentations, enzymic conversions
Purely chemical	Caramelization, alkalinization, acidification (pickling)
Preservation Techniques	
Elevated temperature	Blanching, pasteurization, sterilization, canning, dehydration
Ambient temperature	Salting/brining, irradiation, high-pressure processes, high-intensity field processes
Low temperature	Chilling, freezing, freeze concentration, freeze drying
Others	Packaging, coating, waxing
Separation Techniques	
Solid–solid	Screening, leaching
Solid–gas	Cyclonic separation, air filtration, air classification
Solid–liquid	Sedimentation, centrifugation, filtration, membrane separations
Liquid–liquid	Distillation, extraction

provide consumers with fresh, safe, and nutritious food products. This book reviews and discusses all food processing technologies, whether novel or not, having a common feature of not relying on heat as the food preserving factor.

REFERENCES

Brennan, J.G., Butters, J.R., Cowell, N.D., and Lilly, A.E.V. 1990. *Food Engineering Operations*. London, U.K.: Elsevier.

Fellows, P.J. 2000. *Food Processing Technology: Principles and Practice*. Cambridge, U.K.: Woodhead Publishing Ltd.

Green, D.W. and Maloney, J.O. 1999. *Perry's Chemical Engineers' Handbook*. New York: McGraw-Hill.

McCabe, W.L., Smith, J.C., and Harriot, P. 2001. *Unit Operations in Chemical Engineering*. New York: McGraw-Hill.

Zeuthen, P. and Bogh-Sorensen, L. 2003. *Food Preservation Techniques*. Cambridge, U.K.: Woodhead Publishing Ltd.

2 Overview of Food Safety

Joy G. Waite and Ahmed E. Yousef

CONTENTS

2.1   BURDEN OF FOODBORNE ILLNESS

Over 200 different diseases are known to be transmitted via food and the incidence of foodborne illnesses is increasing worldwide (Bryan 1982, WHO/FOS 2000). In developing countries, diarrheal diseases (primarily food- and waterborne) affect more than 1.5 billion individuals annually with a death toll of 1.8 million (WHO/FOS 2000, World Health Organization 2008a). In the United States, there are an estimated 76 million cases of foodborne illness with 5,000 cases resulting in death annually (Mead et al. 1999). Beyond the direct adverse health impacts, foodborne illnesses cause huge financial burdens. The 2006 United States outbreak of *Escherichia coli* O157:H7 in spinach led to US$100 million losses for the spinach industry due to product recalls and lost sales; the spinach industry is still recovering (Porter and Lister 2006). A single outbreak ($n = 43$) of hepatitis A caused by a single infected food handler would cost US$800,000 (Koopmans et al. 2002). Estimates of annual costs of foodborne illness for the United States are between US$3 and 35 billion, annually. The United Kingdom spends £300–700 million and Australia spends AU$2.6 billion, annually on foodborne illnesses (WHO/FOS 2000).

Increasing incidences of foodborne illnesses are attributed to changes necessary to accommodate globalization of the food supply and changing consumer demographics. Globalization has triggered changes in production, processing and packaging, and distribution systems, thereby encouraging the implementation of intensive agricultural and aquacultural practices. Geographically expansive distribution systems allow for the dissemination of food products as well as the hazards that they may contain, potentially leading to large-scale outbreaks of foodborne illnesses. In developed countries, consumer demographics have changed, resulting in an increased number of consumers who are more susceptible to foodborne illnesses, particularly the elderly and the immunocompromised (WHO/FOS 2000, World Health Organization 2008a). In developing countries, many individuals still have little or no access to safe, uncontaminated water, which also contributes to the contamination of food during production and harvest as well as during final food preparation activities.

This chapter is an introduction to food safety with emphasis on microbiological and toxicological hazards in the food supply. The chapter is organized in five sections, addressing health hazards that (a) are inherent to foods (i.e., natural toxins, allergens, intolerable components), (b) emanate from environmental contaminants (i.e., persistent organic pollutants [POPs], heavy metals, radionuclides), (c) arise from microbiological activity (i.e., infectious agents, bacterial toxins, biogenic amines, mycotoxins, algal phycotoxins), (d) are related to animal production practices (i.e., antibiotic use, hormone treatment, melamine, infectious prions), and (e) result from food processing (i.e., acrylamide, chloropropanols).

2.2   HEALTH HAZARDS INHERENT TO FOODS

Food is derived from organisms during their growth or reproduction. These organisms may naturally produce compounds that are hazardous to the health of consumers. Natural toxins, allergens, and intolerable ingredients present in foods may lead to adverse health conditions. Toxins affect most individuals adversely, whereas allergens and intolerable ingredients only negatively impact sensitive subpopulations.

2.2.1 NATURAL TOXINS

Natural toxins inherent to food are secondary metabolites of plants or mushrooms that upon ingestion lead to adverse symptoms in most, if not all, humans or animals. Over 100,000 different secondary plant metabolites with toxic properties have been identified. Historically, people have utilized these compounds for their sensorial (i.e., flavor, color, aroma, hallucinogenic), medicinal, and toxic qualities. Plant toxins are particularly problematic in livestock production due to the presence of poisonous plants growing in grazing areas throughout the world. Panter (2005) estimated the economic burden of poisonous plants on the United States' livestock industry to exceed US$500 million, annually. Of particular concern in the human food supply are the pyrrolizidine alkaloids that may contaminate grains, milk, honey, and eggs. Most pyrrolizidine poisonings in the United States are due to the consumption of herbal teas or medicinal remedies made from poisonous plants (Walderhaug 1992). Cyanogenic glycoside poisonings are a particular problem in certain geographical regions, due to dependence on crops with high loads of these toxins. In western Africa, cassava is a staple food product that contains high levels of cyanogenic glycosides and must be soaked and cooked properly before it can be safely consumed. Recently, researchers are attempting to reduce the levels of cyanogenic glycosides in cassava by genetic manipulation to create new cultivars (Taylor et al. 2004). Mushrooms also produce a variety of toxins, with several being fatal if consumed. Amatoxins are responsible for the largest number of human fatalities from mushroom consumption. Large-scale outbreaks attributable to natural toxins are summarized in Table 2.1. Comprehensive discussions of natural toxins have been reported by Panter (2005) and Faulstich (2005).

2.2.2 ALLERGENS

Food allergies are adverse immune responses resulting from consumption of specific food proteins. Food allergies affect approximately 6% of children and 3.7% of adults in the United States with 85%–90% of these allergies attributed to milk, eggs, peanuts, tree nuts, soy, wheat, fish, and shellfish (Sampson 2004, Sicherer 2007). The prevalence of allergies to different types of food differs substantially between children and adults (Table 2.2). For example, there is a higher incidence of milk allergies in children and a higher incidence of shellfish allergies in adults. The incidence of food allergies is highest in children under 3 years of age with many individuals developing clinical tolerance as they age (Sampson 2004).

The major food allergens are glycoproteins that range in size between 10 and 70 kDa and are stable to heat, acid, and protease treatments (Sampson 2004). Over 170 different foods have been identified to potentiate an allergic response (Taylor 2000). Symptoms, as a result of contact with allergens, vary from minor to severe. Some patients exhibit minor inconveniences such as oral allergy syndrome, while others may experience anaphylactic shock and potentially

TABLE 2.1

Large-Scale Human Intoxications due to Consumption of Natural Toxins

Year	Geographical Location	Cases	Toxin	Source
1818	Pigeon Creek, Indiana	Unknown	Tremetol	Milk from cows that consumed *Eupatorium rugosum* (white snakeroot)
1950	Uzbekistan	>200 deaths	Pyrrolizidine alkaloids	*Trichodesma incanum*
1951	Jamaica	137 cases	Pyrrolizidine alkaloids	*Crotalaria, Senecio*
1952	Bydgosz, Poland	102 poisonings, 11 deaths	Orellanine	*Cortinarius orellanus* (mushroom)
1954	West Indies	Several hundred deaths	Pyrrolizidine alkaloids	*Crotalaria, Senecio*
1965	USSR	61 cases	Pyrrolizidine alkaloids	*Heliotropium sp.*
1965	France	135 cases, 19 deaths	Orellanine	*Cortinarius orellanus* (mushroom)
1976	Afghanistan	7200 cases	Pyrrolizidine alkaloids	*Heliotropium sp.*
1976	Central India	67 cases	Pyrrolizidine alkaloids	*Crotalaria spp.*
1981	India	67 cases	Pyrrolizidine alkaloids	*Crotalaria nana*
1987	France	26 acute renal failure	Orellanine	*Cortinarius orellanus* (mushroom)
1993	Tajikistan	3900 cases	Pyrrolizidine alkaloids	*Heliotropium*
2001	Australia	3 poisonings	Possibly Orellanine	Mushroom (possibly *Cortinarius sp.*)
2002	United States of America	1072 poisonings, 1 death	Tropane alkaloids	*Datura stramonum* (Jimsom weed)

Sources: Adapted from Bouget, J. et al., *Intensive Care Med.*, 16, 506, 1990; Flick, G.J. and Granata, L.A., *Toxins in Foods*, eds., Dabrowski, W.M. and Sikorski, Z.E., CRC Press, New York, 2005; Huxtable, R.J. and Cooper, R.A., *Natural and Selected Synthetic Toxins: Biological Implications*, eds., Tu, A.T. and Gaffield, W., American Chemical Society, Washington, DC, 2000; Mount, P. et al., *Int. Med. J.*, 32, 187, 2002; Panter, K.E., *Toxins in Food*, eds., Dabrowski, W.M. and Sikorski, Z.E., CRC Press, New York, 2005.

death (Taylor 2000). Food allergens are the most common cause of anaphylaxis outside of the hospital environment and are likely the leading cause of emergency room visits as a result of anaphylaxis (estimated at 29,000 visit per year in the United States) (Bock et al. 2001, Sicherer 2007). Food allergy-induced anaphylaxes are estimated to cause 150 deaths in the United States annually; 90% of these cases are believed to be associated with peanuts and tree nuts (Bock et al. 2001). Sampson (2004) provides a thorough description of the variety of symptoms and disorders associated with food allergies. A comprehensive list of less common food allergens with detailed descriptions of symptoms is reported by Hefle et al. (1996).

TABLE 2.2
Childhood and Adult Food Allergies Rates in the United States

Food	Specific Allergens	Young Children (%)	Adults (%)
Milk	Caseins, whey proteins, serum albumin, immunoglobulins, α-lactalbumin, β-lactoglobulin (25 distinct proteins[113])	2.5	0.3
Egg	Ovomucoid, ovalbumin, livetin (egg yolk), ovotransferrin, lysozyme	1.3	0.2
Peanut	Vicillin, storage seed proteins	0.8	0.6
Tree nuts	Albumins, storage seed proteins	0.2	0.5
Fish	Allergen M, parvalbumins	0.1	0.4
Shellfish	Tropomyosins	0.1	2.0
Soy	Globulin, oleosin		
Wheat	Gliadin, globulins, albumins, α-amylase inhibitor		
Overall		6	3.7

Sources: Adapted from Hefle, S.L., *Impact of Processing on Food Safety*, eds., Jackson, L.S., Knize, M.G., and Morgan, J.N., Kluwer Academic/Plenum Publishers, New York, 1999; Kucharska, E., *Toxins in Food*, eds., Dabrowski, W.M. and Sikorski, Z.E., CRC Press, New York, 2005; Sampson, H.A., *J. Allergy Clin. Immunol.*, 113, 805, 2004; Taylor, S.L. et al , *Food Toxicology*, eds., Helferich, W. and Winter, C.K., CRC Press, New York, 2001.

2.2.2.1 Impact of Processing on Allergens

In theory, thermal food processing could modify the tertiary and quaternary structure of allergens, which are proteins, making them less allergenic. However, most food allergens are relatively stable to thermal and protease treatments. The major allergens, including peanut proteins, generally retain allergenicity during heat treatment of food. Some of the "minor" allergens present in fruits and vegetables are susceptible to heat and microwave treatments (Hefle 1999). Heat treatments have been applied to infant formula in an attempt to reduce allergenicity. Heat treatments effectively reduce the allergenicity of whey proteins, but have no effect on caseins (Lee 1992). Protease treatments have been used to produce hypoallergenic infant formula (Lee 1992, Monti 1994, Hefle 1999). Rice and wheat have also been proteolyzed in attempts to produce additional hypoallergenic products. A cereal manufacturer has effectively reduced allergenicity of psyllium by incorporating an extensive milling process (Beyer et al. 2001).

On the contrary, food processing techniques may increase the allergenicity of certain foods (Hefle 1999). Differences in processing methods may explain the differences in peanut allergy prevalence between the United States (relatively high incidence of peanut allergy) and China (virtual nonexistence of peanut allergy) (Hill et al. 1997, Sampson 2004). Roasted peanuts and peanut butter, made thereof, are the predominant peanut products consumed in the United States. The roasting process increases the allergenicity of peanut proteins via the formation of Maillard products. An informative summary of the formation of allergenic Maillard products is provided by Davis et al. (2001). In China, peanuts are processed and prepared by boiling and frying; both processes have little to no effect on the allergenicity of peanut proteins (Beyer et al. 2001). Not all allergens respond similarly to roasting: Hansen et al. (2003) found the allergenicity of hazelnuts could be reduced by roasting. Maleki (2004) and Sathe et al. (2005) described impacts of food processing methods on the allergenicity of additional food products.

2.2.2.2   Intolerable Ingredients

Food intolerances are nonimmune-mediated reactions that cause adverse symptoms after consumption of specific foods. Particular conditions, often genetic, are believed to cause food intolerance in susceptible individuals (Sicherer 2007). However, many intolerances are poorly characterized and are thus termed idiosyncratic (Taylor et al. 2001). The best characterized intolerances are due to genetic conditions that result in defects in metabolism of specific food components. These include lactose intolerance and favism.

Lactose intolerance results from an inability to metabolize lactose due to a deficiency in the β-galactosidase enzyme. The individual is therefore incapable of digesting and absorbing lactose in the typical manner. Therefore, lactose remains in the intestinal tract and is utilized by the gastrointestinal microbiota resulting in gas production (carbon dioxide and hydrogen) that leads to symptoms of abdominal cramping, flatulence, and frothy diarrhea. The prevalence of lactose intolerance in the United States varies among ethnic and racial groups with estimates ranging between 6% and 12% of the Caucasian population and 60% and 90% of African-Americans, Hispanics, and Asian populations (Taylor et al. 2001).

Favism is an intolerance to fava beans, specifically the oxidants vicine and convicine, which leads to acute hemolytic anemia. Favism is due to a deficiency in glucose-6-phosphate dehydrogenase (G6PDH) in red blood cells. This enzyme is responsible for the recycling of antioxidants (i.e., glutathione and nicotinamide adenine dinucleotide phosphate [NADPH]) to prevent oxidative damage. G6PDH deficiency is common with approximately 100 million people affected worldwide; however, Caucasians and Native Americans are rarely deficient in this enzyme. Favism is a particular problem in the Mediterranean, the Middle East, China, and Bulgaria where dependence on fava beans as a staple crop and high rates of G6PDH deficiencies are common (Taylor et al. 2001).

The idiosyncratic reactions to foods are poorly characterized; however, a few reactions have been associated with certain food components. Sulfites are known to cause asthma in susceptible individuals and aspartame has been proven to cause urticaria (Taylor et al. 2001). The overall prevalence of sulfite intolerance is about 1%–1.5% of the population. Other food additives have been implicated with adverse reactions but the link is considered unproven. These include preservatives (butylated hydroxyanisole, butylated hydroxytoluene, and benzoates), tartrazine (FD&C Yellow No. 5), and monosodium glutamate (Taylor et al. 2001).

2.3 HEALTH HAZARDS ASSOCIATED WITH ENVIRONMENTAL CONTAMINANTS

Food products originate as crops or livestock produced in different geographical locations and climates throughout the world. During the growth of these commodities, they may be exposed to hazardous environmental contaminants present in the local region. POPs, heavy metals, and radionuclides are potential contaminants of concern in the food supply.

2.3.1 PERSISTENT ORGANIC POLLUTANTS

POPs are a class of organic compounds that are produced industrially for use as insecticides and herbicides, as additives in wood and electrical products, and as plasticizers in a variety of products (United States Environmental Protection Agency 2007). Additionally, POPs may be produced as incidental by-products of incineration. These compounds are known for their toxicity, persistence in the environment, and bioaccumulation in the food chain. Long-term exposure to POPs, even at low doses, has been linked to cancers, reproductive disorders, immune system dysfunction, and nerve damage (Schafer and Kegley 2002, United States Environmental Protection Agency 2007). POPs have been found in measurable levels in a variety of animals with blood and breast milk becoming contaminated. These compounds are stable in air and water currents leading to transfer to and accumulation in the northern latitudes, thus global control of the production and use of these compounds is crucial to minimizing risk (Schafer and Kegley 2002).

In May 2004, the Stockholm Convention on Persistent Organic Pollutants Treaty was ratified with signing countries agreeing to commit to the reduction or elimination of 12 POPs with the highest global concern (Table 2.3). Since the ratification, many other countries have also signed this treaty making minimization of POPs a worldwide priority (Anonymous 2008d). The list of POPs includes eight organochlorine pesticides and four industrial chemicals/by-products (Anonymous 2001b). The chemical structures of the 12 POPs included in the treaty are shown in Figure 2.1.

The health risks associated with the 12 POPs are summarized by Schafer (2001). The most current data on POP residues in unprocessed and processed food products may be found in the annual reports of the United States Department of Agriculture's (USDA) Pesticide Data Program and the United States Food and Drug Administration's (U.S. FDA) Total Diet Study, respectively (United States Department of Health and Human Services Food and Drug Administration 2000, United States Department of Agriculture 2007). The most commonly identified POPs in the United States food supply are dichloro-diphenyl-trichloroethane (DDT) and its metabolites and dieldrin (Schafer and Kegley 2002). Dioxin monitoring is neither included in the USDA's Pesticide Data Program, nor in the U.S. FDA's Total Diet Study (United States Department of Health and Human Services Food and Drug Administration 2000, United States Department of Agriculture 2007). Pesticide residues are most often found in fruits, grains and grain products, and vegetables; however, violative residue levels are only found in about 1% of domestic foods in the United States. Imported products have a

TABLE 2.3

United States Regulatory Limits or Recommendations of POPs in Drinking Water and Foodstuffs

POPs	2007 CERCLA[a]: Priority List of Hazardous Substances Rank	U.S. EPA Carcinogen Classification	U.S. EPA Reference Dose	U.S. EPA Maximum Contaminant Level, Drinking Water	U.S. FDA Action Levels and Recommendations
Aldrin	24	Class B2[b]	0.00003 mg/kg/day	0.001 mg/L	0.02–0.1 ppm
Chlordane	20	Class B2 > 20 year latency	0.0005 mg/kg/day	0.002 mg/L	<300 ppb in fruits and vegetables <100 ppb in animal fat and fish
Dieldrin	17	Class B2: No data on latency	0.00005 mg/kg/day	0.002 mg/L	0.02–0.1 ppm
Endrin	41	Class D[c]	0.0003 mg/kg/day	0.002 mg/L	None
DDT (and metabolites)	12, 25, 50, 151	Class B2	0.0005 mg/kg/day	None	0.01–5 ppm depending on food
Heptachlor	34	Class B2	0.0005 mg/kg/day	0.0004 mg/L	0.01–0.02 ppm in raw food crops 0.1 ppm in milk fat 0.3 ppm in seafood
Hexachlorobenzene	93	Class B2	0.0008 mg/kg/day	0.001 mg/L	None
Mirex	N/A	Not listed	0.0002 mg/kg/day	None	<100 ppt in fish
Polychlorinated biphenyls (PCBs) (including Aroclors)	5, 13, 14, 27, 29, 30, 48, 51, 68, 81, 169	Class B2: 5–20 year latency	0.02–0.07 µg/kg/day	0.0005 mg/L	0.2–3 ppm in infant foods, eggs, milk and other dairy products, fish and shellfish, poultry, and red meat
Dioxins	73, 153, 154, 202, 204, 263	Listed individually	Not determined	0.00003 µg/L	<50 ppt in fish and shellfish
Furans	162, 166, 168, 181, 192, 207, 236, 273	Listed individually	Not determined	None	None
Toxaphene	31	Class B2	Not determined	0.005 mg/L	None

Sources: Adapted from United States Department of Health and Human Services Agency for Toxic Substances and Disease Registry 2007, ToxFAQs™. Available online at http://www.atsdr.cdc. gov/toxfaq.html; United States Department of Health and Human Services Food and Drug Administration 2000, Action levels for poisonous of deleterious substances in human food and animal feed. Available online at http://www.cfsan.fda.gov/~lrd/fdaact.html#afla; Ostrowski, S.R. et al., *Toxicol. Ind. Health*, 15, 602, 1999.

[a] CERCLA: Comprehensive Environmental Response, Compensation, and Liability Act.
[b] Probable human carcinogen.
[c] Not classified as a carcinogen.

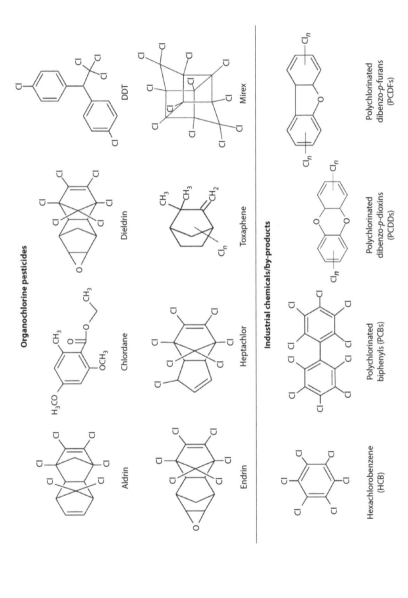

FIGURE 2.1 Chemical structures of POPs targeted for elimination or reduction as agreed to by the Stockholm Convention on POPs. (Adapted from Anonymous, 2001b, Stockholm Convention on Persistent Organic Pollutants. Available online at http://www.pops.int/documents/convtext/convtext_en.pdf.)

higher percentage (4.8%) of violative samples with vegetables and fruits being the predominant violators (Winter 2005).

Worldwide, it is estimated that three million people suffer from pesticide poisoning annually. A majority of the morbidity and mortality of pesticide poisoning occurs in agriculturally intensive developing nations. Approximately 22,000 deaths in India annually are attributable to pesticide poisoning (Broughton 2005). In 1958, 360 people were exposed and 100 died in Kerala, India, following consumption of ethyl parathion-contaminated wheat flour. In the late 1960s, 12 people were sickened due to consumption of aldrin- and hexachlorocyclohexane-contaminated wheat in Madhya Pradesh, India. Symptoms included myoclonic jerks, convulsions, and weakness of limbs. Grand mal seizures were observed in eight individuals following consumption of hexachlorocyclohexane-contaminated wheat in Uttar Pradesh, India. The worst industrial pesticide contamination incident occurred in Bhopal, India in 1984 where 250,000 people were exposed to methylisocyanate vapor, resulting in 8,000 deaths (Gupta 2004).

Between October 1980 and December 1982, the milk supply on Oahu, Hawaii contained high levels of heptachlor (up to 1.2 ppm) (Maskarinec 2005). The milk was contaminated via feeding of dairy cattle with pineapple plant foliage treated with heptachlor. The general population on Oahu was exposed to heptachlor via contaminated milk for 27–29 months. Despite this exposure, there was not an increased risk for birth defects (le Marchand et al. 1986). This population was again analyzed for increased risk of cancer and mortality 20 years after the exposure. There was a slight increase in the rate of cancer between the ages of 15 and 19 for those exposed to the heptachlor as infants; however, there is not a statistically significant increase in cancer rates for the exposed population. At the time of the follow-up study, the case patients were 25–27 years old and carcinogenic effects of heptachlor exposure may take longer for symptoms to develop (Maskarinec 2005).

2.3.2 HEAVY METALS

"Heavy metals" refer to metals, or metalloids, having a specific density greater than $5 \, g/cm^3$. Heavy metals are present in the environment at relatively low levels, nevertheless they are purposefully concentrated for industrial production of building materials, pigments, and pipes (Järup 2003). Heavy metals are used extensively in specific industrial applications (e.g., gold mining, wood preservation, and as fuel additives), thus intensive production and increasing emission have led to increased environmental contamination (Järup 2003). Heavy metal emissions from processing facilities enter the environment and contaminate air, surface waters, and soil which can lead to the contamination of drinking water and food crops.

The major heavy metals of concern in the food supply are lead, cadmium, mercury, and arsenic (a metalloid). Arsenic, lead, and mercury top the 2007 United States priority list of hazardous substances, with cadmium at number seven (United States Department of Health and Human Services Agency for Toxic Substances and Disease Registry 2008). The vicinity of farms to heavy traffic, smelting, electricity plants, and other sources of pollution impacts heavy metal concentrations in food products (Blanusa 1994). Due to the variability in concentrations of heavy metals in foods and differences in dietary habits, average human exposure is difficult to determine (Järup 2003). Direct health impacts due to exposure to heavy metals are shown in Table 2.4. The ability of the body to absorb and excrete heavy metals changes with age, making certain age populations at greater risk for adverse symptoms. For example, adults absorb 10%–15% of the lead present in their diet, whereas children absorb 50%. Additionally, the blood–brain barrier of children and adolescents is more permeable and lead- and mercury-containing complexes may transverse this barrier leading to disease (Järup 2003). Methyl-mercury poisoning is of particular concern due to the damage to fetus in pregnant women (Hamada et al. 1997). While occupational exposure is linked to the highest incidence of acute toxicity to heavy metals, there have been large-scale acute toxic events due to widespread high-level exposure from food. Information on these food poisonings is presented in Table 2.5.

Many of these heavy metals bioaccumulate, especially in aquatic food species, making consumption of them predicative of heavy metal exposure. Fish are a significant source of mercury in the food supply due to bioaccumulation. Fish harvested in the Mediterranean Sea have higher levels of mercury due to proximity to active mercury mining sites, compared to fish harvested in the Atlantic, Pacific, and Indian Oceans. Other smaller water bodies are highly polluted by a variety of industrial processes, potentially leading to excessive exposure of the local coastal population (Blanusa 1994).

2.3.2.1 Impact of Processing on Heavy Metals

Processing can modify the concentration or availability of heavy metals in food. Atta et al. (1997) found that baking and steam-blanching of tilapia that is contaminated with heavy metals (cadmium, lead, copper, and zinc) could moderately reduce the levels of heavy metals in fish flesh. Ersoy et al. (2006) found microwave treatment to increase levels of arsenic and cadmium in sea bass fillets, while significantly reducing lead levels. Frying significantly increased arsenic levels in sea bass; whereas baking reduced lead levels. Nickel was not detected in raw fish; however, it was detected following thermal treatments.

The U.S. FDA has set action levels for the crustaceans at 76 ppm arsenic, 3 ppm cadmium, 1.5 ppm lead, 12 ppm chromium, and 70 ppm nickel. For clams, oysters, and muscles, action levels are 86 ppm arsenic, 4 ppm cadmium, 1.7 ppm lead, 13 ppm chromium, and 80 ppm nickel. Methyl-mercury levels cannot exceed 1.0 ppm in fish (United States Department of Health and Human Services Food and Drug Administration Center for Food Safety and Applied Nutrition 2001). Chromium, selenium, tin, antimony, copper, thallium, fluoride, and zinc also pose potential health threats to consumers and contribute to foodborne illness (Bryan 1982, The American Medical Association, The American Nurses Association, Centers for Disease Control and Prevention, United States Food and Drug Administration Center for Food Safety and Applied Nutrition, and United States Department of Agriculture Food Safety and Inspection Service 2004). High-throughput detection methods for heavy metals in foods have been developed. Inductively coupled plasma–mass

TABLE 2.4

Adverse Health Effects of Heavy Metals

Heavy Metal	Sources of Exposure	Health Effects	Provisional Tolerable Weekly Intake (PTWI)	U.S. FDA Action Levels in Shellfish
Cadmium	Cigarettes and food (potatoes, cereals, shrimp, fish, and shellfish)	Acute gastroenteritis, renal, liver, testicle, and prostate disorders, anemia, hypertension, cardiovascular changes, pregnancy complications, skeletal damage (Itai–itai disease: osteomalacia and osteoporosis), lung cancer (pulmonary adenocarcinomas)	7 µg/kg/week	3 ppm in crustacean
Mercury	Food (fish and shellfish) and dental amalgam	*Inorganic mercury:* lung damage (acute), neurological and psychological symptoms (chronic), kidney damage *Organic mercury (methyl- or ethyl-mercury):* renal, pulmonary, reproductive, and cardiovascular toxicity; nervous system damage, neurotoxic effects, gingivitis, insomnia, memory loss, anorexia, paresthesias, blindness, coma, death *Fetal damage:* mental retardation, cerebral palsy, congenital Minamata disease	5 µg/kg/week (total mercury)	4 ppm in mollusks 1.0 ppm methyl-mercury in all fish
Lead	Food (plants) and air	*Acute lead poisoning:* headache, irritability, abdominal pain, nervous system dysfunction, dark blue lead sulfide line at gingiva, disturbance of hemoglobin synthesis (anemia), kidney damage, death *Lead encephalopathy:* sleeplessness and restlessness, psychosis, confusion, reduced consciousness, memory deterioration Congenital defects *Children:* behavioral problems, learning and concentration difficulties, diminished intellectual capacity	25 µg/kg/week	1.5 ppm in crustacea 1.7 ppm in mollusks
Arsenic	Food (fish and seafood) and drinking water (Bangladesh, Chile, and China)	*Inorganic arsenic:* gastrointestinal symptoms, cardiovascular and central nervous system damage, death, bone marrow depression, hemolysis, hepatomegaly, melanosis, polyneuropathy, encephalopathy, peripheral vascular disease (black foot disease) *Drinking water:* lung, bladder, and kidney cancer; skin cancer and other skin lesions (hyperkeratosis and pigmentation changes)	15 µg/kg/week (inorganic arsenic)	76 ppm in crustacea 86 ppm in mollusks

Sources: Adapted from Bakir, F. et al., *Science*, 181, 230, 1973; Järup, L., *Br. Med. Bull.*, 68, 167, 2003; Protasowicki, M. *Toxins in Foods*, eds., Dabrowski, W.M. and Sikorski, Z.E., CRC Press, New York, 2005; Zukowska. J. and Biziuk, M., *J. Food Sci.*, 73, R21, 2008.

TABLE 2.5

Acute Human Disease Outbreaks due to Consumption of Foods Contaminated with Heavy Metals

Year	Geographical Location	Compound	Food Contaminated	Cause of Contamination	Outcome
Early 1940s	New Zealand Air Force Base	Cadmium	Lemonade, coffee, wine	Cadmium in coatings of containers or freezers leached into product	62 cases
Late 1940s	Lower basin of Jintzu River, Japan	Cadmium	Rice	Cadmium contaminated irrigation water	Large epidemic of Itai–itai disease
1953–1956	Minamata City, Japan (The Minimata catastrophe)	Methyl mercury	Fish	Discharge from acetaldehyde factory	2,259 cases, 46 deaths; ~50 cases of fetal poisoning
1956	Iraq	Methyl mercury	Grain (bread)	Wheat seed treated with ethyl-mercury-*p*-toluene sulfonanilide fungicide	10,000 people poisoned, 6,000 hospitalizations, 450 dead
1960	Iraq	Methyl mercury	Grain (bread)	Wheat seed treated with ethyl-mercury-*p*-toluene sulfonanilide fungicide	1,000 cases, 370 hospitalizations
1961–1970	Canada	Methyl mercury	Fish	Discharge from chlor-alkali plants	Thousands poisoned
1963–1966	Guatemala	Methyl mercury	Grain (bread)	Wheat seed treated with methyl-mercury dicyandiamide fungicide	45 cases, 20 deaths
1964–1966	Niigata City, Japan	Methyl mercury	Fish (zoobenthos)	Discharge from acetaldehyde plant	47 cases, 7 deaths
1972	Iraq	Methyl mercury	Grain (Bread)	Imported wheat treated with methyl-mercurial fungicides	6,530 cases, 459 deaths

Source: Adapted from Jenner, G.G. and Cuningham, J.A.K., *N. Z. Med. J.*, 43, 282, 1944; Protasowicki, M., *Toxins in Foods*, eds., Dabrowski, W.M. and Sikorski, Z.E., CRC Press, New York, 2005.

spectrometry and microchip electrochemistry techniques provide improved sensitivity and capability to analyze additional metals, including antimony, chromium, and tin (Chan et al. 2006, Julshamn et al. 2007, Chailapakul et al. 2008).

2.3.3 RADIONUCLIDES

Development of nuclear weapons and nuclear energy technologies, beginning in the 1940s, led to environmental contamination with radionuclides (Whicker et al. 1999). Above ground testing of nuclear weapons and nuclear accidents have resulted in worldwide contamination of fission products (fallout); of particular importance are strontium-90 and cesium-137. Fallout affects local, intermediate, and global environments due to dispersion by air currents. Geographical regions of the world vary in concentrations of these radioactive isotopes which are known to be present at low levels in groundwater and are incorporated into plant and animal tissues, thus entering the food supply. Fallout concentrates in the temperate latitudes within the hemisphere of detonation, thus the highest levels of contamination due to weapons' testing are in the temperate areas of the northern hemisphere (Langham 1961). Exposure to strontium-90 leads to permanent accumulation in the body and is associated with increased risks of leukemia and bone cancer. Cesium-137 exposure is most often associated with reproductive defects (Langham 1961).

Due to extensive weapons' testing and power facility accidents, thousands of local geographical areas have been measurably contaminated with radionuclides (Whicker et al. 1999). Areas in the United States surrounding Los Alamos, New Mexico, and the Hanford Site, Richland, Washington, are contaminated due to extensive weapons testing in the 1940s–1960s. Bikini Atoll, Marshall Islands, was so heavily contaminated, particularly due to fallout from the nuclear weapon testing in 1954, that adjacent islands (Rongelap and Utirik Atolls) were evacuated, people on adjacent islands and a crew on a nearby Japanese fishing boat (the Fortunate Dragon) experienced radiation sickness, and a high incidence of birth defects has been noted (Alsop and Alsop 1954, Anonymous 1954). Extensive Soviet weapon testing (130 tests) occurred at and around Novaya Zemlya between 1955 and 1990 (Center for Nonproliferation Studies at the Monterey Institute of International Studies 2002). These tests led to worldwide contamination with areas in the Finnish and Swedish Lapplands being extensively contaminated leading to high levels of cesium-137 and strontium-90 in the Finnish and Swedish Lapps population due to consumption of meat from animals that feed on heavily contaminated lichen (Wikland et al. 1990, Skuterud et al. 2005). Civilian nuclear power plant accidents have also resulted in significant environmental contamination. The partial meltdown of the Three Mile Island reactor in Middletown, Pennsylvania on March 28, 1979, resulted in the exposure of two million people to $10\,\mu Sv$ of radiation. The most notable nuclear power facility accident occurred at the Chernobyl facility in Prypiat, Ukraine on April 26, 1986, resulting in a complete meltdown of Reactor No. 4. One hundred thousand people were evacuated from the local contaminated area another 300,000 were evacuated from other contaminated areas of Ukraine, Belarus, and Russia. One thousand square miles surrounding the reactor have been deemed the "Exclusion Zone" and are considered uninhabitable. This meltdown resulted in high levels of radionuclide contamination of parts of northeastern Europe, particularly Poland (Mazur and Jedrorog 1993).

High contamination of the soil with radionuclides poses a significant health risk to people who consume the food produced on these lands. Vegetation absorbs radionuclides during growth in contaminated soils and livestock that consume these will produce milk and meat with high levels of harmful isotopes. Due to their long half-lives (cesium-137: 30.2 years; strontium-90: 28.8 years), radionuclide contamination is a long-term problem (Smith and Beresford 2003). Some remediation efforts have occurred, particularly in the Marshall Islands and the Ukraine, to allow for resettlement in contaminated zones. The most effective means to reduce food supply contamination have been to remove the top layers of soil (30–40 cm) and to treat agricultural land with potassium fertilizers (Seel et al. 1995, Robison et al. 1997). Cleanup of Bikini Atoll began in 1968 and some food crops were planted. Resettlement began with a small number of families, but they were again evacuated in 1978 due to excessive cesium-137 and strontium-90 levels (Robison et al. 1997). Once evacuated in 1978, the United States Department of the Interior stated that resettlement would not be safe until at least 2013 (Anonymous 1978). Through these remediation efforts in Ukraine and evacuation of the Exclusion Zone, most food products available on the open market were below the legal admissible level of radionuclides. However, products from small farms and nonfarm products (e.g., mushrooms, berries, and game) in highly contaminated areas may contain radionuclide levels 100 times higher than acceptable (Matsko and Imanaka 2002). People have illegally returned to the Exclusion Zone and primarily consume food they produced or gathered themselves, which was highly contaminated (Smith and Beresford 2003). Fresh mushrooms from the heavily contaminated Gomel region contain up to 37,000 Bq/kg of cesium-137 (Matsko and Imanaka 2002).

Numerous studies have been completed to determine levels of radionuclides in food products in various countries, including those thought to be heavily contaminated (Strand et al. 1987, Mazur and Jedrorog 1993, Baratta 2003a). Radionuclide monitoring in food products began in 1961 in the United States (Baratta 2003b). The United States' Derived Intervention Levels (DIV) of specific radionuclides are 160 and 1200 Bq/kg for strontium-90 and cesium-(sum of 134 and 137), respectively (United States Department of Health and Human Services Food and Drug Administration 1998). In the first two years following the meltdown at Chernobyl, the United States denied the entry of 23 shipments of food with elevated levels of radionuclides. Most of these products were cheeses, mushrooms, pasta, meat from game (elk or reindeer), and spices (Cunningham et al. 1994). The Total Diet Study conducted by the U.S. FDA measures levels of radionuclides in numerous food products. According to a recent study, low levels of strontium-90 contamination (a maximum of 2.4 Bq/kg in mixed nuts) are found in a large number of foods in the United States. Honey was the only food tested to have measurable levels of cesium-137 (maximum 7 Bq/kg) (United States Department of Health and Human Services Food and Drug Administration Center for Food Safety and Applied Nutrition 2006).

2.4 HEALTH HAZARDS ARISING FROM MICROBIOLOGICAL ACTIVITY

Microorganisms are ubiquitous in nature and therefore are intimately associated with food during production, harvesting, processing, and consumption. Microorganisms are abundant in the field and water where food is produced and harvested and they are commonly associated with the finished product. Most of the associations between food product and microorganisms are not a cause for concern. However, there are many diseases resulting from the consumption of food containing pathogenic microorganisms or their toxic metabolites. Between 1998 and 2002, 2167 foodborne outbreaks of known etiology were reported in the United States. These outbreaks were attributed to bacterial infectious agents (48%), viral agents (33%), bacterial intoxications (7%), biogenic amines (5%), algal phycotoxins (4%), and parasitic infections (1%) (Anonymous 2006c).

2.4.1 INFECTIOUS AGENTS

Numerous bacteria, parasites, and viruses have been identified as causative agents of illness by infective processes. The following sections will highlight specific agents of particular interest to food safety.

2.4.1.1 Viruses

The major foodborne viruses of significance are caliciviruses and hepatitis viruses. Caliciviruses include classic caliciviruses and Norwalk-like or noroviruses.

2.4.1.1.1 Caliciviruses

Classic caliciviruses cause gastroenteritis in adults and infants; however, the incidence is minor compared to that of the Norwalk-like caliciviruses. Norwalk-like viruses (NLV) are responsible for estimated 9.2–23 million cases annually in the United States making these the number one cause of foodborne illness (Mead et al. 1999, Kingsley and Richards 2003). While the occurrence of NLV infections is high, fatal outcomes due to these infections are low (estimated at 7% of foodborne deaths). Humans are the only identified host for NLV and the disease is transmitted by food or water that became contaminated with fecal matter from infected individuals. Ready-to-eat products, such as salads, raw fruits and vegetables, and bakery products, are commonly contaminated by infected food handlers. Shellfish becomes highly contaminated when grown in fecally polluted waters due to filter feeding (Kingsley and Richards 2003).

2.4.1.1.2 Hepatitis Viruses

Hepatitis viruses A (HAV) and E (HEV) cause acute hepatitis with symptoms lasting between 1 and 3 weeks. These viruses are transmitted via the fecal–oral route, person-to-person contact, and contaminated food and water. HAV is a nonenveloped stable virus with the ability to survive in water for weeks to months. Boiling, ultraviolet light, formaldehyde, and chlorine treatments are effective methods to inactivate HAV particles. HAV is a member of the *Picornaviridae* family and is a significant concern with approximately 1.5 million cases worldwide annually; however, the fatality rate is low (0.1% in the United States). There are estimated 83,000 cases of Hepatitis A in the United States each year; however, only 5% are considered to be foodborne (Mead et al. 1999). HAV is considered to be a preventable disease in developed countries due to access to an effective vaccine. HEV is poorly characterized but is classified as a calici-like virus that causes a disease (hepatitis E) with a low-mortality rate (0.5%–3%). Mortality rate due to HEV infection is high in women in the third trimester of pregnancy (20%) (Schoub 2003). Large waterborne outbreaks (>10,000 cases) of HEV have occurred in various countries, whereas only small foodborne outbreaks have been identified. HEV is endemic in Southeast Asia, the Indian subcontinent, and Africa (Koopmans et al. 2002).

Additional viruses are likely transmitted via contamination of food products, including astroviruses and rotaviruses; however, food is estimated to be the vehicle in only 1% of cases (Mead et al. 1999). Poliovirus, avian influenza, adenoviruses, and parvoviruses may also be transmitted via food vehicles.

2.4.1.2   Bacteria

There is a wide variety of bacteria that cause foodborne infections. This discussion will focus on those considered most noteworthy based on prevalence, fatality rate, and classification as emerging pathogens. An incomprehensive list of bacteria that cause foodborne diseases is presented in Table 2.6. Estimates of rates of bacterial foodborne infection vary significantly by region; however, nontyphoid *Salmonella* and *Campylobacter* are believed to be the most prevalent in developed countries. Estimates on rates of *Yersinia* infections indicate a higher prevalence in European countries compared to the United States (Norwegian Institute of Public Health 2008, United States Department of Health and Human Services Centers for Disease Control and Prevention 2008a).

2.4.1.2.1   Nontyphoid Salmonella

Salmonellosis was first identified as a foodborne disease in the late 1800s; however, *Salmonella* is often considered an emerging pathogen due to high prevalence rates of infection throughout the world and the emergence of strains with multidrug resistance. Based on reported salmonellosis cases, the United States Center for Disease Control and Prevention reports a prevalence rate of 14.9 per 100,000 persons; however, many cases are unreported (United States Department of Health and Human Services Centers for Disease Control and Prevention 2008a). Recent estimates predict 3.84 million cases of salmonellosis in the United States annually (Hanes 2003). *Salmonella* is the leading cause of reported bacterial outbreaks in the United States with approximately 117 outbreaks annually (Anonymous 2006a). The prevalence of pathogenic serotypes associated with salmonellosis varies by geographical location. The five most prevalent serotypes in the United States in 2007 were Enteritidis, Typhimurium, Newport, I 4,[5],12:i-, and Javiana (United States Department of Health and Human Services Centers for Disease Control and Prevention 2008a). In Poland in 2001, the most prevalent were Enteritidis, Typhimurium, Hadar, Agona, and Cholerasuis (Hoszokwski and Wasyl 2002), whereas in Thailand between 1993 and 2002, Weltevreden, Enteritidis, Anatum, Derby, and 1,4,5,12:i were the most common (Bangtrakulnonth et al. 2004).

TABLE 2.6

Microbiological Causative Agents of Foodborne and Waterborne Infections

Bacterium	Disease	Year Identified as a Foodborne Pathogen	Associated Foods	Notifiable in United States?	Example Outbreaks
Bacillus anthracis	Gastrointestinal anthrax	1876	Undercooked meat	Yes	Rare
Salmonella enterica serovars Typhi and Paratyphi	Typhoid and enteric fevers	1880	High protein foods, milk, shellfish, fecal–oral contamination of prepared foods, street-vended foods	Yes	1998; Florida, USA; 16 cases; frozen mamey (Olsen et al. 2003)
Mycobacterium tuberculosis	Tuberculosis	1882	Raw milk	Yes	Rare
Vibrio cholerae	Cholera	1883	Contaminated water, raw mussels, shrimp, fish, cucumbers, frozen coconut milk, home-canned palm fruit, shellfish, street-vended food	Yes	2006; Australia; three cases; white bait (Kirk et al. 2008) 2005; Louisiana, USA; two cases; consumption of undercooked seafood posthurricanes (Anonymous 2006d)
Salmonella enterica (nontyphoid)	Salmonellosis	1885	Meat, poultry, eggs, dairy products, fresh produce, spices, condiments, chocolate, coconut, yeast, smoked fish, dried milk, raw milk, unpasteurized juice, cheese, raw fruits and vegetables	Yes	See Table 2.7

Organism	Disease	Year	Foods	Waterborne	Details
Shigella spp.	Shigellosis, Bacilliary dysentery	1888	Contaminated water, moist mixed foods (potato, shrimp, tuna, turkey, and macaroni salads), milk, beans, apple cider, poi, fresh produce	Yes	2007; Australia and Denmark; 55 cases; baby corn imported from Thailand (Kirk et al. 2008)
Brucella spp.	Brucellosis	1928	Raw milk, raw goat milk, cheese, meat	Yes	2008; Thassos, Greece; 55 cases; unpasteurized milk (Vorou et al. 2008) 2002; Andalucia, Spain; 11 cases; unpasteurized goat cheese (Mendez Martinez et al. 2003)
Escherichia coli, enteropathogenic (EPEC)	Infantile diarrhea, watery diarrhea	1940s		No	2000; Nara Prefecture, Japan; 735 cases; boxed lunch (Anonymous 2008a)
Clostridium perfringens (Types A and C)	Type A food poisoning (necrotizing enteritis—malnourished)	1945	Animal proteins, raw meat (beef and pork), poultry, fish, vegetables, meat pie, stew, gravy, cooked meat or poultry (pork, other meat, fish), dried or precooked foods	No	2007; Fukushiyama Prefecture, Japan; 558 cases; boxed lunch (Anonymous 2008a)
Escherichia coli, enteroinvasive (EIEC)	Dysentery-like	1946		No	1985; Houston, Texas, USA; 370 cases; guacamole (Gordillo et al. 1992)
Vibrio parahaemolyticus	Vibriosus	1950s	Raw or undercooked seafood, saltwater fish, mollusks, crustaceans, fish products	Yes	2007; Prefecture, Japan; 620 cases; salted squid; Miyagi (Anonymous 2008a)

(continued)

TABLE 2.6 (continued)
Microbiological Causative Agents of Foodborne and Waterborne Infections

Bacterium	Disease	Year Identified as a Foodborne Pathogen	Associated Foods	Notifiable in United States?	Example Outbreaks
					2006; 10 U.S. states; 177 cases; raw oysters (United States Department of Health and Human Services Centers for Disease Control and Prevention 2007f)
Campylobacter spp.	Campylobac-teriosis	1972	Poultry, raw milk, raw clams, raw beef liver and meat, contaminated water	No	2001; Wisconsin, USA; 75 cases; raw milk (Anonymous 2002b)
Enterobacter sakazakii	Neonatal meningitis, neonatal necrotizing enterocolitis	1961 (1980)	Infant formula	No	2001; Tennessee, USA; 10 cases; infant formula in NICU (Anonymous 2002a)
Escherichia coli, enterotoxigenic (ETEC)	Enteric disease, Traveler's diarrhea	1970s	Contaminated water, foods contaminated with fecal material	No	2000; Tokyo, Japan; 754 cases; boxed lunch (Anonymous 2008a)
Vibrio vulnificus	Vibriosus	1970s	Raw or undercooked seafood, oysters, clams	Yes	1996; Los Angeles, USA; three cases; raw oysters (Anonymous 1996)

Organism	Disease/syndrome	Year	Foods		Outbreak example
Yersinia enterocolitica		1970s	Undercooked pork, poultry, dairy products, seafood, raw milk, chocolate milk, tofu, chitterlings	No	2002: Chicago, USA; nine cases (infants); chitterlings (Anonymous 2003b)
Clostridium botulinum	Infant botulism	1976	Honey, home-canned vegetables and fruits, corn syrup	Yes	Rarely involved in outbreaks
Listeria monocytogenes	Listeriosis (stillbirths, bacteremia, meningitis)	1980s	Meat, dairy products, produce, seafood, ready-to-eat deli meats, hot dogs, unpasteurized cheese, eggs	Yes	2001: North Carolina, USA; 12 cases; homemade Mexican-style cheese (Anonymous 2001a); 1997: Italy, 1566 cases, corn salad (Datta 2003)
Escherichia coli, enterohemorrhagic (EHEC)	Hemolytic uremic syndrome (HUS), hemorrhagic colitis	1982 (O157:H7)	Undercooked hamburgers, lettuce, apples, alfalfa sprouts, raw milk, spinach, unpasteurized juice, raw fruits and vegetables, salami, contaminated water	Yes	See Table 2.8
Escherichia coli, enteroaggregative (EAEC)	Pediatric diarrhea, Traveler's diarrhea	1987		No	1993: Tajimi City, Japan; 2697 cases; school lunch (Itoh et al. 1997)

Sources: Adapted from Bryan, F.L., *Diseases Transmitted by Foods*, Centers for Disease Control, Atlanta, 1982; Entis, P., *Food Safety: Old Habits, New Perspectives*, ASM Press, Washington, DC, 2007; International Commission on Microbiological Specifications for Foods, *Microorganisms in Foods 7: Microbiological Testing in Food Safety Management*, Kluwer Academic/Plenum Publishers, New York, 2002; Meng, J. and Doyle, M.P., *Annu. Rev. Nutr.*, 17, 255, 1997; The American Medical Association, The American Nurses Association, Centers for Disease Control and Prevention, United States Food and Drug Administration Center for Food Safety and Applied Nutrition, and United States Department of Agriculture Food Safety and Inspection Service 2004, Diagnosis and management of food-borne illnesses: A primer for physicians and other health care professionals. Available online at http://www.ama-assn.org/ama/pub/category/3629.html; World Health Organization 2008b, Microbiological risks in food. Available online at http://www.who.int/foodsafety/micro/en/.

Infection with nontyphoid *Salmonella* results in typical gastroenteritis symptoms: abdominal pain, watery diarrhea, nausea, and vomiting with a low-grade fever. Typical infections are self-limiting with symptoms lasting 48 h. More severe symptoms may occur in infants and the elderly and fatality rates are low (~1%). *Salmonella* Typhimurium DT104 is of particular concern due to multidrug resistance and high virulence with fatality rates approaching 10% (Hanes 2003).

A summary of recent *Salmonella* outbreaks is shown in Table 2.7. *Salmonella* infections are most commonly associated with consumption of contaminated eggs and egg products, meat, and poultry. Eggs may become contaminated via horizontal or vertical transmission. Many livestock species naturally harbor *Salmonella* in their gastrointestinal tract with the carcass potentially becoming contaminated during slaughter.

2.4.1.2.2 Salmonella Typhi and Paratyphi

Enteric fevers (typhoid and paratyphoid fevers) are major health problems in developing countries with an annual incidence estimated at 17–20 million cases with 600,000 deaths (Hu and Kopecko 2003b). In Southeast Asia, typhoid fever is endemic (1,000 cases/100,000 population) and is considered the fifth leading cause of death (Hu and Kopecko 2003b). Typhoid fever outbreaks are common in South Africa, particularly in rural areas (Frean et al. 2003). Infection with *Salmonella* Typhi and Paratyphi results in invasion beyond the epithelial layers of the intestine leading to bacteremia and other severe complications including intestinal perforation and death. Some individuals are asymptomatic carriers and may shed these organisms in fecal material for more than 1 year, potentially infecting many others. These carriers may display long-term symptoms such as gallbladder disease and have an increased risk of hepatobiliary cancer. Humans are the only reservoir for *Salmonella* Typhi and Paratyphi and thus are the source of contamination of food and water. Outbreaks have been associated with a variety of foods contaminated by infected food handlers and crops that were contaminated with polluted water. Waterborne outbreaks are common in endemic regions (Hu and Kopecko 2003b).

2.4.1.2.3 Campylobacter spp.

Campylobacter spp. have the second highest prevalence rate of foodborne infection in the United States, causing estimated 2.5 million illnesses per year and 124 deaths (Mead et al. 1999, Hu and Kopecko 2003a). *Campylobacter jejuni* is responsible for 80%–90% of infections caused by this genus and *Campylobacter coli* responsible for 5%–10%. Infection with *Campylobacter* spp. leads to typical gastroenteritis symptoms, including abdominal cramping and diarrhea, rarely accompanied with vomiting. The illness is usually self-limiting with symptoms resolving within 6 days. *Campylobacter* spp. are also capable of causing bacteremia, particularly in the elderly and immunodeficient individuals. *Campylobacter* infections may lead to immune system dysfunction causing additional indirect sequellae, particularly the Guillain–Barré syndrome (Hu and Kopecko 2003a).

The most common sources of *Campylobacter* spp. are undercooked poultry, pork, and beef, raw milk, and contaminated drinking water. Many domesticated animals are infected and serve as the source for these pathogens. A large proportion (50%–70%) of human infections in developed countries is thought to be caused by broiler

TABLE 2.7
Recent Foodborne Salmonellosis Outbreaks

Date	Location	Cases	Serovar	Food Implicated
2008	42 U.S. states, District of Columbia, and Canada	>1000	Saintpaul	Raw tomatoes, fresh cilantro, fresh jalapeño peppers, fresh Serrano peppers, and cilantro are under investigation (United States Department of Health and Human Services Centers for Disease Control and Prevention 2008d)
2008	Denmark	≥636	Typhimurium U292	Pigs (source under investigation) (Ethelberg et al. 2008)
2008	Estonia	94 cases	Enteritidis	Chicken soup (Dontšenko et al. 2008)
2008	15 U.S. states	28 cases	Agona	Puffed cereal (United States Department of Health and Human Services Centers for Disease Control and Prevention 2008c)
2008	16 U.S. states	51 cases	Litchfield	Cantaloupe imported from Honduras (United States Department of Health and Human Services Centers for Disease Control and Prevention 2008e)
2007	47 U.S. states	628 cases	Tennessee	Peanut butter (Anonymous 2007c, United States Department of Health and Human Services Centers for Disease Control and Prevention 2007e)
2007	35 U.S. states	272 cases	I 4,[5],12:i-	Frozen pot pies (United States Department of Health and Human Services Centers for Disease Control and Prevention 2007b)
2007	20 U.S. states	65 cases	Wandsworth and Typhimurium	Puffed vegetable snack (United States Department of Health and Human Services Centers for Disease Control and Prevention 2007d)
2007	Shizuoka Prefecture, Japan	1148 cases	Enteritidis	Boxed lunch (Anonymous 2008a)
2006–2007	Illinois, USA	>85 cases	Newport	Mexican-style aged cheese from inadequately pasteurized milk (Anonymous 2008c)
2006	21 U.S. states	183 cases	Typhimurium	Tomatoes (United States Department of Health and Human Services Centers for Disease Control and Prevention 2006c)

(continued)

TABLE 2.7 (continued)
Recent Foodborne Salmonellosis Outbreaks

Date	Location	Cases	Serovar	Food Implicated
2006	10 U.S. states, Ontario, and Canada	41 cases	Oranienburg	Fruit salad (Anonymous 2007d)
2005	Fukuoka Prefecture, Japan	644 cases	Enteritidis	Cream puff (Anonymous 2008a)
2004	9 U.S. states	31 cases	Typhimurium	Ground beef (Anonymous 2006a)
2002	Australia	55 cases	Montevideo and Tennessee	Tahini imported from Egypt (Kirk et al. 2008)
2001	Australia	55 cases	Stanley and Newport	Dried peanuts imported from China (Kirk et al. 2008)
2001	Australia	23 cases	Typhimurium DT104	Helva imported from Turkey (Kirk et al. 2008)
2000	Iceland	180 cases	Typhimurium DT204b	Iceberg lettuce (Hatakka and Pakkala 2003)
1999	15 U.S. states, 2 Canadian provinces	>200 cases	Muenchen	Orange juice (Richardson and George 1999, Emergency Management Australia 2007)
1999	Aomori Prefecture, Japan	1634 cases	Oranienburg and Chester	Semidried squid snacks (Anonymous 2008a)
1999	South Australia	486 cases	Typhimurium	Orange juice (Ehling-Schulz et al. 2004)
1998	Osaka Prefecture, Japan	1371 cases	Enteritidis	Cake (Anonymous 2008a)
1996	Australia	60 cases	Mbandaka	Peanut butter (Scheil et al. 1998)
1994	41 U.S. states (predominately Minnesota)	593 cases confirmed	Enteritidis	Ice cream (Hennessy et al. 1996)

chickens due to the high rate of contamination of retail raw chicken (at least 60%) (Hu and Kopecko 2003a).

2.4.1.2.4　Vibrio cholerae

Infection with *Vibrio cholerae* serotypes O1 or O139 results in a severe form of excessive rice-watery diarrhea, vomiting, dehydration, and metabolic acidosis. Diarrhea usually subsides within 1–6 days with adequate fluid replacement. Cholera may result in death, primarily due to dehydration. Fecal contamination of water is the primary source of infection during outbreaks; however, seafood has recently been implicated as a vehicle of transmission. Fish, shellfish, crabs, oysters, and clams have been identified as the cause of cholera outbreaks in the United States and

Australia. *V. cholerae* is capable of growth in a variety of foods. However, most are not contaminated unless prepared using contaminated water (Rabbani et al. 2003). After a cholera-free period of 70 years, cholera reappeared in Africa in the 1971. Between 1991 and 1992, African countries reported 153,367 cases of cholera with 14,000 deaths. During the Rwanda crisis in 1994, 500,000 cases of cholera were reported in the refugee population in Zaire, killing 25,000 (Rabbani et al. 2003). In the early 1990s, the seventh pandemic of cholera reached Latin and South America resulting in over one million cases and 11,000 deaths (Franco et al. 2003).

2.4.1.2.5 Listeria monocytogenes

Listeria monocytogenes has the highest fatality rate of the foodborne bacterial pathogens, commonly between 30% and 40%. The incidence rate of *L. monocytogenes* is between 2 and 15 cases per million people worldwide with approximately 2500 cases in the United States annually (Mead et al. 1999, Datta 2003). Listeriosis is classified into two subgroups: neonatal and adult. Neonatal infections may exhibit early- or late-onset. Early-onset infections often lead to abortion, stillbirth, or premature delivery of the fetus. Late-onset infections typically lead to meningitis with a 10% mortality rate (Datta 2003). Adult listeriosis is almost exclusively associated with immunocompromised individuals and begins with flu-like symptoms that may develop into septicemia, meningitis, and endocarditis. Following a period of decline, rates of listeriosis in France began increasing in 2006, particularly in the elderly and immunocompromised individuals (Goulet et al. 2008). *L. monocytogenes* is ubiquitous in the environment and may contaminate a wide variety of food products. Outbreaks have been associated with fresh cheeses, dairy products, and ready-to-eat meats (Datta 2003).

2.4.1.2.6 Enterohemorrhagic Escherichia coli

Twenty-six outbreaks of enterohemorrhagic *E. coli* (EHEC) infections are reported annually in the United States resulting in approximately 290 cases of hemolytic uremic syndrome (Anonymous 2006b, 2008b). However, total illnesses caused by EHEC are estimated to be greater than 73,000 annually, with 85% being transmitted via food (Mead et al. 1999). A summary of recent EHEC disease outbreaks is shown in Table 2.8. Infection with EHEC may cause typical gastroenteritis symptoms, but can lead to severe complications including hemorrhagic colitis, hemolytic uremic syndrome (particularly in children), and thrombotic thrombocytopenic purpura. EHEC strains possess several virulence factors, including Shiga toxins, intimin, and enterohemolysin, that are responsible for the severe outcome of infection. *E. coli* O157:H7 is the predominant serotype of EHEC strains in the United States and developed countries (Ackers et al. 1998); however, other serotypes have been associated with the disease, including O26:H11 and O111:H8 (Eslava et al. 2003).

2.4.1.2.7 Brucella spp.

Brucellosis is a zoonotic disease that has almost been eradicated in many developed countries, but remains a human and animal health concern in many regions of the world. Transmission from animals to humans may be through breaks in the skin, foodborne (commonly raw milk and dairy products), and airborne in abattoirs (Robinson

TABLE 2.8
Disease Outbreaks Caused by Enterohemorrhagic *E. coli*

Date	Location	Cases	Serotype	Food Implicated
2008	7 U.S. states	49 cases	O157	Ground beef (United States Department of Health and Human Services Centers for Disease Control and Prevention 2008b)
2007	8 U.S. states	40 cases	O157	Ground beef (United States Department of Health and Human Services Centers for Disease Control and Prevention 2007c)
2007	10 U.S. states	21 cases	O157	Frozen pepperoni pizza (United States Department of Health and Human Services Centers for Disease Control and Prevention 2007a)
2007	Tokyo, Japan	467 cases	O157	Boxed meals (Anonymous 2008a)
2007	Japan	314 cases	O157	Boxed meals (Anonymous 2008a)
2007	Belgium	5 cases	O145 and O26	Ice cream (De Schrijver et al. 2008)
2006	5 U.S. states	71 cases	O157	Taco Bell (lettuce, cheddar cheese, ground beef) (United States Department of Health and Human Services Centers for Disease Control and Prevention 2006a)
2006	26 U.S. states	199 cases	O157	Spinach (Anonymous 2005, United States Department of Health and Human Services Centers for Disease Control and Prevention 2006a)
2005	Washington and Oregon	18 cases (4 HUS)	O157	Raw milk (Anonymous 2007b)
2004	Japan—military base	3 cases	O157	Ground beef (Anonymous 2005)
1996	Scotland	496 cases (18 deaths)	O157	Ground beef (The Pennington Group 1998)
1996	Sakai City, Japan	7470 cases	O157	White radish (Michino et al. 1999)
1995	Bavaria, Germany	28 HUS cases (children)	O157	Sausages (Ammon et al. 1999)
1995	Montana, USA	92 cases	O157	Leaf lettuce (Ackers et al. 1998)
1995	Southern Australia (Adelaide)	23 HUS cases (children)	O111	Mettwurst (Anonymous 1995)

2003). Fresh goat's milk was the cause of a brucellosis outbreak on a trans-Atlantic voyage of the S.S. Joshua Nicholson in 1905 (Nicoletti 2002). Unpasteurized cow's milk was identified as a vehicle of infection of *Brucella abortus* in 1928 in Rhodesia (Newsom 2007). *Brucella* spp. survive in milk during storage at room temperature for 24 h, in cream held under refrigeration for 4–6 weeks, and in butter and cheese for several months (Memish and Balkhy 2004).

Infection of humans with *Brucella* spp. results in acute fever, chills, night sweats, general fatigue, headache, weight loss, lymphadenopathy, and splenomegaly. Pregnant women are at a high risk to abort their fetus if infected early in their pregnancy (40% terminated). Infection later in the pregnancy results in spontaneous abortion in only 2% of individuals (Memish and Balkhy 2004). The incubation period is commonly 30 days but ranges from 5 days to several months (Anonymous 1994). Infections commonly lead to a significant level of residual disability, often due to the development of arthritis (25% of cases) (Khateeb et al. 1990, Pappas et al. 2006).

There are over half a million cases of brucellosis worldwide annually. Brucellosis is endemic in Mediterranean Europe, north and east Africa, the Near East, India, central Asia, Mexico, and central and South America (Robinson 2003). Syria and Mongolia have the highest reported incidences: 1603.4 and 605.9 cases per 1,000,000 people annually, respectively (Pappas et al. 2006). Rural areas of Greece with pastoral populations may have infection rates as high as 160 cases per 100,000 people annually (Bikas et al. 2003). In Saudi Arabia, 20% of the human population is seropositive for *Brucella* (Robinson 2003). African countries had begun some vaccination programs to combat economical and physical losses; however, due to struggling economies in the region, funding for animal health services has greatly declined (McDermott and Arimi 2002).

In developed countries, the mode of transmission of brucellosis infections has changed from primarily occupational to foodborne transmission. Historically, the majority of human brucellosis cases in Texas occurred in Caucasian males employed in swine production. In the early 1980s, the majority of brucellosis cases transitioned to the Hispanic population that was commonly infected following consumption of raw goat's milk (Taylor and Perdue 1989). Similar demographical changes were noted in brucellosis infections in California during the same time period (Fosgate et al. 2002). In the United States, there are estimated 1500 cases annually with 50% considered foodborne (Mead et al. 1999). Wyoming, Texas, and Hawaii have the highest rates of brucellosis infection at 1.46, 1.38, and 1.09 cases per 1,000,000 people, respectively (Pappas et al. 2006). Counties in the United States that border Mexico have a higher incidence of brucellosis infection (eight times the national rate). This increase is attributed to the increased consumption of imported, unpasteurized dairy products by the large Hispanic subpopulation in the region (Doyle and Bryan 2000).

2.4.1.3 Parasites

Parasites were responsible for 1% of reported foodborne outbreaks in the United States between 1998 and 2002 (Anonymous 2006c). However, parasites contribute largely to foodborne infections in developing countries with millions infected annually.

2.4.1.3.1 Toxoplasma gondii

Toxoplasma gondii causes 225,000 infections in the United States annually and is responsible for 375 foodborne deaths (Mead et al. 1999, United States Department of Health and Human Services Food and Drug Administration 2004). Approximately 16% of the United States human population is seropositive for *T. gondii* with approximately 0.6% of the population suffering from an acute infection annually (Mead et al. 1999, Lake et al. 2002, Jones et al. 2003). A significant proportion (50%–80%) of adults in European countries is seropositive for *T. gondii* (Smith 1991, Lake et al. 2002). Infection often results from consumption of undercooked meat (particularly pork, lamb, mutton, and game), raw milk, and contaminated fruits and vegetables. Infection with *T. gondii* results in toxoplasmosis with any of the following symptoms: enlarged lymph nodes, fatigue, fever, malaise, muscular pain, abdominal pain, headache, rash, and sore throat. Immunocompromised individuals may experience encephalitis, hepatitis, pneumonitis, myocarditis, or pericarditis (Markus 2003).

2.4.1.3.2 Entamoeba histolytica

Entamoeba histolytica is harbored by 50 million people worldwide leading to approximately 70,000 deaths annually. This organism is prevalent in the tropical and subtropical regions of the world with Mexico, western South America, western and southern Africa, areas of the Middle East, and south and Southeast Asia being of the highest risk. In endemic regions, infections are often asymptomatic; however, in symptomatic individuals, the disease is termed amebiasis with symptoms similar to dysentery. Invasive forms of the disease occur in approximately 10% of individuals and include amebic colitis, fulminant amebic colitis, and ámeboma. Liver abscesses due to infection have been reported. Fecal contamination of water and food is the source of infection (Jackson and Reddy 2003).

2.4.1.3.3 Taenia spp.

More than 50 million people worldwide are infected with *Taenia* spp., primarily *Taenia solium* (pork tapeworm) and *Taenia saginata* (beef tapeworm). *T. solium* infection results in two diseases: taeniosis and cysticercosis. Taeniosis is caused by infection with the adult tapeworm; infected individuals may remain asymptomatic or exhibit a loss of appetite, nausea, abdominal discomfort, constipation, or diarrhea. Cysticercosis is caused by infection with the larvae of *T. solium* and affects the central nervous system often leading to epilepsy; this disease results in 50,000 deaths per year worldwide. Survivors are often left with permanent brain damage. *T. solium* infections are endemic in Latin America, sub-Saharan Africa, the Indian subcontinent, and areas of Southeast Asia. Countries with predominantly Muslim inhabitants are almost free from *T. solium* infections due to the lack of pork consumption. Few cases are identified in the United States, with 90% of the cases being in Latin American immigrants (Del Brutto and Sotelo 2003).

Additional parasites contribute to foodborne disease, including the protists: *Giardia*, *Cryptosporidium*, and *Cyclospora*; and the helminthes: *Angiostrongylus cantonensis*, *Anisakis simplex*, *Diphyllobothrium* spp., and *Trichinella spiralis*. Information about each of these parasites can be found in the International Handbook of Foodborne Pathogens (Miliotis and Bier 2003). Infections due to *Giardia* and *Cryptosporidium* are reportable in the United States; however, infections are primarily attributed to recreation water (Mead et al. 1999). Likewise, *Trichinella* infections are reportable with a low incidence in the United States; these are primarily attributed to the consumption of undercooked game, particularly bear meat (Anonymous 2003a).

2.4.1.4   Impact of Processing on Infectious Agents

The majority of food microbiology research focuses on the minimization or elimination of harmful microorganisms in food products by a variety of processing technologies. Thermal, pressure, electrical, chemical, and biological treatments have been investigated to improve the safety of the food supply while maintaining the sensorial and nutritional qualities. The United States' National Advisory Committee on Microbiological Criteria for Foods has established criteria for redefining "pasteurization" for emerging food processing technologies and has summarized the work that must be completed to validate each technology (National Advisory Committee on Microbiological Criteria for Foods 2006). Ongoing research must emphasize the identification of appropriate target and surrogate microorganisms to determine and validate optimum processing conditions. Success of an emerging technology relies on proper identification of target organisms; this requires extensive knowledge about the behavior of potential microorganisms of concern during processing treatments in specific food systems. Before implementation, a new processing technology is commonly validated in the production environment using nonhazardous alternatives to infectious agents (i.e., surrogates). Recent studies have sought to identify surrogates for a variety of foodborne pathogens in specific food systems (Eblen et al. 2005, Niebuhr et al. 2008, Waite et al. 2009). Lado and Yousef (2007) provide a summary of important growth and resistance characteristics of *L. monocytogenes* that must be evaluated to determine an appropriate surrogate organism (Lado and Yousef 2007). Investigators must work with food processors to select the nonpathogenic surrogate microorganisms that do not interfere with the established pathogen testing protocols.

2.4.2   BACTERIAL TOXINS

When bacteria grow in food, they may produce toxic proteins (i.e., toxins). Ingestion of toxin-containing foods causes human diseases that are categorically described as intoxications. Intoxications due to bacterial growth in food include botulism (caused by neurotoxins of *Clostridium botulinum*), emetic syndrome (caused by enterotoxins of *Bacillus cereus*), and staphylococcal gastroenteritis (attributed to enterotoxins of *Staphylococcus aureus*). Incidence of bacterial intoxications is occasionally difficult to determine, due to the short duration of symptoms and self-limiting nature of the disease. One hundred and one outbreaks (2766 cases, 2 deaths) of *S. aureus* intoxication were reported to the United States Centers for Disease Control and Prevention between 1998 and 2002. During the same period, 37 outbreaks (571 cases) of *B. cereus* induced-disease were reported and 12 outbreaks (52 cases) of botulism (Anonymous 2006c). Hundreds of cases of foodborne botulism are reported in developed countries annually, with Poland having a relatively high incidence (an average of approximately 300 cases per year) (Franciosa et al. 2003).

2.4.2.1   Botulism Neurotoxins

During outgrowth of *C. botulinum* spores, neurotoxins are produced within the vegetative cells and are released into the food matrix with cell lysis. Consumption of food containing botulinum neurotoxins leads to botulism, a disease with symptoms that include early diarrhea, late constipation, and descending paralysis, which begins with facial symptoms, followed by limb paralysis and ends with respiratory failure and death. The mortality of botulism in the United States has been reduced from 25% to 6% due to appropriate administration of supportive care and equine antitoxin therapy. The botulinum neurotoxins are categorized as types A–G, with types A, B, and E being primarily responsible for cases of human botulism. Predominant neurotoxin category varies by geographical region (Franciosa et al. 2003). Botulism is primarily associated with improperly home-canned or home-fermented food products, with the commercial canning industry having an excellent track record against *C. botulinum*. Four cases of foodborne botulism linked to commercially canned chili sauce were reported in Texas and Indiana in July 2007. Botulinum neurotoxin A was recovered from bloated cans of chili sauce that were insufficiently processed. These were the first cases associated with commercial canning operations since 1974 (Anonymous 2007a). Botulism also is commonly associated with fermented legumes and fermented fish in China and Japan, respectively (Franciosa et al. 2003).

2.4.2.2   *Bacillus cereus* Emetic Enterotoxin

Growth of *B. cereus* to high levels (greater than 10^5 cells/g) in food products may result in the production of a toxin that upon ingestion leads to an emetic disease. The toxin is a 1.2 kDa water-soluble ionophoric peptide, known as cereulide, with consumption of as little as 30 μg/kg inducing illness (Ehling-Schulz et al. 2004). Improper holding temperatures of food are the leading causes of *B. cereus* intoxication outbreaks. The emetic disease is characterized by a very acute vomiting attack with or without diarrhea. There are estimated 27,000 cases of *B. cereus* intoxications in the United States each year (Mead et al. 1999).

2.4.2.3   Staphylococcal Enterotoxins

It is estimated that 185,000 staphylococcal gastroenteritis illnesses occur in the United States each year (Mead et al. 1999). Growth of *S. aureus* to approximately 10^6 cells/g may result in the production of staphylococcal enterotoxins. These toxins are proteins with a molecular weight ranging from 26 to 29 kDa and exhibit considerable resistance to heat and protease treatments. Staphylococcal enterotoxins are very potent; consumption of 100 ng to 1.0 μg will lead to nausea, vomiting, and abdominal cramping. *S. aureus* is commonly associated with raw animal products, including milk, and is often not a cause for concern. However, processed or prepared products, including meat and meat products, salads, cream-filled bakery products, and dairy products, may become contaminated by food handlers or environmental sources. Subsequent storage of these products at ambient temperature may lead to growth of the pathogen and toxin production, thus causing illness (Bennett and Monday 2003). Improper temperature control may lead to large-scale outbreaks, particularly in food service establishments. For example, in 2005, an outbreak of staphylococcal gastroenteritis affected 862 people who dined on grilled salmon at a restaurant in Shiga Prefecture, Japan. Large outbreaks have also been attributed to food manufacturing facilities, including an outbreak, in the year 2000, caused by staphylococcal entero-toxin A in low-fat milk, leading to 13,000 illnesses in Osaka Prefecture, Japan (Anonymous 2008a).

2.4.3 BIOGENIC AMINES

Biogenic amines are generically defined as biologically formed amines. For the purposes of this chapter, biogenic amines will refer specifically to amines produced during the storage of proteinaceous foods. Biogenic amines form as a result of two distinct processes: (a) endogenous decarboxylation by enzymes inherent to the food or (b) exogenous microbial decarboxylase enzymes released by the product-associated microbiota. Endogenous formation of biogenic amines is believed to be insignificant compared to microbially associated formation. Decarboxylation of specific amino acids leads to relatively high levels of biogenic amines, most notably the conversion of histidine to histamine. Other decarboxylations lead to high levels of putrescine (from ornithine), cadaverine (from lysine), tyramine (from tyrosine), tryptamine (from tryptophan), and β-phenylethylamine (from phenylala-nine). Symptoms of histamine poisoning (also called scombroid fish poisoning) may be gastrointestinal, circulatory, or cutaneous. Gastrointestinal symptoms include nausea, vomiting, and diarrhea. The primary circulatory symptoms are hypoten-sion and palpitations, while cutaneous symptoms include rash, urticaria, tingling, flushing, burning, or itching. Symptoms are noticeable shortly after ingestion of foods containing sufficient levels of biogenic amines and resolve within 24 h (Flick and Granata 2005).

A wide variety of bacteria are capable of histamine production, including: *Acinetobacter* spp., *Aeromonas hydrophila*, *Alteromonas putrefaciens*, *Burkhordelia cepacia*, *Carnobacterium divergens*, *Cedecea* spp., *Citrobacter freundii*, *Clostridium perfringens*, *Enterobacter* spp., *Escherichia* spp., *Hafnia alvei*, *Klebsiella* spp., *Lactobacillus* spp., *Morganella* spp., *Photobacterium* spp., *Pleisomonas shigelloides*, *Proteus* spp., *Providencia* spp., *Pseudomonas* spp., *Serratia* spp., *Shingomonas paucimobilis*, *Staphylococcus xylosus*, *Stenotrophonas maltophilia*, and *Vibrio* spp. (Flick and Granata 2005).

Biogenic amines are of particular concern in fish, particularly scombroid fish (i.e., tunas, bonito, mackerels, bluefish, and saury). Additionally, mahi-mahi, sardines, anchovies, herring, and marlin have also been associated with biogenic amine poisonings. Processing and handling of fish may impact the formation of biogenic amines during storage. In general, whole ungutted fish possesses a faster rate of cadaverine and putrescine formation than fillets from ungutted fish. Conversely, fillets from eviscerated fish are more susceptible to formation of these biogenic amines than whole eviscerated fish. Storage conditions, particularly time and temperature, impact the rates of microbial growth and decarboxylation, and histamine poisoning is most often associated with temperature abused products (Flick and Granata 2005). Biogenic amines are also found in cheese, fermented meat, fresh meat, wine and beer, fruits, vegetables, juice, cocoa beans, and other foods. Of these foods, histamine poisonings have been associated with fish and cheese (Flick and Granata 2005).

In 2006, there were two outbreaks of scombroid fish poisoning associated with tuna steaks in the United States. In November 2006, there were five cases of scombroid poisoning following consumption of tuna steaks in a restaurant in Tennessee. The affected individuals experienced a variety of symptoms including skin rash, headache, diarrhea, and abdominal cramping within 150 min of ingestion. In December 2006, there were six cases of scombroid fish poisoning due to consumption of tuna steaks at a company cafeteria in Louisiana. Affected individuals experienced flushing, rashes, heart palpitations, and diarrhea within 2 h of ingestion (Anonymous 2007e). In attempts to protect consumers from histamine poisoning, regulatory agencies in the United States have introduced guidance levels for histamine content in fresh fish. The U.S. FDA has established a guidance level for histamine of 5 mg/100 g; whereas the European Community has set a maximum average histamine content of 10 mg/100 g (Flick and Granata 2005).

2.4.4 ALGAL PHYCOTOXINS

Dinoflagellates and diatoms are algae that are ubiquitous in marine environments and normally populations of these organisms are constant and produce no adverse impact on other marine species or higher organisms. However, under uncharacterized environmental conditions, the numbers of these organisms may increase rapidly, resulting in algal blooms that may or may not produce characteristic color changes in the affected water (e.g., red tide) (Park et al. 2001, Okaichi et al. 2004). If these blooms progress to the point of causing health problems or environmental damage, they are termed harmful algal blooms (HABs). This proliferation often corresponds with large numbers of killed fish and shellfish (from thousands to millions of fish per bloom) and production of algal toxins (phycotoxins) that accumulate in seafood and, if consumed, lead to a variety of human diseases (Table 2.9). Sixty thousand cases of foodborne illness worldwide are attributed to phycotoxin consumption (Poli 2003). All of the phycotoxin-associated diseases induce gastrointestinal symptoms (diarrhea, vomiting, and nausea). Amnesic, neurotoxic, paralytic, and ciguatera fish poisoning (ASP, NSP, PSP, and CFP, respectively) induce respiratory, cardiac, and neurological symptoms. Onset of symptoms may be as short as 5 min, but is usually less than 24 h postingestion. Most of these diseases resolve within a few days without treatment. However, CFP requires intravenous mannitol treatment and tricyclic depressants, with recovery taking several months. ASP leads to severe neurological symptoms, including paresthesias, confusion, memory loss, disorientation, seizure, and coma; recovery from this disease may take years (Poli 2003). The incidence of HABs has increased and blooms caused by new species are reported in some geographical areas. This upward trend in incidence rate may be partially due to transport of the causative agent in ballast water of ships (Park et al. 2001). A detailed account of phycotoxin-associated diseases and biochemistry has been published by the Food and Agriculture Organization of the United Nations (2004).

Food containing harmful phycotoxins does not appear contaminated as these toxins are odorless and tasteless (Backer et al. 2005). The U.S. FDA and U.S. EPA have established limits on phycotoxins in seafood. Paralytic shellfish toxin cannot exceed

TABLE 2.9

Phycotoxin-Induced Human Diseases

Disease	Phycotoxin	Phycotoxin Producer	Contaminated Food Products	U.S. FDA Action Level	Geographical Locations of Prevalence	Example Outbreaks
Paralytic shellfish poisoning	Saxitoxins	Dinoflagellates: *Alexandrium, Pyrodinium, Gymnodinium*; Algae: *Aphanizomenon, Jania*	Bivalves (scallops, clams, mussels, oysters, cockles); Specific herbivorous fish and crabs	80 µg/100 g meat (clams, mussels, oysters) saxitoxin equivalent	Philippines, Argentina, Japan, Mediterranean Sea, Atlantic coast of Spain, Gulf of California, Gulf of Mexico, Portugal, northeast and west coast of USA	1987; Guatemala; 187 cases, 26 deaths; clam soup (FAO 2004)
Neurotoxic shellfish poisoning	Brevetoxins	Dinoflagellates: *Ptychodiscus (Karenia) brevis*	Bivalves (mussels, clams, scallops)	20 mouse units/100 g shellfish brevetoxin-2 equivalent	Gulf of Mexico, New Zealand	1993; New Zealand; 186 cases (FAO 2004)
Diarrheic shellfish poisoning	Okadaic acid Dinophysistoxins	Dinoflagellates: *Dinophysis* spp., *Prorocentrum lima*	Bivalves (scallops, mussels, clams, oysters)	0.2 ppm okadaic acid and diphysistoxin-1	Europe, Japan, Chile, Indian subcontinent, eastern Canada	2002; Antwerp, Belgium; 403 cases; blue mussels imported from Denmark (FAO 2004)
Amnesic shellfish poisoning	Domoic acids	Diatoms: *Pseudonitzchia* spp., *Nitzchia* spp.	Bivalves (scallops, mussels, clams, oysters); possibly some fish	20 ppm domoic acid 30 ppm domoic acid in viscera of Dungeness Crab	East and west coasts of North America, Spain	1987; Prince Edward Island; 105 cases, 3 deaths; cultured blue mussels (FAO 2004)

(continued)

TABLE 2.9 (continued)
Phycotoxin-Induced Human Diseases

Disease	Phycotoxin	Phycotoxin Producer	Contaminated Food Products	U.S. FDA Action Level	Geographical Locations of Prevalence	Example Outbreaks
Azaspiracid shellfish poisoning	Azaspiracids	Dinoflagellates: *Protoperidinium* spp.	Bivalves	N/A	Europe	2000; United Kingdom; 12–16 cases; frozen mussels harvested from Bantry Bay, Ireland (Twiner et al. 2008)
Ciguatera fish poisoning	Ciguatoxins Maitotoxin, Scaritoxin	Dinoflagellates: *Gambierdiscus toxicus*, possibly others	Large reef fish (barracuda, grouper, red snapper, amberjack)	N/A	Caribbean, subtropical north Atlantic and Pacific Oceans, Pacific islands, Central America, Eastern Africa, Madagascar	1993; Madagascar; 500 cases, 98 deaths; presumed vehicle, shark (FAO 2004)

Sources: Adapted from Backer, L.C. et al., *Toxins in Food*, eds., Dabrowski, W.M. and Sikorski, Z.E., CRC Press, New York, 2005; Okaichi, T. et al., *Red Tides*, Kluwer Academic Publishers, Boston, 2004; Park, D.L. et al., *Food Toxicology*, eds., Helferich, W. and Winter, C.K., CRC Press, New York, 2001; United States Department of Health and Human Services Food and Drug Administration 2000, Action levels for poisonous of deleterious substances in human food and animal feed. Available online at http://www.cfsan.fda.gov/~lrd/fdaact.html#afla.

0.8 ppm (saxitoxin equivalent) in any fish. Neurotoxic shellfish toxins in clams, mussels, and oysters cannot exceed 0.8 ppm (brevetoxin-2 equivalents) and amnesic shellfish toxin cannot exceed 20 ppm (as domoic acid) in any fish; the viscera of Dungeness crab cannot exceed 30 ppm domoic acid (United States Department of Health and Human Services Food and Drug Administration Center for Food Safety and Applied Nutrition 2001).

2.4.5 MYCOTOXINS

Mycotoxins are fungal secondary metabolites that cause a variety of diseases in humans and animals. Mycotoxins are divided into subclasses that differ significantly in chemical structure (Figure 2.2). Aflatoxins, fumonisins, deoxynivalenol, ochratoxin A, and zearalenone are considered the five most important agricultural mycotoxins based on their adverse effects on human and animal health (Shephard 2008). Diseases resulting from ingestion of mycotoxins range from acute (e.g., mycotoxicosis) to chronic (e.g., cancer) conditions. Exposure of humans to mycotoxins varies by geographical location due to the differences in agricultural crops, farming practices, and climate. These variables impact the type of fungi that are present in the field and on harvested products (Bryden 2007). Examples of diseases associated with ingestion of mycotoxin-contaminated commodities are listed in Table 2.10. Acute forms of mycotoxin-induced diseases have been historically associated with disasters such as wars, floods, and famines. Acute mycotoxicoses have been virtually eliminated in developed countries, but are still problematic in the developing nations (Shephard 2008). The association of mycotoxin ingestion with disease symptoms is difficult to establish due to chronic low exposure levels and changes in toxicity that occur with bioactivation following ingestion. However, many commodities including cereal grains contain low levels of mycotoxins and chronic consumption has been associated with health consequences, including cancer, immunosuppression, stunted growth and abnormal development, and premature menarche in girls. Aflatoxin B1, ochratoxin A, rubratoxin B, T-2 toxin, sterigmatocystin, and zearalenone have been shown experimentally to be teratogenic in mammal species (Bryden 2007).

Foods susceptible to mycotoxin contamination and of concern globally include groundnuts, maize, rice, sorghum, spices, and chili. Kumar et al. (2008) reviewed mycotoxin contamination in important agricultural commodities. Presence and level of mycotoxins in food products can be modulated by appropriate industrial handling at the pre- and postharvest stages. Preharvest, risk of mycotoxin production may be minimized by overall improvement of plant health (e.g., cultivar selection, plant vigor, and fungal resistance), sufficient irrigation, and responsible application of insecticides and fungicides. At the postharvest stage, humidity and temperature must be appropriately controlled to minimize fungal growth and toxin production during storage. Antifungal agents may be used when appropriate (Bryden 2007, Kumar et al. 2008). Magan and Aldred (2007) discussed methods of postharvest mycotoxin control.

Mycotoxins generally remain unaltered during food production processes (Bullerman and Bianchini 2007). However, food processes may affect the final

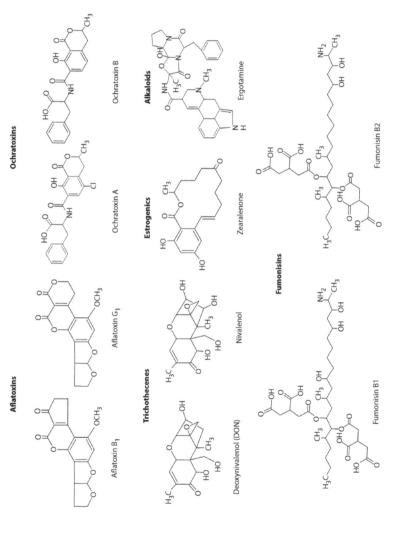

FIGURE 2.2 Classes of mycotoxins and chemical structures of specific mycotoxins of significance to public health.

TABLE 2.10

Human Diseases Associated with Ingestion of Mycotoxin-Contaminated Commodities

Disease	Commodities	Mycotoxin	Fungal Producer
Acute cardiac beriberi (Yellow Rice Disease)	Rice (polished)	Citreoviridin, islanditoxin, luteoskyrin, citrinin, rugolosin	Penicillium spp.
Aflatoxicosis (acute, chronic, and secondary forms)	Maize, rice, cottonseed meal, Brazil nuts, palm kernels, peanuts, soybeans, corn, wheat	Aflatoxins	Aspergillus flavus-ozyrae group
Akakabi-byo disease (Red Mold Toxicosis)	Maize, wheat flour, barley, oats, rice	Trichothecenes (deoxynivalenol)	Fusarium nivale, F. graminearum
Alimentary toxic aleukia	Cereal grains (toxic bread)	Trichothecenes	Fusarium spp.
Arthrinium sugarcane poisoning	Sugarcane	B-nitropropionic acid and possibly fumonisins	Arthrinium spp., Fusarium spp.
Balkan endemic nephropathy (Ochratoxicosis)	Sorghum grain, corn, cereal grains, beans	Ochratoxin A, citrinin	Aspergillus ochraceus, Penicillium spp.
Celery harvester's disease	Celery (pink rot)	Psoralens	Sclerontinia
Deoxynivalenol toxicosis	Wheat flour	Deoxynivalenol, nivalenol, acetyldeoxynivalenol, T-2 toxin	Fusarium spp.
Ergotism (necrotic)/St. Anthony's Fire	Rye meal, bread, oats	Ergot alkaloids (scherotia)	Claviceps purpurea
Ergotism (convulsive)	Wheat, bajra, pearl millet	Ergot alkaloids	Claviceps paspali
Esophageal cancer	Maize	Fumonisins	Fusarium verticilloides, F. proliferatum, F. moniliforme
Fusaritoxicosis	Wheat, maize	Deoxynivalenol	Fusarium spp.
Hyper-estrogenism (F-2 toxicosis, precocious pubertal changes)	Cereal grains	Zearalenone (F-2 toxin)	Fusarium spp.

(continued)

TABLE 2.10 (continued)
Human Diseases Associated with Ingestion of Mycotoxin-Contaminated Commodities

Disease	Commodities	Mycotoxin	Fungal Producer
Kaschin-Beck Disease	Moist grains, bread	Suspected: fusarochromanone, T-2 toxin	*Fusarium* spp.
Kodo Millet Poisoning	Kodo millet	Cytochalasins, Cyclopiazonic acid	*Phomopsis paspali, Aspergillus* spp.
Kwashiorkor	Cereal grains, plantains	Aflatoxins	*Aspergillus flavus, A. parasiticus*
Liver cancer (Hepatocarcinoma)	Cereal grains, peanuts	Aflatoxins	*Aspergillus* spp.
Neural tube defects	Maize	Fumonisins	*Fusarium verticilloides, F. proliferatum*
Onyalai (Thrombocytopenic purpura)	Millet, sorghum	Suspected: Tenuazonic acid salts and moniliformin	*Phoma sorghina, Fusarium* spp.
Oxidative DNA damage	Apples, fruits	Patulin	*Penicillium* spp., *Aspergillus* spp., *Byssochlamys* spp.
Pellagra	Maize	Fumonisins, kojic acid, trichothecenes, zearalenone	*Fusarium* spp.
Reye's syndrome	Cereal grains (grain dust), milk, peanuts	Aflatoxins	*Aspergillus*
Stachybotryotoxicosis	Cereal grains (grain dust)	Satratoxins (trichothecenes)	*Stachybotrys atra*

Sources: Adapted from Bryan, F.L., *Diseases Transmitted by Foods*, Centers for Disease Control, Atlanta, 1982; Bryden, W.L., *Asia Pac. J. Clin. Nutr.*, 16, 95, 2007; Calvo, A.M., *Toxins in Food*, eds., Dabrowski, W.M. and Sikorski, Z.E., CRC Press, New York, 2005; Weidenbörner, M., *Encyclopedia of Food Mycotoxins*, Springer, New York, 2001.

mycotoxin concentration in the end product. Sorting, trimming, cleaning, milling, brewing, cooking, baking, frying, roasting, canning, flaking, alkaline cooling, nixtamalization, and extrusion may impact mycotoxin levels (Bullerman and Bianchini 2007). Thermal processing may or may not affect the level of mycotoxins in the finished product, depending on product characteristics, temperature, humidity, etc. Bullerman and Bianchini (2007) provide a detailed review of the effects of food processing on mycotoxins. Due to the difficulty of reducing mycotoxin levels in the "finished" product, proactive steps must be taken to minimize fungal contamination and growth during production and harvest (Bryden 2007).

International regulations for mycotoxin exposure have been proposed by the Joint Expert Committee on Food Additives. These recommendations would require substantial modification of regulatory limits for aflatoxins in staple food components in developing countries to protect the population from disease (van Egmond et al. 2007).

2.5 HEALTH HAZARDS RESULTING FROM ANIMAL PRODUCTION PRACTICES

Livestock production practices have changed substantially due to the need for intensive farming practices to maximize profitability in a global food market. Some of these changes may impact food safety in a direct or indirect manner. Health hazards related to the following topics remain controversial and require further investigation to determine more accurately the risks to consumers.

2.5.1 ANTIBIOTIC RESIDUES

A number of antibiotics are used in agriculture for therapeutic, prophylactic, and growth promotion reasons. It is estimated that 8–12 million kg of antibiotics are used in livestock production in the United States annually (Katz and Ward 2005). Due to the extensive and potentially excessive use and poor adherence to necessary withdrawal times, it is likely that there are antibiotic residues present in animal products destined for human consumption. Antibiotic residues in animal products pose direct and indirect threats to human health. The consumption of antibiotic residues in animal products may lead to allergic reactions in susceptible individuals, including anaphylactic shock and death. Indirect consequences include the development of antibiotic resistant microorganisms potentially leading to human health hazards due to treatment failures (Dey and Negron 2008).

Beginning in 1982, the United States Food Safety Inspection Service (U.S. FSIS) annually conducts analyses on levels of antibiotic residues and other chemicals of concern in various animal product categories (Cordle 1988). In 2006, 3556 samples were analyzed for antibiotic residues using the 7-plate bioassay. Of these, 173 samples were considered nonviolative positives and 24 were in violation. Bob veal and non-formula-fed veal had the highest percentage of violations, which were 3.9% and 3.0%, respectively. The bob veal category has consistently had violative samples for neomycin residues every year from 2003 to 2006 (United

States Department of Agriculture Food Safety and Inspection Service 2007). Bob veal is a product from cull calves, usually bull calves from dairy herds, between a few days old up to a weight of 150 lb. These calves are often fed with medicated milk replacer that may contain high levels of neomycin and/or oxytetracycline and there are no legally required antibiotic withdrawal times for this product category (Quigley 2004). Samples from dairy cows and heavy calves were also found to violate limits for antibiotic residues. Violations were most commonly due to neomycin (14 violations) and gentamicin (8 violations) residues. Single violations were due to high levels of oxytetracycline (bob veal) and penicillin (dairy cow). Animals ($n = 3006$) were also tested for sulfonamide residues and 14 samples were in violation due to sulfamethazine residue levels and 3 due to sulfadimethoxine (United States Department of Agriculture Food Safety and Inspection Service 2007). Dey and Negron provide several suggestions to minimize or eliminate violative levels of antibiotics in food animals (Dey and Negron 2008). Additional information is available specifically for reducing residue levels in bob veal (Quigley 2004).

2.5.2 HORMONE RESIDUES

There are concerns that consumption of exogenous hormones may lead to adverse health outcomes. Animal products naturally contain estradiols and related metabolic products in varying concentrations depending on the physiological status of the animal and tissue (Collier 2000, Aksglaede et al. 2006). Hormone use to enhance growth rate for food production has been banned in Europe since 1989. Likewise, Europe does not allow the importation of hormonally treated meat (Katz and Ward 2005). However, hormones are used in animal production in the United States, Canada, New Zealand, Australia, and Argentina to enhance the growth rates, particularly of beef cattle (Aksglaede et al. 2006). Natural hormones (estradiol, progesterone, and testosterone) and synthetic hormones (trenbolone, zeranol, and melengestrol) can be used in specific classes of cattle and other livestock species to improve yield and efficiency. The Code of Federal Regulations, Title 21, Part 556, describes limits for hormone residues present in animals treated relative to untreated animals (United States Government 2008). The U.S. FSIS monitors levels of trenbolone, zeranol, and melengestrol acetate in cattle. In 2006, there were no violations for zeranol or melengestrol hormone residues; however, there were two violations (1.1%) for high levels of trenbolone in nonformula fed veal (United States Department of Agriculture Food Safety and Inspection Service 2007). An extensive review of steroid use in the animal production industry is provided by Lone (1997).

Hormone treatments are commonly administered by ingestion of pellets or, more commonly, by ear implantation (subcutaneous). With proper dosage and adherence to withdrawal times, the risk for high levels of hormone residues in the finished product is low (Katz and Ward 2005). However, several reports indicate that there are significantly higher levels of estrogen residues in the edible tissues of animals treated with hormone, compared to those that are untreated (Daxenberger et al. 2001, Partsch and Sippell 2001). Accurate quantification of hormone residues is difficult due to the large number of potentially active metabolites that may elicit estrogenic effects on the consumer. Daxenberger et al. (2001) reviewed the literature available on hormone-treated and untreated cattle and other animals to determine overall differences in hormone concentrations in different tissues. Based on predicted consumer daily intakes of animal products, consumption of untreated animal products would deliver 4.30 ng of 17-β estradiol per day, whereas consumption of treated animals would deliver 19.98 ng/day (Daxenberger et al. 2001). Consumer safety depends on strict adherence of withdrawal periods and legal uses of hormone administration for production purposes. Illegal treatment with hormones, incorrect administration protocols, or inappropriate withdrawal periods could lead to high hormone levels in the meat. There have been observations of illegal intermuscular hormone implants and a lack of removal of implants prior to slaughter, indicating a lack of withdrawal (Lone 1997, Katz and Ward 2005).

Partsch and Sippell (2001) and Aksglæde et al. (2006) provide detailed reviews on the effects of exogenous estrogens on humans, particularly children. There has been a large-scale incidence of hormone-associated illnesses; however, this outbreak could not be confirmed as foodborne. During 1977–1978, there was a large outbreak of breast enlargement and thelarche of boys and girls at a specific grade school in Milan, Italy. Fara et al. (1979) suspected that meat, possibly veal or poultry, served at the school cafeteria was contaminated with high levels of estrogen (Fara et al. 1979). Cases included 213 boys (aged 3–14) and 110 girls (aged 3–7) that exhibited early secondary sexual characteristics. Slightly elevated levels of 17-β estradiol were found in case subjects. During this time period, it was not uncommon for animals to be illegally treated with diethylstilbestrol (DES) via intramuscular or subcutaneous implantation without a withdrawal period prior to slaughter. It is speculated that the remaining, unmetabolized DES present in the animal (at the site of implantation) became homogenized in processed meat products, such as hamburger and meatballs, resulting in higher levels of hormone residues (Loizoo 2003). Twenty-two years following the exposure, there was a clinical follow-up study on the Milan cases (Sampson 2004). The men who had been exposed in the late 1970s had significantly fewer children than controls, indicating a reduction in fertility that was clinically associated with a reduced testicular volume. One of the cases had developed testicular cancer, but it is unclear whether this was related to the earlier exposure. The subjects were still young (25–36 years of age) at the time of the follow-up study; therefore potential observations of associated cancers would be unlikely, necessitating follow-up in later years (Sampson 2004).

2.5.3 MELAMINE

Melamine is a trimer of cyanic acid that has been suggested as a nonprotein nitrogen source for livestock. Compared to other feeds, melamine is not effectively utilized by ruminants. Melamine scrap has been added to animal feed to increase the perceived protein content; however, it is not approved for use in food or feeds in the United States (United States Department of Health and Human Services Food and Drug Administration 2007).

In 2007, pet foods were recalled from the U.S. markets due to high levels of white granular melamine added as an adulterant to wheat gluten by a feed supplier in China. Consumption of the adulterated pet food led to the deaths or serious illnesses of hundreds of dogs and cats due to renal failure (Swaminathan 2007). Crystalline melamine was found in the kidneys and urine of the deceased animals. Further investigation into melamine prevalence in pet food identified additional manufacturers from China and within the United States that had been adulterating feed with melamine. This product recall prompted an investigation into the prevalence of melamine in animal feeds and the potential risk to humans that consume meat from these animals. The investigation resulted in the indictment of two Chinese businesses and a United States company for their roles in the melamine adulteration scheme (United States Department of Health and Human Services Food and Drug Administration 2008). The U.S. FDA concluded that consumption of meat from animals fed melamine would pose little risk to consumers (Lang 2007).

2.5.4 INFECTIOUS PRIONS

Accumulation of abnormal isoforms of prion protein in nerve cells is associated with transmissible spongiform encephalopathies (TSEs) in animals (Takemura et al. 2004). TSEs are neurological diseases that occur in a variety of animals due to the misfolding of normal prion proteins within brain tissue leading to neurodegenerative disorders (United States Department of Health and Human Services Centers for Disease Control and Prevention 2006b). These diseases are caused by unconventional, infectious agents that induce a shift in the normal prion protein from a predominantly α-helical secondary structure to a β-sheet conformation (Belay and Schonberger 2005). TSEs have been identified in several animal species, including humans. Of particular importance to food safety are bovine spongiform encephalopathy (BSE) in cattle, scrapie in sheep, and chronic wasting disease (CWD) in deer and elk (United States Department of Health and Human Services Centers for Disease Control and Prevention 2006b).

A primary food safety concern is the relationship between BSE, also known as mad cow disease, in beef cattle and TSE in humans. Mad cow disease was first identified in cattle following a large outbreak in the United Kingdom in 1985–1986. The cattle were thought to become infected following ingestion of feed containing sheep parts (meat-and-bone meal) from animals infected with scrapie (Belay and Schonberger 2005). In 1996, a new human TSE was identified in the United Kingdom and termed variant Creutzfeldt–Jakob Disease (vCJD). This disease is characterized initially by psychiatric symptoms including anxiety and depression, followed by neurological symptoms of ataxia and involuntary movements (Vicari et al. 2003). vCJD is believed to be caused by consumption of BSE-infected cattle by-products with onset of symptoms typically greater than 7 years following consumption (United States Department of Health and Human Services Centers for Disease Control and Prevention 2007g). The disease is always fatal and no treatment options are available (Vicari et al. 2003). It has been estimated that between 870,000 and 1.6 million head of cattle infected with BSE entered the food chain in the 1980s and 1990s. Millions of consumers in the United Kingdom were exposed to BSE via consumption of contaminated by-products (Belay and Schonberger 2005).

Cattle and their by-products were exported from the United Kingdom during the BSE outbreak and BSE disease spread throughout parts of Europe and to

Israel and Japan. Single cases have been identified in other countries, including Canada and the United States. Meat is considered to be a low risk for transmission of TSEs, whereas brain, spinal cord, and small intestine are considered high risk and have been banned in feed for ruminant animals (Belay and Schonberger 2005). As of January 2007, 208 cases of vCJD in 11 countries have been reported (United States Department of Health and Human Services Centers for Disease Control and Prevention 2007g). Host genotype is believed to contribute to the disease progression of vCJD, specifically only individuals that are homozygous for methionine at the 129th codon of normal prion protein have developed vCJD (Vicari et al. 2003).

Prions are resistant to protease treatments, disinfectants, radiation, and standard sterilization protocols (Vicari et al. 2003, Belay and Schonberger 2005). Rendering processes in Europe have been modified to treat ruminant protein waste with 133°C at a pressure of 3 bar for 20 min. Treatment with 20,000 ppm available chlorine or 1–2 N sodium hydroxide at 20°C for 1 h is also effective to inactivate TSE agents (Vicari et al. 2003). In the United Kingdom, use of mammalian proteins in livestock feed is banned (Belay and Schonberger 2005).

2.6 HEALTH HAZARDS EMERGING FROM FOOD PROCESSING

Food processing methods are typically designed to enhance food safety; however, these processes may lead to the production of hazardous chemical compounds in the finished product. Polycyclic aromatic hydrocarbons (PAHs) and heterocyclic amines have been well studied for their association with health problems. Products of lipid and protein oxidation may cause health hazards ranging from decreased nutritional capacity to carcinogenic potential. Nitropolycyclic aromatic hydrocarbons have been recently identified in smoked sausages and roasted coffee beans (Sikorski 2005). The potential production and health implications of acrylamide and chloropropanols will be briefly discussed.

2.6.1 ACRYLAMIDE

Acrylamide is formed in starchy foods as a Maillard reaction product during high temperature thermal treatments (i.e., grilling, baking, or frying). Fried potato products have the highest reported levels of acrylamide and many factors can be modified to reduce acrylamide formation. Taeymans et al. (2004) and Claeys et al. (2005) discussed the factors that contribute to acrylamide formation, from cultivar selection to processing conditions, and how manufacturers may modify processes to minimize the concentration in the finished product. Surface area and frying time are known to significantly impact acrylamide formation in fried potato products (Taubert et al. 2004). Average dietary intake of acrylamide from food is estimated between 0.28 and 1.4 g/kg$_{bw}$/day. Acrylamide is a known neurotoxin as is a suspected carcinogen and reproductive toxin (low fertility, testicular toxin) (Tritscher 2004).

2.6.2 CHLOROPROPANOLS

Chlorinated propanols are produced in foods treated with high temperature combined with concentrated hydrochloric acid. The derivative, 3-monochloroproanediol, was originally identified in acid hydrolyzed vegetable protein, along with smaller concentrations of other chloropropanols. Further investigation into similarly processed products revealed the presence of chloropropanols in soya sauce (and similar sauces), modified starches, and malted cereals, including roasted barley malt (Schlatter et al. 2003). Dietary exposure to chloropropanols varies geographically, primarily due to differences in diet, but estimated consumptions range between 140 µg/person/day in the United States and 540 µg/person/day in Japan (Tritscher 2004). Toxic effects of chloropropanols, including renal tubule hyperplasia, have been revealed in animal studies (Schlatter et al. 2002, Tritscher 2004). Currently, there are no data to document the hazards associated with consumption of chloropropanols to human health.

2.7 CONCLUSION

Foodborne illnesses constitute a huge burden on individuals in the developed and developing nations of the world. Different illnesses are prevalent in different geographical regions due to environmental, economic, and agricultural variables. However, with the globalization of the food supply, many illnesses are being spread beyond historical boundaries associated with these problems. This chapter is an attempt to communicate a summary on the variety of foodborne hazards important in the local, regional, national, and global food supplies.

REFERENCES

Ackers, M.-L., Mahon, B.E., Leahy, E. et al. 1998. An outbreak of *Escherichia coli* O157:H7 infections associated with leaf lettuce consumption. *Journal of Infectious Diseases* 177: 1588–1593.

Aksglaede, L., Juul, A., Leffers, H., Skakkebæk, N.E., and Andersson, A.-M. 2006. The sensitivity of the child to sex steroids: Possible impact of exogenous estrogens. *Human Reproduction Update* 12: 341–349.

Alsop, J. and Alsop, S. 1954. The H-bomb fallout: Lessons of the "Dragon." *Time Magazine* August 23, 1954.

Ammon, A., Petersen, L.R., and Karch, H. 1999. A large outbreak of hemolytic uremic syndrome caused by an unusual sorbitol-fermenting strain of *Escherichia coli* O157:H. *Journal of Infectious Diseases* 179: 1274–1277.

Anonymous. 1954. Polite complaint. *Time Magazine* May 24, 1954.

Anonymous. 1978. Blunder on Bikini Island. *Time Magazine* April 3, 1978.

Anonymous. 1994. Brucellosis outbreak at a pork processing plant—North Carolina, 1992. *MMWR* 43: 113–116.

Anonymous. 1995. Community outbreak of hemolytic uremic syndrome attributable to *Escherichia coli* O111:NM-South Australia. *MMWR* 44: 550–551, 557–558.

Anonymous. 1996. *Vibrio vulnificus* infections associated with eating raw oysters—Los Angeles, 1996. *MMWR* 45: 621–624.

Anonymous. 2001a. Outbreak of listeriosis associated with homemade Mexican-style cheese—North Carolina, October 2000–January 2001. *MMWR* 50: 560–562.

Anonymous. 2001b. Stockholm Convention on Persistent Organic Pollutants. Available online at http://www.pops.int/documents/convtext/convtext_en.pdf.

Anonymous. 2002a. *Enterobacter sakazakii* infections associated with the use of powdered infant formula—Tennessee, 2001. *MMWR* 51: 298–300.

Anonymous. 2002b. Outbreak of *Campylobacter jejuni* infections associated with drinking unpasteurized milk procured through a cow-leasing program—Wisconsin, 2001. *MMWR* 51: 548–549.

Anonymous. 2003a. Trichinellosis surveillance—United States, 1997–2001. *MMWR* 55: 1–8.

Anonymous. 2003b. *Yersinia enterocolitica* gastroenteritis among infants exposed to chitterlings—Chicago, Illinois, 2002. *MMWR* 52: 956–958.

Anonymous. 2005. *Escherichia coli* O157:H7 infections associated with ground beef from a U.S. military installation—Okinawa, Japan, February 2004. *MMWR* 54: 40–42.

Anonymous. 2006a. Multistate outbreak of *Salmonella typhimurium* infections associated with eating ground beef—United States, 2004. *MMWR* 55: 180–182.

Anonymous. 2006b. Ongoing multistate outbreak of *Escherichia coli* serotype O157:H7 infections associated with consumption of fresh spinach—United States, September 2006. *MMWR* 55: 1045–1046.

Anonymous. 2006c. Surveillance of foodborne-disease outbreaks—United States, 1998–2002. *MMWR* 55: 1–34.

Anonymous. 2006d. Two cases of toxigenic *Vibrio cholerae* O1 infection after hurricanes Katrina and Rita—Louisiana, October 2005. *MMWR* 55: 31–32.

Anonymous. 2007a. Botulism associated with commercially canned chili sauce—Texas and Indiana, July 2007. *MMWR* 56: 767–769.

Anonymous. 2007b. *Escherichia coli* O157:H7 infection associated with drinking raw milk—Washington and Oregon, November–December 2005. *MMWR* 56: 165–167.

Anonymous. 2007c. Multistate outbreak of *Salmonella* serotype Tennessee infections associated with peanut butter—United States, 2006–2007. *MMWR* 56: 521–524.

Anonymous. 2007d. *Salmonella* Oranienburg infections associated with fruit salad served in health-care facilities—northeastern United States and Canada, 2006. *MMWR* 56: 1025–1028.

Anonymous. 2007e. Scombroid fish poisoning associated with tuna steaks—Louisiana and Tennessee, 2006. *MMWR* 56: 817–819.

Anonymous. 2008a. Bacterial food poisoning in Japan, 1998–2007. *IASR* 29: 213–214.

Anonymous. 2008b. Notifiable diseases/deaths in selected cities weekly information. *MMWR* 57: 749–761.

Anonymous. 2008c. Outbreak of multidrug-resistant *Salmonella enterica* serotype Newport infections associated with consumption of unpasteurized Mexican-style aged cheese—Illinois, March 2006–April 2007. *MMWR* 57: 432–435.

Anonymous. 2008d. Status of ratification: Stockholm convention on persistent organic pollutants. Available online at http://chm.pops.int/Countries/StatusofRatification/tabid/252/language/en-US/Default.aspx.

Anonymous. 2008e. Summary of notifiable diseases—United States 2006. *MMWR* 55: 1–94.

Atta, M.B., El-Sebaie, L.A., Noaman, M.A., and Kassab, H.E. 1997. The effect of cooking on the content of heavy metals in fish (*Tilapia nilotica*). *Food Chemistry* 58: 1–4.

Backer, L.C., Schurz-Rogers, H., Fleming, L.E., Kirkpatrick, B., and Benson, J. 2005. Marine phycotoxins in seafood. In *Toxins in Food*, eds. W.M. Dabrowski and Z.E. Sikorski, pp. 155–189. New York: CRC Press.

Bakir, F., Damluji, S.F., Amin-Zaki, L. et al. 1973. Methylmercury poisoning in Iraq. *Science* 181: 230–241.

Bangtrakulnonth, A., Pornreongwong, S., Pulsrikarn, C. et al. 2004. *Salmonella* serovars from humans and other sources in Thailand, 1993–2002. *Emerging Infectious Diseases* 10: 131–136.

Baratta, E.J. 2003a. Determination of radionuclides in foods from Minsk, Belarus from Chernobyl to present. *Czechoslovac Journal of Physics* 53: A31–A37.

Baratta, E.J. 2003b. Radionuclides in foods: American perspectives. In *Food Safety: Contaminants and Toxins*, ed. J.P.F. D'Mello, pp. 391–407. Wallingford, U.K.: CABI Publishing.

Belay, E.D. and Schonberger, L.B. 2005. The public health impact of prion diseases. *Annual Review of Public Health* 26: 191–212.

Bennett, R.W. and Monday, S.R. 2003. *Staphylococcus aureus*. In *International Handbook of Foodborne Pathogens*, eds. M.D. Miliotis and J.W. Bier, pp. 41–59. New York: Marcel Dekker, Inc.

Beyer, K., Morrow, E., Li, L.X.-M. et al. 2001. Effects of cooking methods on peanut allergenicity. *Journal of Allergy and Clinical Immunology* 107: 1077–1081.

Bikas, C., Jelastopulu, E., Leotsinidis, M., and Kondakis, X. 2003. Epidemiology of human brucellosis in a rural area of northwestern Peloponnese in Greece. *European Journal of Epidemiology* 18: 267–274.

Blanusa, M. 1994. Heavy metal dietary intake: A European comparison. In *Chemical Safety: International Reference Manual*, ed. M. Richardson, pp. 171–181. Weinhem, Germany: VCH Publishers.

Bock, S.A., Muñoz-Furlong, A., and Sampson, H.A. 2001. Fatalities due to anaphylactic reactions to foods. *Journal of Allergy and Clinical Immunology* 107: 191–193.

Bouget, J., Bousser, J., Pats, B. et al. 1990. Acute renal failure following collective intoxication by *Cortinarius orellanus*. *Intensive Care Medicine* 16: 506–510.

Broughton, E. 2005. The Bhopal disaster and its aftermath: A review. *Environmental Health* 4: 6–12.

Bryan, F.L. 1982. *Diseases Transmitted by Foods*. Atlanta: Centers for Disease Control.

Bryden, W.L. 2007. Mycotoxins in the food chain: Human health implications. *Asia Pacific Journal of Clinical Nutrition* 16: 95–101.

Bullerman, L.B. and Bianchini., A. 2007. Stability of mycotoxins during food processing. *International Journal of Food Microbiology* 119: 140–146.

Calvo, A.M. 2005. Mycotoxins. In *Toxins in Food*, eds. W.M. Dabrowski and Z.E. Sikorski, pp. 215–235. New York: CRC Press.

Center for Nonproliferation Studies at the Monterey Institute of International Studies. 2002. Russia: Central Test Site, Novaya Zemlya. Available online at http://www.nti.org/db/nisprofs/russia/weafacl/othernuc/novayaze.htm.

Chailapakul, O., Korsrisakul, S., Siangproh, W., and Grudpan, K. 2008. Fast and simultaneous detection of heavy metals using a simple and reliable microchip-electrochemistry route: An alternative approach to food analysis. *Talanta* 74: 683–689.

Chan, K.C., Yip, Y.C., Chu, H.S., and Sham, W.C. 2006. High-throughput determination of seven trace elements in food samples by inductively coupled plasma–mass spectroscopy. *Journal of AOAC International* 89: 469–479.

Claeys, W.L., de Vleeschouwer, K., and Hendrickx, M.E. 2005. Quantifying the formation of carcinogens during food processing: Acrylamide. *Trends in Food Science and Technology* 16: 181–193.

Collier, R. 2000. Regulations of rbST in the U.S. *AgBioForum* 3: 156–163.

Cordle, M.K. 1988. USDA regulation of residues in meat and poultry products. *Journal of Animal Science* 66: 413–433.

Cunningham, W.C., Anderson, D.L., and Baratta, E.J. 1994. Radionuclides in domestic and imported foods in the United States, 1987–1992. *Journal of AOAC International* 77: 1422–1427.

Datta, A.R. 2003. *Listeria monocytogenes*. In *International Handbook of Foodborne Pathogens*, eds. M.D. Miliotis and J.W. Bier, pp. 105–121. New York: Marcel Dekker, Inc.

Davis, P.J., Smales, C.M., and James, D.C. 2001. How can thermal processing modify the antigenicity of proteins. *Allergy* 56: 56–60.

Daxenberger, A., Ibarreta, D., and Meyer, H.H. 2001. Possible health impact of animal oestrogens in food. *Human Reproduction Update* 7: 340–355.

De Schrijver, K., Buvens, G., Possé, B. et al. 2008. Outbreak of verocytotoxin-producing *E. coli* O145 and O26 infections associated with the consumption of ice cream produced at a farm, Belgium, 2007. *EuroSurveillance* 13: Article 5.

Del Brutto, O.H. and Sotelo, J. 2003. *Taenia solium*. In *International Handbook of Foodborne Pathogens*, eds. M.D. Miliotis and J.W. Bier, pp. 525–538. New York: Marcel Dekker, Inc.

Dey, B.P. and Negron, E. 2008. Food safety and antimicrobial residues in food animals. Available online at www.fsis.usda.gov/oppde/animalprod/Presentations/Residue/Residue.ppt.

Dontšenko, I., Võželevskaja, N., Põld, A., Kerbo, N., and Kutsar, K. 2008. Outbreak of salmonellosis in a kindergarten in Estonia, May 2008. *EuroSurveillance* 13: Article 1.

Doyle, T.J. and Bryan, R.T. 2000. Infectious disease morbidity in the U.S. region bordering Mexico, 1990–1998. *Journal of Infectious Diseases* 182: 1503–1510.

Eblen, D.R., Annous, B.A., and Sapers, G.M. 2005. Studies to select appropriate nonpathogenic surrogate *Escherichia coli* strains for potential use in place of *Escherichia coli* O157:H7 and *Salmonella* in pilot plant studies. *Journal of Food Protection* 68: 282–291.

Ehling-Schulz, M., Fricker, M., and Scherer, S. 2004. *Bacillus cereus*, the causative agent of an emetic type of food-borne illness. *Molecular Nutrition & Food Research* 48: 479–487.

Emergency Management Australia. 2007. EMA Disasters Database: Adelaide, SA: Salmonella Outbreak. Available online at http://www.ema.gov.au/ema/emadisasters.nsf/54273a46a9c753b3ca256d0900180220/fa63b3eb05875f76ca256d330005ae45?OpenDocument.

Entis, P. 2007. *Food Safety: Old Habits, New Perspectives*. Washington, DC: ASM Press.

Ersoy, B., Yanar, Y., Küçükgülmez, A., and Çelik, M. 2006. Effects of four cooking methods on the heavy metal concentrations of sea bass fillets (*Dicentrarchus labrax* Linne, 1785). *Food Chemistry* 99: 748–751.

Eslava, C., Villaseca, J., Hernandez, U., and Cravioto, A. 2003. *Escherichia coli*. In *International Handbook of Foodborne Pathogens*, eds. M.D. Miliotis and J.W. Bier, pp. 123–135. New York: Marcel Dekker, Inc.

Ethelberg, S., Wingstrand, A., Jensen, T. et al. 2008. Large ongoing outbreak of infection with *Salmonella* Typhimurium U292 in Denmark—February–July 2008. *EuroSurveillance* 13: Article 1.

FAO. 2004. *Marine Biotoxins, FAO Food and Nutrition Paper 80*. Rome: Food and Agriculture Organization of the United Nations.

Fara, G.M., del Corvo, G., Bernuzzi, S. et al. 1979. Epidemica of breast enlargement in an Italian school. *The Lancet* 2: 295–297.

Faulstich, H. 2005. Mushroom toxins. In *Toxins in Food*, eds. W.M. Dabrowski and Z.E. Sikorski, pp. 65–83. New York: CRC Press.

Flick, G.J. and Granata, L.A. 2005. Biogenic amines in foods. In *Toxins in Foods*, eds. W.M. Dabrowski and Z.E. Sikorski, pp. 122–154. New York: CRC Press.

Fosgate, G.T., Carpenter, T.E., Chomel, B.B., Case, J.T., DeBess, E.E., and Reilly, K.F. 2002. Time-space clustering of human brucellosis, California, 1973–1992. *Emerging Infectious Diseases* 8: 672–678.

Franciosa, G., Aureli, P., and Schechter, R. 2003. *Clostridium botulinum*. In *International Handbook of Foodborne Pathogens*, eds. M.D. Miliotis and J.W. Bier, pp. 61–90. New York: Marcel Dekker, Inc.

Franco, B.D.G.M., Landgraf, M., Destro, M.T., and Gelli, D.S. 2003. Foodborne diseases in Southern South America. In *International Handbook of Foodborne Pathogens*, eds. M.D. Miliotis and J.W. Bier, pp. 733–743. New York: Marcel Dekker, Inc.

Frean, J., Keddy, K., and Koornhof, H. 2003. Incidence of foodborne illness in Southern Africa. In *International Handbook of Foodborne Pathogens*, eds. M.D. Miliotis and J.W. Bier, pp. 703–710. New York: Marcel Dekker, Inc.

Gordillo, M.E., Reeve, G.R., Pappas, J., Mathewson, J.J., DuPont, H.L., and Murray, B.E. 1992. Molecular characterization of strains of enteroinvasive *Escherichia coli* O143, including isolates from a large outbreak in Houston, Texas. *Journal of Clinical Microbiology* 30: 889–893.

Goulet, V., Hedberg, C., Le Monnier, A., and de Valk, H. 2008. Increasing incidence of listeriosis in France and other European countries. *Emerging Infectious Diseases* 14: 734–740.

Gupta, P.K. 2004. Pesticide exposure—Indian scene. *Toxicology* 198: 83–90.

Hamada, R., Arimura, K., and Osame, M. 1997. Maternal–fetal mercury transport and fetal methylmercury poisoning. In *Metal Ions in Biological Systems*, eds. H. Siegel and A. Siegel, pp. 405–420. New York: CRC Press.

Hanes, D. 2003. Nontyphoid *Salmonella*. In *International Handbook of Foodborne Pathogens*, eds. M.D. Miliotis and J.W. Bier, pp. 137–149. New York: Marcel Dekker, Inc.

Hansen, K.S., Ballmer-Weber, B.K., Lüttkop, D. et al. 2003. Roasted hazelnuts—allergenic activity evaluated by double-blind placebo-controlled food challenge. *Allergy* 58:132–138.

Hatakka, M. and Pakkala, P. 2003. Incidence of foodborne infections in Northern Europe. In *International Handbook of Foodborne Pathogens*, eds. M.D. Miliotis and J.W. Bier, pp. 669–683. New York: Marcel Dekker, Inc.

Hefle, S.L. 1999. Impact of processing on food allergens. In *Impact of Processing on Food Safety*, eds. L.S. Jackson, M.G. Knize, and J.N. Morgan, pp. 107–119. New York: Kluwer Academic/Plenum Publishers.

Hefle, S.L., Nordlee, J.A., and Taylor, S.L. 1996. Allergenic foods. *Critical Reviews in Food Science and Nutrition* 36: S69–S89.

Hennessy, T.W., Hedberg, C.W., Slutsker, L. et al. 1996. A national outbreak of *Salmonella enteritidis* infections from ice cream. *New England Journal of Medicine* 334: 1281–1286.

Hill, D.J., Hosking, C.S., Zhie, C.Y. et al. 1997. The frequency of food allergy in Australia and Asia. *Environmental Toxicology and Pharmacology* 4: 101–110.

Hoszokwski, A. and Wasyl, D. 2002. *Salmonella* serovars found in animals and feeding stuffs in 2001 and their antimicrobial resistance. *Bulletin of the Veterinary Institute in Pulawy* 46: 165–178.

Hu, L. and Kopecko, D.J. 2003a. *Campylobacter* Species. In *International Handbook of Foodborne Pathogens*, eds. M.D. Miliotis and J.W. Bier,. pp. 181–198. New York: Marcel Dekker, Inc.

Hu, L. and Kopecko, D.J. 2003b. Typhoid *Salmonella*. In *International Handbook of Foodborne Pathogens*, eds. M.D. Miliotis and J.W. Bier, 151–165. New York: Marcel Dekker, Inc.

Huxtable, R.J. and Cooper, R.A. 2000. Pyrrolizidine alkaloids: physicochemical correlates of metabolism and toxicity. In *Natural and Selected Synthetic Toxins: Biological Implications*, eds. A.T. Tu and W. Gaffield, pp. 100–117. Washington, DC: American Chemical Society.

International Commission on Microbiological Specifications for Foods. 2002. *Microorganisms in Foods 7: Microbiological Testing in Food Safety Management*. New York: Kluwer Academic/Plenum Publishers.

Itoh, Y., Nagano, I., Kunishima, M., and Ezaki, T. 1997. Laboratory investigation of enteroaggregative *Escherichia coli* O untypeable:H10 associated with a massive outbreak of gastrointestinal illness. *Journal of Clinical Microbiology* 35: 2546–2550.

Jackson, T.F.H.G. and Reddy, S.G. 2003. *Entamoeba histolytica*. In *International Handbook of Foodborne Pathogens*, eds. M.D. Miliotis and J.W. Bier, pp. 459–471. New York: Marcel Dekker, Inc.

Järup, L. 2003. Hazards of heavy metal contamination. *British Medical Bulletin* 68: 167–182.

Jenner, G.G. and Cuningham, J.A.K. 1944. An outbreak of cadmium poisoning. *New Zealand Medical Journal* 43: 282–283.

Jones, J.L., Kruszon-Moran, D., and Wilson, M. 2003. *Toxoplasma gondii* infection in the United States, 1999–2000. *Emerging Infectious Diseases* 9: 1371–1374.

Julshamn, K., Maage, A., Norli, H.S., Grobecker, K.H., Jorhem, L., and Fecher, P. 2007. Determination of arsenic, cadmium, mercury, and lead by inductively coupled plasma/ mass spectrometry in foods after pressure digestion: NMKL interlaboratory study. *Journal of AOAC International* 90: 844–856.

Katz, S.E. and Ward, P.-M.L. 2005. Antibiotic and hormone residues in foods and their significance. In *Toxins in Food*, eds. W.M. Dabrowski and Z.E. Sikorski, pp. 270–284. New York: CRC Press.

Khateeb, M.I., Araj, G.F., Majeed, S.A., and Lulu, A.R. 1990. *Brucella* arthritis: A study of 96 cases in Kuwait. *Annals of Rheumatic Diseases* 49: 994–998.

Kingsley, D.H. and Richards, G.P. 2003. Calciviruses. In *International Handbook of Foodborne Pathogens*, eds. M.D. Miliotis and J.W. Bier, pp. 1–13. New York: Marcel Dekker, Inc.

Kirk, M., Musto, J., Gregory, J., and Fullerton, K. 2008. Obligations to report outbreaks of foodborne disease under the International Health Regulations (2005). *Emerging Infectious Disease* [Electronic publication ahead of print]:http://www.cdc.gov/eid/content/14/9/pdfs/08-0468.pdf.

Koopmans, M., von Bonsdorff, C.-H., Vinjé, J., de Medici, D., and Monroe, S. 2002. Foodborne viruses. *FEMS Microbiology Reviews* 26: 187–205.

Kucharska, E. 2005. Food allergies and food intolerance. In *Toxins in Food*, eds. W.M. Dabrowski and Z.E. Sikorski, pp. 105–120. New York: CRC Press.

Kumar, V., Basu, M.S., and Rajendran, T.P. 2008. Mycotoxin research and mycoflora in some commercially important agricultural commodities. *Crop Protection* 27: 891–905.

Lado, B.H. and Yousef, A.E. 2007. Characteristics of *Listeria monocytogenes* important to food processors. In *Listeria, Listeriosis, and Food Safety*, 3rd edition, eds. E.T. Ryser and E.H. Marth, pp. 157–214. Boca Raton, FL: CRC Press.

Lake, R., Hudson, A., and Cressey, P. 2002. Risk profile: *Toxoplasma gondii* in red meat and meat products. Available online at http://www.nzfsa.govt.nz/science/risk-profiles/toxoplasma-gondii-in-red-meat.pdf.

Lang, L. 2007. FDA issues statement on diethylene glycol and melamine food contamination. *Gastroenterology* 133: 5–6.

Langham, W.H. 1961. Some considerations of present biosperic contamination of radioactive fallout. *Journal of Agricultural and Food Chemistry* 9: 91–95.

le Marchand, L., Kolonel, L.N., Siegel, B.Z., and Dendle, W.H. 1986. Trends in birth defects for a Hawaiian population exposed to heptachlor and for the United States. *Archives of Environmental Health* 41: 145–148.

Lee, Y.-H. 1992. Food-processing approaches to altering allergcnic potential of milk-based formula. *Journal of Pediatrics* 121: S47–S50.

Loizoo, A. 2003. Epidemic of breast enlargement in some Italian schools: Twenty years later. *Italian Journal of Pediatrics* 29: 4–5.

Lone, K.P. 1997. Natural sex steroids and their xenobiotic analogs in animal production: Growth, carcass quality, pharmokinetics, metabolism, mode of action, residues, methods, and epidemiology. *Critical Reviews in Food Science and Nutrition* 37: 93–209.

Magan, N. and Aldred, D. 2007. Post-harvest control strategies: Minimizing mycotoxins in the food chain. *International Journal of Food Microbiology* 119: 131–139.

Maleki, S.J. 2004. Food processing: Effects on allergenicity. *Current Opinion in Allergy and Clinical Immunology* 4: 941–945.

Markus, M.B. 2003. *Toxoplasma gondii*. In *International Handbook of Foodborne Pathogens*, eds. M.D. Miliotis and J.W. Bier, pp. 511–523. New York: Marcel Dekker, Inc.

Maskarinec, G. 2005. Mortality and cancer incidence among children and adolescents in Hawaii 20 years after a heptachlor contamination episode. *Journal of Environmental Pathology Toxicology and Oncology* 24: 235–249.

Matsko, V.P. and Imanaka, T. 2002. Content of radionuclides of Chernobyl origin in food products for the Belarusian population. *KURRI-KR* 79: 103–111.

Mazur, G. and Jedrorog, S. 1993. Radioactive food contamination in Warsaw in the fifth year after the Chernobyl accident. *Polish Journal of Food Science* 2: 87–91.

McDermott, J.J. and Arimi, S.M. 2002. Brucellosis in sub-Saharan Africa: Epidemiology, control and impact. *Veterinary Microbiology* 90: 111–134.

Mead, P.S., Slutsker, L., Dietz, V. et al. 1999. Food-related illness and death in the United States. *Emerging Infectious Diseases* 5: 607–625.

Memish, Z.A. and Balkhy, H.H. 2004. Brucellosis and international travel. *Journal of Travel Medicine* 11: 49–55.

Mendez Martinez, C., Paez Jimenez, A., Cortés-Blanco, M. et al. 2003. Brucellosis outbreak due to unpasteurized raw goat cheese in Andalucia (Spain), January–March 2002. *EuroSurveillance* 8: Article 3.

Meng, J. and Doyle, M.P. 1997. Emerging issues in microbiological food safety. *Annual Reviews in Nutrition* 17: 255–275.

Michino, H., Araki, K., Minami, S. et al. 1999. Massive outbreak of *Escherichia coli* O157:H7 infection in schoolchildren in Sakai City, Japan, associated with consumption of white radish sprouts. *American Journal of Epidemiology* 150: 787–796.

Miliotis, M.D. and Bier, J.W. 2003. *International Handbook of Foodborne Pathogens*. New York: Marcel Dekker, Inc.

Monti, J.C. 1994. Milk based hypoallergenic infant formulas. *Dairy Products in Human Health and Nutrition: Proceedings of the First World Conference*, 407–410.

Mount, P., Harris, G., Sinclair, R., Finlay, M., and Becker, G.J. 2002. Acute renal failure following ingestion of wild mushrooms. *International Medicine Journal* 32: 187–190.

National Advisory Committee on Microbiological Criteria for Foods. 2006. Requisite scientific parameters for establishing the equivalence of alternative methods of pasteurization. *Journal of Food Protection* 69: 1190–1216.

Newsom, S.W.B. 2007. Brucellosis, or Mediterranean fever. *British Journal of Infection Control* 8: 13–16.

Nicoletti, P. 2002. A short history of brucellosis. *Veterinary Microbiology* 90: 5–9.

Niebuhr, S.E., Laury, A., Acuff, G.R., and Dickson, J.S. 2008. Evaluation of nonpathogenic surrogate bacteria as process validation indicators for *Salmonella enterica* for selected antimicrobial treatments, cold storage, and fermentation in meat. *Journal of Food Protection* 71: 714–718.

Norwegian Institute of Public Health. 2008. EpiNorth: A Co-operation Project for Communicable Disease Control in Northern Europe. Available online at http://www.epinorth.org/.

Okaichi, T., Fukuyo, Y., Hata, Y. et al. 2004. *Red Tides*. Boston: Kluwer Academic Publishers.

Olsen, S.J., Bleasdale, S.C., Magnano, A.R. et al. 2003. Outbreaks of typhoid fever in the United States, 1960–99. *Epidemiology and Infection* 130: 13–21.

Ostrowski, S.R., Wilbur, S., Chou, C.-H.S.J. et al. 1999. Agency for toxic substances and disease registry's 1997 priority list of hazardous substances. Latent effects—carcinogenesis, neurotoxicology, and developmental deficits in humans and animals. *Toxicology and Industrial Health* 15: 602–644.

Panter, K.E. 2005. Natural toxins of plant origin. In *Toxins in Food*, eds. W.M. Dabrowski and Z.E. Sikorski, pp. 11–63. New York: CRC Press.

Pappas, G., Papadimitriou, P., Akritidis, N., Christou, L., and Tsianos, E.V. 2006. The new global map of human brucellosis. *The Lancet Infectious Diseases* 6: 91–99.

Park, D.L., Ayala, C.E., Guzman-Perez, S.E., Lopez-Garcia, R., and Trujillo, S. 2001. Microbial toxins in foods: Algal, fungal, and bacterial. In *Food Toxicology*, eds. W. Helferich and C.K. Winter, pp. 93–135. New York: CRC Press.

Partsch, C.-J. and Sippell, W. 2001. Pathogenesis and epidemiology of precocious puberty. Effects of exogenous oestrogens. *Human Reproduction Update* 7: 292–302.

Poli, M.A. 2003. Foodborne marine biotoxins. In *International Handbook of Foodborne Pathogens*, eds. M.D. Miliotis and J.W. Bier, pp. 445–458. New York: Marcel Dekker, Inc.

Porter, D.V. and Lister, S.A. 2006. CRS report for congress: Food safety: Federal and state response to the spinach *E. coli* outbreak. Available online at http://www.nationalaglaw-center.org/assets/crs/RL33722.pdf.

Protasowicki, M. 2005. Heavy metals. In *Toxins in Foods*, eds. W.M. Dabrowski and Z.E. Sikorski, pp. 237–249. New York: CRC Press.

Quigley, J. 2004. Calf note #106: calves and antibiotic residues. Available online at http://www.calfnotes.com/pdffiles/CN106.pdf.

Rabbani, G.H., Sack, D.A., and Choudhury, M.R. 2003. *Vibrio cholerae*. In *International Handbook of Foodborne Pathogens*, eds. M.D. Miliotis and J.W. Bier, pp. 217–235. New York: Marcel Dekker, Inc.

Richardson, K. and George, B. 1999. *Salmonella* in unpasteurised orange juice–U.S. *Food Safety and Hygiene: A Bulletin for the Australian Food Industry*. November 1999.

Robinson, A. 2003. Guidelines for coordinated human and animal brucellosis surveillance. *FAO Animal Production and Health*. Paper No. 156.

Robison, W.L., Bogen, K.T., and Conrado, C.L. 1997. An updated dose assessment for resettlement options at Bikini Atoll—a U.S. nuclear test site. *Health Physics* 73: 100–114.

Russo, G., Guarneri, M.P., Garancini, M.P. et al. 2003. Epidemic of breast enlargement in a school population: A twenty-year follow up. *Italian Journal of Pediatrics* 29: 57–60.

Sampson, H.A. 2004. Update of food allergy. *Journal of Allergy and Clinical Immunology* 113: 805–819.

Sathe, S.K., Teuber, S.S., and Roux, K.H. 2005. Effects of food processing on the stability of food allergens. *Biotechnology Advances* 23: 423–429.

Schafer, K.S. 2001. Nowhere to hide: Persistent toxic chemicals in the U.S. food supply. Available online at http://www.panna.org/resources/documents/nowhereToHideAvail.dv.html.

Schafer, K.S. and Kegley S.E. 2002. Persistent toxic chemicals in the U.S. food supply. *Journal of Epidemiology and Community Health* 56: 813–817.

Scheil, W., Cameron, S., Dalton, C., Murray, C., and Wilson, D. 1998. A South Australian Salmonella Mbandaka outbreak investigation using a database to select controls. *Australian and New Zealand Journal of Public Health* 22: 536–539.

Schlatter, J., Baars, A.J., DiNovi, M., Lawrie, S., and Lorentzen, R. 2002. 3-Chloro-1,2-propane-diol. *Proceedings of the Fifty-Seventh Meeting of the Joint FAO/WHO Expert Committee on Food Additives (JECFA) of the Safety Evaluation of Certain Food Additives and Contaminants*, 401–432.

Schoub, B.D. 2003. Hepatitis. In *International Handbook of Foodborne Pathogens*, eds. M.D. Miliotis and J.W. Bier, pp. 15–25. New York: Marcel Dekker, Inc.

Seel, J.F., Whicker, F.W., and Adriano, D.C. 1995. Uptake of ^{137}Cs in vegetable crops grown on a contaminated lakebed. *Health Physics* 68: 793–799.

Shephard, G.S. 2008. Impact of mycotoxins on human health in developing countries. *Food Additives and Contaminants* 25: 146–151.

Sicherer, S.H. 2007. Food allergy. In *Handbook of Nutrition and Food*, eds. C.D. Berdanier, J.Dwyer, and E.B. Feldman, pp. 1111–1123. New York: CRC Press.

Sikorski, Z.E. 2005. The effect of processing on the nutritional value and toxicity of foods. In *Toxins in Food*, eds. W.M. Dabrowski and Z. E. Sikorski, pp. 285–312. New York: CRC Press.

Skuterud, L., Gwynn, J.P., Gaare, E., Steinnes, E., and Hove, K. 2005. ^{90}Sr, ^{210}Po, and ^{210}Pb in lichen and reindeer in Norway. *Journal of Environmental Radioactivity* 84: 441–456.

Smith, J.L. 1991. Foodborne toxoplasmosis. *Journal of Food Safety* 12: 17–57.

Smith, J.T. and Beresford, N.A. 2003. Radionuclides in foods: The post-Chernobyl evidence. In *Food Safety: Contaminants and Toxins*, ed. J.P.F. D'Mello, pp. 373–390. Wallingford, U.K.: CABI Publishing.

Soderholm, C.G., Otterby, D.E., Linn, J.G. et al. 1988. Effects of recombinant bovine somatotropin on milk production, body composition, and physiological parameters. *Journal of Dairy Science* 71: 355–365.

Strand, T., Strand, P., and Baarli, J. 1987. Radioactivity in foodstuffs and doses to the Norwegian population from the Chernobyl fallout. *Radiation Protection Dosimetry* 20: 221–229.

Swaminathan, N. 2007. Special report: The poisoning of our pets. *Scientific American* March 28, 2007.

Taeymans, D., Wood, J., Ashby, P. et al. 2004. A review of acrylamide: An industry perspective on research, analysis, formation, and control. *Critical Reviews in Food Science and Nutrition* 44: 323–347.

Takemura, K., Kahdre, M., Joseph, D., Yousef, A., and Sreevatsen, S. 2004. An overview of transmissible spongiform encephalopathies. *Animal Health Research Reviews* 5: 103–124.

Takizawa, Y., Kosaka, T., and Sugai, R. 1972. Studies on the cause of the Niigata episode of Minamata disease outbreak. *Acta Medica et Biologica* 19: 193–206.

Taubert, D., Harlfinger, S., Henkes, L., Berkels, R., and Schömig, E. 2004. Influence of processing parameters on acrylamide formation during frying of potatoes. *Journal of Agricultural and Food Chemistry* 52: 2735–2739.

Taylor, J.P. and Perdue, J.N. 1989. The changing epidemiology of human brucellosis in Texas, 1977–1986. *American Journal of Epidemiology* 130: 160–165.

Taylor, N., Chavarriaga, P., Raemakers, K., Siritunga, D., and Zhang, P. 2004. Development and application of transgenic technologies in cassava. *Plant Molecular Biology* 56: 671–688.

Taylor, S.L. 2000. Emerging problems with food allergens. In *Food, Nutrition, and Agriculture*, 26, ed. J. Albert, pp. 14–23. Rome: Food and Agriculture Organization of the United Nations.

Taylor, S.L., Hefle, S.L., and Gauger, B.J. 2001. Food allergies and sensitivities. In *Food Toxicology*, eds. W. Helferich and C.K. Winter, pp. 1–36. New York: CRC Press.

The American Medical Association, The American Nurses Association, Centers for Disease Control and Prevention, United States Food and Drug Administration Center for Food Safety and Applied Nutrition, and United States Department of Agriculture Food Safety and Inspection Service. 2004. Diagnosis and management of foodborne illnesses: A primer for physicians and other health care professionals. Available online at http://www.ama-assn.org/ama/pub/category/3629.html.

The Pennington Group. 1998. Report on the circumstances leading to the 1996 outbreak of infection with *E.coli* O157 in Central Scotland, the implications for food safety and the lessons to be learned. Available online at http://www.scotland.gov.uk/library/documents-w4/pgr-00.htm.

Tritscher, A.M. 2004. Human health risk assessment of processing-related compounds in food. *Toxicology Letters* 149: 177–186.

Twiner, M.J., Rehmann, N., Hess, P., and Doucette, G.J. 2008. Azaspiracid shellfish poisoning: A review on the chemistry, ecology, and toxicology with an emphasis on human health impacts. *Marine Drugs* 6: 39–72.

United States Department of Agriculture. 2007. Pesticide data program: Annual summary, calendar year 2006. Available online at http://www.ams.usda.gov/AMSv1.0/getfile?dDocName = STELPRDC5064786.

United States Department of Agriculture Food Safety and Inspection Service. 2007. 2006 FSIS National Residue Program Data. Available online at http://www.fsis.usda.gov/PDF/2006_Red_Book_Intro.pdf, http://www.fsis.usda.gov/PDF/2006_Red_Book_Results_pp25-58.pdf, http://www.fsis.usda.gov/PDF/2006_Red_Book_Results_pp59-end.pdf.

United States Department of Health and Human Services Agency for Toxic Substances and Disease Registry. 2007. ToxFAQs™. Available online at http://www.atsdr.cdc.gov/toxfaq.html.

United States Department of Health and Human Services Agency for Toxic Substances and Disease Registry. 2008. 2007 CERCLA Priority List of Hazardous Substances. Available online at http://www.atsdr.cdc.gov/cercla/07list.html.

United States Department of Health and Human Services Centers for Disease Control and Prevention. 2006a. Multistate Outbreak of *E. coli* O157 Infections, November–December 2006. Available online at http://www.cdc.gov/ecoli/2006/december/121406.htm.

United States Department of Health and Human Services Centers for Disease Control and Prevention. 2006b. Prion Diseases. Available online at http://www.cdc.gov/ncidod/dvrd/prions/.

United States Department of Health and Human Services Centers for Disease Control and Prevention. 2006c. Salmonellosis—Outbreak Investigation, October 2006. Available online at http://www.cdc.gov/ncidod/dbmd/diseaseinfo/salmonellosis_2006/110306_outbreak_notice.htm.

United States Department of Health and Human Services Centers for Disease Control and Prevention. 2006d. Update on Multi-State Outbreak of *E. coli* O157:H7 Infections From Fresh Spinach, October 6, 2006. Available online at http://www.cdc.gov/foodborne/ecolispinach/100606.htm.

United States Department of Health and Human Services Centers for Disease Control and Prevention. 2007a. Investigation of Outbreak of Human Infections Caused by *E. coli* O157:H7. Available online at http://www.cdc.gov/ecoli/2007/october/103107.html.

United States Department of Health and Human Services Centers for Disease Control and Prevention. 2007b. Investigation of Outbreak of Human Infections Caused by *Salmonella* I 4,[5],12:i:-. Available online at http://www.cdc.gov/Salmonella/4512eyeminus.html.

United States Department of Health and Human Services Centers for Disease Control and Prevention. 2007c. Multistate Outbreak of *E. coli* O157 Infections Linked to Topp's Brand Ground Beef Patties. Available online at http://www.cdc.gov/ecoli/2007/october/100207.html.

United States Department of Health and Human Services Centers for Disease Control and Prevention. 2007d. Salmonella Wandsworth Outbreak Investigation, June–July 2007. Available online at http://www.cdc.gov/salmonclla/wandsworth.htm.

United States Department of Health and Human Services Centers for Disease Control and Prevention. 2007e. Salmonellosis—Outbreak Investigation, February 2007. Available online at http://www.cdc.gov/ncidod/dbmd/diseaseinfo/salmonellosis_2007/030707_outbreak_notice.htm.

United States Department of Health and Human Services Centers for Disease Control and Prevention. 2007f. Summary of human *Vibrio* isolates reported to CDC, 2006. Available online at http://www.cdc.gov/nationalsurveillance/PDFs/CSTEVibrio2006website.pdf.

United States Department of Health and Human Services Centers for Disease Control and Prevention. 2007g. vCJD (Variant Creutzfeldt–Jakob Disease). Available online at http://www.cdc.gov/ncidod/dvrd/vcjd/factsheet_nvcjd.htm.

United States Department of Health and Human Services Centers for Disease Control and Prevention. 2008a. FoodNet Facts and Figures—number of infections and incidence per 100,000 persons. Available online at http://www.cdc.gov/FoodNet/factsandfigures/incidence.html.

United States Department of Health and Human Services Centers for Disease Control and Prevention. 2008b. Investigation of Outbreak of Human Infections Caused by *E. coli* O157:H7. Available online at http://www.cdc.gov/ecoli/june2008outbreak/.

United States Department of Health and Human Services Centers for Disease Control and Prevention. 2008c. Investigation of Outbreak of Infections Caused by *Salmonella* Agona. Available online at http://www.cdc.gov/salmonella/agona/.

United States Department of Health and Human Services Centers for Disease Control and Prevention. 2008d. Investigation of Outbreak of Infections Caused by *Salmonella* Saintpaul. Available online at http://www.cdc.gov/Salmonella/saintpaul/.

United States Department of Health and Human Services Centers for Disease Control and Prevention. 2008e. Investigation Update: Outbreak of *Salmonella* Litchfield Infections, 2008. Available online at http://www.cdc.gov/salmonella/litchfield/.

United States Department of Health and Human Services Centers for Disease Control and Prevention. 2008f. Nationally notifiable infectious disease. Available online at http://www.cdc.gov/ncphi/disss/nndss/PHS/infdis2008.htm.

United States Department of Health and Human Services Food and Drug Administration. 1998. Accidental radioactive contamination of human food and animal feeds: recommendations for state and local agencies. Available online at http://www.fda.gov/cdrh/dmqrp/1071.pdf.

United States Department of Health and Human Services Food and Drug Administration. 2000. Action levels for poisonous of deleterious substances in human food and animal feed. Available online at http://www.cfsan.fda.gov/~lrd/fdaact.html#afla.

United States Department of Health and Human Services Food and Drug Administration. 2003. Food and Drug Administration Pesticide Program Residue Monitoring 2003. Available online at www.cfsan.fda.gov/~acrobat/pes03rep.pdf.

United States Department of Health and Human Services Food and Drug Administration. 2004. Physicians, Nurses and U.S. Government Release New Foodborne Illness Guide. Available online at http://www.fda.gov/bbs/topics/news/2004/NEW01047.html.

United States Department of Health and Human Services Food and Drug Administration. 2005. National Shellfish Sanitation Program Guide for the Control of Molluscan Shellfish 2005 IV.II.04.Action Levels, Tolerances and Guidance Levels for Poisonous or Deleterious Substances in Seafood. Available online at http://www.cfsan.fda.gov/~ear/nss3-42d.html.

United States Department of Health and Human Services Food and Drug Administration. 2007. Transcript of FDA Press Conference on the Pet Food Recall. Available online at http://www.fda.gov/oc/opacom/hottopics/petfood/transcript040507.pdf.

United States Department of Health and Human Services Food and Drug Administration. 2008. Pet food recall (melamine)/Tainted animal feed. Available online at http://www.fda.gov/oc/opacom/hottopics/petfood.html.

United States Department of Health and Human Services Food and Drug Administration Center for Food Safety and Applied Nutrition. 2001. FDA & EPA safety levels in regulations and guidance. In *Fish and Fisheries Products Hazards and Controls Guidance*, 285–288, Washington, DC.

United States Department of Health and Human Services Food and Drug Administration Center for Food Safety and Applied Nutrition. 2006. Total Diet Study Statistics on Radionuclide Results. Available online at www.cfsan.fda.gov/~acrobat/tdsbyrn.pdf.

United States Environmental Protection Agency. 2007. Persistant Organic Pollutants (POPs). Available online at http://www.epa.gov/pesticides/international/pops.htm.

United States Environmental Protection Agency. 2008. Integrated risk information system. Available online at http://cfpub.epa.gov/ncea/iris/index.cfm.

United States Government. 2008. Code of Federal Regulations: Title 21, Part 556: Tolerances for residues of new animal drugs in foods. Available online at http://www.gpoaccess.gov/CFR/INDEX.HTML.

van Egmond, H.P., Schothorst, R.C., and Jonker, M.A. 2007. Regulations relating to mycotoxins in food. *Analytical and Bioanalytical Chemistry* 389: 147–157.

Vicari, A.S., Hueston, W.D., and Travis, D.A. 2003. Variant Creutzfeldt–Jakob disease and bovine spongiform encephalopathy. In *International Handbook of Foodborne Pathogens*, eds. M.D. Miliotis and J.W. Bier, pp. 659–667. New York: Marcel Dekkcer, Inc.

Vorou, R., Gkolfinopoulou, K., Dougas, G., Mellou, K., Pierroutsakos, I.N., and Papadimitriou, T. 2008. Local brucellosis outbreak on Thassos, Greece: a preliminary report. *EuroSurveillance* 13: Article 1.

Waite, J.G., Lado, B.H., and Yousef, A.E. 2009. Selection and identification of a *Listeria monocytogenes* surrogate for ultra-high pressure and pulsed electric field process validation. *Journal of Food Protection* (submitted).

Walderhaug, M. 1992. Foodborne pathogenic microorganisms and natural toxins handbook: pyrrolizidine alkaloids. In *Bad Bug Book*, Washington, DC: United States Food and Drug Administration Center for Food Safety and Applied Nutrition.

Weidenbörner, M. 2001. *Encyclopedia of Food Mycotoxins*. New York: Springer.

Wheatley, B., Barbeau, A., Clarkson, T.W., and Lapham, L.W. 1979. Methylmercury poisoning in Canadian Indians—the elusive diagnosis. *Canadian Journal of Neurological Sciences* 6: 417–422.

Whicker, F.W., Shaw, G., Voigt, G., and Holm, E. 1999. Radioactive contamination: State of the science and its application to predictive models. *Environmental Pollution* 100: 133–149.

Wikland, K., Holm, L.-E., and Eklund, G. 1990. Cancer in Swedish lapps who breed reindeer. *American Journal of Epidemiology* 132: 1078–1082.

Winter, C.K. 2005. Pesticides in food. In *Toxins in Food*, eds. W.M. Dabrowski and Z.E. Sikorski, pp. 252–267. New York: CRC Press.

World Health Organization. 2008a. Food safety. Available online at http://www.who.int/foodsafety/en/.

World Health Organization. 2008b. Microbiological risks in food. Available online at http://www.who.int/foodsafety/micro/en/.

World Health Organization Food Safety Programme (WHO/FOS). 2000. Foodborne illness: An increasing public health concern. Available online at http://www.wpro.who.int/fsi_guide/files/foodborne_illness.ppt.

Zukowska, J. and Biziuk, M. 2008. Methodological evaluation of method for dietary heavy metal intake. *Journal of Food Science* 73: R21–R29.

3 Food Quality Perception

Russell S.J. Keast

CONTENTS

3.1   INTRODUCTION

The concept of food quality differs at each stage of the food system. What defines quality at the production level is not the same as quality at the manufacturing level, and quality will differ once again in the eyes of the consumer. Food quality perception is primarily involved with the consumer, and judgment of quality will vary significantly among people.

Food quality perception is a multidimensional concept that cannot be defined quantitatively. Perceived quality is determined within the context of sensory and nonsensory factors which in turn are subject to individual differences and situational factors. Consumers use cues to predetermine food quality prior to consumption. During and after consumption consumer's experiences will govern future purchases and expectations over food products. This chapter provides a basic overview of the factors involved in food quality perception.

Food quality perception: three descriptive words, each with multiple meanings. Before discussing the meaning of the collective term, an understanding of each individual word will complement the overall concept of food quality perception.

"Food" is any material that is eaten or taken into the body for nourishment. It includes all unprocessed, semiprocessed, or processed items, which are intended for use as food or drinks including any ingredients incorporated into food or drink, and any substances that come into contact with food during production, processing, or treatment (Alli 2003).

"Quality" comprises both subjective and objective characteristics and can be defined in many ways and from many different perspectives. Indeed quality is a theoretical construct, not a physical entity with a fixed time or position. In food science, quality concerns a product's "fitness for consumption." A product is considered to be of quality if it satisfies all the needs and requirements relating to the characteristics of food as determined by the producer, manufacturer, or consumer (Alli 2003). The term quality has been defined as the totality of features relevant to the ability of a product to fulfill its requirements (Muller and Steinhart 2006).

"Perception" is the central cognitive process in which information is selected, organized, and interpreted through input from our sensory systems. Perception is shaped by cognitive processes, understandings, and experiences. Perception is individualistic and influenced by memories, personality, mood, knowledge, and expectations.

The concept of food quality differs at each stage of the food system (Muller and Steinhart 2006). What defines quality at the production level is not the same as quality at the manufacturing level, and quality will differ once again in the eyes of the consumer. Whereas the production definition of quality requires more objective measures and is aimed at obtaining consistency among like products. While this chapter covers food quality perception, it is nevertheless important to briefly mention the objective measures of quality in terms of food quality perception.

3.2   QUALITY AND THE FOOD SYSTEM

3.2.1   PRODUCTION, MANUFACTURING, AND QUALITY OF FOOD

Food quality at the production level is judged by "value for sale" of the product. Values are defined by grading/measuring systems which have standards for size, shape, appearance, damage, and composition of foods such as muscle size of meat, fat percentage of canola seed, or the alcoholic strength of beer (Harrison and Hester 2001, Brunso et al. 2005). Product- and manufacturing-oriented quality constitutes the objective qualities of food products (Bredahl 2003). Objective qualities include all factors which can be measured analytically, are repeatable, and can be quantified and include weight, fat content, tenderness, pH, and specific gravity (Harrison and Hester 2001). In some cases objective characteristics can be used to infer subjective experiences such as cuts of meat with less visible fat being perceived as higher quality because they are thought to taste better and/or be healthier (Bredahl 2003). At the manufacturing level quality is also defined by composition as well as production methods and sensory characteristics thought to be valued by consumers (Brunso et al. 2005).

Manufacturers can refrain from using additives or genetically modified ingredients, reduce fat, or farm organically to increase consumer appeal (Bredahl 2003). Quality at the manufacturing level does not necessarily effect the physical properties of food, but relies on subjective consumer perception of quality (Brunso et al. 2005). The individual differences in consumer preferences are evident by the diverse range of food products currently available. Options such as gluten free, high fiber, or salt reduced are just some examples of the variety. These many options are a result of the food producers and manufacturers being well aware of subjective processes which drive quality perception, allowing them to tailor production and manufacturing processes, appealing to certain groups of consumers (Grunert 1997).

3.2.2 Food Safety and Food Quality

Assurances that foods will not cause harm or include harmful substances within or on foods prior to consumption, alongside correct handling, preparation, and storage methods are essential (Alli 2003). It is generally assumed that all food available for human consumption has met prior safety standards and requirements ensuring consumption will not cause harm (Harrison and Hester 2001). In most cases food safety is an implicit part of food quality—it is what we expect of all food purchased. While this may hold true in most cases, the safety and reliability of the food supply has been shattered in recent times by scares which threaten human health. Outbreaks of Bovine Spongiform Encephaly (BSE), bird flu, or swine fever alongside occurrences of food contamination and food poisoning all result in long lasting damage to market demand and credibility of food companies, brands, and regulatory bodies (Grunert 2002).

Food safety is integrated within food quality assurance and quality control programs. Quality control and quality assurance are processors concerned with meeting standards, regulations, and maintaining consistency of predetermined products and may reflect quality at the manufacturing or production level, but not necessarily at the consumer level as quality will be subject to one's own personal set of standards (Harrison and Hester 2001). Food quality assurance sets targets for policies, programs, systems, and procedures concerning the food (Muller and Steinhart 2006). Quality assurance is concerned with assurance and consistency within food products. Food quality control is closely tied to the manufacturing processes ensuring raw materials meet quality specifications.

The presence of pesticides, allergens, contaminants, residues, physical hazards, biological and pathogenic organisms as well as other undesirable contents such as trans-fat all disrupt quality perception and expectations (Muller and Steinhart 2006). However, food safety is not solely the responsibility of producer or manufacturer. Correct handling, preparation, and storage of foods are also the responsibility of the intermediate consumer (restaurants and retailers) and the end consumer (Alli 2003).

Consumers can partially judge food safety prior to purchase with indirect cues such as color, expiry date, smell, and condition of products (Grunert 2002). These sensory and labeling cues provide consumers some guidance for choosing safe foods, but our judgments of quality are subjective and based on personal experiences, expectations, and perceptions about food (Brunso et al. 2005). Subjective factors and food quality perception will be the focus of this review.

3.3   FLAVOR AS A DIMENSION OF QUALITY

Flavor is a key factor in quality perception of foods. Perception of flavor is an easy task. This ease is due to the sophistication of our sensory systems which continuously sample the local environment and our brain which processes millions of pieces of information, while we consciously expend no energy or attention to the entire process. The perceived flavor of the food is derived when three independent sensory systems are activated: (1) taste, (2) olfaction, and (3) oro-nasal somatosensations (irritation, thermal, and texture) (Keast et al. 2004). In addition to sensory system activation, societal and lifestyle factors influence judgments on the perceived flavor, what one person believes extremely bitter and unpleasant another may perceive very little bitterness and pleasant. There is variation in perception and liking from myriad sources, including sensory adaptation, background noise in the nervous system, a variety of peripheral receptors being activated during the course of eating, an individual's sociocultural heritage, and the local environment, to name a few.

Cognitive processes which occur in the brain vary from individual to individual (Williams 1992). While it is clear that individuals analyze, process, and respond to sensory data differently due to diverse cultural backgrounds, nutritional, social, and attitudinal requirements, evolutionarily it is suspected initial encoding and arrangement of sensation schemata in the brain is similar among all (Lawless 1995). Affective coding of flavor in the brain is highly individualized and thought to be a result of personal learning experiences, beliefs, and genetics (Reed et al. 2006).

Our impressions of quality are formed before we consume a food. The mere sight or smell of a specific food is sometimes enough to evoke physiological responses in the mouth modifying further sensory experiences such as flavor, taste, and texture (Williams 1992). Flavor release, texture, and taste can all be altered by mouth pH, saliva levels, enzyme activity, food temperature, and chewing rate. Once inside the mouth, food chemicals activate receptors where further sensory information is relayed to the brain. Food chemicals and properties activate sensory systems resulting in individual responses to acceptance, rejection, and quality ratings of specific foods. Recognition of food components at receptor sites can be masked or affected by competing food components such as flavor enhancers like citric acid, or flavor blockers such as zinc ions (Keast 2003).

As the brain is constantly upgrading, reorganizing, and learning new information, sensory perception too is in a state of constant flux. Changes occur with changing lives, needs, and situations. Before a discussion on external factors affecting perceived quality, it is important to outline the sensory processes essential for collecting the information on foods that are used to guide our perception of quality.

3.4 PERIPHERAL SENSORY INFLUENCES ON QUALITY PERCEPTION

3.4.1 TASTE

Tastants from food dissolve in saliva and are transported to receptor sites throughout the mouth. Taste is elicited when saliva-soluble compounds stimulate taste receptor cells in the mouth and throat areas. The majority of taste receptor cells are organized into rosette-like structures called taste buds, which are embedded in folds or lingual bumps called papillae. These papillae are located on the tongue; fungiform papillae occur at the anterior of the tongue, vallate papillae occupy the posterior sides of the tongue, and circumvallate papillae are located at the posterior dorsal surface of the tongue. In addition, there are large numbers of taste buds on the soft palate and pharynx in humans, but these do not occur in papillae. Taste buds contain taste cells/receptor mechanisms responsible for detecting chemical moieties on compounds, once activated electrical signals are sent via nerve fibers to processing regions of the brain and we experience the appropriate taste quality (e.g., sweet, sour, salty, bitter, and umami) (Chandrashekar et al. 2006). Variation at the level of the receptor between peoples will result in variation of taste perception.

3.4.2 THE SENSE OF SMELL

Given the small number of tastes, it is not surprising that the immense diversity of flavors we associate with foods is primarily derived from olfactory input, via the volatile compounds that are released in the oral cavity when food or liquids are chewed and swallowed. The importance of the sense of smell is exemplified by the approximately 1000 genes encoding olfactory receptors (ORs), the largest gene family in the entire genome (Buck 1992). However, human reliance on the sense of smell for survival is not as crucial as the rest of the animal world, and over half of these genes in humans (~65%) are nonfunctional pseudo-genes. We have approximately 350 functional genes that encode for ORs. The ORs are housed high in the nasal cavity, and odorants must travel through an aqueous mucus layer to reach them. Once the odorant has reached the ORs, it may activate one or more of the 350 different receptors, similarly, any one OR may be activated by multiple odorants. The odors we perceive after activation of the sense of smell are a result of a pattern of activation of ORs that decoded by the cognitive centers of the brain (Firestein 2001).

There are two routes for activation of the ORs, and both routes are used when tasting wine. First, a wine taster will sniff the wine, the volatile chemicals will travel through the nostrils and activate the ORs; this is termed orthonasal olfaction and is a result of active or passive sniffing of the local environment. Second, the wine is tasted by taking a sample in the mouth and swishing it around. The volatile chemicals travel up the nasal pharynx to activate the ORs; this is termed retronasal olfaction and is directly related to the release of volatiles from food or beverages that are in the mouth. Variations in type and quantity of ORs between individuals will cause variation in flavor perception, as will variation in volatile compounds' ability to access the ORs, for example when we have congested airways, e.g., cold/flu symptoms (Keast et al. 2004). When odorants fail to reach or activate ORs, foods seem tasteless or lack flavor, so people who have constricted airways will perceive less aroma than people with higher nasal airflow.

3.4.3   Chemesthesis—Chemical Irritation

Free endings of individual nerve fibers innervating both the oral and nasal mucosa have specialized sensory receptors that respond to both noxious heat and noxious cold both of which evoke thermal and pain sensations. The oro-nasal nerve fibers are not independent sensory systems, but a component of the pain and temperature fibers that occur throughout the skin. A common feature of oro-nasal chemical irritation is the long latency of sensation relative to that of taste or smell, due to the time taken for the chemicals to diffuse through tight junctions or epithelium to engage receptors on the nerve fibers (Green and Lawless 1991). The disparity in temporal onset between the odor/taste and chemesthetic sensations adds complexity to the perception of flavor.

There are a number of chemicals that are capable of activating irritant sensations and different adjectives to describe the sensations: the burn of chili pepper, the warmth of ethanol, the tingle of CO_2, the pungency of wasabi, etc. However, the terminology associated with oro-nasal irritant sensations lacks the qualitative breadth of olfactory sensations and may be influenced by the taste and smell coelicited with the irritation. For example, the perceived irritant burn from chili pepper or pungency from wasabi may be the same irritant sensation, but when chili pepper or wasabi are eaten they appear perceptually distinguishable. This may be due to the interactions of irritation with the taste and odor components rather than the quality of irritation per se.

3.4.4   Food Texture

Texture includes all mechanical, geographical, and surface attributes of food products (Szcesniak 2002). Texture can be determined by visual, tactile, and in some cases auditory attributes of a food. Texture can be assessed both before and after consumption of the food: avocadoes can be squeezed to test ripeness, a tough steak is hard to cut, and cooked potatoes will be easy to mash.

Mechanical textural attributes of food include their tenderness, fracturability, viscosity, adhesiveness, and elasticity. Different food purposes may call for different mechanical properties of the same food product. Cream can be either viscous (as when whipped) or runny, dependent on its intended purpose. Mechanical properties of foods can change making the food undesirable, as when biscuits become stale, or desirable, as in overripe bananas which are perfect for cakes.

Geometric properties of foods refer to the size, shape, and arrangement of particles with in the food. Textures can be dry like plain crackers, moist as in cake, or juicy in oranges. The temperature of foods at times alters these properties; consider the difference in texture of a cooked potato when served hot and when cold.

Once in the mouth oral evaluation of the food begins. Some foods may exhibit textural changes once inside the mouth. Ice cream and chocolate melt, crunchy foods such as biscuits become softened through mixing with saliva and mouth enzymes. It has been postulated that the change in texture during mastication is partially responsible for the enjoyment of certain foods (Hyde and Witherly 1993).

Unpleasant texture attributes may lead to food aversions and dislikes. Tough meat can be perceived as low quality; slimy foods such as oysters can cause involuntary gagging; spinach not properly washed can be gritty. All negative experiences can lead to dislike of specific foods. In addition, foods with a distinctive texture such as "crispness" in apples may be disliked if their texture is soft and mushy when consumed (Harker et al. 2002).

3.4.5 APPEARANCE

Appearance covers all visible cues to food quality, which include pigmentation, shape, color, size, translucency, and gloss. Visual cues are important for assessment of the food quality. Bruised fruit, moldy bread, or curdled milk may indicate low quality (Kennedy et al. 2005). The most important qualities for meat appear to be color, freshness, visible fat, price, and presentation (Acebron and Dopico 2000). Quality perception increases with attractive appearance and freshness and decreases with amounts of visible fat (Acebron and Dopico 2000). Fat is the most commonly used quality cue; unfortunately, the way consumers use it is dysfunctional. Consumers expect meats with less visible fat or marbling to taste better and be of higher quality. As fat promotes tenderness, juiciness, and flavor, cuts with higher marbling and more fat tend to taste better and result in pleasurable sensations and experiences, contrary to consumer expectations (Brunso et al. 2005).

Color is used to perceive ripeness (in fruits and vegetables), the extent of cooking (with meat), and the degree of strength (tea, juices, or sauce) and is an important cue concerning quality (Harrison and Hester 2001). Even though color may help consumers avoid food that is off or unnatural, it also can lead to incorrect expectations (Schröder 2003). Pigmentation of chicken meat differs between animals fed on different diets. Corn-fed chicken meat has a yellow hue, whereas wheat-fed chicken a tan hue. Traditionally, a yellowish hue was perceived as high quality meat; these days, even though corn-fed chicken is more flavorsome, tender, and tasty than wheat fed, when shown raw meat consumers perceive it as "rancid" or "overly processed" (Kennedy et al. 2005).

The appearance of food may well be the first sensory evaluation made on its quality, leaving long lasting and significant impacts on quality perception. Given the large number of food colors which are added to processed foods to increase acceptability, it is little wonder consumers question foods with unfamiliar colors.

3.4.6 Auditory Texture

Distinctive sounds are associated with consumption of specific foods. When we eat corn chips sounds are emitted as our teeth break the chip, similarly the sizzle of meat or the crunch of a fresh apple can add to our perception of food quality. The auditory texture of foods is an important experience cue, and when auditory texture deviates from expectations, liking of the product can decrease (Harrison and Hester 2001).

3.4.7 Integration of Taste and Smell

Taste and odor form the unique flavors of foods. It has been suggested that tastes and odors are encoded in the brain as part of a unique perceptual system in the form of discrete flavor entities (Prescott 1999). When paired together tastes and odors form flavor memories in the brain. Taste and aroma both influence our perception of each other, meaning taste qualities can be attributed to odors and vice versa. The theory that taste and odor operate together is suggested from experiments in which sweet substances smelled prior to or during consumption of foods enhanced the perception of sweetness (Frank and Byram 1988). Likewise other smells can suppress tastes in food when the odor is not usually associated with the taste, such as pairing salty flavors (like peanut butter) with sweet flavors (like caramel) (Frank et al. 1993).

The aroma and taste of food are powerful sensations, decreasing appetite and flavor perception with nasal congestion and colds or producing saliva and preparing the mouth for mastication when you walk past an aromatic restaurant.

With each new flavor experience quality perception changes and may affect previous and future expectations and experiences. Both humans and animals are cautious when exposed to novel foods or flavors but once it has been established that food is safe to consume, acceptance normally follows. In most cases more than one exposure to novel foods and flavors are required for appreciation (Martins and Pliner 2005). Take for example a child stating an extreme dislike for the flavor of beer. In a few years time, that same child (now an adult) willingly and regularly purchases that same brand of beer. It is not only the flavor of beer we become accustomed to, but also our liking is modified by the postingestive influences of the beer.

3.4.8 Postingestive Effects of Foods:
Influence on Quality Perception

The postingestive effects of foods are a determining factor in the perceived quality of a food. The development of flavor preferences and aversions is established when individuals associate (unconsciously) a food/flavor with its postingestive consequences. Indeed caffeine is a good example to illustrate this point. The mode of action of caffeine in developing flavor preference is not immediate (Yeomans et al. 2000) as, for example, we experience with a sucrose solution (sweet and appetitive). Caffeine may elicit no perceived flavor or bitterness in the mouth depending on concentration (Keast and Roper 2007), but the positive affects occur postconsumption with increased vigilance and attention, enhanced mood, and arousal as well as enhanced motor activity. Behavioral studies have shown that the consumption of caffeine promotes a dependence that is reinforced with repeat consumption (Hughes et al.

1993, Schuh and Griffiths 1997, Garrett and Griffiths 1998). The common method of repeat caffeine consumption is via caffeinated foods such as coffee, tea, cocoa, and soft drinks which are hedonically pleasant to drink. It is well known that beverages containing caffeine, such as tea, coffee, and cola soft-drinks, are amongst the highest consumed beverages worldwide and we speculate this is due to caffeine enhancing the overall liking of these beverages (Keast and Riddell 2007).

While specific chemicals or nutrients can influence our perception of quality, so do our culture, personal experiences and myriad other influences that affect our conscious lives. What follows is an overview of the cognitive processes involved in quality perception.

3.5 COGNITIVE INFLUENCES ON QUALITY PERCEPTION

Advanced cognitive pathways are in operation when choosing the products, preparing them, and finally experiencing their quality (Grunert 1997). While sensory input is important in developing impressions of quality, the cognitive processing of sensory intensity and affective valence happens at different regions of the brain (Small et al. 2003) meaning there is a functional segregation between sensory input and liking. Therefore, just because we like sweetness in food does not ensure that increasing the level of sweetness will increase the preference (or quality perception) of the food. Individual preferences for foods manifest based on personal needs, knowledge, beliefs, and prior experiences (Rozin 2007). Consumers use information available to them from myriad sources to form a perception of quality. Consumer perception of food quality is subjective and formed as consumers assimilate and interpret information surrounding a product prior to purchase; after consumption quality can be reassessed by experience (Acebron and Dopico 2000, Brunso et al. 2005).

The dimensions driving consumer oriented or perceived quality are numerous and complex. For example, we may say that quality at the retail level is based on how quickly a product disappears from the shelf, and is based on consumer demand (Harrison and Hester 2001). This is a simplistic view of quality and while volume of sales is very important to drive profits, it is not always the most purchased item that is the best quality (Lawless 1995). Consider the popularity of Budweiser Beer which is mass produced, yet its quality merits probably lie with its availability, consistency of flavor, and price (it is cheap). Most people would argue that beers produced from boutique breweries are of higher quality (e.g., Emerson's India Pale Ale), but at the same time may not be available at bars, or suitable for all occasions (hot day after mowing the lawn) or affordable to those on low incomes.

The nonsensory cues which consumers use to interpret food quality will now be discussed. As nonsensory factors can alter the way in which food quality is perceived, they are important for any discussion on food quality perception. The interactions between both sensory and the nonsensory factors will together form the final decision on food choices and quality perception.

3.6 NONSENSORY FACTORS IN QUALITY PERCEPTION

Nonsensory factors influencing quality perception direct people where to shop, which brands to choose, and which products are most suited to their intended purpose.

In addition, ethical and cultural beliefs, price, convenience, processing, credence attributes, traditions, and the context in which the food is consumed form the non-sensory factors influencing perceived quality (Jaeger 2006).

3.6.1 PRICE

The price of food is considered both an important contributor toward value-for-money quality as well as perceived quality. In general price is positively related to quality, foods that are higher in price are expected to be of higher quality (Harker et al. 2002). How much money one is willing to spend on a particular product will depend on what characteristics the food poses, how important they are to the consumer, and whether the additional cost is reasonable for the added benefits (Stefani et al. 2006).

The emergence and popularity of organic foods, which tend to cost more than conventionally farmed products, acknowledges people's willingness to pay more for a perceived higher quality product (Torjusen et al. 2001). Choosing to pay more for organic produce rather than conventionally farmed products is considered superior for both health and environmentally centered reasons, thereby increasing the perceived quality (Grunert et al. 2003).

3.6.2 CONVENIENCE

Reliance on convenience at all levels including purchasing, preparation, and consumption has been a growing trend since the nineteenth century (Jaeger 2006). Ready-to-eat meals, snacks, and precooked foods are emerging as an important quality trait as women enter the work force leaving less time for traditional meal preparation and eating times. Meals can be prepared in less time and be consumed "on the go" while walking or driving. Foods with extended shelf life are also important to time-poor consumers, as they are easily stored and prepared when needed (Muller and Steinhart 2006).

3.6.3 BRANDING

Brand names, generic marks, and labeling all provide additional information and cues to consumers, and can improve quality perception at the point of purchase (Brunso et al. 2005). Branding benefits both the consumer and the manufacturer, not only allowing consumers to easily identify products, but also allowing manufacturers to market products with specific qualities (Brunso et al. 2005). Brands allow consumers to draw on previous experiences or information with specifically marketed products (Jaeger 2006). As food is bought and consumed on a regular basis, and at times in a scripted, subconscious fashion, branding contributes to the ease of routine in food purchasing, decreasing time needed for product evaluation (Bredahl 2003).

Companies can change brand names or sell their produce at specific outlets to appeal to different consumers. Pilchards and sardines are the same small fish, yet pilchards are associated with cheap, tinned fish, and sardines are associated with Mediterranean cuisine. The Pilchard Works in the United Kingdom renamed their product from "pilchards" to "Cornish Sardines," which are sold exclusively at high class supermarkets (Jaeger 2006).

The relevance of branding is a function of the information visible to the consumer (Brunso et al. 2005, Jaeger 2006). Labels stating the level of hygiene practiced during production, for example, are generally not of interest to consumers and brands must be considered as trustworthy in regard to both history and sponsors. Once again individual preferences will dictate brand appeal, so not all brands or labels will be of interest to all consumers (Brunso et al. 2005).

3.6.4 FOOD PROCESSING

Food processing techniques are important indicators of quality for religious, ethical, and environmental reasons (Schröder 2003). Religion may dictate food quality without prior experience through beliefs and traditions. People who follow the Muslim religion are only allowed meat which is "halal," referring to the specific slaughter technique, products which do not meet the criteria are immediately excluded (Schröder 2003).

The emergence and public rejection of genetically modified food products has brought to light consumers' interest in food production technologies, its importance to quality perception, and also consumers' lack of trust regarding science (Grunert 2002). Consumers appear to associate high risk with food products containing genetically modified foods, even though most experts dismiss these concerns and praise the merits these products have to offer (Grunert 2002). An obvious counterpoint and benefactor is organic foods and environmental concerns. Organics once again highlight consumers' interest in environmentally conscious farming and land sustainability.

3.6.5 ETHICAL CONCERNS AND ANIMAL WELFARE

Ethical concerns regarding animal welfare have become important considerations, especially in regard to meat purchases (Bernues et al. 2003). Organic farming techniques, free-range chickens, and barn-laid eggs challenge conventional agriculture by practicing natural, environmentally friendly as well as livestock friendly farming techniques thereby raising the perceived quality and the price of many food products (Torjusen et al. 2001).

3.6.6 CREDENCE ATTRIBUTES

Under most circumstances credence qualities cannot be ascertained by the consumer, their presence in foods is a matter of credibility and trust (Brunso et al. 2005). Credence characteristics include nutritional value, health benefits, production techniques, and ingredients, their role toward quality perception is becoming increasingly important (Grunert 2002, Brunso et al. 2005). The emergence of chronic diet-related diseases such as obesity and type 2 diabetes along with degenerative pathologies which are related increase the appeal of credence qualities which promise better health, longevity, or weight loss (Bellisle 2003).

As credence qualities are not experienced during consumption, repeat purchases are driven by the ability of credence attributes to fulfill consumer requirements, and to the extent they are used as a basis for inferences to perceived quality. A bread with added omega-3 fatty acids can be marketed as lowering the risk of cardiovascular disease, which is a long-term effect not easily measured by the consumer. This credence attribute may add perceived quality to that bread if the consumer has a family history of cardiovascular disease. It could equally decrease quality perception for others who prefer traditional bread, or those who are ambivalent about their health.

3.6.7   CULTURAL DIFFERENCES

The perception of food attributes which predict quality varies depending on culture. Food manufacturers are well aware of this and will modify products to suit their market. Foods can be produced sweeter, saltier, or packaged differently to meet specific cultural values or preferences (Nielsen et al. 1998). Indeed the quality of the same product can be considered high, but for a range of different reasons in different countries. In Denmark, the most important quality of olive oil was health related; in France, quality referred to taste and traditional ingredients; and in England, olive oil is considered quality in regard to price and versatility (Nielsen et al. 1998). Olive oil is just one example of how food quality perception differs amongst cultures.

3.6.8   CONTEXTUAL INFLUENCES

The context in which food is bought, prepared, and consumed influences food experiences and perceptions. Contextual concepts include the meal, the environment where it is consumed, and social interactions surrounding the meal (King et al. 2007).

Purpose and company of the meal are important determinants in food quality perception. Food that is prepared for special guests may appear more impressive, possesses more desirable sensory attributes, or be presented and served in nice crockery, increasing quality perception (Meiselman et al. 2000). The purpose or use of a food can also alter quality perception; overripe tomatoes may be considered low quality for a salad, but are ideal for soup or stews (Ophius et al. 1995).

The social context surrounding a meal can influence quality perception. Positive effects on food acceptance and perceived quality occur when the people eating know each other (Meiselman et al. 2000). Additionally, people may choose different foods when dining alone or with company (King et al. 2007).

Where and when the food is consumed is equally important. Food that needs to be eaten on the run has different quality dimensions than food that will be eaten at the table. Food on the run needs to be transportable, disposable, and convenient (King et al. 2007). The quality of fast food is judged differently from home-cooked meals, and differently again, from restaurant meals (Jaeger 2006).

The environments in which food is served or consumed can affect quality perception. Food served in a restaurant is thought to be of better quality than food served in a cafeteria or institution (Meiselman et al. 2000). Manipulations of the environments such as in ethnic food restaurant increase quality perception when the surroundings reinforce the culture and traditions of the food being served (King et al. 2007).

Time of day and type of food are also important, a fresh crisp flavorsome apple may be considered a high quality snack between meals, but not satisfactory for a main meal (Jaeger 2006).

3.6.9 FOOD TRADITIONS

Food traditions and traditional food practices contribute to quality perception in a variety of ways. Traditional foods can be associated with special events such as birthdays, Christmas, and other occasions. Religious beliefs may also reflect food traditions, at times prohibiting foods and at others associating signature dishes with specific calendar events and celebrations. Traditional foods also relate to local and artisan ingredients, place of origin, methods of production, as well as know-how (Cayot 2006).

Traditional food, food practices, and methods of production are becoming increasingly lost as technology, safety, convenience, and health emerge as trends dictating the food supply. The next question is to what extent do these modern issues detract from quality?

The ban on exports of cheese made from raw milk in many countries is an example of traditional practices which need to be changed to ensure food safety (Hersleth et al. 2005). Cheese maturation and sensory experiences differ between traditional methods which use raw milk and conventional methods which use pasteurized milk (Cayot 2006). The shift from raw to pasteurized milk decreases the growth and proliferation of unknown microorganisms which reside in raw milk, and presumably is less risky to the consumer (Teng et al. 2004). The microorganisms that live in raw milk are essential for flavor development in artisan cheese. By allowing only pasteurized milk to be used in cheese you dramatically reduce the quality and diversity of flavor. You could argue such processing quality controls actually reduce the quality of foods. Committees such as the Appellation d'Origine Controlle (AOC) and other institutionalized systems of quality certification seek to maintain the traditions associated with certain foods and beverages. Products associated with this organization need to abide by traditional methods of farming, processing, and ingredients to maintain the high quality that is associated with these products. AOC cheeses are exclusively made with raw milk; additives and processing are strictly controlled (Bertozzi and Panari 1993).

Low-fat cheese occupies 20% of the cheese market in the United States where it is perceived as a quality product defined by health (Banks 2004). In other countries such as Italy and France, low-fat cheese appeals only to a small percentage of consumers as cheese quality in these countries is rooted in authenticity, traditional methods of farming, and production as well as taste quality (Banks 2004). The debate over what product holds more quality is up to the reader.

Place of origin is also related to tradition and is a signal of quality, shaped by values and customs (Stefani et al. 2006). Country of origin relates authenticity and tradition to foods whose production has been embedded in a specific environment with consolidated production methods which cannot be reproduced with technology (Bertozzi and Panari 1993). In some cases, place of origin of foods has become famous for the food produced there (e.g., Emmental, Parmesan, and Champagne) (Bertozzi and Panari 1993).

3.7 INTEGRATION OF SENSORY AND NONSENSORY FACTORS IN QUALITY PERCEPTION

Food quality perception occurs as a result of the perceptual process which integrates all information surrounding foods in order to form judgments, and as such is a psychological construct. Consumers form subjective impressions over foods through psychological processes which are shaped by cognitive competencies, previous knowledge, and prior experiences.

Consumers use both sensory (sight, hearing, taste, smell, and touch) and nonsensory dimensions such as convenience and health to form quality perceptions (Bredahl 2003). Information regarding food products can be evaluated both prior to purchase by touching, smelling, inspecting foods or studying labels, health claims and brands, as well as after preparation and consumption by sensations and other postingestional outcomes or convenience dimensions experienced during preparation (Holm and Kildevang 1996).

Perception is situation specific and not static over time, attributes of specific foods are considered desirable for different occasions, at different times and as our knowledge and exposure to food increases so does our perception of quality. A bagel with cream cheese could be considered quality for a picnic or a food be eaten "on the go." Conversely a poached filet of sole is more suited to consumption when seated in a high class restaurant (Meiselman et al. 2000, King et al. 2007). Both foods are considered to have quality attributes for the specific settings.

Specific product attributes are irrelevant when judged on their own. Contributions and importance of attributes to quality depend on consumer beliefs over the desirable effects which may occur with, during, or after consumption. The more important specific attributes or characteristics of food are to the individual the higher they will rate the quality of the food. These cognitive processes which occur when evaluating food product quality depend on the associations between the consumers' personal values, product attributes, and self-relevant consequences of purchase (Nielsen et al. 1998). For example, consumers may infer concrete characteristics such as low-fat yoghurt to abstract concepts such as fewer calories, being slimmer, more socially acceptable, and confident (Grunert 1997). This can result in an increase in perceived quality of the product.

Many attributes of food cannot be ascertained before purchase or consumption such as taste, mouth feel, and tenderness; other qualities such as health benefits may never be ascertained (Grunert 1997). In order to make logical purchase decisions, consumers need to form quality expectations leading to quality purchases. The Total Food Quality Model proposed by Grunert (1997) illustrates how consumers gather and use information to form judgments and expectations over food products before purchase and then how their experiences relate to expectations. Perceived quality is decreased when expectation fails to meet the actual experience. As consumers base expectancies on prior experience, knowledge, health claims, and cues which may or may not indicate quality, foods will not always fulfill their expectations. Experiences after consumption influence future purchases and expectations on foods (Grunert 2002).

3.8 CONCLUSION

Food quality perception is a multidimensional concept that cannot be defined quantitatively. Objective measures of food quality may play a role in quality assurance and quality control, but have little effect on consumers' food choices. Quality in food science is best described as "fitness for consumption," the factors, attributes, and characteristics which define fitness will be shaped by individual opinions, views, and beliefs and may be rooted in ethical, cultural, or religious constructs.

Perceived quality is determined within the context of sensory and nonsensory factors which in turn are subject to individual differences and situational factors. Consumers use cues to predetermine food quality prior to consumption. During and after consumption consumer experiences will govern future purchases and expectations over food products.

Emerging food trends such as lowered fat products are viewed as quality by some, yet despised by others who perceive quality not through fat content but prefer foods embedded within traditions and taste. Food quality perception is a subjective experience which is definable, only by the individual. What rates high on one's quality list will not appear on another's.

ACKNOWLEDGMENTS

I wish to thank Jessica Stewart for assistance in preparing this chapter. This work was supported by The Institute of Biotechnology, Deakin University, and the RPA cluster funding from the Health, Medicine, Nursing and Behavioural Sciences Faculty, Deakin University.

REFERENCES

Acebron, B.L. and Dopico, C.D. 2000. The importance of intrinsic and extrinsic cues to expected and experienced quality: An empirical application to beef. *Food Quality and Preference* 11: 229–238.

Alli, I. 2003. *Food Quality Assurance: Principles and Practices*. Boca Raton, FL: CRC Press.

Banks, J.M. 2004. The technology of low-fat cheese manufacture. *International Journal of Dairy Technology* 57: 199–207.

Bellisle, F. 2003. Why should we study human food intake behaviours? *Nutritional, Metabolic and Cardiovascular Disorders* 13: 189–193.

Bernues, A., Olaizola, A., and Corcoran, K. 2003. Extrinsic attributes of red meat as indicators of quality in Europe: An application for market segmentation. *Food Quality and Preference* 14: 265–276.

Bertozzi, L. and Panari, G. 1993. Cheeses with Appellation d'Origine Controlee (AOC): Factors that affect quality. *International Journal of Dairy Technology* 3: 297–312.

Bredahl, L. 2003. Cue utilisation and quality with perception with regard to branded beef. *Food Quality and Preference* 15: 65–75.

Brunso, K., Bredahl, L., Grunhert, K.G., and Scoholder, J. 2005. Consumer perception of the quality of beef resulting from various fattening regimes. *Livestock Production Science* 94: 83–93.

Buck, L.B. 1992. The olfactory multigene family. *Current Biology* 2: 467–473.

Cayot, N. 2006. Sensory quality of traditional foods. *Food Chemistry* 102: 445–453.

Chandrashekar, J., Hoon, M.A., Ryba, N.J., and Zuker, C.S. 2006. The receptors and cells for mammalian taste. *Nature* 444: 288–294.

Firestein, S. 2001. How the olfactory system makes sense of scents. *Nature* 413: 211–218.

Frank, R.A. and Byram, J. 1988. Taste–smell interactions are tastant and odorant dependent. *Chemical Senses* 13: 445–455.

Frank, R.A., van der Klaauw, N.J., and Schifferstein, H.N.J. 1993. Both perceptual and conceptual factors influence taste–odour and taste–taste interactions. *Perception and Psychophysics* 54: 343–354.

Garrett, B.E. and Griffiths, R.R. 1998. Physical dependence increases the relative reinforcing effects of caffeine versus placebo. *Psychopharmacology (Berl)* 139: 195–202.

Green, B.G. and Lawless, H.T. 1991. The psychophysics of somatosensory chemoreception in the nose and mouth. In *Smell and Taste in Health and Disease*, eds. T.V. Getchell, L.M. Bartoshuk, R.L. Doty, and J.B. Snow, pp. 235–253. New York: Raven Press.

Grunert, K.G. 1997. What's in a steak? A cross-cultural study on the quality perception of beef. *Food Quality and Preference* 8: 157–174.

Grunert, K.G. 2002. Current issues in the understanding of consumer food choice. *Trends in Food Science and Technology* 13: 275–285.

Grunert, K.G., Bredahl, K., and Brunso, K. 2003. Consumer perception of meat quality and implication for product development in the meat sector—A review. *Meat Science* 66: 259–272.

Harker, R.F., Gunson, R.F., Brookfield, P.L., and White, A. 2002. An apple a day: The influence of memory on consumer judgment of quality. *Food Quality and Preference* 13: 173–179.

Harrison, R.M. and Hester, R.E. 2001. *Food Safety and Food Quality Issues in Environmental Science and Technology*. Cambridge: Royal Society of Chemistry.

Hersleth, M., Ilseng, M.A., Martens, M., and Naes, T. 2005. Perception of cheese: A comparison of quality scoring, descriptive analysis and consumer responses. *Journal of Food Quality* 28: 333–349.

Holm, L. and Kildevang, H. 1996. Consumers' views on food quality. A qualitative interview study. *Appetite* 27: 1–14.

Hughes, J.R., Oliveto, A.H., Bickel, W.K., and Higgins, S.T. 1993. Indication of caffeine dependence in a population based study. In *Problems of Drug Dependence 1993. Proceedings of the 55th Annual Scientific Meeting*. Washington DC: U.S. Government Printing Office.

Hyde, R.J. and Witherly, S.A. 1993. Dynamic contrast: A sensory contribution to palatability. *Appetite* 21: 1–16.

Jaeger, S.R. 2006. Non-sensory factors in sensory science research. *Food Quality and Preference* 17: 132–144.

Keast, R.S.J. 2003. The effect of zinc on human taste perception. *Journal of Food Science* 68: 1871–1877.

Keast, R.S.J. and Riddell, L.J. 2007. Caffeine as a flavor additive in soft-drinks. *Appetite* 49: 255–259.

Keast, R.S.J. and Roper, J. 2007. A complex relationship among chemical concentration, detection threshold, and suprathreshold intensity of bitter compounds. *Chemical Senses* 32: 245–253.

Keast, R.S.J., Dalton, P.H., and Breslin, P.A.S. 2004. Flavour interactions at the sensory level. In *Flavor Perception*, eds. A.J. Taylor and D.D. Roberts, pp. 228–255. Oxford, U.K.: Blackwell Publishing.

Kennedy, O.B., Stewart-Knox, B.J., Mitchell, P.C., and Thurnham, D.I. 2005. Flesh colour dominates consumer preference for chicken. *Appetite* 44:181–186.

King, S.C., Meiselman, H.L., Hottenstein, A.W., Work, T.M., and Cronk, V. 2007. The effects of contextual variables on food acceptability: A confirmatory study. *Food Quality and Preference* 18: 58–65.

Lawless, H. 1995. Dimensions of Sensory quality: A critique. *Food Quality and Preference* 6: 191–199.

Martins, Y. and Pliner, P. 2005. Human food choices: An examination of the factors underlying acceptance/rejection of novel and familiar animal and nonanimal foods. *Appetite* 45: 214–224.

Meiselman, H.L., Johnson, J.L., Reeve, W., and Crouch, J.E. 2000. Demonstrations of the influence of the eating environment on food acceptance. *Appetite* 35: 231–237.

Muller, A. and Steinhart, H. 2006. Recent developments in instrumental analysis for food quality. *Food Chemistry* 102: 436–444.

Nielsen, N.A., Larsen, B.T., and Grunert, K.G. 1998. Consumer purchase motives and product perceptions: A laddering study on vegetable oil in three countries. *Food Quality and Preference* 9: 455–466.

Ophius Oude, P.A.M. and Van Trip, H.C.M. 1995. Perceived quality: A market driven and consumer oriented approach. *Food Quality and Preference* 6: 177–185.

Prescott, J. 1999. Flavour as a psychological construct: Implications for perceiving and measuring the sensory qualities of foods. *Food Quality and Preference* 10: 349–356.

Reed, D.R., Tanaka, T., and McDaniel, A.H. 2006. Diverse tastes: Genetics of sweet and bitter perception. *Physiology and Behavior* 88: 215–226.

Rozin, P. 2007. Food choice: An introduction. In *Understanding Consumers of Food Products*, eds. L. Frewer and H. van Trijp, Chapter 1. Cambridge: Woodhead Publishing.

Schröder, M.J.A. 2003. *Food Quality and Consumer Value: Delivering Food that Satisfies.* New York: Springer.

Schuh, K.J. and Griffiths, R.R. 1997. Caffeine reinforcement: The role of withdrawal. *Psychopharmacology (Berl)* 130: 320–326.

Small, D.M., Gregory, M.D., Mak, Y.E., Gitelman, D., Mesulam, M.M., and Parrish, T. 2003. Dissociation of neural representation of intensity and affective valuation in human gustation. *Neuron* 39: 701–711.

Stefani, G., Romano, D., and Cavicchi, A. 2006. Consumer expectations, liking and willingness to pay for speciality foods: Do sensory characteristics tell the whole story? *Food Quality and Preference* 17: 53–62.

Szcesniak, A.S. 2002. Texture is a sensory property. *Food Quality and Preference* 13: 215–225.

Teng, D., Wilcock, A., and Aung, M. 2004. Cheese quality at farmers markets: Observation of vendor practices and survey of consumer perceptions. *Food Control* 15: 579–587.

Torjusen, H., Leiblein, G., Wandel, M., and Francis, C.A. 2001. Food system orientation and quality perception amongst consumers and producers of organic foods in Hedmark County, Norway. *Food Quality and Preference* 12: 207–216.

Williams, A.A. 1992. Flavour quality—understanding the relationship between sensory responses and chemical stimuli. What are we trying to do? The data, approaches and problems. *Food Quality and Preference* 5: 3–16.

Yeomans, M.R., Jackson, A., Lee, M.D. et al. 2000. Acquisition and extinction of flavour preferences conditioned by caffeine in humans. *Appetite* 35: 131–141.

Part II

Evaluating the Safety and
Quality of Foods

4 Models for Microorganism Inactivation: Application in Food Preservation Design

Stella M. Alzamora, Sandra N. Guerrero,
Aurelio López-Malo, Enrique Palou,
Cielo D. Char, and Silvia Raffellini

CONTENTS

4.1   INTRODUCTION

The complexity of the preharvest and postharvest environments makes it impossible to control all potential sources of microbial contamination. Main efforts at prevention and control are so implemented throughout processing systems.

Preservation techniques currently act in one of three ways: preventing microorganisms' access to foods, inactivating them should they gain access, or preventing or slowing their growth. Traditionally, the most popular preservation technologies for the inactivation of microbial populations of food have relied on thermal treatments. Unfortunately, heating induces physical and chemical changes in the food, and efforts to develop many alternative processing technologies that result in products that have minimal process-induced changes in sensory and nutritional characteristics are being investigated. Most new and emerging preservation techniques (high hydrostatic pressure [HHP], ultrasound, high voltage gradient pulses, electron beam and gamma irradiation, light pulses, and ultraviolet, among others) aim to inactivate microorganisms in foods rather than inhibit them.

The introduction of any alternative technology or combination of various alternative/traditional technologies as well as the optimization of traditional methods require scientific data about microbial responses. In particular, kinetic parameters and models are essential to develop food preservation processes that ensure safety (López Malo et al. 2002, McMeekin and Ross 2002, Mafart 2005, McMeekin et al. 2006, McMeekin 2007). The parameters also allow comparison of the ability of different process technologies to reduce microbial populations. There is no question that quantitative microbiology is one of the most useful approaches not only in determining the microbiological impact of the different steps associated with the production, distribution, and retailing of a food but in determining the optimum conditions for many preservation processes. A predictive microbiological approach is needed to select factors and levels with a statistical sense in the development of alternative preservation techniques (Alzamora et al. 2000, Alzamora and López-Malo 2002). An accurate model prediction of survival curves would be beneficial to the food industry in selecting the optimum combinations of lethal agents and environmental factors as well as exposure times to obtain desired levels of inactivation while minimizing production costs and maintaining a maximum degree of sensory and nutrient qualities. Potential combinations of preservation factors are numerous, but until now not much quantitative microbiological information about combinations of alternative processes with traditional constraints is available. Predictive microbiology provides the tools to compare the impact of different factors on the reduction of microbial populations and is an aid to understanding biological system behavior, assisting in the development of mechanistic models that will explain microbial activity in terms of properties and events at the cellular and molecular levels.

This chapter is intended to briefly review some selected mathematical models proposed by researchers to describe microbial inactivation/decline, and to discuss the applications of these models in the design of food preservation techniques.

The lethality of a stress factor is strongly affected by the presence/intensity of other lethal/inhibitory factors. Examples are here presented where predictive inactivation models allow quantification of the microbial response in treatments with conventional/unconventional lethal agents combined with other environmental stress factors. These models would improve both microbial safety and the quality of foods.

4.2   INACTIVATION/SURVIVAL MICROBIAL BEHAVIOR

In thermal and nonthermal inactivation of microorganisms, four common types of semilogarithmic survival curves (Figure 4.1) are usually found: linear curves (first-order inactivation), curves with a shoulder or initial lag period, curves with a tailing (or biphasic curves), and sigmoid curves (Peleg and Cole 1998, Xiong et al. 1999).

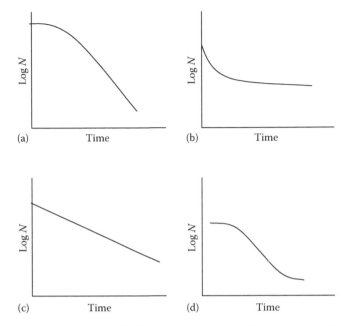

FIGURE 4.1 Different inactivation/decline curves: (a) with a shoulder, (b) with a tail, (c) log-linear, and (d) sigmoid. (From Raffellini, S. et al., *Journal of Food Safety*, 28: 514. With permission.)

Traditionally, microbial inactivation has been considered a process that follows first-order kinetics (Figure 4.1c). It has been implicitly assumed that all cells and spores have identical resistance to a lethal agent and each microorganism has the same probability of dying (van Boekel 2002). Following this mechanistic approach, the equation proposed by Chick (1908) has been widely used to calculate sterility in thermal preservation methods:

$$\frac{dN(t)}{dt} = -k(T)N(t) \tag{4.1}$$

From this equation, after integration and transformation into a decimal logarithm, the well-known concept of D value or decimal reduction time was derived:

$$\log\left(\frac{N_0}{N(t)}\right) = \frac{t}{D(T)} \tag{4.2}$$

where
   N_0 is the initial number of cells
   N is the number of cells at time t
   $k(T)$ is a temperature-dependent rate constant, with $D(T) = 2.303/k(T)$

The temperature dependence of $D(T)$ is expressed using the so-called z value, that is, the increase in temperature that would reduce the decimal reduction time by a factor of 10:

$$\frac{D(T_1)}{D(T_2)} = 10^{\frac{T_2 - T_1}{z}} \qquad (4.3)$$

The z value can be related to the activation energy E_a as follows:

$$z = \frac{2.303RT^2}{E_a} \qquad (4.4)$$

If the temperature history $T(t)$ is known, the lethality of the thermal process is calculated as

$$F_0(t) = \int_0^t 10^{[T(t) - T_{ref}]/z} \, dt \qquad (4.5)$$

and the survival ratio results:

$$\log S(t) = -F_0(t)/D_{T=T_{ref}} \qquad (4.6)$$

where T_{ref} is the reference temperature, usually being 250°F for the sterilization of low-acid foods or 212°F for the sterilization of acid foods.

This interpretation has dominated the field of quantitative microbiology for many years, but this analysis has appeared insufficient because the decrease of the population does not usually follow first-order kinetics (Peleg 2006). Forcing a straight line through a concave downward semilogarithmic survival curve or a concave upward semilogarithmic survival curve can result in an over- or underestimate of the organism resistance, respectively, and so in overprocessing in the first case or in underprocessing in the second case (Peleg 1999). The common explanation for shoulders and tails in the survival curves has been to consider the presence of subpopulations of different resistances, each one with its own first-order inactivation kinetics.

In 2003, 40 world-renowned scientists participated in the Institute of Food Technologists Second Research Summit to advance the understanding of microbial inactivation kinetics and models for non-log-linear survivor curves and identify needs for further research (Heldman and Newsome 2003). Two different approaches to explain microbial inactivation were discussed, among others:

- Mechanistic: The decay is produced by some molecular or physical mechanism (i.e., a monomolecular transformation or an enzyme-catalyzed reaction). This approach is deterministic in nature since all cells behave in the same way and the death of a single cell is due to one single event.

- Vitalistic: In line with other areas of biological studies, this approach is based on the assumption that the individual microorganisms in a population do not have identical resistances, and that microbial sensitivity to lethal agents is distributed (i.e., biological variability) (Booth 2002). Shoulders and tails are due to underlying physiological reactions of the cells/spores to lethal conditions.

Although there were a few issues where consensus was not reached, the summit resolution stated: "Since there is significant evidence that microbial survivor curves can be described by non-log-linear kinetic expressions, the scientific and technical community should recognize alternative models and parameters for description and communication of the survival of microbial populations when exposed to various lethal agents." The concept of "the number of log cycles of reduction rather than the classical D value" to communicate the performance of food preservation processes was also identified as one of the implications of this new approach in modeling survival curves.

A variety of alternative models (mechanistic or empirical, deterministic or probabilistic) have been developed to describe curvilinear semilogarithmic inactivation curves, and a number of excellent articles, books, and book chapters are available on the subject, including the following: Xiong et al. (1999), Albert and Mafart (2005), Chmielewsky and Frank (2004), McKellar and Lu (2004), McMeekin et al. (1993), Peleg (2006), Sapru et al. (1993), Mafart et al. (2002), Geeraerd et al. (2000), Shadbolt et al. (1999), and ter Steeg and Ueckert (2002). Some of these alternative models show accuracy but are either overparameterized (mechanistic models) or have parameters without any physical or biological significance (empirical models). Moreover, the complexity of many of these models hinders their application in actual process calculations. Other models, which consider the survival curve as a cumulative form of the temporal distribution of a lethality event distribution, presented a probabilistic approach. In addition, to enhance the use of inactivation models by the food industry and food microbiologists, a user-friendly interface, GInaFiT (Geeraerd and Van Impe Inactivation Model Fitting Tool), for testing nine different types of microbial survival models on user-specific experimental data relating the evolution of the microbial population with time was recently developed by researchers at the Katholieke Universiteit Leuven (Geeraerd et al. 2005). The software covers all survivor curve shapes observed until now in the literature for vegetative cells and is distributed under a free license agreement. The nine model types allowed fitting classical log-linear curves; curves displaying a so-called shoulder before a log-linear decrease are apparent; curves displaying a so-called tail after a log-linear decrease; survival curves displaying both shoulder and tailing behaviors; concave curves; convex curves; convex/concave curves followed by tailing; biphasic inactivation kinetics; and biphasic inactivation kinetics preceded by a shoulder.

Process models must account for the agent/level used for microbial inactivation and the potential for multiple environmental parameters as well as for multiple agents in combination. Some selected and relatively simple mathematical models that allow quantification of the effects of various lethal agents on microbial inactivation and their application to food process design are described in the following section.

4.3 DESCRIPTION OF SELECTED INACTIVATION/ SURVIVAL MODELS

4.3.1 MODIFIED GOMPERTZ MODEL

The Gompertz equation and its modified forms have been used primarily in modeling the asymmetrical sigmoid shape of microbial growth curves (McMeekin et al. 1993). Nonlinear survival curves could be modeled using a modified Gompertz equation as described as follows:

$$\log [N_t/N_0] = C \exp (-\exp(A + Bt)) - C \exp (-\exp(A)) \qquad (4.7)$$

where

$\log [N_t/N_0]$ represents the logarithmic surviving fraction

Three parameter estimates (A, B, C) represent the different regions of the survival curve (Figure 4.2): the initial shoulder $(A, [\min])$; the maximum death rate $(B, [\min^{-1}])$; and the overall change in the survivor number $(C, [-])$

However, the parameters of the modified Gompertz equation have no direct link with microbial death kinetics when the logarithmic number of microorganism cells is used. Ideally, it should be used for the cell's concentration rather than for its logarithms (Baranyi et al. 1993).

This equation has been demonstrated to be particularly suitable for sigmoid survival curves (an initial shoulder followed by an exponential phase and a tailing region) and survival curves with tails or shoulders and has the property of an asymptotic nonzero probability of cell (the lethality approaches $- C$) when time approaches infinity (Juneja and Marks 2003).

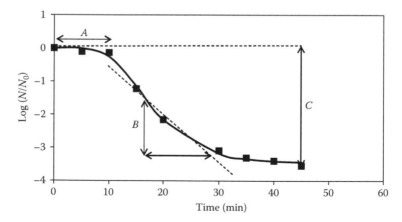

FIGURE 4.2 Graphical representation of a typical sigmoid curve fitted by the modified Gompertz model: the shoulder region (A), the maximum death rate (B), and the overall change in the survivor number (C).

Badhuri et al. (1991) were the first to demonstrate that the modified equation of the Gompertz model could describe the nonlinear survival curves for *Listeria monocytogenes* heated in sausage slurry. Subsequently, Linton et al. (1995a,b) successfully used it to fit nonlinear survival curves for *L. monocytogenes* Scott A and reported that it was effective in modeling survival curves for both linear curves and those containing a shoulder and tail.

In thermal studies, it was also likely to provide a more accurate estimate of a microorganism's resistance than the first-order kinetic model (Bhaduri et al. 1991). It was successfully tested to describe bacterial heat inactivation (Linton et al. 1995a,b). Char et al. (2009) studied the response of *L. innocua*, a surrogate of *L. monocytogenes*, in orange juice to combined treatments involving heating at moderate temperatures (57°C–61°C) and the addition of different levels of vanillin (0–1100 ppm). The modified Gompertz equation accounted for more than 98% of the variation in surviving counts since the adjusted determination coefficient (R^2_{adj}) ranged between 98.3% and 99.9%. In general, high vanillin concentrations in combination with high heating temperatures turned out in inactivation curve shapes closer to linearity with a smaller or absent shoulder region (with consequent lower *A* values) and a larger maximum death rate characterized by the *B* parameter. The decrease in the *B* value reflected a decrease in the resistance of the cells to the combined treatment, as was suggested by Linton et al. (1995a). This behavior was verified in most systems, except for the system containing 1100 ppm vanillin treated at 59°C and the system with 900 ppm vanillin heated at 60°C. But, in these systems, the severity of the combined treatment limited the number of experimental data points, diminishing the predictive power of the Gompertz equation. Parameter *C* exceeded six log-cycle reductions for all tested conditions, indicating that all treatments were effective in destroying microbial populations. Since most curves did not show a noticeable tailing region, the modified Gompertz model overestimated the overall change in survivor value with respect to the observed values. These results agreed with those reported by Linton et al. (1995a) when used to evaluate the heat resistance of *L. monocytogenes* in infant formula at 50°C, 55°C, and 60°C; pH values of 5.0, 6.0, and 7.0; and 0%, 2%, or 4% NaCl concentration. These authors observed that as the heating temperature increased from 50°C to 60°C, microbial cell death was accelerated, generating survival curves without or with scarce shoulder or tailing regions. Chhabra et al. (2002) obtained similar responses when studying the effects of growth conditions (different pH and milk fat levels) on the thermal resistance of *L. monocytogenes* (thermally treated at 55°C, 60°C, or 65°C). They also found that the modified Gompertz model was very versatile in describing the different types of inactivation curves (nearly linear curves, curves with lag periods, and sigmoid curves) with high correlation coefficients.

The modified Gompertz model was also used to describe nonthermal inactivation. Koutsoumanis et al. (1999) evaluated the performance of the Baranyi model, a nonautonomous, first-order differential equation (Baranyi and Roberts 1994) and the modified Gompertz function, as suggested by Gibson et al. (1988), to predict the death of *Salmonella enteritidis* in homemade tarama salad supplemented with oregano essential oil (0.0%, 0.5%, 1.0%, and 2.0% v/w) under different conditions of controlling factors (storage temperatures of 0°C, 5°C, 10°C, 15°C, and 20°C; pHs of 4.3, 4.8, 5.0, and 5.3). The modified Gompertz equation gave a short lag phase (small shoulder) in most cases, which was not in agreement with the experimental data points, and greater maximum death rates, possibly due to the fact that the Gompertz curve shows a curvature while the Baranyi model is a straight line in the exponential zone. These authors demonstrated that when the microorganism death was low, there was no difference between model performances, but for higher death rates, the Gompertz equation lost goodness of fit, probably because the lag phase is not separated from the exponential and the stationary phases (Koutsoumanis et al. 1999), indicating that it is more adequate for completely sigmoid rather than linear or semisigmoid curves (curves with shoulder or tail).

More recently, some authors used this model to characterize microorganism resistance to other emerging factors. Guerrero et al. (2004) studied the resistance of *Sacchromyces cerevisiae* cells to the action of ultrasound (20 kHz, 95.2 μm wave amplitude) at 45°C in laboratory media (pH 5.6) containing low weight chitosan (1000 ppm). The presence of such an additive and the time of contact greatly modified the inactivation curve shape rendering to sigmoid survival curves. Experimental data were successfully modeled by using the modified Gompertz equation (Figure 4.3a). The obtained regression parameters were greatly useful in generating predictions for the yeast behavior at these times (Table 4.1). The addition of chitosan enhanced inactivation due to ultrasound. Incubating the yeast with a positively charged polysaccharide for several minutes prior to sonication increased the inhibitory effect of the chitosan and ultrasound. After 30 min of exposure to the polysaccharide, approximately one log-cycle reduction of viable yeast cells was obtained, leading to a final reduction of three log cycles at the end of the ultrasonic treatment. A further increase in the time of exposure resulted in little significant change in the level of yeast reduction (Figure 4.3a).

Ferrante (2004) used the modified Gompertz equation to model *L. monocytogenes* survival in sonicated (600 W, 20 kHz, 95.2 μm wave amplitude, 45°C) fresh, squeezed orange juice (pH 3.5) containing vanillin (0, 1000, and 1500 ppm). The modified Gompertz model successfully represented *L. monocytogenes* survival curves, which were completely sigmoid and semisigmoid depending on the applied treatment (Figure 4.3b). The *A*, *B*, and *C* parameters well characterized the overall curve for the shoulder region, the maximum death rate, and the tailing region, respectively. Statistical parameters indicated that the predicted and observed values for all treatments were not significantly different from each other. The fraction of the variation that was explained by the model ranged between 97.8 and 99.6 (Table 4.1). The estimated values for the parameters showed, in general, that the combined emerging treatment of vanillin addition/sonication was more effective than the individual treatments. Increasing the concentration of vanillin resulted mainly in a significant decrease in *A* values (representing the shoulder region) and *C* values (related to the population that remained at the end of the treatment). These authors also described an overestimation of the *C* parameter for the most severe conditions (1000 and 1500 ppm of vanillin addition plus ultrasonic treatment) since the death rate of the microbial population was fast, giving a survival curve without a tailing region and following an almost first-order kinetics (Ferrante et al. 2004).

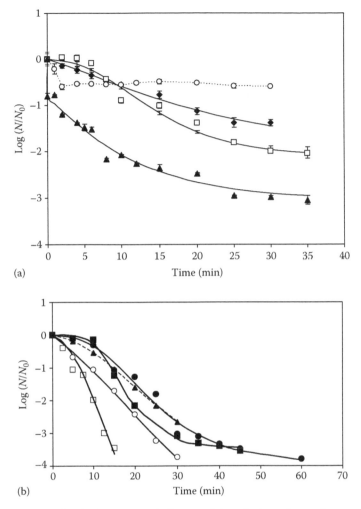

FIGURE 4.3 (a) Survival curves of *S. cerevisiae* during ultrasonic treatments in Sabouraud broth: ultrasonic treatment (-♦-); 1000 ppm chitosan without sonication (⋯ O ⋯); 1000 ppm chitosan and sonication (-□-); 1000 ppm chitosan, incubation 30 min and sonication (-▲-); predicted values (—); standard deviation (I), initial number of yeast cells (N_0:CFU/mL); and yeast survivor count (*N*:CFU/mL). (Adapted from Guerrero, S.N. et al., *J. Food Control*, 16, 131, 2004.) (b) Survival curves (points) and predicted modified Gompertz model (lines) of *L. monocytogenes* during ultrasonic treatments (pH 3.5, 95.2 μm wave amplitude, 45°C) in orange juice fitted by Equation 4.1 as a model: ultrasonic treatment (--▲--); 1000 ppm vanillin without sonication (-●-); 1000 ppm vanillin and sonication (-O-); 1500 ppm vanillin without sonication (-■-); 1500 ppm vanillin and sonication (-□-); predicted Gompertz model values (—); initial number of cells (N_0:CFU/mL); and bacteria survivor count (*N*:CFU/mL). (Adapted from Ferrante, S., Utilización combinada de ultrasonido y antimicrobianos naturales para inhibir el desarrollo de *Listeria monocytogenes* en jugos frutales. Thesis. Universidad Católica Argentina, Buenos Aires, Argentina, 2004.)

TABLE 4.1

Estimates of Gompertz Model Parameters for Microorganism Survival in Laboratory Media or Orange Juice Added with Natural Antimicrobials with and without Ultrasonic Treatment (95.2 µm Wave Amplitude, 45°C)

System	Initial Shoulder A (min)	Std. Error	Maximum Death Rate B (min⁻¹)	Std. Error	Overall Change in Survivor Number C (–)	Std. Error	R^2_{adj}	F
L. monocytogenes in Orange Juice, pH 3.5								
US	1.4	0.2	−0.07	0.02	−4.4	1.3	98.9	501**
1000 ppm vanillin	1.8	0.3	−0.09	0.01	−3.9	0.2	98.1	506**
US/1000 ppm vanillin	1.0	0.3	−0.07	0.02	−5.3	1.0	99.1	1015**
1500 ppm vanillin	2.6	0.2	−0.17	0.02	−3.5	0.06	99.6	2037***
US/1500 ppm vanillin	1.5	0.2	−0.11	0.02	−8.6	0.9	98.2	389**
S. cerevisiae in Laboratory Media, pH 5.6								
US	0.51	0.26	−0.07	0.03	−2.27	0.79	98.8	590**
US/1000 ppm chitosan	1.65	0.36	−0.15	0.03	−2.11	0.17	98.0	210**
US/1000 ppm chitosan (30 min preincubation)	−0.45	0.06	−0.10	0.02	−3.16	0.84	99.8	400**

Sources: Adapted from Ferrante, S., Utilización combinada de ultrasonido y antimicrobianos naturales para inhibir el desarrollo de *Listeria monocytogenes* en jugos frutales. Thesis. Universidad Católica Argentina, Buenos Aires, Argentina, 2004; Guerrero, S.N. et al., *J. Food Control*, 16, 131, 2004.

** Significant at 1% level.
*** Significant at 5% level.

An innovative modified Gompertz model was recently used to predict microbial inactivation under time varying temperature conditions (Gil et al. 2006). It is well known that temperature may vary throughout the complete process. Consequently, the kinetic parameters, estimated under time varying temperature conditions, may differ from those estimated under isothermal conditions. Gil et al. (2006) designed a model that included the variations of temperature along the process by differentiating a modified Gompertz equation with respect to time:

$$\log N = \log N_0 - \log\left(\frac{N_0}{N_f}\right) \times \exp\left(-\exp\left[\frac{k\exp(1)}{\log\left(\frac{N_0}{N_f}\right)}(L-t)+1\right]\right) \tag{4.8}$$

where

N represents the microbial cell density at a certain process time (t)
L is the time parameter (or shoulder)
k is the maximum inactivation rate constant
N_0 and N_f are the initial and residual microbial cell densities, respectively

The final expression considering time varying temperature conditions is

$$\log N_0 - \int_0^t \left[k\exp(1)\exp\left(\frac{k\exp(1)}{\log\frac{N_0}{N_f}}(L-t')+1\right) \right.$$
$$\left. \times\exp\left(-\exp\left(\frac{k\exp(1)}{\log\frac{N_0}{N_f}}(L-t')+1\right)\right) \right] dt' \tag{4.9}$$

The parameters k and L are time–temperature relying. These authors used an Arrhenius expression to describe the relationship between L and temperature as follows:

$$L = a\exp\left(b\left(\frac{1}{T}-\frac{1}{T_{ref}}\right)\right) \tag{4.10}$$

where

a and b are parameters
T_{ref} is a fixed reference temperature

In relation to the rate constant, the expression used is (Ratkowsky et al. 1982)

$$k = c(T-d)^2 \tag{4.11}$$

where c and d are regression parameters.

If expressions 4.10 and 4.11 are included in Equation 4.9, a mathematical model that describes the microbial content related to time and temperature is obtained. Three main assumptions were made: (a) no microbial growth occurs during the come-up time of the nonisothermal heat treatment; (b) there is a limit of temperature below which no inactivation is observed; and (c) the temperature history has not a significant effect on the microbial heat resistance, thus meaning that the use of the actual temperature in expressions 4.10 and 4.11 is adequate. Gil et al. (2006) successfully applied this modified equation over pseudoexperimental data (computer-generated data using model parameters) calculated on the basis of experimental inactivation data of *L. monocytogenes* Scott A published by Casadei et al. (1998). These authors supposed two different heating regimes typically used in surface pasteurization processes and compared the results with isothermal conditions. The authors concluded that the heating period performance (slow or fast) seriously affected bacterial inactivation and the determination of the lowest temperature of inactivation was critical. Although these considerations may be more realistic in a time varying temperature process, the high number of parameters of the model (log N_f, a, b, c, and d) could make the convergence procedure difficult.

In conclusion, various modifications of the Gompertz model can be found in the literature to characterize survival data. All of the previously discussed modified Gompertz models seem to adequately represent the experimental data when sigmoid or semisigmoid survival curves were obtained but failed representing microorganism behavior for the most severe conditions of stress factors when a tail is absent, since these models overestimated the overall change in survivors.

4.3.2 WEIBULL DISTRIBUTION OF RESISTANCES MODEL

The Weibull distribution function takes biological variation into account and is used to describe the spectrum of resistances of the population to a lethal agent under different conditions (Peleg and Cole 1998, Peleg 2000, 2006, von Boekel 2002). Survival patterns are explained without assuming the validity of any kinetic model. As a matter of fact, the Weibull function has been widely used to explain the probabilistic distribution of fracture stresses in brittle materials as well as yielding stresses in ductile materials. The main advantages of the model based on the Weibull distribution are its simplicity and its capability of modeling survival curves that are linear and those that contain shoulder or tailing regions (Peleg 1999).

In contrast to what is assumed in the first-order approach, a lethal event is considered a probabilistic event rather than deterministic. The inactivation mechanism at the molecular level may vary from cell to cell. It is unlikely that all cells in the population behave in the same way and that the death of a single cell is due to one single event. So there will be a distribution of inactivation times and the survival curve should not be treated in kinetic terms but should be considered the cumulative form of the temporal distribution of lethal events (Peleg and Penchina 2000). If each individual of a microbial population is inactivated at a specific time (t_{ci}) and considering t_{ci} as having a continuous distribution, the survival ratio $S(t)$ can be written as

$$S(t) = \int_0^1 f\left[t, t_c\left(\phi\right)\right] \mathrm{d}\phi \qquad (4.12)$$

where
  t_c is the time at which the microorganism dies or loses its viability
  $S(t)$ is the survival fraction
  N/N_0 (N_0 is the initial number of cells and N is the number of cells at
    time t)
  $f[t, t_c(\phi)]$ is a function of the exposure time t and the fraction of organisms ϕ
    that share any given t_c

Assuming t_c has a Weibull-type distribution

$$\frac{d\phi}{dt_c} = bnt_c^{n-1} \exp(-b't_c^n) \tag{4.13}$$

Operating, the cumulative form of Weibull distribution results:

$$S(t) = (N/N_0) = \exp(-b't^n) \tag{4.14}$$

Since microbiologists use logarithms of numbers, Equation 4.14 can be written as
follows:

$$\log_{10} S(t) = -bt^n \tag{4.15}$$

where $b = b'/\log_e 10$.
  Although this transformation distorts the error structure of data, it induces a normal distribution of errors, stabilizes the variance, and assigns adequate weight to the
data point in the tailing region (Peleg 1999, van Boekel 2002).
  The Weibull model has two parameters: the scale or nonlinear rate parameter b
(with dimensions time $^{-n}$) and the dimensionless shape parameter n, which indicate
the overall steepness and the shape of the survival curves, respectively. The survival curves can be fitted using Equation 4.15 and Weibull parameters (n, b) can be
obtained for each experimental condition. These values are then used to calculate
the frequency distribution of resistances or sensitivities using Equation 4.13, and
the distribution's mode, t_{cm}; the mean, $\bar{t}_c$; the variance, σ_{tc}^2; and the coefficient of
skewness, v_1:

$$t_{cm} = \left[(n-1)/nb\right]^{1/n} \tag{4.16}$$

$$\bar{t}_c = \left\{\Gamma\left[(n+1)/n\right]\right\}\big/ b^{1/n} \tag{4.17}$$

$$\sigma_{tc}^2 = \left\{\Gamma\left[(n+2)/n\right] - \left(\Gamma\left[(n+1)/n\right]\right)^2\right\}\big/ b^{2/n} \tag{4.18}$$

and

$$v_1 = \frac{\left[\Gamma(n+3/n)\big/b^{3/n}\right]}{\left[\Gamma(n+2/n)\big/b^{2/n}\right]^{3/2}} \tag{4.19}$$

where Γ is the gamma function.

The distribution mode, t_{cm}, represents the treatment time at which the majority of the population dies or inactivates. The mean, $\bar{t}_c$, corresponds to the inactivation time on average with its variance, σ_{tc}^2. The coefficient of skewness, v_1, represents the skew of the distribution.

According to this model, the log S vs t plot has a downward concavity whenever $n > 1$ and an upward concavity whenever $n < 1$, while when $n = 1$ the survival curve is linear and the Weibull distribution reduces to an exponential distribution. A coefficient of skewness >1 indicates a skew to the right, and <1 indicates a skew to the left. A distribution with a coefficient of skewness = 1 is symmetric.

Although the Weibull distribution is an empiric model, a link can be made with the physiological effects involved in microbial inactivation according to the concavity of the curve (von Boekel 2002). If $n < 1$, it means that the remaining cells have less probability of dying or perhaps adapting to stress. If $n > 1$, cumulative damage occurs and the remaining cells become increasingly susceptible to the lethal agent. If $n = 1$, the probability of dying does not depend on time and each cell is equally susceptible no matter how long the inactivation treatment lasts. Peleg (2006) also postulated that both mechanisms, accumulative damage and sturdier survivors, can coexist. If both modes are balanced, the survival curve would appear as log-linear. If one of them dominates the first stages of inactivation and the other the last stages, sigmoid survival curves would be obtained.

As microbial mortality is increased by lowering the organism's resistance to the treatment (for instance, when an additive or synergistic combination of lethal agents is used and/or the severity of a lethal agent is increased), the mode and the mean of the distribution are lowered. From the point of view of preservation system design, the combination of factors/levels to choose are those that also decrease the spread or variance of the distribution (Peleg 2000). The weibull model can also be used to quantify the influence of the environmental factors on inactivation behavior.

An interesting example is given in Figures 4.4 through 4.6 to illustrate the effect of hydrogen peroxide solutions at different concentrations (0%–3% w/v) and pH values (3.0–7.2) on the inactivation of *Escherichia coli* ATCC 35218 at 25°C (Raffellini et al. 2008). Figure 4.4 shows the experimental survival data and the curves fitted using the cumulative Weibull distribution function. Table 4.2 shows the underlying regression parameters, calculated from Equations 4.15 through 4.19. Most semilogarithmic survival curves were clearly nonlinear. Experimental curves were highly correlated to the predicted data, obtaining very significant adjusted coefficients of determination R_{adj}^2 (0.946–0.998). The survival pattern varied according to the concentration of the lethal agent and the pH of the medium, reflecting resistance distributions with different modes, variances, and skewness. Decreasing pH and

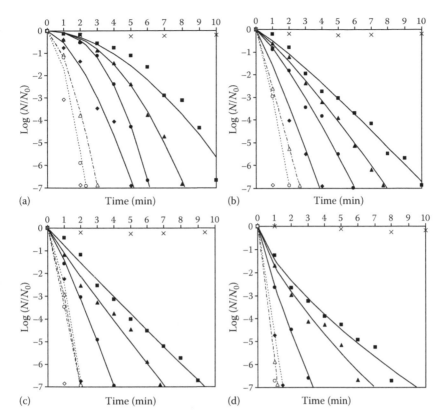

FIGURE 4.4 Effect of H_2O_2 concentration and pH on semilogarithmic survival curves of *E. coli* at 25°C. Experimental (points) and fitted values derived from the Weibullian model (lines). Control (×), 0.50% w/v H_2O_2 (■), 0.75% w/v H_2O_2 (▲), 1.00% w/v H_2O_2 (•), 1.50% w/v H_2O_2 (♦), 2.00% w/v H_2O_2 (△), 2.50% w/v H_2O_2 (○), 3.00% w/v H_2O_2 (♦): (a) pH 7.2, (b) pH 5.8, (c) pH 4.4, and (d) pH 3.0. (Adapted from Raffellini, S. et al., *J. Food Safety*, 28, 514, 2008.)

increasing the antimicrobial agent concentration generally improved the effectiveness of the treatment. The parameter *n* or shape parameter was >1 for the concave downward survival curves (all curves obtained at pH 7.2 and some curves obtained at pH 5.8), indicating that a cumulative damage occurred decreasing the resistance of the remaining cells to the treatment. The inactivation pattern at pH 4.4 mainly exhibited *n* values close to 1 as well as some curves at pH 5.8, meaning that the probability of dying did not depend on treatment time. On the contrary, at pH 3.0, the shape parameter was <1 for most of the inactivation curves, evidencing that sensitive members were destroyed while the remaining cells had less probability of dying and were the sturdy ones. For sanitizer concentrations higher than 1% w/v, the number of experimental points was too small to ascribe a specific inactivation behavior, because of the rapid reduction in *E. coli* population in those conditions. Overall, at a given pH, the parameter *b* or scale parameter gradually increased as the H_2O_2

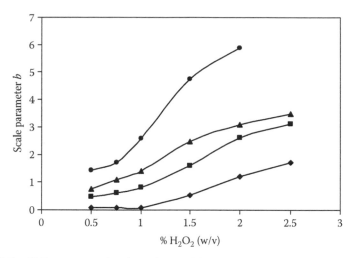

FIGURE 4.5 H_2O_2 concentration dependence of the scale parameter b of *E. coli*, fitted with the Weibullian model, for different pH values. ●: 3.0, ▲: 4.4, and ■: 5.8, ♦: 7.2. (Adapted from Raffellini, S. et al., *J. Food Safety*, 28, 514, 2008.)

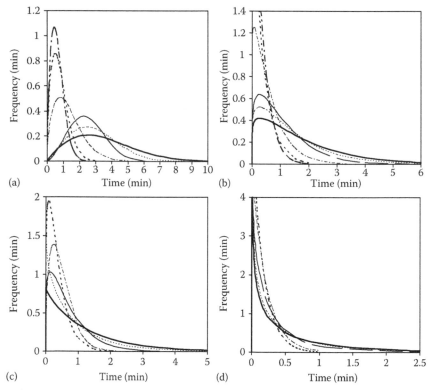

FIGURE 4.6 Frequency distributions of resistances of *E. coli* to H_2O_2 obtained by applying Equation 4.11 at different H_2O_2 concentrations: (—) 0.50% w/v, (....) 0.75% w/v, (-· -· -) 1.00% w/v, (– –) 1.50% w/v, (- -) 2.00% w/v, (— · — · —) 2.50% w/v: (a) pH 7.2, (b) pH 5.8, (c) pH 4.4, and (d) pH 3.0. (Adapted from Raffellini, S. et al., *J. Food Safety*, 28, 514, 2008.)

TABLE 4.2

Weibull Distribution Parameters of *E. coli* under Treatments with Different H$_2$O$_2$ Concentrations at pHs 7.2, 5.8, 4.4, and 3.0 (25°C)

pH	H$_2$O$_2$ Conc. (% w/v)	b (min^{-n})	N	R^2_{adj}	t_{cm} (min)	$\bar{t}_c$ (min)	$v_1(-)$	σ^2_{tc} (min^2)
7.2	0.50	0.08	1.84	0.946	2.6	3.5	1.4	3.8
	0.75	0.08	2.14	0.993	2.4	2.9	1.3	2.1
	1.00	0.08	2.47	0.995	2.3	2.5	1.2	1.1
	1.50	0.53	1.57	0.97	0.8	1.3	1.5	0.8
	2.00	1.21	1.59	0.99	0.5	0.8	1.5	0.3
	2.50	1.72	1.67	0.99	0.4	0.6	1.4	0.2
5.8	0.50	0.49	1.13	0.987	0.3	1.8	1.9	2.5
	0.75	0.63	1.16	0.996	0.3	1.4	1.9	1.5
	1.00	0.82	1.20	0.987	0.3	1.1	1.8	0.9
	1.50	1.60	1.09	0.959	0.1	0.6	1.9	0.3
	2.00	2.61	1.01	0.99	0.1	0.4	2.1	0.1
	2.50	3.10	1.17		0.1	0.4	1.8	0.1
4.4	0.50	0.77	0.99	0.990	—	1.3	2.1	1.8
	0.75	1.10	0.94	0.976	—	0.9	2.2	1.0
	1.00	1.40	1.16	0.998	0.1	0.7	1.9	0.4
	1.50	2.47	1.44		0.2	0.5	1.6	0.1
	2.00	3.07	1.19		0.1	0.4	1.8	0.1
	2.50	3.48	1.01		0.1	0.3	2.1	0.1
3.0	0.50	1.44	0.70	0.978	—	0.8	3.3	1.2
	0.75	1.72	0.72	0.976	—	0.6	3.2	0.7
	1.00	2.57	0.84	0.998	—	0.4	2.6	0.2
	1.50	4.73	1.00		—	0.2	2.1	0.1
	2.00	5.89	0.99		—	0.2	2.1	0.1

Source: Adapted from Raffellini, S. et al., *J. Food Safety*, 28, 514, 2008.

Notes: b, n = constants of Weibullian model; R^2_{adj} = statistic; t_{cm} = distribution's mode; $\bar{t}_c$ = distribution's mean; σ^2_{tc} = variance; and v_1 = coefficient of skewness.

level increased, except in the H$_2$O$_2$ concentration range 0.5%–1.0% w/v at pH 7.2, where the b value was approximately constant (Figure 4.5). The relative increase in b with H$_2$O$_2$ level was more pronounced at the highest pH. For a given agent concentration, the parameter b increased when the pH decreased. Frequency distribution profiles markedly changed with H$_2$O$_2$ concentration and pH (Figure 4.6). In general, the greater the H$_2$O$_2$ concentration and the lower the pH, the narrower the distribution and the lower the mean, as well as the mode, and the variance values. At pH 7.2, frequency shapes with a considerable spread of data, heavy tails, large mode, mean and variance values, indicated that an important fraction of the microorganism population survived after these treatments. At a given pH, the largest variance corresponded to the lowest H_2O_2 level. The frequency distributions corresponding to pH 3.0 did not have a peak and were skewed to the right (with high skewness coefficients and no mode). The combinations of pH 3.0 and H_2O_2 concentrations greater or equal to 1% w/v showed a frequency distribution shape with the smallest mean and without a tail (put on evidence by the small variance values obtained). Accordingly, with this behavior, the majority of the microorganisms were destroyed in a short time of exposure. The population was not only more sensitive on average but it had a more uniform sensitivity to the treatment.

The Weibull distribution is a flexible and simple model and has been used extensively to describe thermal and nonthermal microorganism inactivation. Van Boekel (2002) evaluated the applicability of the Weibull model to describe the thermal inactivation of microbial vegetative cells in 55 case studies taken from the literature. In 53 cases, the curves were found to be nonlinear and could be described very well by the Weibull model. In 39 cases, $n > 1$; in 14 cases, $n < 1$; and only in 2 cases was a true linear curve observed. The temperature dependence of the b parameter could be described by a power law relationship as for the traditional D value, while in 47 cases the n parameter seemed to be independent of temperature. He concluded that this distribution could be used to model nonlinear survival curves and may be helpful to pinpoint relevant physiological effects caused by heating.

The influence of the electric field strength, the treatment time, the total specific energy, and the conductivity of the treatment medium (McIlvaine buffer of pH 7.0) on *L. monocytogenes* inactivation by a pulsed electric field (PEF) have been investigated by Alvarez et al. (2003). A mathematical model based on the Weibull distribution was fitted to the experimental data when the field strength (15–28 kV/cm), treatment time (0–2000 µs), and specific energy (0–3490 kJ/kg) were used as PEF control parameters. A linear relationship was obtained between the $\log_{10}$ of the scale factor b and the electric field strength while the shape parameter n obtained at all the field strengths investigated was the same. That treatment time and specific energy necessary to achieve a given level of inactivation decreased at higher electric field strengths. Therefore, from an industrial application point of view of this technology, higher electric field strengths would be more suitable due to lower residence times and energy consumption. When PEF treatment was performed in the same medium at different pHs (3.5–7.0), the general shape of survival curves was concave upward but n decreased as the pH decreased (Gomez et al. 2005). At each pH, the shape parameter did not depend on the electric field but the parameter b varied with this variable and the pH. This model based on the Weibull distribution was successfully validated by the authors when they analyzed the inactivation of *L. monocytogenes* by PEF application in apple juice.

Peleg and Penchima (2000), Peleg and Cole (2000), Campanella and Peleg (2001), Peleg (2006), and corradini and peleg (2004) proposed an alternative method to calculate commercial thermal processes when the microbial population exhibits isothermal survival curves described by the Weibullian power law model. This method was based on the following assumptions: (a) growth and damage repair did not occur over the pertinent timescale; (b) all the survival curves could be described by the same primary model; (c) the temperature dependence of the parameters of the primary model (n and b) could be described mathematically; (d) the temperature history of the coldest point in the food could be described mathematically; (e) the mortality rate was a function of only the momentary agent intensity and of the organism's (or spore's) survival fraction, but was unaffected by agent application rate; and (f) the environmental factors (pH, food composition, etc.) were practically unchanged during the process.

The isothermal survival curves are described by the Weibullian power law relationship

$$\log_{10} S(t) = -b(T)t^{n(T)} \tag{4.20}$$

where $b(T)$ and $n(T)$ are temperature-dependent parameters. The momentary time-dependent isothermal inactivation rate is given by

$$\left. \frac{d\log_{10} S(t)}{dt} \right|_{T=\text{const}} = -b(T)n(T)t^{n(T)-1} \tag{4.21}$$

and the time that corresponds to the momentary survival ratio is

$$t^* = \left[\frac{-\log_{10} S(t)}{b(T)} \right]^{\frac{1}{n(T)}} \tag{4.22}$$

Combining Equations 4.21 and 4.22, we get the nonisothermal inactivation rate:

$$\frac{d\log_{10} S(t)}{dt} = -b[T(t)]n[T(t)]\left\{ \frac{-\log_{10} S(t)}{b[T(t)]} \right\}^{\frac{n(T)-1}{n(T)}} \tag{4.23}$$

The momentary inactivation rate in nonisothermal conditions is that which corresponds to the momentary temperature $T(t)$, at the time corresponding to the momentary survival ratio, $\log_{10} S(t)$, that is, t^*. This differential equation can be solved numerically by different mathematical software for any temperature profile. Equation 4.23 has been demonstrated to predict the survival patterns of *L. monocytogenes*, *Salmonella*, and *Clostridium botulinum* spores during nonisothermal heat treatments. The proposed procedure uses mathematical expressions directly derived from the experimental survival data without assuming any kinetic model and could be extended for lethal agents with varying intensities other than temperature (pressure, voltage, radiation, etc.).

Couvert et al. (2005) studied the influence of environmental factors on Weibull model parameters corresponding to the thermal inactivation of *Bacillus pumilus* spores. The environmental heating and recovery conditions did not present clear and regular influences on the shape parameter n and could not be described by any model tried. Conversely, the scale parameter b depended on the heating temperature and heating and recovery medium pH. The authors checked the independence of the parameters b and n and found a structural correlation between them (i.e., an error on b will be balanced by an error on n in the same way). So they proposed to estimate a single n value from the whole set of kinetics for eliminating the structural correlation between the b and n parameters. Getting a single overall estimation of n values per strain, regardless of the environmental conditions of heat treatment and recovery, seems to be enough for bacterial food predictive modeling and canning process calculations. Despite a slight loss of goodness of fit, this modification has lead to an improvement of the robustness of the model.

The Weibull model was successfully used by Avsaroglu et al. (2006) in describing the lactococcal bacteriophage inactivation by HHP. Four lactococcal bacteriophages (fLl6-2, fLl35-6, fLd66-36, and fLd67-42) in M17 broth were pressurized at 300 and 350 MPa at room temperature. The semilogarithmic survival curves were clearly nonlinear (monotonic upward concavity). The Weibull model produced a better fit to the data than the traditional linear model for all of the studied phages.

4.3.3 Dose–Response: Fermi Model

As stated before, four common types of semilogarithmic survival curves are usually found: linear curves, curves with a shoulder or initial lag period, curves with a tailing, and sigmoid curves. If the number of survivors expressed as a fraction, as a function of the applied dose, have the later shape and a symmetric or approximately symmetric form around the inflection point (Figure 4.7), the Fermi function can be used to describe and model the microbial response. The Fermi function is the inverted form

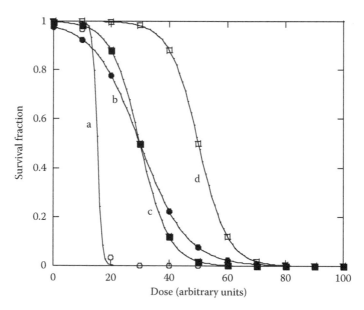

FIGURE 4.7 Simulated dose–response curves of a sensitive (a), intermediate (b and c), and resistant (d) microorganisms using the Fermi equation as a model.

of the logistic equation (Peleg 1996) and its convenience has been verified in modeling microbial survival exposed to PEFs (Peleg 2006), HHP (Palou et al. 1999), and antimicrobial agents (López-Malo et al. 2000), among other lethal factors.

The inactivation or inhibition pattern of microorganisms exposed to lethal or inhibitory agents or environments can be described by a dose–response curve. A dose–response curve, also known as a survival curve, is a graphic representation of the survival microbial fraction as a function of the applied dose of the antimicrobial factor or agent (Figure 4.7). Within this category of lethal factors or agents, several preservation factors (i.e., temperature, pressure, ultrasound intensity, UV radiation), disinfectant agents (chlorine, ozone, and organic acids), and antimicrobials (sorbates, benzoates, and naturally occurring antimicrobials) can be included. In experiments to determine dose–response curves, time is usually not treated as a variable, so the survivor microbial population is determined before and after the application or the exposure to the antimicrobial agent or factor under specific conditions.

When the microbial response is plotted as a function of the intensity of the preservation factor (Peleg et al. 1997, Peleg 2006), the dose–response curve typically has a sigmoid shape, characterized by a region of marginal or unnoticeable inhibition or mortality followed by a clearly linear region, representing an exponential decay (Peleg 1996, Palou et al. 1998, Alzamora and López-Malo 2002, Peleg 2006). This behavior has been traditionally separated into two regions: one with marginal or no effect at low exposure intensities of the preservation factor and a second region with a linear (after a logarithmic transformation) behavior, which is modeled using first-order reaction kinetics (Alzamora and López-Malo 2002).

An alternative approach to modeling microbial dose–response is to consider it continuous for the entire range of the applied preservation factor, without abrupt changes in response kinetics but with a gradual transition from no or marginal effects at relatively low intensities to inhibitory or lethal effects at high intensities of the preservation factor (Figure 4.7). The Fermi equation can be written as follows:

$$S(P) = \left[\frac{1}{1 + \exp\left(\dfrac{P - P_c}{k} \right)} \right] \tag{4.24}$$

where
 $S(P)$ is the survival fraction
 P is the intensity of the applied factor or the lethal agent concentration
 P_c is the critical level of P where $S(P)$ is 0.5
 k is a constant indicating the steepness of the dose–response curve around P_c

Since about 90% of the inhibition occurs within $P_c \pm 3k$ (Peleg 1996), a large value of k means a wide span, while a small k value means a very steep decline. Parameter values of the Fermi equation can be typically obtained by nonlinear regression. The generated models can be used to predict the minimal concentration of the lethal agent or dose required of the factor or antimicrobial that causes a specific reduction in the microbial population, i.e., 90%, 99%, or 99.9% reduction, and can be compared with those experimentally obtained or previously published. The Fermi function can describe curves with short or long delays as shown in Figure 4.7a and d, respectively, and also for a gradual decrease in the quantity of survivors (Figure 4.7b and c). The microbial population declines as a result of exposure to lethal agents, i.e., high temperature, PEFs, high pressure treatments, radiation or antimicrobial dose, can be described by the Fermi function (Peleg 1995, 1996, Peleg et al. 1997, Palou et al. 2007, Alzamora and López-Malo 2002). The particular application of the Fermi function is that the extent of its constants (P_c and k) can be directly related to the survival curve shape, which makes possible the comparison among microorganisms or the evaluation of factor intensity that determine microbial survival.

As an example of the applicability of dose–response modeling using the Fermi equation, yeast survival after HHP treatments is described next.

The extent of microbial inactivation achieved at a particular high pressure treatment depends on a number of interacting factors including type and number of microorganisms, magnitude and duration of pressure treatment, temperature, and composition of the suspension media or food (Palou et al. 1997a,b,c, 1999). Several authors reported first-order inactivation kinetics for selected microorganisms. However, Earnshaw et al. (1995) stated that there is clear evidence that pressure inactivation kinetics is not first order and survivor curves often show pronounced tails. In the application of HHP treatments to foods, it has been reported that, while an increase in pressure increases microbial inactivation, increasing the holding time does not necessarily increase the lethal effect of pressure (Palou et al. 1999). For each microorganism, there is a pressure threshold above which increasing the exposure time does not cause a significant reduction in the initially inoculated microbial cells (Palou et al. 1999). The pressure sensitivity of microorganisms varies among species, within strains of the same species, and, in general, cells in the exponential phase are more sensitive to pressure treatments than cells in the stationary phase of growth (Earnshaw 1995, Mackey et al. 1995, Patterson et al. 1995a,b, Palou et al. 1999).

Determination of the microbial pressure inactivation kinetics depends on pressure increase and decrease rates, and includes a decompression step for sampling. Therefore, sampling, treatment evaluation, and survivor enumeration are not continuous. The pressure increase and decrease are not always reported and the come-up time to reach the pressure (or pressure buildup velocity) is not always taken into account in the logarithmic representation of the survivors (Palou et al. 1999). In many reported HHP survival curves, it is not clear if time zero experiments are controls without pressure treatment or are pressure treatments that only take into account the come-up time to reach the working pressure. Palou et al. (2007) investigated the effects of the come-up time at different pressures on *S. cerevisiae* and *Zygosaccharomyces bailii* viability and model yeast responses using the Fermi function.

The authors (Palou et al. 1999) used laboratory media adjusted to selected water activities (0.98 or 0.95), inoculated with 1×10^5 CFU/mL of yeast cells. For *Z. bailii*, the inoculum was taken from the exponential or stationary growth phase when grown at a_w 0.98, and from the stationary phase when grown at a_w 0.95. For *S. cerevisiae*, the inoculum was taken from the exponential growth phase. The yeasts were exposed to HHP treatments for the time needed to reach the desired pressure at 21°C. The studied pressures ranged between 50 and 689 MPa.

In this case, the Fermi function parameters were defined by Palou et al. (2007) as follows: $S(P)$ is the fraction of surviving microorganisms, P is the pressure applied (MPa), P_c is a critical level of P where the survival fraction is 0.5 (i.e., $S(P)$ inflection point), and k is a constant (MPa) indicating the steepness of the survival curve around P_c.

The applicability of the Fermi equation to the HHP survivor curves of Z. *bailii* and S. *cerevisiae* is demonstrated in Figures 4.8 and 4.9. The regression parameters are listed in Table 4.3. As judged by statistical criteria ($r > 0.996$, $0.008 < \chi^2 < 0.030$), the fit was highly satisfactory; hence the Fermi equation was successful in capturing the features of the survivor curves. Pressure come-up times exert an important effect on S. *cerevisiae* and Z. *bailii* survival fractions, decreasing counts as the pressure is increased.

Differences in inactivation patterns can be expressed quantitatively in terms of the magnitude of the Fermi equation parameters (P_c and k). The larger P_c value calculated for S. *cerevisiae* than for Z. *bailii* indicates that the former is more pressure resistant. However, both yeasts present similar sensitivities to pressure treatment (similar k values). The progressive increase of lethality as a result of increasing a_w

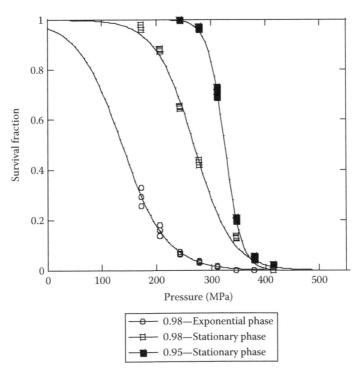

FIGURE 4.8 Survivor curves of Z. *bailii* suspended in selected a_w media (0.95 or 0.98) and from different stages of growth (stationary or exponential), exposed to pressure treatments only for the time needed to reach the desired pressure. Open circles, squares, and diamonds are experimental data; solid lines are the fit of the Fermi equation. (Adapted from Palou, E. et al., *J. Food Prot.*, 61, 1657, 1998.)

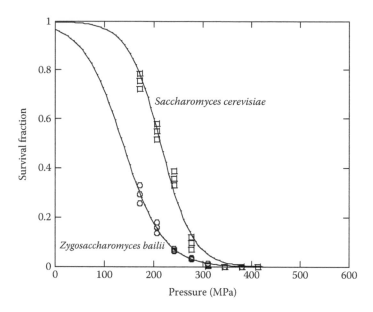

FIGURE 4.9 Survivor curves of yeasts from the exponential growth phase suspended in a_w 0.98 media, exposed to pressure treatments only for the time needed to reach the desired pressure. Open circles and squares are experimental data; solid lines are the fit of the Fermi equation. (Adapted from Palou, E. et al., *J. Food Prot.*, 61, 1657, 1998.)

TABLE 4.3
Survival Regression Parameters[a] Using Fermi Equation as a Model for Z. *bailii* and S. *cerevisiae* Exposed to High Hydrostatic Pressure Treatments Only for the Time Needed to Reach the Desired Pressure

Yeast	a_w	Growth Phase	r[b]	P_c (MPa)	k (MPa)
Z. *bailii*	0.98	Exponential	0.997	145.0 ± 2.8	34.3 ± 2.1
S. *cerevisiae*	0.98	Exponential	0.999	213.6 ± 1.7	33.1 ± 1.8
Z. *bailii*	0.98	Stationary	0.998	297.8 ± 1.0	28.7 ± 0.9
	0.95	Stationary	0.999	324.0 ± 0.7	15.8 ± 0.6

Source: Adapted from Palou, E. et al., *J. Food Prot.*, 61, 1657, 1998.

[a] Figures ± are the standard deviations.

[b] Regression coefficient.

of the suspension media was primarily expressed in a lower critical pressure (P_c) for a_w 0.98 than for a_w 0.95. These results confirm the baroprotective effects of low a_w for Z. *bailii* reported by Palou et al. (1997a,b,c, 2007).

The Fermi equation is a phenomenological or empirical model. Consequently, it can only be used to describe and compare inactivation patterns. The Fermi equation allows for a smooth transition from a nonlethal to a lethal regime, and at the same time accounts for a "lag" of any length as well as for both very steep and moderate lethality patterns (Peleg 1996, 2006). However, the applicability of this equation is limited to survivor curves that are symmetric or approximately symmetric around the inflection point. Apart from being able to describe very sharp drops in survival, the special appeal of the Fermi equation is that P_c and k can be used for comparison between microorganisms and to assess the effects of environmental conditions or the efficacy of lethal agents or treatments. Obviously, it would be interesting to learn how biophysical mechanisms at the tissue, cellular, and/or molecular levels affect the magnitude of P_c and k. The Fermi equation cannot be related to any specific destruction mechanism when used to model antimicrobial or preservation factors effects on microbial survival at the physiological/molecular level. However, once such a mechanism is known or understood, its manifestation can, at least in principle, be expressed by equation mathematical structure and the magnitude of its parameters as expressed by Peleg (1995, 1996, 2006) and peleg et al. (1997).

4.4 FINAL REMARKS

Inactivation models can add considerable value to product and process designs and evaluation as well as essential decision making in industry.

Information provided by the models can be used advantageously in the design of preservation technologies in many ways (Alzamora and López-Malo 2002):

- Although most of techniques are preserved using more than one preservation factor, there is a lack of quantitative data available to predict the adequate and necessary levels of each preservation factor. Thus, if we can predict with accuracy the decay kinetics for an identified target microorganism under several combinations of factors, the selection of such factors can be made on a sound basis, and the selected preservation factors can be kept at their minimum doses.
- Modeling has applications in products where the perceived "naturalness" prevents the use of large intensities of chemical preservation and/or solutes and/or acids and/or heating and/or other nonthermal factors. Sensory selection of preservation factors and their levels may be done between several "safe" equivalent combinations of interactive effects determined by the models.
- While much relevant information is available in the scientific literature concerning factors/interactions of factors that influence microbial activities in foods, it is seldom usable in formulating combined techniques. Many times, the information involves only data from traditional challenge testing

in particular food conditions or microbial presence or absence tests. These isolated results do not allow us to compare quantitatively what happens in a food system when the levels of the independent preservation factors are changed. Neither can the sensitivity of the key microorganisms to the different factors be inferred.

ACKNOWLEDGMENTS

We acknowledge financial support from Universidad de Buenos Aires (Project X197), CONICET, and ANPCyT-BID (Project 13955) of Argentina, as well as from Universidad de las Américas, Puebla, and CONACyT (Projects 44088 and 52653-Z) of Mexico.

REFERENCES

Albert, I. and Mafart, P. 2005. A modified Weibull model for bacterial inactivation. *International Journal of Food Microbiology* 100: 197–211.

Alvarez, I., Pagán, R., Condón, S., and J. Raso, J. 2003. The influence of process parameters for the inactivation of *Listeria monocytogenes* by pulsed electric fields. *International Journal of Food Microbiology* 87: 87–95.

Alzamora, S.M. and López-Malo, A. 2002. Microbial behavior modeling as a tool in the design and control of minimally processed foods. In *Engineering and Food for the 21st Century*, eds., J. Welti-Chanes, G. Barbosa-Cánovas, and J.M. Aguilera, pp. 631–650. Boca Raton, FL: CRC Press.

Alzamora, S.M., López-Malo, A., and Tapia, M.S. 2000. Overview. In *Minimally Processed Fruits and Vegetables. Fundamental Aspects and Applications*, eds., S.M. Alzamora, M.S. Tapia and A. López-Malo, 1–9. Gaithersburg, MD: Aspen Publishers, Inc.

Avsaroglu, M.D., Buzrul, S., Alpas, H., Akcelik, M., and Bozoglu, F. 2006. Use of the Weibull model for lactococcal bacteriophage inactivation by high hydrostatic pressure. *International Journal of Food Microbiology* 108: 78–83.

Baranyi, J. and Roberts, T.A. 1994. A dynamic approach to predicting bacterial growth in food. *International Journal of Food Microbiology* 23: 277–294.

Baranyi, J., Roberts, T.A., and McClure, P. 1993. A non-autonomous differential equation to model bacterial growth. *Food Microbiology* 10: 43–59.

Bhaduri, S., Smith, P.W., Palumbo, S.A. et al. 1991. Thermal destruction of *L. monocytogenes* in liver sausage slurry. *Food Microbiology* 8: 75–78.

Booth, I.R. 2002. Stress and the single cell: Intrapopulation diversity is a mechanism to ensure survival upon exposure to stress. *International Journal of Food Microbiology* 78: 19–30.

Campanella, O.H. and Peleg, M. 2001. Theoretical comparison of a new and the traditional method to calculate *Clostridium botulinum* survival during thermal inactivation. *Journal of the Science of Food and Agriculture* 81: 1069–1076.

Casadei, M.A., Esteves de Matos, R., Harrison, S.T., and Gaze, J.E. 1998. Heat resistance of *Listeria monocytogenes* in dairy products as affected by the growth medium. *Journal of Applied Microbiology* 84: 234–239.

Char, C.D., Guerrero, S., and Alzamora, S.M. 2009. Survival of *Listeria innocua* in thermally processed orange juice as affected by vanillin addition. *Food Control* 20: 67–74.

Chhabra, A.T., Carter, W.H., Linton, R.H., and Cousin, M.A. 2002. A predictive model that evaluates the effect of growth conditions on the thermal resistance of *Listeria monocytogenes*. *International Journal of Food Microbiology* 78: 235–243.

Chick, H. 1908. An investigation of the laws of disinfection. *Journal of Hygiene Cambridge* 8: 92–158.

Chmielewski, R.A.N. and Frank, J.F. 2004. A predictive model for heat inactivation of *Listeria monocytogenes* biofilm on stainless steel. *Journal of Food Protection* 67: 2712–2718.

Corradini, M.G. and Peleg, M. 2004. Demonstration of the applicability of the Weibull-log-logistic survival model to the isothermal and nonisothermal inactivation of *Escherichia coli* K-12 MG1655. *Journal of Food Protection* 67: 2617–2621.

Couvert, O., Gaillarda, S., Savyb, N., Mafarta, P., and Leguérinel, I. 2005. Survival curves of heated bacterial spores: Effect of environmental factors on Weibull parameters. *International Journal of Food Microbiology* 101: 73–81.

Earnshaw, R.G., Appleyard, J., and Hurst, R.M. 1995. Understanding physical inactivation processes: Combined preservation opportunities using heat, ultrasound and pressure. *International Journal of Food Microbiology* 28: 197–219.

Ferrante, S. 2004. Utilización combinada de ultrasonido y antimicrobianos naturales para inhibir el desarrollo de *Listeria monocytogenes* en jugos frutales. Thesis. Universidad Católica Argentina, Buenos Aires, Argentina.

Geeraerd, A.H., Herremans, C.H., and van Impe, J.F. 2000. Structural model requirements to describe microbial inactivation during a mild heat treatment. *International Journal of Food Microbiology* 59: 185–209.

Geeraerd, A.H., Valdramidis, V.P., and van Impe, J.F. 2005. GInaFit, a free tool to assess non-log-linear microbial survivor curves. *International Journal of Food Microbiology* 102: 95–105.

Gibson, A.M., Bratchell, N., and Roberts, T.A. 1988. Predicting growth responses of salmonellae in a laboratory medium as affected by pH, sodium chloride and storage temperature. *International Journal of Food Microbiology* 6: 155–178.

Gil, M.M., Brandao, T.R.S., and Silva, C.L.M. 2006. A modified Gompertz model to predict microbial inactivation under time-varying temperature conditions. *Journal of Food Engineering* 76: 89–94.

Gómez, M., García, D., Alvarez, I., Condón, S., and Raso, J. 2005. Modelling inactivation of *Listeria monocytogenes* by pulsed electric fields in media of different pH. *International Journal of Food Microbiology* 103: 199–206.

Guerrero, S.N., Tognon, M., and Alzamora, S.M. 2004. Response of *Saccharomyces cerevisiae* to the combined action of ultrasound and low weight chitosan. *Journal of Food Control* 16: 131–139.

Heldman, D.R. and Newsome, R.I. 2003. Kinetic models for microbial survival during processing. *Food Technology* 57: 40–46, 100.

Juneja, V.K. and Marks, H.M. 2003. Mathematical description of non-linear survival curves of *Listeria monocytogenes* as determined in a beef gravy model. *Innovative Food Science and Emerging Technologies* 4: 307–317.

Koutsoumanis, K., Lambropoulou, K., and Nychas, G.J.E. 1999. A predictive model for the non-thermal inactivation of *Salmonella enteritidis* in a food model system supplemented with. *International Journal of Food Microbiology* 49: 63–74.

Lebert, I. and Lebert, A. 2006. Quantitative prediction of microbial behaviour during food processing using an integrated modelling approach: A review. *International Journal of Refrigeration* 29: 968–984.

Linton, R.H., Carter, W.H., Pierson, M.D., and Hackney, C.R. 1995a. Use of a modified Gompertz equation to model nonlinear survival curves for *Listeria monocytogenes* Scott A. *Journal of Food Protection* 58: 946–954.

Linton, R.H., Carter, W.H., Pierson, M.D., Hackney, C.R., and Eifert, J.D. 1995b. Use of modified Gompertz equation to predict the effects of temperature, pH and NaCl on the inactivation of *Listeria monocytogenes* Scott A heated in infant formula. *Journal of Food Protection* 59: 16–23.

López-Malo, A., Alzamora, S.M., and Guerrero, S.N. 2000. Natural antimicrobials from plants. In *Minimally Processed Fruits and Vegetables. Fundamental Aspects and Applications*, eds., S.M. Alzamora, M.S.Tapia, and A. López-Malo, pp. 237–263. Gaithersburg, MD: Aspen Publishers, Inc.

Mackey, B.M., Forestiere, K., and Isaacs, N.S. 1995. Factors affecting the resistance of *Listeria monocytogenes* to high hydrostatic pressure. *Food Biotechnology* 9: 1–11.

Mafart, P. 2005. Food engineering and predictive microbiology: On the necessity to combine biological and physical kinetics. *International Journal of Food Microbiology* 100: 239–251.

Mafart, P., Couvert, O., Gaillard, S., and Leguerinel, I. 2002. On calculating sterility in thermal preservation methods: Application of the Weibull frequency distribution model. *International Journal of Food Microbiology* 72: 107–113.

McKellar, R.C. and Lu, X. 2004. *Modeling Microbial Responses in Food*. Boca Raton, FL: CRC Press.

McMeekin, T.A. 2007. Predictive microbiology: Quantitative science delivering quantifiable benefits to the meat industry and other food industries. *Meat Science* 77: 17–27.

McMeekin, T.A. and Ross, T. 2002. Predictive microbiology: Providing a knowledge-based framework for change management. *International Journal of Food Microbiology* 78: 133–153.

McMeekin, T.A., Olley, J.N., Ross, T., and Ratkowsky, D.A. 1993. *Predictive Microbiology: Theory and Application*. Great Britain: John Wiley & Sons Inc.

McMeekin, T.A., Baranyi, J., Bowman, J. et al. 2006. Information systems in food safety management. *International Journal of Food Microbiology* 112: 181–194.

Palou, E., López-Malo, A., Barbosa-Cánovas, G.V., Welti-Chanes, J., and Swanson, B.G. 1997a. Combined effect of high hydrostatic pressure and water activity on *Zygosaccharomyces bailii* inhibition. *Letters in Applied Microbiology* 24: 417–420.

Palou, E., López-Malo, A., Barbosa-Cánovas, G.V., Welti-Chanes, J., and Swanson, B.G. 1997b. High hydrostatic pressure as a hurdle for *Zygosaccharomyces bailii* inactivation. *Journal of Food Science* 62: 855–857.

Palou, E., López-Malo, A., Barbosa-Cánovas, G.V., Welti-Chanes, J., and Swanson, B.G. 1997c. Kinetic analysis of *Zygosaccharomyces bailii* by high hydrostatic pressure. *Lebensmittel-Wissenschaft und Technology* 30: 703–708.

Palou, E., López-Malo, A., Barbosa-Cánovas, G.V., Welti-Chanes, J., Davidson, P.M., and Swanson, B.G. 1998. High hydrostatic pressure come-up time and yeast viability. *Journal of Food Protection* 61: 1657–1660.

Palou, E., López-Malo, A., Barbosa-Cánovas, G.V., and Swanson, B.G. 2007. High-pressure treatment in food preservation. In *Handbook of Food Preservation*, ed. M.S. Rahman, 2nd ed., pp. 815–853. New York: CRC Press.

Patterson, M.F., Quinn, M., Simpson, R., and Gilmour, A. 1995a. Effects of high pressure on vegetative pathogens. In *High Pressure Processing of Foods*, eds., D.A. Ledward, D.E. Johnston, R.G. Earnshaw, and A.M.P. Hastings, pp. 47–63. Nottingham, U.K.: Nottingham University Press.

Patterson, M.F., Quinn, M., Simpson, R., and Gilmour, A. 1995b. Sensitivity of vegetative pathogens to high hydrostatic pressure treatment in phosphate-buffered saline and foods. *Journal of Food Protection* 58: 524–529.

Peleg, M. 1995. A model of microbial survival after exposure to pulsed electric fields. *Journal of the Science of Food and Agricultural* 67: 93–99.

Peleg, M. 1996. Evaluation of the Fermi equation as a model of dose–response curves. *Applied Microbiology and Biotechnology* 46: 303–306.

Peleg, M. 1999. On calculating sterility in thermal and non-thermal preservation methods. *Food Research International* 32: 271–278.

Peleg, M. 2000. Modeling and simulating microbial survival in foods subjected to a combination of preservation methods. In *Innovations in Food Processing*, eds., G.V. Barbosa-Cánovas, and G.W. Gould, pp. 163–182. Boca Raton, FL: CRC Press.

Peleg, M. 2006. *Advanced Quantitative Microbiology for Foods and Biosystems—Models for Predicting Growth and Inactivation*. Boca Raton, FL: CRC Press/Taylor & Francis.

Peleg, M. and Cole, M.B. 1998. Reinterpretation of microbial surival curves. *Critical Reviews in Food science* 38: 353–380.

Peleg, M. and Cole, M.B. 2000. Estimating the survival of *Clostridium botulinum* spores during heat treatments. *Journal of Food Protection* 63: 190–195.

Peleg, M. and Penchina, C.M. 2000. Modeling microbial survival during exposure to a lethal agent with varying intensity. *Critical Reviews in Food Science and Nutrition* 40: 159–172.

Peleg, M., Normand, M.D., and Damrau, E. 1997. Mathematical interpretation of dose–response curves. *Bulletin of Mathematical Biology* 59: 747–761.

Raffellini, S., Guerrero, S., and Alzamora, S.M. 2008. Effect of hydrogen peroxide concentration and pH on inactivation kinetics of *Escherichia coli*. *Journal of Food Safety* 28: 514–533.

Ratkowsky, D.A., Olley, J, McMeekin, T.A., and Ball, A. 1982. Relationship between temperature and growth rate of bacterial cultures. *Journal of Bacteriology* 149: 1–5.

Sapru, V., Smerage, G.H., Teixeira, A.A., and Lindsay, J.A. 1993. Comparison of predictive models for bacterial spore population resources to sterilization temperatures. *Journal of Food Science* 58: 223–228.

Shadbolt, C.T., Ross, T., and McMeekin, T.A. 1999. Nonthermal death of *Escherichia coli*. *International Journal of Food Microbiology* 49: 129–138.

ter Steeg, P.F. and Ueckert, J.E. 2002. Debating the biological reality of modeling preservation. *International Journal of Food Microbiology* 73: 409–414.

van Boekel, M.A.J.S. 1999. Testing of kinetic model: Usefulness of the multiresponse approach as applied to chlorophyll degradation in foods. *Food Research International* 32: 261–269.

van Boekel, M.A.J.S. 2002. On the use of Weibull model to describe thermal inactivation of microbial vegetative cells. *International Journal of Food Microbiology* 74: 139–159.

Xiong, R., Xie, G., Edmonson, A.E., and Sheard, M.A. 1999. A mathematical model for bacterial inactivation. *International Journal of Food Microbiology* 46: 45–55.

5 Kinetics of Chemical Reactions and the Case of Monitoring Phenolic Compounds by High-Performance Liquid Chromatography Coupled with Mass Spectrometry

Erika Salas-Muñoz and Enrique Ortega-Rivas

CONTENTS

5.1 INTRODUCTION

Food quality, in a broad sense, means satisfying the expectation of the consumer in terms of a series of attributes, such as color, presence or absence of certain flavor compounds, texture features, etc. Food quality related issues within the food industry are normally the responsibility of chemists, since the general subject of undesirable changes in the quality of food products has been traditionally related to alterations in chemical structure. The first thing a food technologist should do, regardless of whether he/she is a chemist or not, is to relate quality changes to chemical and physical processes taking place in the food. There are several key reactions of different sorts known to have effects on food quality, and they can be summarized as follows:

- Chemical reactions: these are mainly due to oxidation or Maillard-type reactions.
- Microbial reactions: microorganisms can grow in foods, and this can be desirable as in fermentation, or undesirable when causing spoilage or health risk in the case of pathogens.
- Biochemical reactions: many foods contain endogenous enzymes that can potentially catalyze reactions leading to quality loss, although in some fermentation processes enzymes can be utilized to actually improve quality.
- Physical reactions: many foods are heterogeneous and contain particles, which are normally unstable and can coalesce, aggregate, and in turn sediment, leading to quality loss. Some changes in texture can also be considered as physical reactions, although the underlying mechanism causing such changes may be of a chemical nature.

Food materials are unstable in a thermodynamic sense, i.e., they tend to change from a low-entropy, high-enthalpy state to a high-entropy, low-enthalpy state. There are some barriers to overcome this tendency so that foods can be in a kinetically stable state. Thermodynamics indicates the direction of change but not the rate at which such a change occurs. Chemical kinetics may be used for modeling of quality changes such as degradation of compounds (e.g., vitamins), formation of compounds (e.g., acrylamide), kinetics of aggregation in texture formation, kinetics of enzyme inactivation, or kinetics of sedimentation and crystallization. Models based on kinetics can be used to describe quality changes, bearing in mind that food materials are quite complex and that many interactions can occur. Thus, models can be used as guidelines in controlling and predicting food quality attributes and their changes, but they do not eliminate the need for experimental corroboration in most of the cases.

The most important reactions taking place in food processing as well as the reactivity of the main components in foods are presented in Tables 5.1 and 5.2, respectively (van Boekel 2007). As stated before, these tables indicate the many interactions possible in reaction kinetics during food processing. Therefore, when proposing models it is important to consider all these possible interactions in order to minimize errors in interpreting results.

TABLE 5.1
Main Types of Reactions Affecting Quality in Foods

Example	Type	Some Consequences
Nonenzymic browning	Chemical reaction (Maillard reaction)	Affects color, taste, aroma, and nutritive value; formation of toxicologically suspect compounds
Fat oxidation	Chemical reaction	Loss of essential fatty acids, rancid flavor, formation of toxicologically suspect compounds
Fat oxidation	Biochemical reaction (lipoxygenase)	Developing off-flavors due to aldehydes and ketones
Hydrolysis	Chemical reaction	Change in flavor and vitamin content
Lipolysis	Biochemical reaction (lipase)	Formation of free fatty acids, rancid taste
Proteolysis	Biochemical reaction (protease)	Formation of amino acids and peptides, bitter taste
Enzymic browning	Biochemical reaction of polyphenols	Browning

Source: Adapted from van Boekel, M.A.J.S. *Food Product Design: An Integrated Approach*, eds., Linneman, A.R. and van Boekel, M.A.J.S., Wageningen Academic Publishers, Wageningen, the Netherlands, 2007.

TABLE 5.2
Reactions of Key Components in Foods

Compound	Reaction	Some Consequences
Proteins	Denaturation	Gelation, precipitation, solubility
	Hydrolysis	Formation of peptides and amino acids
	Maillard reaction	Crosslinking, loss of nutritional value, browning
Lipids	Oxidation	Loss of essential fatty acids, rancidity
	Fat hardening	Formation of trans-fatty acids
	Hydrolysis	Formation of free fatty acids, soapy off-flavor
Mono- and disaccharides	Maillard reaction	Nonenzymic browning
	Caramelization	Taste and flavor changes
	Hydrolysis	Sugar inversion
Polysaccharides	Hydrolysis	Softening of tissue, texture changes
	Gelatinization	Staling of bread
Polyphenols	Enzymic polymerization	Browning
	Interaction with proteins	Crosslinking, gelation
Vitamins	Oxidation	Loss of nutritional value

Source: Adapted from van Boekel, M.A.J.S. *Food Product Design: An Integrated Approach*, eds., Linneman, A.R. and van Boekel, M.A.J.S.,Wageningen Academic Publishers, Wageningen, the Netherlands, 2007.

5.2　USE OF MODELS BASED ON KINETICS

As suggested in the previous discussion, many changes in food processing can be followed by kinetics and described by mathematical models containing characteristic kinetic parameters such as activation energies and rate constants. According to Haefner (2005), modeling in science can serve the goals of understanding, predicting, and controlling. These goals can be achieved fairly straightforward when dealing with pure reactants, for example, in the chemical industry. Connecting to fundamental reaction mechanisms and associated kinetic parameters yields other parameters valid for elementary reactions in simple, usually dilute, ideal systems.

Foods are far from being dilute or ideal chemical systems and the observed changes in processing may be due to many interacting, complex reaction mechanisms and not to a single elementary step. In some applications, when composition of raw materials is complex like in foods, models can be derived by the use of model systems rather than real systems. Food materials are so incredibly complex that there is significant risk in applying models directly to foods, when they have been based on fundamental reactions studied in model systems. The alternative would be to study kinetics directly on real food systems, but the derived parameters cannot be immediately interpreted as they will be in a well-defined dilute ideal system. Therefore, the derived models will result empirical, or at best semiempirical, meaning that extra precautions should be taken when applying such models outside the parameter region on which it was based.

Mathematical models consist, of course, of equations providing an output (such as vitamin content) based on a set of input data (time or temperature, for example). It can be considered a concise way to express physical behavior in mathematical terms. Perhaps one of the best examples of mathematical modeling in food technology is represented by the one used to predict the inactivation of microorganisms as a function of heating time and temperature. This model is effectively a first-order model and has been quite successful in thermal process calculations. Nevertheless, it is constantly updated and modified based on research and new trends in food processing, and even some alternatives have been proposed (van Boekel 2002, Peleg 2006).

As previously stated, modeling food quality attributes refers to modeling changes because quality of food products is, normally and inevitably, continuously changing with time. Food quality modeling could be, therefore, taken as kinetic modeling. Differential equations frequently form the basis for mathematical models, but their analytical solution is not relatively easy even with all the software and spreadsheets available. It has been also emphasized that not even the best solved mathematical model will be able to substitute crossing experimental information, when dealing with effects on food processing. The following discussion covering the main aspects of chemical and enzymic kinetics should be, therefore, used with extreme care and confirmatory experimental work is always recommended in food process engineering.

5.2.1　CHEMICAL REACTIONS MODELING

The typical chemical reaction between two molecules A and B to yield products C and D can be, of course, represented by

$$A + B \rightarrow C + D \tag{5.1}$$

The rate of reaction r is defined by

$$r = -\frac{d[A]}{dt} = -\frac{d[B]}{dt} = \frac{d[C]}{dt} = \frac{d[D]}{dt} = k[A][B] \tag{5.2}$$

The proportionality constant, k, is known as the rate constant. For molecules to react they must first come together and this happens via diffusion. If the encounter frequency is rate limiting, a reaction is called diffusion limited and such reactions take place rapidly.

In most cases the actual reaction step will be rate limiting rather than the encounter rate, and a simpler equation possible, known as the general rate law, can be represented by

$$r = -\frac{dc}{dt} = kc^n \tag{5.3}$$

where
c is the concentration of a single reactant
power n represents the order of reaction

Equation 5.3 can be integrated with respect to time to obtain the course of the concentration as a function of time, i.e.,

$$c^{1-n} = c_0^{1-n} + (n-1)kt \quad \text{for } n \neq 1 \tag{5.4a}$$

$$c = c_0(e)^{-kt} \quad \text{for } n = 1 \tag{5.4b}$$

where c_0 is the initial concentration.

In food science, quality changes are normally modeled by means of a zero-, first-, or second-order reaction, and thus if $n = 0$ for a decomposition reaction, according to Equation 5.4a

$$-\frac{dc}{dt} = k \tag{5.5}$$

and integration of Equation 5.5 leads to

$$c = c_0 - kt \tag{5.6}$$

Zero-order reactions are usually reported for changes in foods when the amount of product formed is only a small fraction of precursors present, or also for decomposition reactions where a small amount of product is formed from a reactant. In this case, since the reactant is in such large excess, its concentration remains effectively constant during the observation period. Therefore, the rate appears to be independent of the concentration.

First-order reactions are also reported for foods ($n = 1$), so substituting again into Equation 5.4a, the resultant relation is

$$-\frac{dc}{dt} = kc \tag{5.7}$$

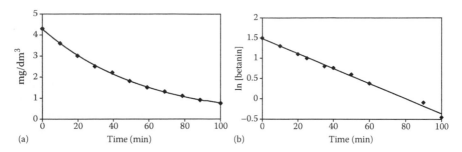

FIGURE 5.1 Example of a first-order reaction for the degradation of betanin at 75°C: (a) untransformed concentration plotted using Equation 5.4b and (b) logarithmically transformed using Equation 5.8. (Adapted from Saguy, I. et al., *J. Agric. Food Chem.*, 26, 360, 1978.)

Integration of Equation 5.7 will result, of course, in the expression given by Equation 5.4b. The logarithmic form of such an equation can also be used to obtain

$$\ln c = \ln c_0 - kt \tag{5.8}$$

Equation 5.4b is nonlinear but it is transformed to linear in its logarithmic form represented by Equation 5.8. An example of linearization of first-order reactions in food is presented in Figure 5.1. It concerns the heat-induced degradation of betanin, a natural color compound found in red beets. Figure 5.1a shows the first-order plot for untransformed data according to Equation 5.4b, while Figure 5.1b shows the plot for logarithmically transformed data according to Equation 5.8. A logarithmic plot resulting in a straight line is generally interpreted as a proof of a first-order reaction taking place.

The equation for a second-order reaction, i.e., $n = 2$, is thus

$$-\frac{dc}{dt} = kc^2 \tag{5.9}$$

Integration of Equation 5.9 leads to

$$c = \frac{c_0}{1 + c_0 kt} \tag{5.10}$$

and in linearized form it becomes

$$\frac{1}{c} = \frac{1}{c_0} + kt \tag{5.11}$$

Second-order reactions are sometimes reported for changes of amino acids involved in the Maillard reaction. An example that can be cited is the loss of lysine in sterilized milk due to this reaction. Therefore, a plot of the inverse of concentration of lysine as a function of time will be described as a straight line as illustrated in Figure 5.2.

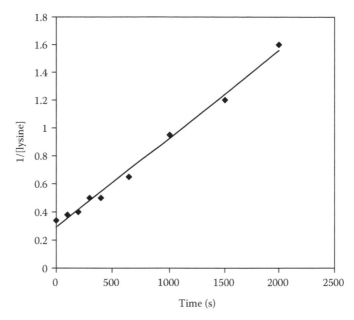

FIGURE 5.2 Reduction of lysine concentration in milk treated at 160°C, as a function of time and according to a second-order model. (Adapted from Horak, F.P., Über die Reaktionskinetic der Sporenabtötung und chemischer Veränderungen bei der thermischen Haltbarmachung von Milch zur Optimieruung von Erhitzungverfahren. PhD thesis. Germany: Technical University of Munich, 1980.)

Although theoretical concepts of kinetics like those described above can be applied in food science, reactions are usually much more complicated and cannot be simply described by the equations discussed. Many chemical reactions in foods, for example, fat oxidation and the Maillard reaction, cannot be given in one simple equation but in a network of linked reaction steps. Unraveling of such a network can be carried out by multiresponse mechanistic modeling (van Boekel 2000), which is useful in scientific explanations, but probably less suitable for practical applications.

5.2.2 TEMPERATURE DEPENDENCE OF CHEMICAL REACTIONS

Arrhenius' law was empirically derived to describe the temperature dependence of simple chemical reactions, and it relates the rate constant, k, of a reaction to absolute temperature T. Its mathematical expression is as follows:

$$k = A(e)^{-E_a/RT} \tag{5.12}$$

The linearized form is

$$\ln k = \ln A - \frac{E_a}{RT} \tag{5.13}$$

where
    A is the preexponential factor
    E_a is the activation energy
    R is the universal gas constant
    T is the absolute temperature

The activation energy can be taken as the energy barrier that molecules need to cross in order to be able to react. The proportion of molecules likely to react increases with temperature, which explains in a quantitative manner the effect of temperature on rates. The physical meaning of A is that it represents the rate constant at which all molecules count on enough energy to react ($E_a = 0$).

The parameters described so far in the kinetics models discussed, i.e., orders, rate constants, and activation parameters, are all that are needed in chemical kinetics. In food chemistry kinetics, however, some other parameters are commonly used. They are normally associated with kinetics of microbial inactivation and are analogously used in food chemistry. The parameter, Q_{10}, describes the temperature dependence of a reaction as the factor by which the reaction rate is changed when the temperature is increased by 10°C. It is described by

$$Q_{10} = \frac{k_T + 5}{k_T - 5} \approx \frac{k_T + 10}{k_T} \tag{5.14}$$

If the Arrhenius equation holds, it can be shown that

$$Q_{10} = e^{(10 E_a / RT^2)} \tag{5.15}$$

Another parameter to describe temperature dependence is represented by Z, and expresses the increase in temperature that would produce an increase in rate by a factor of 10. It is related to Q_{10} by the relationship

$$Z = \frac{2.303 RT^2}{E_a} = \frac{10}{\log Q_{10}} \tag{5.16}$$

The parameters, Q_{10}, and Z, can be linked to the more fundamental kinetic parameters but they are somewhat proper to food systems and they are usually estimated in real foods. As such, they reflect appropriately a time–temperature dependence characteristic that can be used for engineering purposes.

5.2.3 Enzyme Reactions Modeling

Biochemical reactions are important for food quality because virtually all foods are biological materials, and as such they contain enzymes. Some enzymes may be desirable, for example, for cheese ripening, but most of them need to be deactivated because their action will cause deterioration of food quality. For example, enzymic browning of apples, potatoes, and cauliflower is due to polyphenoloxidase (PPO), while formation of a soapy or rancid taste in raw milk is due to the action of lipases. A model often used to describe enzyme kinetics is known as the Michaelis–Menten model and is basically represented by the equation

$$v = v_{max} \frac{[S]}{[S] + K_M} \tag{5.17}$$

where

    v is the initial rate of the reaction
    v_{max} is the maximum rate of the enzyme under the condition of study
    [S] is the substrate concentration
    K_M is a constant known as the Michaelis constant

By knowing v_{max} and K_M, the parameters of the equation, the rate of the enzymic reaction can be predicted.

In order to prevent the action of enzymes, inactivation kinetics is used. Enzymes are proteins and inactivation is due to unfolding of the protein. A general model can be given by the relation

$$N \underset{k_2}{\overset{k_1}{\leftrightarrow}} D \overset{k_3}{\rightarrow} 1 \tag{5.18}$$

where

    N is the native protein
    D is the denatured protein
    l is the inactivated protein
    k_1, k_2, k_3 are the rate constants for each step

The importance of the equilibrium between N and D is that proteins can refold after denaturation, with enzyme activity restored upon eliminating the cause of denaturation. In most cases in food materials, the cause of denaturation is heat and if the denatured protein is subjected to further reaction, they may lead to the inactivated form l, in which case the enzyme cannot return to its active form. In most cases, a first-order model, as given in Equation 5.4b, appears to be applicable to describe enzyme inactivation. This implies that the third step in Equation 5.8 is rate determining. An example of first-order inactivation rate in food processing is the inactivation kinetics of pectin methylestarase in tomato, as shown in Figure 5.3.

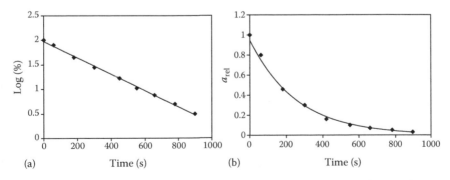

FIGURE 5.3 Plots of first-order heat induced inactivation kinetics of pectin methylesterase from tomato at 69.8°C: (a) logarithmic plot and (b) relative activity plot. (Adapted from Anthon, G.E. et al., *J. Agric. Food Chem.*, 50, 6, 2002.)

5.3 CASE STUDY: MONITORING PHENOLIC COMPOUNDS BY HIGH-PERFORMANCE LIQUID CHROMATOGRAPHY COUPLED WITH MASS SPECTROMETRY

The compounds present in food undergo many changes during processing and storage; the changes related to the finished product first involve the selective extraction of different compounds in the raw material by different operations (like pressing for fruit juices and maceration or skin contact for wine making) and second involve the elaboration process itself, as it could be the use of high temperatures in processing and enzyme utilization for enhancing extraction, creating as a consequence a modification in the raw material native composition. Some reactions take place: (1) Formation of new compounds is due to the presence of native compounds that were located in different compartments or clusters in the fruit and are now extracted in the juice and are able to react with each other. Diffusion starts after the crushing of the fruit, and in the case of winemaking it continues until the red wine is separated from the seeds and skins by pressing. Diffusion kinetics depend both on the solubility of the compounds and on their localization in the fruit. (2) Degradation of the compounds present in the raw material is caused by the increase of temperature (pasteurization of fruit juices and fermentation of wines) and light exposure; another key element in the degradation of native compounds present in raw materials is the enzymes that are naturally present in fruit, for example, PPO. This enzyme shows two activities: cresolase (hydroxylating monophenols to o-diphenols) and catecholase (oxidizing o-diphenols to o-quinones), using molecular oxygen as cosubstrate. PPO is the enzyme responsible for browning in apples or any other fruit with a high phenolic content, as well as fruit juices. Some of the preferred substrates of PPO are phenolic acids like chlorogenic acid (present in apples) and caffeoyl tartaric acid (present in grapes). Another transformation due to enzymes is the secondary activity of many commercial preparations; these enzymes are generally added in processing to enhance extraction and as clarification and filtration agents. The enzymes most commonly used in the fruit juice and winemaking industry are pectinases and β-glucanases. The secondary β-glucosidase activity of commercial enzymes could result in deglucosylation of anthocyanins (phenolic pigments present in red fruits) leading to anthocyanidins, which are more thermally instable and prone to degradation than anthocyanins (Piffaut et al. 1994). The degradation of the natural pigments present in the fruit causes the discoloration of fruit juice.

These types of reactions will produce changes in the raw material native composition and the newly formed compounds will have different physicochemical properties than their predecessors; therefore, the organoleptic characteristics (especially color and taste) of the beverage will change.

In complex food matrices, it is complicated to follow the disappearance of native compounds and the appearance of newly formed ones. Analytical techniques used to separate and/or fractionate compounds with similar structures are, therefore, quite useful to identify chemical compounds formed by different reactions.

5.3.1 High-Performance Liquid Chromatography

High-performance liquid chromatography (HPLC) is an analytical tool commonly used in the analysis of complex mixtures due to its sensibility and reproducibility. This technique allows, under high pressure, to separate molecules very similar to each other (even isomers) in columns where the stationary phase is constituted of beads with a size of 5 μm (the most common size). These fine beads allow achieving a large number of theoretical plates, leading to an enhanced separation of molecules. HPLC can be used in an analytical, semipreparative, or preparative scale. One of the stationary phases most commonly used in food analysis is reversed phase (RP-HPLC), which is a nonpolar stationary phase (C18 = octadecylsilyl) and uses aqueous and increasing amounts of methanol or acetonitrile as a mobile phase. In this stationary–mobile phase system, polar compounds elute more rapidly than the nonpolar ones.

5.3.2 Electrospray Ionization Mass Spectrometry

Coupling HPLC with mass spectrometry (MS) equipped with mild ionization techniques, such as electrospray ionization (ESI-MS), allows the study of complex mixtures by the association of one separation method and one identification method. In ESI-MS, the ionization process is done at room temperature and at atmospheric pressure. ESI-MS also allows the detection of multicharged ions in the positive ion mode $[M + zH]^{z+}/z$ or negative ion mode $[M - zH]^{z-}/z$. Besides, increasing the orifice voltage in the ion source favors the formation of fragment ions (MS^2), which gives supplementary information about the structure of the molecule. Fragment ions can be used as tool for determining molecular structure as they arise from the well-determined breaking of chemical bonds.

Tandem mass spectrometry (MS-MS) and ion trap mass spectrometry (IT-MS, MS^n), which allows fragmentation patterns to be obtained on selected individual ions, are progressively replacing the classical ESI-quadrupole mass spectrometers in HPLC-MS coupling. This facilitates the structural elucidation of unknown compounds. In addition, the utilization of the select ion monitoring mode increases considerably the sensitivity. The use of HPLC-MS enables the study of phytochemicals present in plants and food, such as phenolic compounds.

5.3.3 Polyphenols

Phenolic compounds are a very important group of compounds present in natural products of plant origin; polyphenols are responsible for some of the main organoleptic properties of fruits (and fruit-derived beverages) like color and astringency. Among phenolic compounds, two groups, namely flavonoids, based on a common C6–C3–C6 skeleton (Figure 5.4) and nonflavonoids, are classically distinguished. Each of these groups is further divided into several families. Two types of flavonoids, namely anthocyanins and flavanols, are particularly important for the quality of red wines and other fruit-derived beverages. Particularly, anthocyanins are responsible

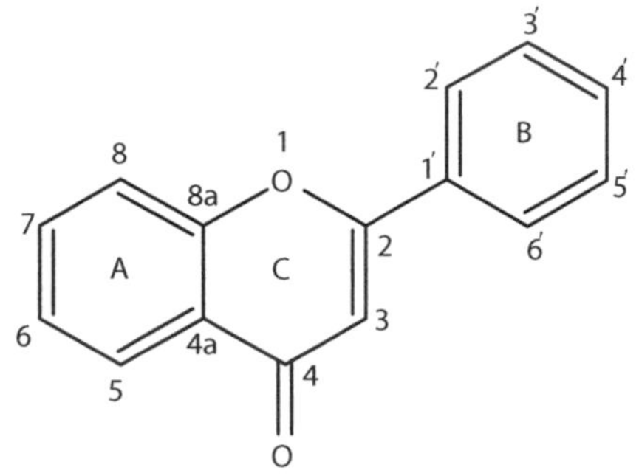

FIGURE 5.4 Flavonoid skeleton.

for red wine color (Ribéreau-Gayon 1964) and flavanols are usually associated with bitterness and astringency (Lea 1978). During winemaking and aging, the ensemble of phenolic compounds evolves due to their reactivity. With storage time, the color of young red wine changes from a reddish blue to the reddish brown color of matured wines and astringency decreases. Color changes are due to gradual conversion of the anthocyanin pigments extracted from red grape into various derivatives through different reaction mechanisms. Anthocyanins are usually localized in the berry skins of the red grape varieties (and other red fruits including berries as strawberry, blueberry, etc.). The anthocyanins present in red grapes are glucosylated derivatives of five aglycones (or anthocyanidins), namely cyanidin, peonidin, petunidin, delphinidin, and malvidin.

Flavanols are encountered in grape as monomers and as oligomers and polymers, called proanthocyanidins because they release red anthocyanidins when heated in acidic medium. Proanthocyanidins show a great diversity of structures, from the rather simple dimers to extremely complex molecules, due to the occurrence of several constitutive units that can be linked through different positions and make up chains of variable length.

Anthocyanins extracted from grape skins during winemaking are responsible for the color of red wines. The red color of anthocyanins is due to their cationic flavylium form (AH^+), which is predominant only in acidic media (pH <2). The flavylium cation is in equilibrium with various other forms through proton transfer, hydration, and tautomerization reactions. At wine pH, the nucleophilic attack of water at position 2 or 4 of the pyrylium nucleus leads to the colorless hemiketal form (AOH), which is, itself, partly converted into chalcones (C). AOH and C correspond to the thermodynamic products of anthocyanin structural transformations in aqueous solutions (Brouillard et al. 1977). Concurrently, fast deprotonation of the flavylium cation leads to the kinetic products, the neutral quinonoid bases (A) in mildly acidic solutions, which further deprotonate into the anionic quinonoid base (A^-) in neutral to mildly alkaline solutions (Figure 5.5). Native anthocyanins are hydrated at 50% at a pH value around 2.6. Hence, in red wine (pH 3.2–4.0), anthocyanins are expected to be largely under colorless hydrated forms (>70%). Thus, some color-stabilizing mechanisms must take place.

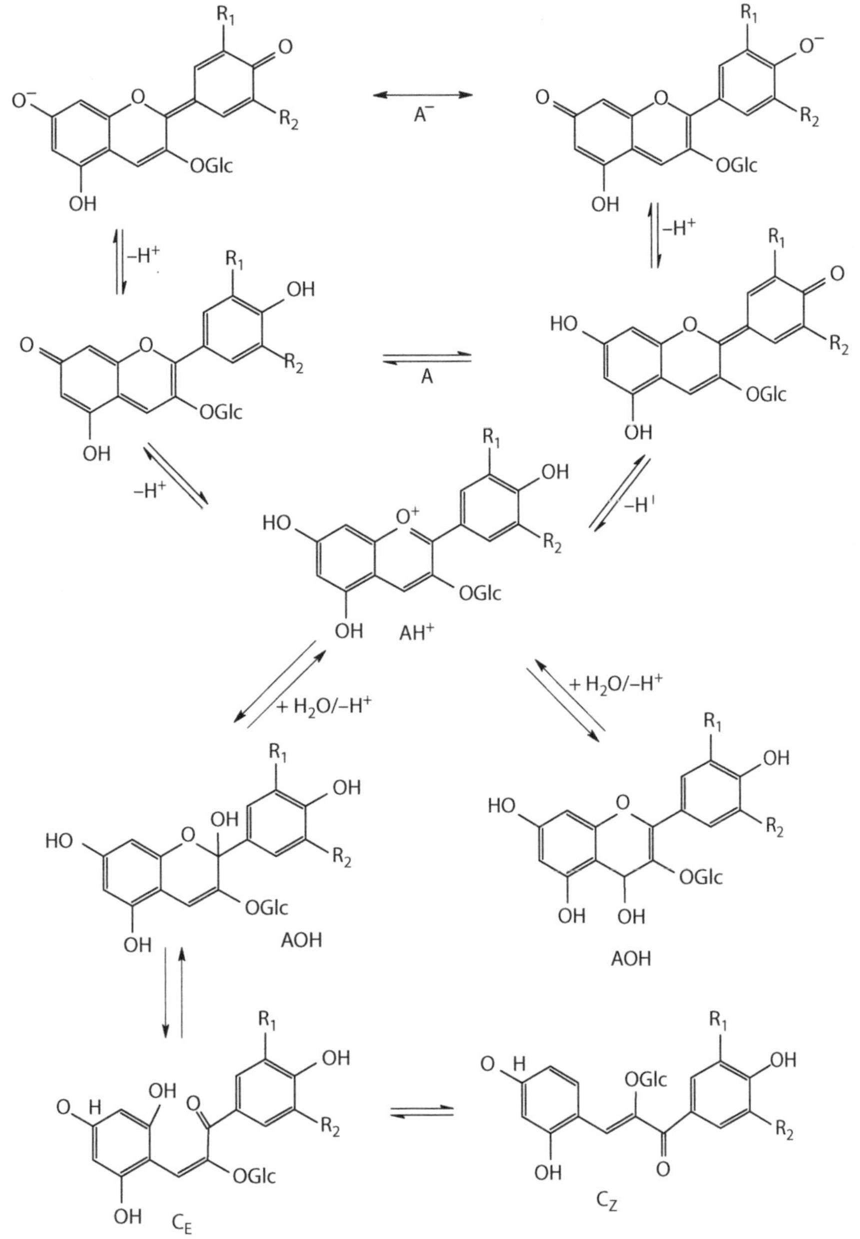

FIGURE 5.5 Structural transformations of an anthocyanin in aqueous solutions.

One of the color-stabilizing mechanisms, especially important in wine, is the conversion of native anthocyanins into more stable pigments through various reactions involving other wine components such as tannins and yeast metabolites. Several pathways have been proposed to explain the formation of more stable pigments in wines. One of the mechanisms is a direct reaction between anthocyanins and flavanols (F), leading to flavanol–anthocyanin (F–A⁺) and anthocyanin–flavanol (A⁺–F) adducts (Salas et al. 2003, 2004a).

Studies of phenolic reactions in wine and in food involve two complementary approaches. On one hand, wine analysis demonstrates the occurrence of new compounds that can be isolated and identified. Mechanisms generating them are then postulated and checked in wine-like model systems containing their potential precursors. On the other hand, model solution studies yield new products that can then be searched for in wine. In this study, it was worked with model solutions because it is easier to follow the kinetics of the reactions, meaning the disappearance of the precursors and the apparition of the new formed products. The presence of some of the reaction products of anthocyanins was then compared with wine (Salas et al. 2004a, 2005).

In this work, the reaction between an anthocyanin, (malvidin 3-*O*-glucoside), and a procyanidin dimer, Ec-EcG ((−)-epicatechin-(4β-8)-(−)-epicatechin 3-*O*-gallate), was monitored by HPLC-ESI-MS (Figure 5.6). The study was done at two different pHs, as the influence of pH on anthocyanin and tannin reactivity was also investigated, by comparing two pH values, 2 and 3.8. The former was chosen to favor acidic cleavage of the interflavanic bond and reactions involving the flavylium cation. The latter was selected, as it is a typical pH value encountered in red wine. Incubations were carried out at pH 2 and 3.8. For each pH condition, polyphenols were dissolved in the corresponding media in order to obtain final concentrations of 1.4 mM for Mv3glc and of 0.7 mM for Ec-EcG. The reaction was initiated by mixing the two polyphenol solutions. Each solution was distributed in aliquots of 0.5 mL for each point of the kinetics (0, 6, 14, 24, 31, 45, and 64 days). Each 1 mL screw cap vial was purged with argon and incubated at 30°C in the dark. Control solutions containing Mv3glc or Ec-EcG alone at the concentration used in mixture model solutions were incubated under the same conditions.

Mv3glc and Ec-EcG were isolated from grape skin and grape seed, respectively (Ricardo da Silva et al. 1991, Sarni et al. 1995). The identity and purity of both compounds were checked by MS and HPLC. Mv3glc was quantified from the area of its

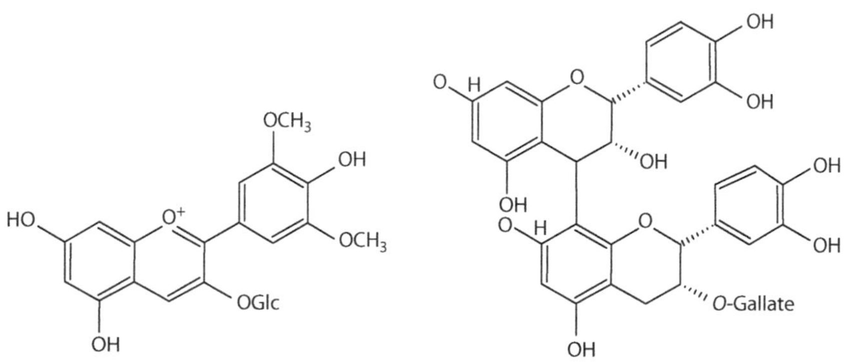

FIGURE 5.6 Malvidin 3-*O*-glucoside (Mv3glc) and procyanidin dimer Ec-EcG.

peak at 520 nm by HPLC using multiple point calibration with an external standard of commercial Mv3glc. Mv3glc was 90% pure, the other 10% corresponding to other anthocyanins, namely peonidin 3-O-glucoside, malvidin 3,5-di-O-glucoside, and malvidin 3-O-acetylglucoside. Ec-EcG was quantified from the area of its peak at 280 nm by HPLC using multiple point calibration with an external standard purified by semipreparative HPLC from a grape seed fraction.

5.3.4 Reaction Kinetics

Compositional changes were monitored by HPLC-MS. Peak areas were measured at 520 nm for Mv3glc and at 280 nm for Ec-EcG. The disappearance of Mv3glc and Ec-EcG was modeled through first-order kinetics for each solution. The observed reaction rate constants k were determined using linear regression to plot the ln of the concentration at different time intervals (after 6, 14, 24, 31, 45, and 64 days) as shown in Table 5.3. The disappearance of both species was much faster at pH 3.8 than at pH 2; meaning, on one hand, that the hemiketal form of Mv3glc is less stable than its flavylium form, on the other hand, that reactions involving acid-catalyzed cleavage of the interflavanic bond in the procyanidin dimer (Ec-EcG) are not predominant. Mv3glc disappeared two times faster in the presence of Ec-EcG at both pH values, suggesting that Mv3glc reacts with Ec-EcG. The disappearance of Ec-EcG was slightly affected by the presence of Mv3glc at pH 3.8, but not at pH 2. The difference of k values for Ec-EcG at pH 3.8 (between control and mixture solutions) suggests that Mv3glc protects Ec-EcG from degradation.

The degradation of Mv3glc is due to the fact that anthocyanins are relatively more stable than their aglycones because of the substitution at position 3. In case of enzymatic attack (β-glucosidase), the aglycon is released and can be degraded by cleavage of the heterocycle of the malvidin chalcone form, which is in equilibrium with the hemiketal form (Brouillard et al. 1977), leading to syringic acid (from the B ring) and 2,4,6-trihydroxybenzaldehyde (from the A ring) (Piffaut

TABLE 5.3
Rate Constants of Mv3glc and Ec-EcG Disappearance

pH 2	$K = -\Delta \ln C/\Delta T$ (days^{-1})	R^2
Mv3glc	0.0153	0.8478
Ec-EcG	0.0251	0.9217
Mv3glc control	0.0076	0.7925
Ec-EcG control	0.025	0.9612
pH 3.8		
Mv3glc	0.0639	0.9866
Ec-EcG	0.0456	0.9962
Mv3glc control	0.0294	0.7685
Ec-EcG control	0.0586	0.9664

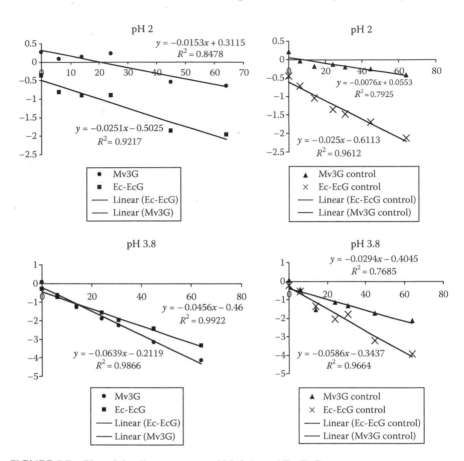

FIGURE 5.7 Plot of the disappearance of Mv3glc and Ec-EcG.

et al. 1994) (Figure 5.7). Heat or oxidative agents such as H_2O_2 can cause this cleavage. Thermal degradation of anthocyanins follows first-order kinetics (Cemeroglu et al. 1994). Color losses are often attributed to anthocyanin degradation. The cleavage of the heterocycle, due to heat, is favored by the instability of the open chalcone form at high temperature (Piffaut et al. 1994). The exact mechanism of thermal degradation of anthocyanins is still hypothetical even though the passing through the chalcone form seems necessary (Jurd 1972) (Figure 5.8). The mechanism of degradation of anthocyanins agrees with that described by Hrazdina and Franzese (1974). The anthocyanin goes by a malvone that breaks into a phenolic acid (B ring) and, as an effect of pH, into a coumarin or a phenolic acid (A ring) (Figure 5.9).

Syringic acid was chosen as a marker for anthocyanin (Mv3glc) degradation and its concentration at pH 3.8 was also plotted and k calculated (Table 5.4). The observed rate of syringic acid apparition was 50% lower in the presence of Ec-EcG than in the control even though Mv3glc disappeared two times faster suggesting that other reactions besides degradation are taking place (Figure 5.10).

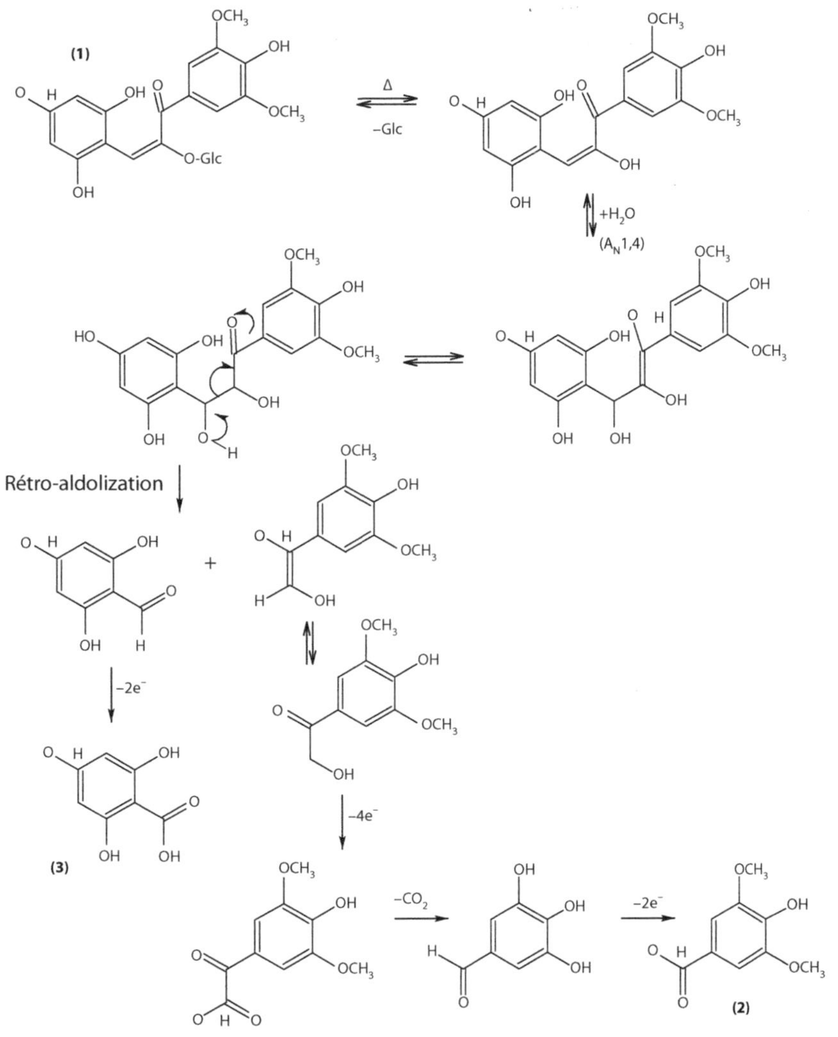

FIGURE 5.8 Hypothetical thermal degradation mechanism of anthocyanins (1) malvidin in the chalcone form, (2) syringic acid, and (3) 2,4,6-trihydroxybenzenic acid.

Disappearance of Mv3glc was almost total after 64 days of incubation at pH 3.8, and at pH 2 it remained around 40% of the initial Mv3glc concentration after 64 days.

5.3.5 MONITORING THE CHANGES IN THE REACTION MEDIA BY HPLC-MS

At pH 2, the appearance of a peak with a mass signal at m/z 441 (epicatechin gallate, EcG) indicated that the acidic cleavage of the interflavanic bond of the Ec-EcG

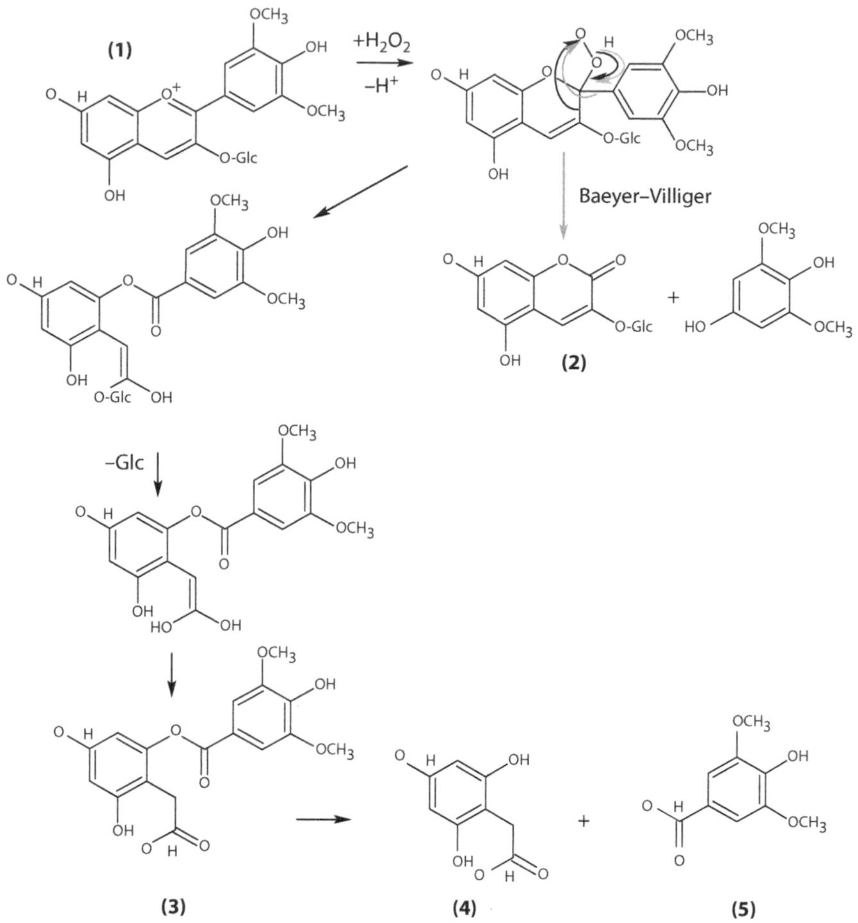

FIGURE 5.9 Hypothetical oxidative degradation mechanism of anthocyanins (1) malvidin in the flavylium form, (2) coumarin, (3) malvone, (4) 2,4,6-trihydroxyphenilacetic acid, and (5) syringic acid.

TABLE 5.4
Rate Constants of Syringic Acid Formation

pH 3.8	$k = \Delta \ln C / \Delta T$ (days^{-1})	R^2
Syringic acid	0.0166	0.749
Syringic acid control	0.0247	0.8837

dimer had taken place. At pH 3.8, although the disappearance of Ec-EcG was faster than at pH 2, no EcG was detected, indicating that acid-catalyzed C–C bond breaking process did not take place or yielded different products to those observed at pH 2 (Figure 5.11).

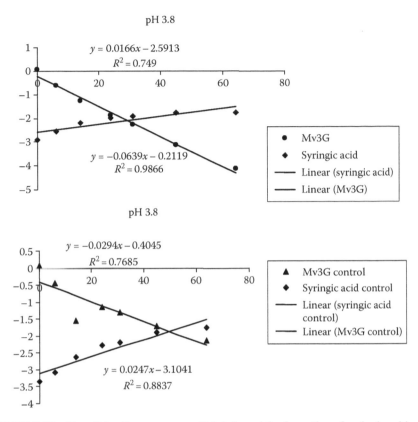

FIGURE 5.10 Plot of the disappearance of Mv3glc and the formation of syringic acid.

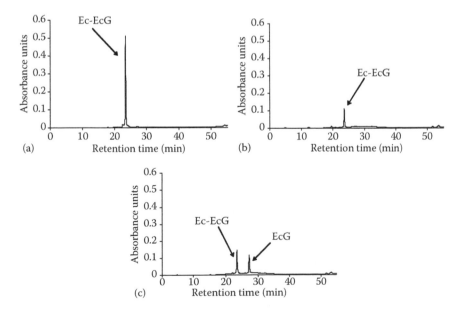

FIGURE 5.11 HPLC profile at 280 nm of the solution containing Ec-EcG at 0 days (a) and after 31 days of incubation at pH 3.8 (b) and at pH 2 (c).

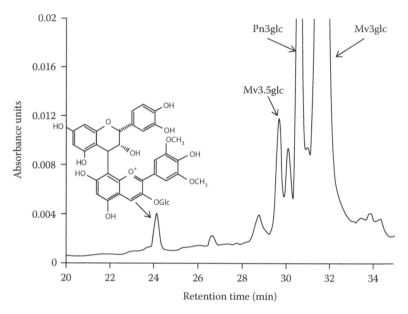

FIGURE 5.12 HPLC profile at 520 nm of the solution containing Mv3glc and Ec-EcG at pH 2.0 after 24 days of incubation.

A new compound showing a mass signal at m/z 781 in the positive ion mode and a UV–visible spectrum similar to that of Mv3glc gradually accumulated until 24 days of incubation in the solution containing Mv3glc and Ec-EcG at pH 2 (Figure 5.12) and decreased afterwards. This compound may correspond to the dimer of one flavanol (epicatechin, Ec) and Mv3glc under its flavylium form: the Ec-Mv3glc (F–A$^+$) dimer. This newly formed dimer is issued from the following mechanism: procyanidin dimer (Ec-EcG) is affected by acid-catalyzed cleavage of its interflavanic bond, releasing the intermediate carbocation Ec$^+$ (Haslam 1980), which acts as an electrophile, whereas the anthocyanin, in its hydrated hemiketal form (AOH), acts as a nucleophile (Ribéreau-Gayon 1982). This yields the colorless dimer (F-AOH), which dehydrates to the red flavylium form (F–A$^+$) (Figure 5.13). The MS/MS analysis of the molecular ion at m/z 781 gave two signals at m/z 763 and m/z 619, which correspond, respectively, to the loss of a water molecule [M − 18]$^+$ as described by Friedrich et al. (2000), and to that of the anthocyanin glucose moiety [M − 162]$^+$. MS^3 fragmentation of the ion at m/z 619 (Figure 5.14) gave four ions detected at m/z 601, 493, 467, and 373. The ion at m/z 601 corresponds to the loss of a water molecule [M − 18]$^+$. The ion at m/z 493 [M − 126]$^+$ can be interpreted as resulting from loss of fragment $C_6H_6O_3$ (A ring). This fragmentation scheme indicates that the flavanol is the upper unit and the anthocyanin is the lower unit; since the fragment ion at m/z 493 (M − 126) cannot arise from the anthocyanin A ring and, according to Friedrich et al. (2000), it is specific of the upper unit of dimers. The ion at m/z 467 [M − 152]$^+$ results from the retro Diels–Alder (RDA) decomposition of the

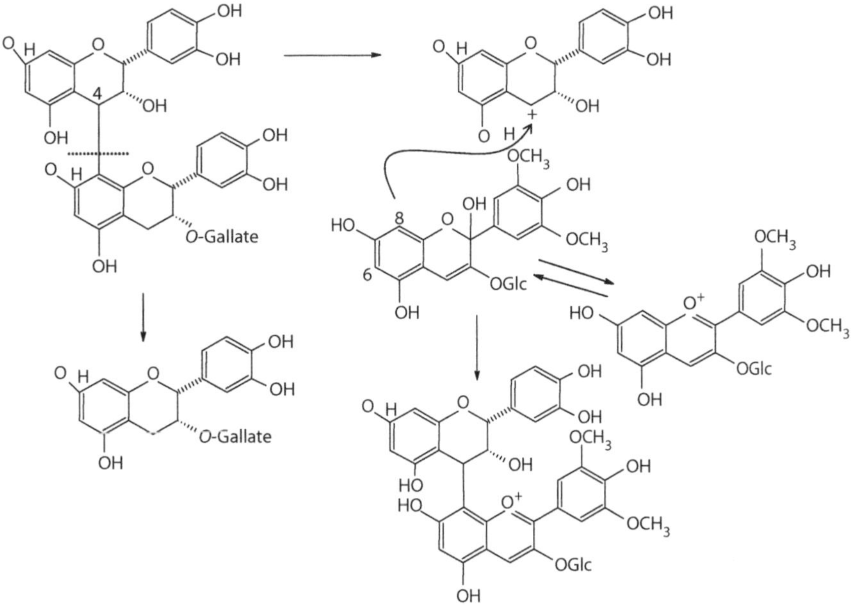

FIGURE 5.13   Mechanism of flavanol–anthocyanin (F–A⁺) formation.

flavanol unit and provides no information on its position in the dimer. Nevertheless, the fragmentation pattern agrees with the postulated Ec-Mv3glc structure, which arises from nucleophilic addition of Mv3glc onto the carbocation Ec⁺, issued from the acidic cleavage of Ec-EcG.

A similar F–A⁺ dimer synthesized with catechin instead of epicatehin was characterized by NMR and confirmed the postulated structure of the anthocyanin being the upper unit and the flavanol the lower unit (Salas et al. 2004b).

Incubation of Mv3glc and Ec-EcG at pH 3.8 did not give rise to the Ec-Mv3glc (F–A⁺) dimer but yielded another product with a mass signal at m/z 1221, in the positive ion mode, and a maximum wavelength of absorbance at 545 nm (Figure 5.15), which was not found in the control solution of Mv3glc at pH 3.8 and corresponds probably to Mv3glc⁺–Ec-EcG. This trimer gradually accumulated until 14 days of incubation and disappeared afterwards. It is expected to result from the following mechanism (Figure 5.16): the Mv3glc is in the flavylium form (A⁺) and acts as an electrophile. The hydroxyl groups in C-5 and C-7 have a mesomer effect and confer the flavanol (Ec-EcG dimer) a nucleophilic character in C-6 or C-8. Nucleophilic addition of the flavanol onto the flavylium cation leads to the colorless flavene (A–F), which can either be oxidized to the red flavylium (A⁺–F) (Jurd 1969, Somers 1971, Liao 1992) or proceed to a colorless cyclic condensation product with an A type bond (A-(4–8, 2-O-7)-F). The latter has recently been detected in wine and identified in a model solution (Remy-Tanneau et al. 2003). Ion extraction of the molecular ion at m/z 1221 showed the presence of two trimers and MS/MS analysis of both signals

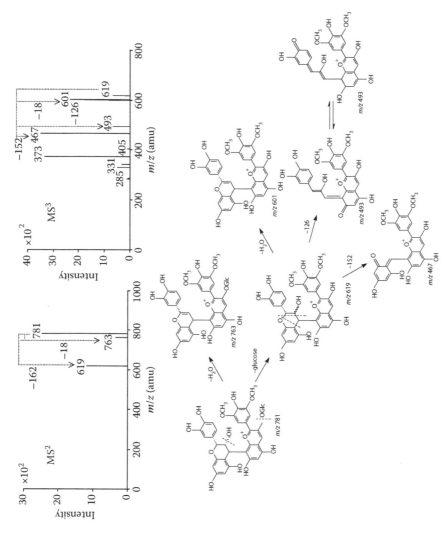

FIGURE 5.14   MS² and MS³ analysis of the Ec-Mv3glc molecular ion (*m/z* 781).

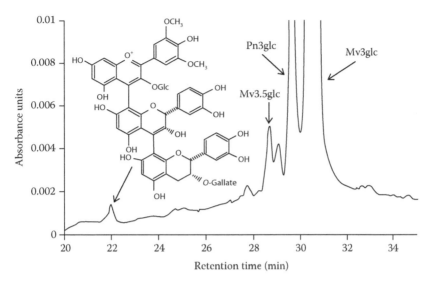

FIGURE 5.15   HPLC profile at 520 nm of the solution containing Mv3glc and Ec-EcG at pH 3.8 after 14 days of incubation.

showed different fragmentation patterns, suggesting different position linkages of the anthocyanin to the Ec-EcG dimer (Salas et al. 2003).

Because of the high reactivity of polyphenols, many other reactions like degradation, oxidation, and formation of new compounds took place in this rather simple model solution system. The apparition kinetics rate (k) values were not calculated for the newly formed dimers and trimers since these compounds did not follow a linear behavior. These values show that there are other reactions besides the postulated A^+–F and F–A^+ -studied mechanisms. These unknown mechanisms are predominant at pH 3.8.

5.4   CONCLUSIONS

A thorough understanding of chemical kinetics is necessary for designing, optimizing, and operating food processes in industry. Rates of chemical and enzyme reactions need to be quantified in food systems, and then integrated with mass and heat transfer phenomena principles to derive models useful in the operation of food-processing plants. Such an integration is not an easy task because foods are complex heterogeneous mixtures of multiple compounds that affect the rate of reactions. For this reason, most kinetic studies of foods remain empirical, and experimental confirmation of models derived by principles of kinetics of chemical reactions must be considered a crucial step. Matching and corroboration of theory and practice in chemical kinetics pertaining to food processing is facilitated by the use of instrumental analytical chemistry. An example would be the monitoring of phenolic compounds described in the chapter. Numerous reactions of flavonoids occur in the course of winemaking and aging, producing a huge variety of colored and colorless products. Their relative quantity depends on many factors (Fulcrand et al. 2004). The

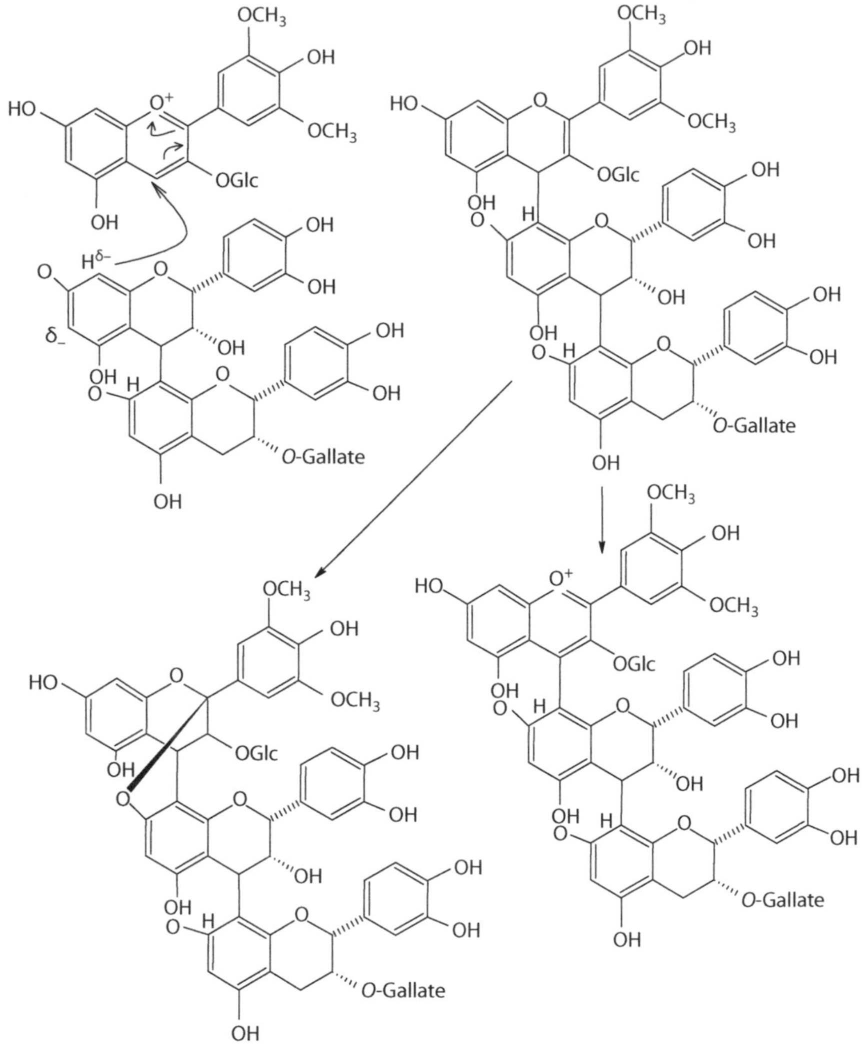

FIGURE 5.16 Mechanism of anthocyanin–flavanol (A⁺–F) formation.

pH determines the percentage of the flavylium cations compared with the hydrated forms of anthocyanins. Development of analytical techniques has unraveled some of the structures derived from tannins and/or anthocyanins in wine and enabled determination of the mechanisms involved in their formation. Each of these species is present in very small amount; even combined they represent only a small proportion of total wine phenolics and make a limited contribution to wine composition. However, most may be regarded as markers of numerous related structures similarly formed from the complete series of anthocyanins and tannins extracted from grape. Furthermore, some of these derivatives are also unstable and undergo cleavage and polymerization reactions leading to an increased diversity of structures. In spite of progress in recent years, determination of wine phenolic composition remains one of the major challenges for wine and food research. However, the availability of marker molecules representing some of the groups of derived products has made it possible to study their sensory properties.

REFERENCES

Anthon, G.E., Sekine, Y., Watanabe, N., and Barrett, D.M. 2002. Thermal inactivation of pectin metylesterase, polygalacturonase, and peroxidase in tomato juice. *Journal of Agricultural and Food Chemistry* 50: 6–9.

Brouillard, R. and Delaporte, B. 1977. Chemistry of anthocyanin pigments. 2. Kinetic and thermodynamic study of proton transfer, hydration, and tautomeric reactions of malvidin 3-glucoside. *Journal of the American Chemical Society* 99: 8461–8468.

Cemeroglu, B., Velioglu, S., and Isik, S. 1994. Degradation kinetics of anthocyanins in sour cherry juice and concentrate. *Journal of Food Science* 6: 1216–1218.

Friedrich, W., Eberhardt, A., and Galensa, R. 2000. Investigation of proanthocyanidins by HPLC with electrospray ionization mass spectrometry. *European Food Research and Technology* 211: 56–64.

Fulcrand, H., Atanasova, V., Salas, E., and Cheynier, V. 2004. The fate of anthocyanins in wine: Are there determining factors? In *Red Wine Color: Revealing the Mysteries*, eds. A.L. Waterhouse and J.A. Kennedy, pp. 68–85. Washington DC: American Chemical Society.

Haefner, J.W. 2005. *Modeling Biological Systems: Principles and Applications*. New York: Springer.

Haslam, E. 1980. In vino veritas: Oligomeric procyanidins and the ageing of red wines. *Phytochemistry* 19: 2577–2582.

Horak, F.P. 1980. Über die Reaktionskinetic der Sporenabtötung und chemischer Veränderungen bei der thermischen Haltbarmachung von Milch zur Optimieruung von Erhitzungverfahren. PhD thesis. Germany: Technical University of Munich.

Hrazdina, G. and Franzese, A.J. 1974. Oxidation products of acylated anthocyanins under acidic and neutral conditions. *Phytochemistry* 13: 231–234.

Jurd, L. 1969. Review of polyphenol condensation reactions and their possible occurrence in the aging of wines. *American Journal of Enology and Viticulture* 20: 197–195.

Lea, A. 1978. The phenolics of cider: Oligomeric and polymeric procyanidins. *Journal of the Science of Food and Agriculture* 58: 471–477.

Liao, H., Cai, Y., Haslam, E., 1992. Polyphenol interactions. Anthocyanins: Co-pigmentation and color changes in red wines. *Journal of the Science of Food and Agriculture* 59: 299–305.

Peleg, M. 2006. Time to revise thermal processing theories. *Food Technology* 60: 92–96.

Piffaut, B., Kader, F., Girardin, M., and Metche, M. 1994. Comparative degradation pathways of malvidin3,5-diglucoside after enzymatic and thermal treatments. *Food Chemistry* 59: 115–120.

Remy-Tanneau, S., Guerneve, C.L., Meudec, E., and Cheynier, V. 2003. Characterization of a colorless anthocyanin-flavan-3-ol dimer containing both carbon–carbon and ether inter-flavanoid linkages by NMR and mass spectrometry. *Journal of Agricultural and Food Chemistry* 51: 3592–3597.

Ribéreau-Gayon, P. 1964. Les composés phénoliques du raisin et du vin II. Les flavonosides et les anthocyanosides. *Annales de Physiologie Végétale* 3: 211–242.

Ribéreau-Gayon, P. 1982. *The Anthocyanins of Grapes and Wines*. New York: Academic Press.

Ricardo da Silva, J.M., Rigaud, J., Cheynier, V., Cheminat, A., and Moutounet, M. 1991. Procyanidin dimers and trimers from grape seeds. *Phytochemistry* 30: 1259–1264.

Saguy, I., Kopelman, I.J., and Mizrahi, S. 1978. Thermal kinetic degradation of betanin and betalamic acid. *Journal of Agricultural and Food Chemistry* 26: 360–362.

Salas, E., Fulcrand, H., Meudec, E., and Cheynier, V. 2003. Reactions of anthocyanins and tannins in model solutions. *Journal of Agricultural and Food Chemistry* 51: 7951–7961.

Salas, E., Atanasova, V., Poncet-Legrand, C., Meudec, E., Mazauric, J., and Cheynier, V. 2004a. Demonstration of the occurrence of flavanol–anthocyanin adducts in wine and in model solutions. *Analytica Chimica Acta* 513: 325–332.

Salas, E., Le Guernevé, C., Fulcrand, H., Poncet-Legrand, C., and Cheynier, V. 2004b. Structure determination and color properties of a newly synthesized direct-linked flavanol–anthocyanin dimer. *Tetrahedron Letters* 45: 8725–8729.

Salas, E., Dueñas, M., Schwarz, M., Winterhalter, P., Cheynier, V., and Fulcrand. H. 2005. Characterization of pigments from different high speed countercurrent chromatography wine fractions. *Journal of Agricultural and Food Chemistry* 53: 4536–4546.

Sarni, P., Fulcrand, H., Souillol, V., Souquet, J.-M., and Cheynier, V. 1995. Mechanisms of anthocyanin degradation in grape must-like model solutions. *Journal of the Science of Food and Agriculture* 69: 385–391.

Somers, T. C., 1971. The polymeric nature of wine pigments. *Phytochemistry* 10: 2175–2186.

van Boekel, M.A.J.S. 2000. Kinetic modelling in food science: A case study on chlorophyll degradation in olives. *Journal of the Science of Food and Agriculture* 80: 3–9.

van Boekel, M.A.J.S. 2002. On the use of the Weibull model to describe thermal inactivation of microbial vegetative cells. *International Journal of Food Microbiology* 74: 139–150.

van Boekel, M.A.J.S. 2007. Key reactions in foods and ways to model them. In *Food Product Design: An Integrated Approach*, eds. A.R. Linneman and M.A.J.S. van Boekel, pp. 67–106. Wageningen, the Netherlands: Wageningen Academic Publishers.

6 Use of Survival Analysis Statistics in Analyzing the Quality of Foods from a Consumer's Perspective

Guillermo Hough

CONTENTS

6.1 QUALITY AND THE CONSUMER

Quality has been defined in a number of ways, but implicit in each definition is the fact that it is the consumer who finally decides if the quality is acceptable or not. Muñoz et al. (1992) emphasize the importance of defining quality in terms of consumer perceptions. Generic definitions of quality imply the satisfaction of needs; thus it should be made clear that it is the person whose needs are being satisfied who determines quality, that is, the consumer.

The importance of the consumer is recognized in the definition given for standard on sensory shelf life by the ASTM (2005) as "Sensory shelf life is the time period during which the products' sensory characteristics and performance are as intended by the manufacturer. The product is consumable or usable during this period, providing the end-user with the intended sensory characteristics, performance, and benefits." In defining other aspects inherent to food quality such as the optimum concentration of an ingredient, the upper limit of an off-flavor, the ideal cooking temperature, or the optimum ripening time, the consumer is the one to ask. He/she is the one to say if the color of a yogurt is too strong or if the tomato should have been picked later. In this last example, there may be a relationship between ripeness as perceived by the consumer and soluble sugar contents (expressed as degrees Brix), which could allow the researcher to say that a crop of tomatoes needs a few more days on the plant because the °Brix are too low; but the researcher can only say it with confidence if a previous study on consumers' perception has been completed.

Correctly estimating the shelf life of a product is an important step in setting up the quality of a food product. Shelf life is defined as the time during which the food remains suitable for consumption from a sanitary perspective, keeping sensory, functional, and nutritional characteristics within the range of those previously established as acceptable (Hough and Wittig 2005). Sensory evaluation is the key factor for determining the shelf life of many food products. Microbiologically stable foods, such as biscuits or mayonnaise, will have their shelf life defined by the changes in their sensory properties. Many fresh foods, such as yogurt or pasta, after relatively prolonged storage, may be microbiologically safe to eat but rejected due to changes in their sensory properties.

This chapter will deal mostly with survival analysis methodology applied to sensory shelf life as seen by the consumer. It is in this area where there has been most published research. Other applications of this methodology in the estimation of optimum concentrations and cooking temperatures will also be mentioned.

6.2 INTRODUCTION TO SURVIVAL ANALYSIS

Survival analysis (Klein and Moeschberger 1997, Meeker and Escobar 1998) is a branch of statistics used extensively in clinical studies, epidemiology, biology, sociology, and reliability studies. It encompasses a group of statistical procedures which analyze data that have time between two events as the response variable. A characteristic of these types of data is an event of interest and the time in which it occurs. Table 6.1 illustrates this concept for different types of events. Originally this methodology was developed in biology, medicine, and epidemiology; but as observed in Table 6.1 it is used in many other fields such as durability of industrial products, social events and it has even been used to estimate how many years go by before a building is perceived as historical. Durability of industrial products is the field of sensory shelf life of foods. However, survival analysis has been applied to a wide range of products such as car tires, electronic equipment, and machinery (Nelson 1990, Meeker and Escobar 1998) and only in the recent years it has been applied to shelf life of foods.

Gacula and Singh (1984) introduced the Weibull model, derived from survival analysis, to food shelf-life studies. The model was applied in later studies

TABLE 6.1
Events of Interest and Measured Time

Event of Interest	Measured Time
Survival after cancer diagnosis	Since application of medication till death
Infection with HIV	From day of taking intravenous drugs till infection
Burn out of a lightbulb	From connection to burn out
Going back to prison	Since coming out of prison after first sentence till going back for second sentence
Getting bored of eating chocolate ice cream	Till consumer declares he is bored of eating chocolate ice cream for dessert every dinner
Consumer rejecting a food	Since the food leaves the factory till it is consumed

without considering censoring phenomena (see Section 6.3.3) inherent in the data (Hough et al. 1999, Cardelli and Labuza 2001, Duyvesteyn et al. 2001). Hough et al. (2003) applied the concept of interval censoring and survival analysis tools to food shelf life.

Traditionally food shelf life has been centered on the product. For example, in a recent study on the use of ionizing radiation to extend the shelf life of green onion leaves (Fan et al. 2003), a panel of three assessors measured the overall quality of the product using a 9-point scale, with 1 being unusable and 9 excellent. The authors chose the 6 of this scale as commercialization limit. Under the studied conditions, they determined that the shelf life of the onion leaves was 9 days. This is what we call shelf life centered on the product. An interesting point is what consumers would think about these onion leaves kept for 9 days. It is very probable that a very demanding consumer would find the product stored for 6 days unacceptable and that another, less demanding consumer would be quite happy consuming the onion leaves stored 12 days. Thus, from a sensory point of view, the onion leaves do not have a shelf life of their own; rather this will depend on the interaction of the food with the consumer. In survival analysis, the cumulative distribution function or rejection function $F(t)$ is defined as the probability of an individual failing or dying before time t (Meeker and Escobar 1998). Referring this definition to the sensory shelf life of the onion leaves, the "individual" would not be the vegetable itself, but rather the consumer, that is the rejection function would be defined as the probability of a consumer rejecting a product stored for less than time t. The hazard would not be focused on the product deteriorating, rather on the consumer rejecting the product.

6.3 SHELF-LIFE METHODOLOGY

6.3.1 EXPERIMENTAL DATA USED TO ILLUSTRATE THE METHODOLOGY

Actual experimental data are used to illustrate the shelf-life methodology based on survival analysis is from a study on shelf life of yogurt (Curia et al. 2005). Stirred fat-free strawberry flavored yogurt was obtained from a dairy company. Bottles

(1000 mL) from different batches were stored at 10°C in such a way as to have samples with different storage times ready on the same day. Storage times at 10°C were 0, 14, 28, 42, 56, 70, and 84 days. All batches were made with the same formulation and were checked to be similar to the previous batch by consensus among three expert assessors. Once samples had reached the storage time at 10°C, they were refrigerated at 2°C until they were tasted. Previous microbiological tests (coliforms, yeasts, molds, and *Staphylococcus aureus*) were performed to ensure there were no health risks.

Eighty regular consumers of fat-free yogurt tested samples. Each consumer received the seven yogurt samples (corresponding to each storage time at 10°C) monadically in random order. Fifty grams of each sample were served in 70 mL plastic cups. Water was available for rinsing. For each sample, subjects had to answer the question, "Would you normally consume this product? Yes or No." It was explained that this meant that if they bought the product to eat it or if it was served to them at their homes, would they consume it or not? Table 6.2 shows the data for 5 of the 80 consumers who did the test.

6.3.2 BASIC CONCEPTS OF THE MODEL

In shelf-life studies, samples with different storage times (hours, days, weeks, etc.) are presented to consumers. The random variable T is defined as the time at which a consumer rejects the sample. It is supposed that there is time when the shelf-life testing begins, a zero time point which can be when the product is manufactured, when the ingredients equilibrate, when the product is put into storage conditions, or when the consumer is first likely to see it. For this zero time point, it is taken that all consumers accept the product. If a consumer rejects the fresh sample, he/she probably does not like the product to start with and their data would not be considered in the study. Each individual's rejection time can have random variations, and thus formally, T is a nonnegative random variable. The variations are intrinsic to the

---

TABLE 6.2

Acceptance/Rejection Data for Five Subjects Who Tasted Yogurt Samples with Different Storage Times at 10°C

Consumer	Storage Time (Days)							Censoring
	0	14	28	42	56	70	84	
1	Yes	Yes	Yes	Yes	No	No	No	Interval: 42–56
2	Yes	Yes	Yes	Yes	Yes	Yes	Yes	Right: >80
3	Yes	Yes	No	Yes	No	No	No	Interval: 14–56
4	Yes	No	No	No	No	No	No	Left: ≤14
5	No	No	Yes	Yes	Yes	Yes	No	Not considered

Source: Adapted from Curia, A. et al., *J. Food Sci.*, 70, S442, 2005. With permission.

individual and are produced by many nonmeasurable factors such as the physical state of the individual, the foods and beverages he/she has consumed prior to the test, their general mood, etc. To interpret the random time variable, the rejection function $F(t)$ can be used. This function is defined as the probability of a consumer rejecting a product before time t, and is represented by $F(t) = P(T \leq t)$. Formally $F(t)$ is the cumulative distribution function of T, and is the complement of the survival function $S(t) = 1 - F(t)$.

The $F(t)$ graph, percentage of rejection versus storage time, is the rejection curve. Time can go from zero to infinity, and $F(t)$ from 0% to 100%. When the rejection curve is drawn from experimental data, a step-like curve is obtained. Percentage of rejection is calculated counting the number of rejections at each storage time, and working out the percentage over the total number of answers. In the present example, these were 74 as there were six consumers whose data were not considered due to their rejection of the fresh sample. Table 6.3 shows the percentage of rejection for each storage time, and Figure 6.1 is the corresponding rejection curve. An approximation to the yogurt's shelf life can be obtained from Figure 6.1. If a 50% rejection is adopted to define the shelf life of the product, estimated storage time is 70 days. This value is not statistically reliable it is only a naïve estimation. As observed in Figure 6.1 the estimated storage time of 70 days has a wide rejection interval going from 34% to 72%. As shall be seen further ahead, a more precise parametric estimation of shelf life at 50% rejection was 48 days.

6.3.3 DATA CENSORING

A difficulty of shelf-life data is that the information on the time a consumer takes to reject a sample depends on the storage times at which he/she tastes the product. The rejection time T is not observed exactly leading to what is called censored data. Censoring can be of different types (Meeker and Escobar 1998):

TABLE 6.3
Experimental Percentage of Rejection for Each Storage Time

Storage Time (Days)	Percentage of Rejection
0	0
14	18
28	17
42	23
56	34
70	72
84	72

Source: Adapted from Curia, A. et al., *J. Food Sci.*, 70, S442, 2005. With permission.

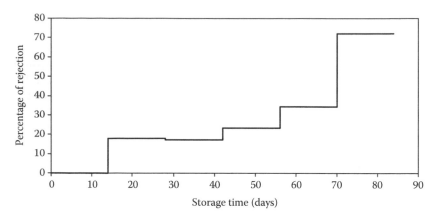

FIGURE 6.1 Percentage of rejection of consumers for yogurts stored at different times. (Adapted from Curia, A. et al., *J. Food Sci.*, 70, S442, 2005. With permission.)

- Right censoring: This is when the consumer does not reject the sample with the longest storage time (t_{last}). In this case the consumers' rejection time is higher than t_{last}.
- Interval censoring: This is when the consumer rejects a sample stored between two storage times. He/she accepts a sample stored at time t_j and rejects a sample stored at time t_k. In this case, T is between t_j and t_k.
- Left censoring: It is a particular case of interval censoring which occurs when a consumer rejects a sample at the first storage time t_1. In this case, T is between 0 and t_1.

Table 6.2 illustrates the nature of censored data for 5 of the 80 consumers who tasted yogurt samples (Curia et al. 2005). Storage times were 0, 14, 28, 42, 56, 70, and 84 days:

Consumer 1 behaved as expected in a shelf-life study; that is, he/she accepts samples up to a certain storage time and after this rejects them consistently. Data are interval censored because exact rejection time between 42 and 56 days is unknown. In the study, 40 consumers presented these types of data.

Consumer 2 accepted all samples. Supposedly for a prolonged storage time ($T >$ 84 days), samples would be rejected; thus data were right censored. Two consumers presented these types of data.

Consumer 3 was inconsistent: He/she rejected the sample with 28 days storage time, accepted the 42 day sample and rejected the 56 day and following samples. This inconsistency led to a widening of the interval censorship. A number of 14 consumers presented these types of data.

Consumer 4 rejected the sample at the first storage time and thus his/her data were left censored. A group of 18 consumers presented these types of data.

Consumer 5 rejected the fresh sample. Causes for this answer could have been that he/she (a) was recruited by mistake, that is, he/she did not like yogurt; (b) preferred the stored product to the fresh one; or (c) did not understand the task. It would not be reasonable to consider the results of these consumers in establishing the products' shelf life. For example, a company would have to produce a yogurt with a different flavor profile for those consumers that prefer the stored to the fresh product, and not encourage them to consume the aged product. In a shelf-life study of sunflower oil, Ramírez et al. (2001) found that 9 of the 60 consumers preferred the stored samples to the fresh sample and their results were not included in failure calculations. In the present yogurt study, a total of six consumers rejected the fresh sample.

6.3.4   SHELF-LIFE CALCULATIONS FROM CONSUMER DATA

6.3.4.1   Rejection Function Estimation

Hough et al. (2003) presented the model for shelf-life estimation based on survival analysis statistics. The likelihood function, which is used to estimate the failure function, is the joint probability of the given observations of n consumers (Meeker and Escobar 1998):

$$L = \prod_{i \in R}(1 - F(r_i)) \prod_{i \in L} F(l_i) \prod_{i \in I}(F(r_i) - F(l_i)) \tag{6.1}$$

where
   R is the set of right-censored observations
   L is the set of left-censored observations
   I is the set of interval-censored observations

Equation 6.1 shows how each type of censoring contributes differently to the likelihood function.

If an appropriate distribution for the data can be assumed, the use of parametric models furnishes more precise estimates of the survival function and other quantities of interest than nonparametric estimators. Usually, survival times are not normally distributed instead their distribution is often right skewed. Usually, a log-linear model ($Y = \ln(T) = \mu + \sigma W$) is chosen, where W is the error term distribution. That is, instead of the survival time T, its logarithmic transformation is modeled. In Klein and Moeschberger (1997), different possible distributions for T are presented, for example the log-normal or the Weibull distribution. In case of the former, W is the standard normal distribution, in case of the Weibull distribution, is the smallest extreme value distribution.

If the log-normal distribution is chosen for T, the failure function is given by

$$F(t) = \Phi\left(\frac{\ln(t) - \mu}{\sigma}\right) \tag{6.2}$$

where
   $\Phi(\cdot)$ is the standard normal cumulative distribution function
   μ and σ are the model's parameters

The failure function of a Weibull distribution is given by

$$F(t) = F_{sev}\left(\frac{\ln(t) - \mu}{\sigma}\right) \tag{6.3}$$

where
   $F_{sev}(\cdot)$ is the distribution function of the smallest extreme value distribution, that is, $F_{sev}(w) = 1 - \exp(-e^w)$
   μ (location parameter) and σ (shape parameter) are the model's parameters of interest

The parameters of the log-linear model are obtained by maximizing the likelihood function (Equation 6.1). The likelihood function is a mathematical expression which describes the joint probability of obtaining the data actually observed on the subjects in the study as a function of the unknown parameters of the model being considered. To estimate μ and σ for the log-normal or the Weibull distribution, the likelihood function can be maximized by substituting $F(t)$ into Equation 6.1 by the expressions given in Equation 6.2 or 6.3, respectively.

Once the likelihood function is formed for a given model, specialized software can be used to estimate the parameters (μ and σ) that maximize the likelihood function for the given experimental data. The maximization is obtained by numerically solving the following system of equations using methods like the Newton–Raphson method (Gómez 2004): ($\partial \ln L(\mu, \sigma)/\partial \mu = 0$) or ($\partial \ln L(\mu,\sigma)/\partial \sigma = 0$).

For more details on likelihood functions Klein and Moeschberger (1997) or Meeker and Escobar (1998) can be consulted. Different statistical software packages have facilities to deal with interval censored data as generated in food shelf-life studies. An option is S-PLUS (Insightful Corporation, Seattle, WA). The first step in using this software is transforming the data as shown in Table 6.2 to the data as shown in Table 6.4. This last type of data is processed in S-PLUS using the CensorReg procedures. A useful set of S-PLUS procedures to transform and process shelf-life data were developed by Garitta et al. (2004) in the form of a tutorial. This tutorial presents instructions and procedures to perform the following:

TABLE 6.4

Time Intervals and Type of Censoring Corresponding to Four Consumers Who Tasted Yogurts with Different Storage Times

Consumer	Lower Time Interval (Days)	Upper Time Interval (Days)	Type of Censoring
1	42	56	Interval
2	80	80	Right
3	14	56	Interval
4	14	14	Left

Source: Adapted from Curia, A. et al., *J. Food Sci.*, 70, S442, 2005. With permission.

- Importing raw data of the type shown in Table 6.2, from Excel into S-PLUS
- Estimating nonparametric terms of the rejection curve
- Selecting the best-adjusting parametric model
- Calculating the parametric model's parameters μ and σ, with their corresponding confidence intervals
- Obtaining the rejection curve graph
- Computing percentiles with confidence intervals

The mentioned tutorial can be downloaded free of cost from www.desa.edu.ar.

The experimental data for the yogurt study previously described (Curia et al. 2005), consisting of a matrix of 74 rows (one for each consumer who did not reject the fresh sample) and 4 columns (consumer number, lower time interval, higher time interval and censoring code), were submitted to S-PLUS software obtaining the following results:

- Weibull distribution was found adequate to model the data.
- Weibull's μ and σ parameters were 3.98 (3.90, 4.07) and 0.31 (0.24, 0.39), respectively; upper and lower 95% confidence limits are in brackets.

With these μ and σ parameters, Equation 6.3 could be applied to draw the rejection curve shown in Figure 6.2. This graph would be used to obtain approximate values of percentage of rejection for different storage times, or inversely, storage times for different values of percentage of rejection. These data can be obtained analytically using Equation 6.3; confidence intervals can be calculated with the S-PLUS software.

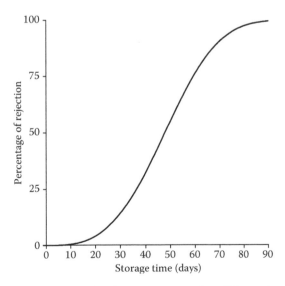

FIGURE 6.2 Percentage of rejection for yogurts stored at different times as modeled by the Weibull distribution. (Adapted from Curia, A. et al., *J. Food Sci.*, 70, S442, 2005. With permission.)

To estimate shelf life, the probability of a consumer rejecting a product (i.e., $F(t)$) must be chosen. Gacula and Singh (1984) mentioned a nominal shelf-life value considering 50% rejection, and Cardelli and Labuza (2001) used this criterion in calculating the shelf life of coffee. Curia et al. (2005) used 25% and 50% rejection for a yogurt study. These percentages are somewhat arbitrary. To have a clear guide as to what percentage to choose, the consumption distribution over the shelf life of the product must be known. That is, what percentage of the consumers will eat the product in the first 5 days of storage, how many between 5 and 10 days storage, and so on. Many companies have this information but it is propriety. In the Instituto Superior Experimental de Tecnologia Alimentaria, Argentina, research is being conducted to measure distribution times of consumption of various food products. This will serve as a basis for establishing reliable percentages of rejection values useful in estimating shelf lives. For the present it is suggested calculating shelf lives corresponding to 10%, 25%, and 50% rejection. From these values, and the knowledge of the product under study, conclusions can be drawn.

Table 6.5 shows shelf lives different percentages of rejection for the yogurt data. The company that made this yogurt was giving it 35 days shelf life, so this means that if a consumer tasted a yogurt with 35 days storage, there was approximately a 25% probability that he/she would reject the product. This was considered adequate, especially as it was known that the turnover of this yogurt was relatively quick: very few consumers actually tasted the product with more than 20 days storage. S-PLUS software can be used to calculate the estimated percentage of rejection (95% confidence limits) corresponding to 20 days storage: 4% (1.5, 10). Thus this shelf-life study confirmed that the stamped shelf life of 35 days was adequate, and that with the current turnover there was a small probability (4%) of consumers rejecting the product.

An important aspect of this methodology is that experimental sensory work is relatively simple. In this yogurt example, 80 consumers each tasted seven yogurt samples with different storage times, answering yes or no to whether they would consume the samples. This information was sufficient to model the probability of consumers accepting the products with different storage times, and from the models shelf-life estimations were made.

TABLE 6.5
Yogurt Shelf-Life Estimations with their Lower and Upper 95% Confidence Intervals for Different Values of Percentage of Rejection

Percentage of Rejection	Shelf Life (Days)	Lower Limit	Upper Limit
10	27	22	33
25	36	31	42
50	48	43	53

Source: Adapted from Curia, A. et al., *J. Food Sci.*, 70, S442, 2005. With permission.

6.4 DETERMINING THE OPTIMUM CONCENTRATION OF AN INGREDIENT

This section is based on an article published by Garitta et al. (2006). For shelf-life studies described above, there is only one reason for rejection: sample deterioration due to prolonged storage time. Hough et al. (2004) applied survival analysis concepts and calculations to consumers' acceptance/rejection data of samples with different levels of sensory defects in ultrahigh temperature (UHT) pasteurized milk. For example, for the dark color defect consumers were asked if they accepted or rejected the appearance of each one of a series of samples with different intensities of dark color. That is, consumers either found the sample appropriate or too dark. In this case the dark color was considered a defect and was rejected for this reason. But when the color is desirable such as the case of red in strawberry yogurt the consumer can find the color too light, appropriate, or too dark. The rejection of the product can be for two reasons: too light or too dark. In shelf-life or sensory defect studies, there is only one event of interest, time to failure or concentration limit, respectively. When a consumer opens a pot or bottle of yogurt he/she can find the yogurt too light, appropriate, or too dark. Thus there are two events of interest: the transition from too light to appropriate, and the transition from appropriate to too dark.

As discussed above, from a sensory point of view food products do not have shelf lives of their own, but they will depend on the interaction of the food with the consumer. In optimizing the color intensity of a yogurt, there will not be an ideal color of the product on its own and which will be considered acceptable by all consumers. What one consumer finds too light, another will find too dark and vice versa. The objective will be to find an optimum color that maximizes the proportion of consumers who find the color appropriate. At this optimum there will be a group of consumers who will find it appropriate, another group who will find it too light, and still another group who will find it too dark.

6.4.1 Experimental Data Used to Illustrate the Methodology

A range of seven colors of liquid strawberry yogurt were used. For the medium color, a leading commercial product was chosen. The lighter colors were obtained by mixing the yogurt with whole milk. The darkest colors were obtained by adding red coloring (Color 646-Christian Hansen, Quilmes, Buenos Aires, Argentina) to the commercial yogurt. Detailed preparation of samples is in Table 6.6. The color intensity of the above samples was measured with a Color-Tec PCM/PSMTM colorimeter (Color-Tec, Clinton, NJ) using daylight (D65) as illuminant option and calibrated with its color standards. The color measurement was expressed in the CIELAB system using the a^* parameter as representative of the yogurt's red color (Hutchings 1994), these values are presented in Table 6.6. Sixty children between 10 and 12 years old from the town of Nueve de Julio, Argentina, and common consumers of liquid yogurt, were recruited. Each consumer received the seven samples corresponding to the seven colors (Table 6.6) monadically in random order. Fifty milliliters of each sample was presented in a 70 mL plastic glass. For each sample they had to look at

TABLE 6.6
Preparation of Samples with Different Color Intensities and the Parameter a^* Values of Color Space CIELAB

Sample	Preparation (% Ingredients)	Parameter a^*
1	0% yogurt/100% whole milk	0.2
2	20% yogurt/80% whole milk	3.9
3	25% yogurt/75% whole milk	6.3
4	50% yogurt/50% whole milk	10.9
5	100% yogurt/0% whole milk	15.8
6	0.5 mL colorant/L of yogurt	21.4
7	1 mL colorant/L of yogurt	25.3

Source: Adapted from Garitta, L. et al., *J. Food Sci.*, 71, S526, 2006. With permission.

TABLE 6.7
Acceptance and Rejection Data for Five Consumers and their Corresponding Censoring

	Red Color a^*							Censoring	
Consumer	0.2[a]	3.9	6.3	10.9	15.8	21.4	25.3	Light Color Rejection	Dark Color Rejection
1	1[b]	1	0	0	0	2	2	Interval: 3.9–6.3	Interval: 15.8–21.4
2	1	1	1	0	2	0	2	Interval: 6.3–10.9	Interval: 10.9–25.3
3	1	1	0	0	0	0	0	Interval: 3.9–6.3	Right: >25.3
4	1	1	1	1	1	1	1	Right: >25.3	Right: >25.3
5	0	1	1	1	2	1	0	Not considered	

Source: Adapted from Garitta, L. et al., *J. Food Sci.*, 71, S526, 2006. With permission.

[a] Sample preparation is in Table 6.6.

[b] 1, rejection due to too light; 0, color is ok; and 2, rejection due to too dark.

it and mark one of the following options: "too light," "ok," or "too dark." Table 6.7 shows the data for 5 of the 60 consumers who did the test.

6.4.2 DATA CENSORING CONSIDERATIONS

Data from five consumers will be used to illustrate the type of censoring that can occur with two-event data. Consumer 1 behaved as expected. He/she rejected the very light yogurts, accepted those with intermediate colors, and rejected those that were very dark. The exact color below which this consumer rejected the yogurts because they were too light was unknown; it was between $a^* = 3.9$ and $a^* = 6.3$ and his/her data were thus interval censored for light color rejection. Analogously, the exact color above which this consumer rejected the yogurts because they were too dark was unknown; it was between $a^* = 15.8$ and $a^* = 21.4$, and his/her data were thus interval censored for dark color rejection. Consumer 2 behaved similarly to Consumer 1, the only difference is that there was an inconsistency in his/her dark color rejection, leading to a wider censoring interval. Consumer 3 rejected the very light yogurts, and then accepted all yogurts. He/she was interval censored for light color rejection between $a^* = 3.9$ and $a^* = 6.3$. The color above which this consumer rejected the yogurts because they were too dark was $a^* > 25.3$ and thus his/her data were right censored for dark color rejection. Consumer 4 rejected all yogurts because they were too light. Thus his/her data were right censored for light and dark color rejection.

Completely inconsistent data such as those of Consumer 5 were not considered in the calculations. Eleven of the 60 consumers presented these types of data, most probably because they did not concentrate on the task or did not understand it. Some of the groups of children performing the task were rather boisterous and it was difficult to get them to concentrate. This could explain the relatively large number of inconsistent consumers in this data set.

6.4.3   Two-Event Models Using Survival Analysis

Let A be the random variable representing the optimum color value for a consumer. Assume that A is absolutely continuous with distribution function F. For each value of color a^*, there will be two rejection functions:

- $R_1(a^*)$ = probability of a consumer (or proportion of consumers) rejecting a yogurt with a color = a^* because it is too light, that is, $R_1(a^*) = P(A > a^*) = 1 - F(a^*)$
- $R_d(a^*)$ = probability of a consumer (or proportion of consumers) rejecting a yogurt with a color = a^* because it is too dark, that is, $R_d(a^*) = P(A < a^*) = F(a^*)$

The consumer language may be defined in the sense that "too light" means low intensity of red color (low values of a^*), and "too dark" means high intensity of red color (high values of a^*).

As for shelf-life studies described above, the likelihood function, which is used to estimate the failure function, is the joint probability of the given observations of the n consumers (Klein and Moeschberger 1997). In this two-event model, there are two likelihood functions, namely L_1 (light colors) and L_d (dark colors):

$$L_1 = \prod_{i \in R} R_1(r_i) \prod_{i \in L} (1 - R_1(l_i)) \prod_{i \in I} (R_1(l_i) - R_1(r_i)) \tag{6.4a}$$

$$L_{\rm d} = \prod_{i \in R}(1 - R_{\rm d}(r_i)) \prod_{i \in L} R_{\rm d}(l_i) \prod_{i \in I}(R_{\rm d}(r_i) - R_{\rm d}(l_i)) \qquad (6.4b)$$

Both in Equation 6.4a and b, R is the set of right-censored observations, L is the set of left-censored observations, and I is the set of interval-censored observations. Equation 6.4a and b shows how each type of censoring contributes differently to the likelihood functions.

In shelf-life studies, failure values are not normally distributed, but they tend to be represented by right-skewed distributions. For this reason, a log-linear type function, that is, $Y = \ln(T) = \mu + \sigma W$ where W is the error term distribution, is selected. In other words, instead of the color a^* failure intensities, its logarithmic transformation is modeled.

For shelf-life studies, the log-normal and Weibull distributions were presented. If the normal distribution is chosen, the rejection functions are given by

$$R_{\rm l}(a^*) = 1 - \Phi\left(\frac{\ln(a^*) - \mu_{\rm l}}{\sigma_{\rm l}}\right) \qquad (6.5a)$$

$$R_{\rm d}(a^*) = \Phi\left(\frac{\ln(a^*) - \mu_{\rm d}}{\sigma_{\rm d}}\right) \qquad (6.5b)$$

Both in Equation 6.5a and b, $\Phi(\cdot)$ is the standard normal cumulative distribution, and $\mu_{\rm l}$, $\mu_{\rm d}$ and $\sigma_{\rm l}$, $\sigma_{\rm d}$ are the model's parameters.

The parameters of the log-linear model are obtained by maximizing the likelihood functions (Equation 6.4a and b). To estimate μ and σ for the normal distribution, the likelihood functions are maximized by substituting $R_{\rm l}(a^*)$ and $R_{\rm d}(a^*)$ in Equation 6.4a and b by the expressions given in Equation 6.5a and b, respectively. Once the likelihood function is formed, as discussed in the previous shelf-life study, specialized software such as S-PLUS can be used to estimate the parameters (μ and σ) that maximize the likelihood function for the given experimental data.

6.4.4 DETERMINATION OF THE OPTIMUM FOR A TWO-EVENT MODEL

The normal distribution showed a better fit to these color data than the Weibull distribution. Figure 6.3a and b shows rejection probability for the light and dark colors, respectively, using the normal function. Looking at Figure 6.3a it could be decided to minimize rejection due to light color and thus a high value of a^* could be chosen; but if the a^* value is too high there is an increase in the probability of rejection due to too dark. Analogously if it is wanted to minimize rejection due to dark color, a low value of a^* (see Figure 6.3b) could be chosen; but if the a^* value is too low there is an increase in the probability of rejection due to too light. This discussion leads to the necessity of obtaining an optimum color where the sum of percentage of rejection

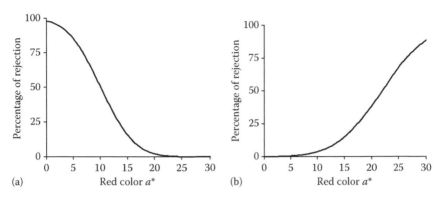

FIGURE 6.3   Percentage of consumers rejecting the yogurt versus the red color a^* for (a) R_1, light color rejection and (b) R_d, dark color rejection. (Adapted from Garitta, L. et al., *J. Food Sci.*, 71, S526, 2006. With permission.)

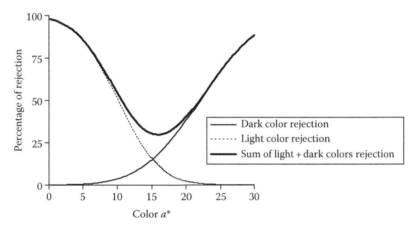

FIGURE 6.4   Optimum color (sum of light color rejection + dark color rejection). (Adapted from Garitta, L. et al., *J. Food Sci.*, 71, S526, 2006. With permission.)

due to too light + percentage of rejection due to too dark is a minimum. Figure 6.4 shows the sum of Figure 6.3a and b and the minimum of the resulting curve is the desired optimum $a^* = 16.0$.

In order to get an estimate of the variability of the optimum a^*, the following formula can be applied:

$$OC \pm Z_{1-\alpha/2} \, (se_{average}) \tag{6.6}$$

where
   OC is the optimum color
   $Z_{1-\alpha/2}$ is the $(1 - \alpha/2)$ coordinate of the standard normal distribution

$se_{average}$ is the average standard error, calculated by the following equation:

$$se_{average} = \frac{1}{2}\sqrt{se_1^2 + se_d^2} \tag{6.7}$$

where se_1 and se_d are the standard errors of the optimum color calculated from the rejection due to too light and rejection due to too dark curves, respectively. Using the above formulas the 95% confidence interval for the optimum color was 16 ± 1.5. At this optimum color there is still a 30% rejection: 12% of consumers will find this color too light, and 18% will find it too dark; this means that 15 of the 49 consumers will reject the optimum color ($49 \times 0.30 \approx 15$).

Similarly to what was mentioned for the shelf-life study, an important aspect of the methodology for determining optimal concentration of a food ingredient is that experimental sensory work is relatively simple. In this work, 60 consumers each looked at seven yogurt samples with different red color intensities, responding to whether they found the color too light, ok, or too dark. This information was sufficient to determine an optimum color with an acceptable confidence interval.

6.5   FURTHER RESEARCH ON SURVIVAL ANALYSIS METHODOLOGY APPLIED TO FOOD QUALITY

A number of articles have been published on the use of survival analysis methodology to shelf life and other food quality studies. Following is a list of these publications for the interested reader to pursue:

- Basic initial study: Hough et al. (2003) presented the survival analysis model to estimate shelf life of foods based on consumers' acceptance/rejection of stored samples.
- Bayesian modeling: Calle et al. (2006) applied Bayesian modeling to shelf-life estimations as a way of being able to use prior information from previous experience and thus obtain better parameter estimates.
- Concentration limits of sensory defects: Hough et al. (2004) extended the basic model where the explanatory variable of the rejection curve is storage time to a model where the explanatory variable was concentration of a sensory defect. They applied this model to sensory defects in UHT milk.
- Effect of covariates on shelf-life estimations: Curia et al. (2005) presented the necessary equations to study the effect of covariates such as product formulation or consumer demographics on sensory shelf life. They applied the model to shelf life of yogurts with different flavors and fat content.
- Accelerated shelf-life modeling: Hough et al. (2006) developed an accelerated shelf life, which allowed estimating an activation energy of how acceptance/rejection behavior of consumers' varied as a function of storage time and temperature. As a case study they applied the model to appearance of minced beef.

REFERENCES

ASTM. 2005. *E2454–05: Standard Guide for Sensory Evaluation Methods to Determine the Sensory Shelf Life of Consumer Products.* West Conshohocken, PA: ASTM International.

Calle, M., Hough, G., Curia, A., and Gómez, G. 2006. Bayesian survival analysis modelling applied to sensory shelf life of foods. *Food Quality and Preference* 17: 307–312.

Cardelli, C. and Labuza, T.P. 2001. Application of Weibull hazard analysis to the determination of the shelf life of roasted and ground coffee. *Lebensmittel-Wissenschaft und Technology* 34: 273–278.

Curia, A., Aguerrido, M., Langohr, K., and Hough, G. 2005. Survival analysis applied to sensory shelf life of yogurts. I: Argentine formulations. *Journal of Food Science* 70: S442–S445.

Duyvesteyn, W.S., Shimoni, E., and Labuza, T.P. 2001. Determination of the end of shelf-life for milk using Weibull hazard method. *Lebensmittel-Wissenschaft und Technology* 34:143–148.

Fan, X., Niemira, B.A., and Sokorai, K.J.B. 2003. Use of ionizing radiation to improve sensory and microbial quality of fresh-cut green onion leaves. *Journal of Food Science* 68: 1478–1483.

Gacula, M.C. and Singh, J. 1984. *Statistical Methods in Food and Consumer Research.* New York: Academic Press.

Garitta, L., Gómez, G., Hough, G., Langohr, K., and Serrat, C. 2004. *Estadística de Supervivencia Aplicada a la Vida Util Sensorial de Alimentos.* Tutorial for S-Plus. Madrid: CYTED Program.

Garitta, L., Serrat, C., Hough, G., and Curia, A. 2006. Determination of optimum concentrations of a food ingredient using survival analysis statistics. *Journal of Food Science* 71: S526–S532.

Gómez, G. 2004. *Análisis de Supervivencia.* Barcelona: Ahlens, SL.

Hough, G. and Wittig, E. 2005. Introducción al análisis sensorial. In *Estimación de la Vida Util Sensorial de los Alimentos*, eds. G. Hough and S. Fiszman, pp. 13–16. Madrid: CYTED Program.

Hough, G., Puglieso, M.L., Sánchez, R., and Mendes da Silva, O. 1999. Sensory and microbiological shelf-life of a commercial Ricotta cheese. *Journal of Dairy Science* 82: 454–459.

Hough, G., Langohr, K., Gómez, G., and Curia, A. 2003. Survival analysis applied to sensory shelf life of foods. *Journal of Food Science* 68: 359–362.

Hough, G., Garitta, L., and Sánchez, R. 2004. Determination of consumer acceptance limits to sensory defects using survival analysis. *Food Quality and Preference* 15: 729–734.

Hough, G., Garitta, L., and Gómez, G. 2006. Sensory shelf life predictions by survival analysis accelerated storage models. *Food Quality and Preference* 17: 468–473.

Hutchings, J.B. 1994. *Food Color and Appearance.* London, U.K.: Chapman & Hall.

Klein, J.P. and Moeschberger, M.L. 1997. *Survival Analysis, Techniques for Censored and Truncated Data.* New York: Springer-Verlag.

Meeker, W.Q. and Escobar, L.A. 1998. *Statistical Methods for Reliability Data.* New York: John Wiley & Sons.

Muñoz, A.M., Civille, G.V., and Carr, B.T. 1992. *Sensory Evaluation in Quality Control.* New York: Van Nostrand Reinhold.

Nelson, W. 1990. *Accelerated Testing. Statistical Models, Test Plans and Data Analyses.* New York: John Wiley & Sons.

Ramírez, G., Hough, G., and Contarini, A. 2001. Influence of temperature and light exposure on sensory shelf-life of a commercial sunflower oil. *Journal of Food Quality* 24: 195–204.

Part III

Effects of Specific Technologies on Safety and Quality

7 Safety and Quality of Thermally Processed Foods in Hermetically Sealed Containers

Romeo T. Toledo

CONTENTS

7.1 INTRODUCTION

High moisture thermally processed food products are the most convenient ready-to-eat foods currently available. These foods have quality attributes that mimic their home-kitchen-cooked counterparts. They do not require refrigeration, and they can be eaten directly out of the container with or without preheating. Although canned food quality has improved in recent years, eating quality, particularly of products containing vegetables, falls far short of that prepared in a home kitchen. One feature of commercially canned foods that make them different from fresh, frozen, home kitchen, or institutionally cooked products is their low risk of food borne illness. Safety against food poisoning microorganisms has been assured through adequate thermal inputs to inactivate food poisoning microorganisms and the application of modern techniques of quality assurance that ensures maintenance of hermetic seal integrity after the packaged product has been thermally processed to achieve commercial sterility. Commercial sterility, the main feature of canned foods, is defined as a condition where microorganisms capable of growing in the food product during storage and distribution at ambient temperatures are inactivated during processing. Commercially sterile canned foods are safe and stable when stored on a dry shelf. These products are high moisture foods packaged in hermetically sealed containers. Containers can be made of glass, metal, or multilayered polymers and laminates with adequate mechanical resistance to abuse and adequate oxygen barrier properties. This chapter discusses the requirements for thermal processes to produce commercially sterile foods, kinetics of thermal degradation of food quality attributes, and techniques for optimization of thermal treatments to maximize processed food quality.

7.2 SAFETY OF COMMERCIALLY STERILE FOODS

7.2.1 FOOD SAFETY RISKS FROM CANNED FOODS

The food borne illness associated with canned foods is botulism caused by toxin production when the anaerobic spore forming microorganism *Clostridium botulinum* grows in the food prior to consumption. Food products at risk for botulism are those where the water activity is above 0.85 and the pH values of the contents of a container prior to thermal processing are above 4.6. Some products that have been acidified to a pH of 4.6 and below but was originally at a high pH before acidification may also carry some risk for botulism. These food products are considered low acid or acidified foods, respectively. The former requires thermal processing at temperatures above 120°C in high pressure steam or water to inactivate spores of bacteria. The latter requires relatively mild temperatures at around 100°C but care must be taken to ensure that product pH inside the container must be at 4.6 or below within 24 h after thermal processing. Advances in thermal processing technology have resulted in virtual elimination of food poisoning outbreaks from commercial thermally processed foods. Although there are still about 20–30 cases of botulism outbreaks per year in the United States from 1978 to 1994, these outbreaks are mainly due to foods prepared in the kitchen and improperly handled after cooking or home-caned foods that are inadequately processed (Bell and Kyriakides 2000). The last botulism outbreak in the United States from commercially processed canned soup which resulted in fatalities occurred in 1982. Two nationwide recalls of canned products necessitated by the threat of botulism occurred in 2000 and 2007 (FDA/CFSAN 2008). The 2000 recall involved canned creamed soups and the 2007 recall involved several types of canned chili and canned French cut green beans.

7.2.2 U.S. REGULATIONS TO ENSURE FOOD SAFETY FROM CANNED FOODS

In 1982, the U.S. Food and Drug Administration (U.S.FDA) codified regulations 21:CFR 113 (CFR 2008a) requiring the registration of food establishments that process low acid foods packed in hermetically sealed containers. These establishments are required to file with the U.S.FDA scheduled thermal processes for these foods before they are made available for sale and consumption by the general public. These regulations apply to U.S. and foreign manufacturers marketing their products for sale in the United States. Implementation of 21:CFR 113 required processors to be cognizant of factors that affect heat penetration into food inside a container and to demonstrate that data used to determine the thermal process were developed by a competent processing authority. For foods that are naturally low acid but are acidified to prevent growth and toxin production by *C. botulinum*, the provisions of 21:CFR 114 (CFR 2008b) ensured that these foods are free from risks of botulism. In order to assist manufacturers in complying with these regulations, the U.S.FDA sanctioned Better Process Control School using a curriculum developed by the Food Processor's Institute (FPI 1995) and taught as a short course at universities across the United States by local and FPI instructors in the presence of an FDA representative. Processing establishments must ensure that operators of retorts or aseptic processing systems are certified by passing the examination after the Better Process Control School. Canned meat and poultry products are regulated by the United States Department of Agriculture and the corresponding regulations are codified in 9:CFR 318 (CFR 2008c).

7.2.3 PRODUCTION RECORDS AND ENSURING PRODUCT SAFETY

The process of collecting and maintaining production records not only instills diligence on the part of production personnel in following established production protocols but also stresses the importance of accurate measurements of critical factors. Records must not only be maintained but they must be regularly reviewed. Recording charts of time–temperature used on each retort batch must be filed with identifying marks such that the record can be traced to production. In addition, records of number of containers processed per batch, number of batches processed, and container closure examination data must also be maintained. Records of values of critical factors specified in the filed process must also be kept. Well-maintained records will assist in the determination of possible causes of potential food safety threats in order to avoid these threats in future production.

7.2.3.1   U.S.FDA Filing of Scheduled Thermal Processes

All processing establishments that have registered with the U.S.FDA and issued a Food Canning Establishment number may file a process using FDA form no. 2541. All process filings may be done online using the electronic forms and instructions available online at the U.S.FDA web site (U.S.FDA 2008). The fundamental food safety principle behind process filing is making responsible individuals in a process-ing firm aware of the relationships between the processing system, the container type, the critical factors of container fill, product composition and component properties, and the method of container placement in the retort as they affect microbial inactiva-tion rates, heat penetration into the critical point in the container, and adequacy of thermal treatment received by all containers in a retort batch. A scheduled process designed to produce a commercially sterile product must be specified. Processing the product named in a filed process outside the submitted parameters is illegal and could result in detention and destruction of such products, customs seizure of foreign products entering the United States, and closure of a plant in the United States.

7.2.3.2   Establishing a Scheduled Process

A scheduled process must meet an adequate level of lethality towards microorgan-isms. The lethal value of the heat treatment is expressed as the F_o value defined as the time the microbial suspension is held at the reference temperature of 121.1°C to achieve a specified reduction in the number of viable organisms:

$$F_o = D_o \log\left(\frac{N_o}{N}\right) \tag{7.1}$$

where
  D_o is the decimal reduction time or the time at the reference temperature (T_{ref}) of 121.1°C needed to reduce the viable microbial population by a factor of 10
  N_o and N are initial and final numbers of viable organisms, respectively

Equation 7.1 is based on a log-linear change in viable microbial numbers with heating time. The decimal reduction time is temperature dependent and decreases logarith-mically with an increase in temperature. Thus, a plot of log D against temperature is considered to be linear and the value of the slope is the negative reciprocal of the z value:

$$\log\left(\frac{D}{D_o}\right) = -\frac{1}{z}(T - T_{ref}) \tag{7.2}$$

The log-linear relationship of the D value with temperature is valid over the tem-perature range over which the D values were measured. However, extrapolation of thermal inactivation rate beyond the range of the measurement should be avoided since Equation 7.2 may deviate from the log-linear relationship over a very wide temperature range.

There is a strong evidence that the microbial inactivation curve deviates from linearity particularly when initial microbial numbers are small, there is also a long track record of canned food safety for processes based on the log-linear inactivation curve. Pflug (1987a) pointed out that the uncertainties inherent in the kinetics of microbial inactivation could be in the order of $\pm 10\%$–20%; therefore, enough safety margin must be included in the selection of the F_0 value used as a basis for a scheduled process. The minimum F_0 value is generally considered to be that equivalent to a 12 decimal reduction of *C. botulinum*. However, Pflug (1987a) argues that the number of decimal reductions should be related to the final probability of a *C. botulinum* spore surviving the heating process of 10^{-9} Thus, if the spore load in a product to be processed is 100 per can, the minimum process is only 11D rather than 12D. A very unlikely high spore load of 1000 per can will have to be present to require a 12D process. Since the highest D_0 value of *C. botulinum* is 0.3 min, an 11D reduction, i.e., $\log(N_0/N) = 11$, then the minimum F_0 value of a thermal process should be 3.3 min. Other spoilage organisms are more heat resistant than *C. botulinum*; therefore, processes based on the minimum F_0 would not be adequate to achieve commercial sterility. Pflug (1987b) recommends 9D reduction of a mesophilic spore forming organism having a D_0 value of 0.5 min or an F_0 value of 4.5 min as a minimum heat treatment to avoid product spoilage. If the product is exposed to conditions that favor thermophilic spoilage, Pflug (1987b) also recommended as a minimum F_0 a value equivalent to 4D reduction of an organism with a D_0 of 1.5 or an F_0 of 6.0 min. Actual thermal processes used in the industry, however, are carried out using F_0 values much higher than the minimum. In reality, processors utilize F_0 values much higher than the minimum to compensate for uncertainties in retort temperature distribution, heat transfer rates, microbial load levels, and variations in the number and heat resistance of the microorganisms to be inactivated by the thermal process.

Products such as leafy greens and asparagus, which are highly unacceptable when processed using high F_0 values, are processed to an F_0 value close to the minimum. However, when using F_0 values close to the minimum, process safety will need to be validated using inoculated packs.

7.2.4 OPERATOR PROCESS TIME

Since the F_0 value is the time the organism to be inactivated is exposed to a constant temperature of 121.1°C, and since the container with product when heated at a constant retort temperature will exhibit a slowly increasing temperature with time at the critical point, the actual processing time to achieve commercial sterility will depend upon the rate of heating of the food inside the container. The scheduled process will consist of a processing temperature (T_r), a process time (B_b), an initial product temperature (T_0) at the critical point in the containers at the time the containers are loaded, the retort closed, and steam is introduced. B_b is measured starting from the time where the retort temperature reaches the specified processing temperature to the time when steam is turned off. B_b is also referred to as the "operator process time."

7.2.5   DETERMINATION OF THERMAL PROCESS LETHALITY

A process carried out at a specified retort temperature for a specified time can be converted to a lethal value of the heat treatment received by microorganisms in the container equivalent to a time of instantaneous exposure to a constant reference temperature of 121.1°C. This lethal value is called the process F_o and is calculated as follows:

$$F_o = \int_0^t 10^{\frac{T-T_{ref}}{z}}\, dt \tag{7.3}$$

where

     T is the recorded temperature at the critical point (slowest heating point) inside a container at any time during a thermal process

     z is the temperature dependence of microbial inactivation rate as defined in Equation 7.2

z is also defined as the change in temperature that would change the inactivation rate by a factor of 10. The F_o is determined from a curve of temperature at the critical point in a container time. The value of the exponential term inside the integral sign is calculated at different times, plotting the values against time and calculating the area under the curve using Simpson's rule. When specifying a scheduled process, it is imperative that the process lethality calculated using Equation 7.3 equals or exceeds the specified microbial lethality calculated using Equation 7.1.

7.2.5.1   Heating Curve Parameters

Equation 7.3 requires a curve of temperature at the critical point (T) against processing time (t). The critical point is defined as the slowest heating point inside the container. An empirical curve obtained by placement of a thermocouple at the critical point in the container and recording the temperature reading as a function of time can be used to determine the F_o value according to Equation 7.3. However, in practice, it is not convenient to measure the temperature in each batch of containers in the retort during a thermal process. The T vs. t curves are obtained by plotting the logarithm of the dimensionless temperature ratio against time and calculating the heat penetration parameters as follows:

$$\log\left[\frac{T_r - T}{j_h\,(T_r - T_o)}\right] = -\frac{t}{f_h} \tag{7.4}$$

A rearranged Equation 7.4 results in a form that will permit plotting $\log(T_r - T)$ against t to obtain the heating curve parameters:

$$\log(T_r - T) = \log\left[j_h * (T_r - T_o)\right] - \frac{1}{f_h}t \tag{7.5}$$

In Equation 7.5, t is measured as the time from steam introduction into the retort. The log-linear dependence of the temperature difference $(T_r - T)$ with time of heating will hold if T_r is constant after T_r has leveled off at the processing temperature. A retort sued for heat penetration studies should have accurate temperature controls to maintain a constant retort temperature quickly after steam introduction. Since retort temperature is not at T_r until some time has elapsed after steam introduction, a pseudo-initial time calculated as 0.6 times the retort come-up time is considered to be the start of constant temperature heating.

The term $(T_r - T_o)$ is I, the initial temperature difference. The term $(T_r - T)$ is the unaccomplished temperature difference, g. Thus, Equation 7.4 may be written as

$$t = f_h[\log(jI) - \log(g)] \tag{7.5a}$$

If a semilogarithmic plot ($\log[g]$ vs. t) is constructed, the ratio of the value of g at the pseudo-initial time to the value of I will give the intercept factor or lag factor of the heating curve, j_h. The lag factor is an index of how fast the temperature at the critical point in the container responds to the applied retort temperature to assume the log-linear curve represented by Equation 7.4 or 7.5. The slope of the linear segment of the log-linear plot is the negative reciprocal of the slope index f_h. If the value of g is adequate to meet a safe value of F_o the value of t calculated using Equation 7.5 is equivalent to B_b, the required operator process time.

7.2.5.2 Heating Curve Parameters Calculated Using a Spreadsheet

Determination of the heating and cooling curve parameters can be done consistently and accurately by importing directly into a spreadsheet time and temperature data collected in heat penetration runs. Table 7.1 shows an example of heating and cooling data for a canned product at $T_r = 120°C$ and a retort come-up time of 3 min. A plot of $\log(T_r - T)$ vs. heating time is constructed first by converting can temperature data to $\log(T_r - T)$ and the plot is used to determine the heating time at the start and end of the linear segment. The corresponding values of $\log(T_r - T)$ and t within this line segment are entered into a regression and values of the slope and intercept are then determined. The regression equation is then solved for the value of $\log(T_r - T)$ at $t = 0.6$ times the retort come-up time. The antilog of this value is $j_h * I$. When this value is divided by $(T_r - T_o)$, the quotient is the value of the lag factor for heating, j_h. The slope index, f_h, is the negative reciprocal of the slope. In Figure 7.1, a regression of the data points between 14 and 45 min of heating gives $\log(T_r - T) = -0.04 * t + 2.067$ at $t = 0.6 * 3 = 1.8$ min, $\log(T_r - T) = -0.04 * 1.8 + 2.067 = 1.995$ and $j_h * I = 10^{1.995} = 98.8$. $I = 120 - 38 = 82$ and $j_h = 98.8/82 = 1.2$. The slope index $f_h = -1/\text{slope} = -1/0.04 = 25$ min.

7.2.5.3 Cooling Curve Parameters

Residual heat in the container after the start of cooling may contribute significantly to the F_o value. Products that heat slowly will also cool slowly. The most significant contribution of the cooling curve to process lethality occurs during the first few minutes of cooling since the critical point in the container will remain at the

TABLE 7.1

Heating and Cooling Data, Canned Food Processed at 120°C, Come-Up Time 3 min, Cooling Water Temperature 15°C

Heating Time (min)	Temperature (°C)	Cooling Time (min)	Temperature (°C)
0	38	0	118
3	39	2	117
5	44	4	116
10	71	5	114
15	91	10	79
20	102	15	56
25	108	20	41
30	113	25	32
35	116		
40	117		
45	118		

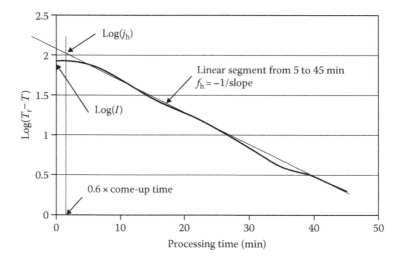

FIGURE 7.1 Semilogarithmic plot of simple heating curve showing how heating curve parameters are obtained using a spreadsheet.

same temperature after termination of heating for some time after introduction of the cooling water. The equation representing the cooling curve is as follows:

$$\log\left(\frac{(T - T_c)}{j_c * (T_g - T_c)}\right) = -\frac{t_c}{f_c} \qquad (7.6)$$

where
    T is the can temperature
    T_c is the cooling water temperature

T_g is the can temperature at the start of cooling

t_c is the cooling time

j_c and f_c are the lag factor and slope index of the cooling curve, respectively

As with the heating curve, the cooling curve can be represented by the following equation:

$$\log(T - T_c) = \log\left[j_c{}^*\log(T - T_c)\right] = \log\left[j_c{}^*(T_g - T_c)\right] - \frac{1}{f_c}*t_c \qquad (7.7)$$

Equation 7.7 shows that a semilogarithmic plot of $(T - T_c)$ against t_c will be log linear and the intercept at $t_c = 0$ will be $\log[j_c*(T_g - T_c)]$. Thus, the value of the lag factor for the cooling curve can be calculated from the antilog of the intercept divided by the value of $(T_g - T_c)$.

A plot of $\log(T - T_c)$ against t_c can be made and the values of t_c that spans the linear segment can be used in a regression of $\log(T - T_c)$ against t_c. The negative reciprocal of the slope is the value of f_c and the antilog of the intercept divided by $(T_g - T_c)$ is the value of j_c.

Cooling data for a canned product with cooling water at 15°C are shown in Table 7.1. Corresponding values of $\log(T - T_c)$ and cooling time, t_c, are plotted. Figure 7.2 shows the cooling curve showing a linear segment at cooling times from 5 to 25 min. A regression of the data points between 5 and 25 min gives the regression equation: $\log(T - T_c) = -0.0384*t_c + 2.188$. The value of $\log(j_cI_c)$ is 2.188, which is the intercept of the regression equation at $t_c = 0$. Thus, the value of j_cI_c is $10^{2.188} = 154$, $I_c = 103$, the difference between can temperature at the end of heating (118°C) and the cooling water temperature (15°C). Thus, $j_c = 154/103 = 1.5$.

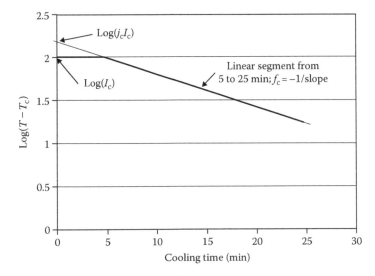

FIGURE 7.2 Semilogarithmic plot of cooling curve showing how cooling curve parameters are obtained.

The values of f_h, j_h, and j_c are used to calculate the scheduled process time at a specified processing temperature and specified initial product temperature. j_c is used to enter the $g f_h/U$ tables to obtain the value of g used in Equation 7.5 to calculate the required process time B_b. U used in the f_h/U vs. g tables is the time at the processing temperature equivalent to the value of F_o.

7.2.5.4 Process Time Calculations Using the f_h/U vs. g Tables

Tabulated values of f_h/U g are shown in Table 7.2 (values of g in °F from Toledo [2006]). Using the heating and cooling curve parameters obtained for the data in Table 7.1, a process time can be calculated. To determine a process time for a target, F_o value of 6 min, $U = 6^*(10^{(120-121.1)/10} = 4.7$. $f_h/U = 25/4.7 = 5.32$. From Table 7.2, for $z = 18$°F and $j_c = 1.5$, the values of g (in °F) are 6.2 for $f_h/U = 5$ and 7.2 for $f_h/U = 6$. Thus, interpolating for $f_h/U = 5.32$, $g = 6.4$°F or 3.6°C. A process time at 120°C and an initial can temperature of 38°C are calculated as follows:

$$B_b = 25^* [\log(1.2 * 82) - \log(3.6)] = 36\,\text{min}$$

Unless otherwise specified, the standard F_o value is based on a z value of 18°F (10°C). Thus, U is calculated at temperatures other than 250°F (121.1°C) using $z = 18$°F (10°C) and values from the f_h/U g table are obtained at $z = 18$°F (10°C).

7.2.5.5 Broken Heating Curve

Some products exhibit multiple linear segments in the heating curve. This condition will occur when fluid viscosity changes or when solids absorb some of the liquid placed inside the can before sealing, or when the solids settle down to form a compact mass at the bottom of the container. When physical changes in the product reduce the rate of heat transfer towards the critical point, the heating rate slows down. Thus, the heating curve will be manifested by an initial rapid heating rate with a lower value of the slope index followed by a slower heating rate with a higher value of the slope index. Table 7.2 shows heat penetration data for a canned food processed at a temperature of 120.5°C. A plot of $\log(T_r - T)$ vs. t is shown in Figure 7.3. The plot shows a first linear segment between 8 and 22 min of heating and a second linear segment between 25 and 55 min of heating. The calculated values of j and slope indices f_1 and f_2 are 1.3, 24, and 30, respectively. The time of heating at the point where the two linear segments intersect is t_{bh}. From Figure 7.3, this value is 22 min. The unaccomplished temperature difference $(T_r - T)$ at the point of intersection of the two line segments of the heating curve is g_{bh}. The equations of the line segments of the heating curves are

$$T = T_r - j * (T_r - T_o) * [10]^{\frac{-t}{f_1}} \quad \text{for } t < t_{bh} \tag{7.8a}$$

$$T = g_{bh} [10]^{\frac{-(t-t_{bh})}{f_2}} \quad \text{for } t > t_{bh} \tag{7.8b}$$

TABLE 7.2

f_h/U vs. g Values Used to Calculate Scheduled Processes Using Stumbo's Procedure, Tabular Values Are for $j_c = 1.0$, Use Interpolating Value $\Delta g/\Delta j$ for $j_c \neq 1.0$

	$z = 14$		$z = 18$		$z = 22$	
f_h/U	g	$\Delta g/\Delta j$	G	$\Delta g/\Delta j$	g	$\Delta g/\Delta j$
0.2	0.000091	0.0000118	0.0000509	0.0000168	0.0000616	0.0000226
0.3	0.00175	0.00059	0.0024	0.00066	0.00282	0.00106
0.4	0.0122	0.0038	0.0162	0.0047	0.020	0.0067
0.5	0.0396	0.0111	0.0506	0.0159	0.065	0.0197
0.6	0.0876	0.0224	0.109	0.036	0.143	0.040
0.7	0.155	0.036	0.189	0.066	0.25	0.069
0.8	0.238	0.053	0.287	0.103	0.38	0.105
0.9	0.334	0.07	0.400	0.145	0.527	0.147
1.0	0.438	0.009	0.523	0.192	0.685	0.196
2.0	1.56	0.37	1.93	0.68	2.41	0.83
3.0	2.53	0.70	3.26	1.05	3.98	1.44
4.0	3.33	1.03	4.41	1.34	5.33	1.97
5.0	4.02	1.32	5.40	1.59	6.51	2.39
6.0	4.63	1.56	6.25	1.82	7.53	2.75
7.0	5.17	1.77	7.00	2.05	8.44	3.06
8.0	5.67	1.95	7.66	2.27	9.26	3.32
9.0	6.13	2.09	8.25	2.48	10.00	3.55
10.0	6.55	2.22	8.78	2.69	10.67	3.77
15.0	8.29	2.68	10.88	3.57	13.40	4.60
20.0	9.63	2.96	12.40	4.28	15.30	5.50
25.0	10.7	3.18	13.60	4.80	16.90	6.10
30.0	11.6	3.37	14.60	5.30	18.20	6.70
35.0	12.4	3.50	15.50	5.70	19.30	7.20
40.0	13.1	3.70	16.30	6.00	20.30	7.60
45.0	13.7	3.80	17.00	6.20	21.10	8.00
50.0	14.2	4.00	17.70	6.40	21.90	8.30
60.0	15.1	4.30	18.90	6.80	23.20	9.00
70.0	15.9	4.50	19.90	7.10	24.30	9.50
80.0	16.5	4.80	20.80	7.30	25.30	9.80
90.0	17.1	5.00	21.60	7.60	26.20	10.1
100.0	17.6	5.20	22.30	7.80	27.00	10.4
150.0	19.5	6.10	25.20	8.40	30.30	11.4
200.0	20.8	6.70	27.10	9.10	32.70	12.1

Source: Adapted from Toledo, R.T., *Fundamentals of Food Process Engineering*, 3rd edn., Springer, New York, 2006. With permission.

Notes: To use for values of j other than 1, solve for g_j as follows:

$$g_j = j_{j=1} + (j-1)(\Delta g/\Delta j)$$

Example: g for $(f_h/U) = 20$ and $j = 1.4$ at $z = 18$

$$g_{j=1.4} = 12.4 + (0.4)(4.28) = 14.11.$$

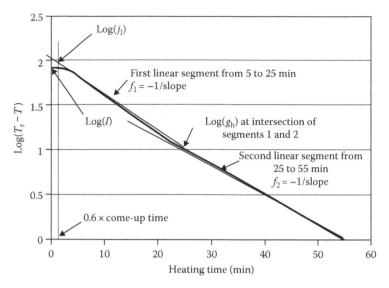

FIGURE 7.3   Semilogarithmic plot of product that exhibits a broken heating curve.

Equation 7.8a can be used to calculate g_{bh} from the value of t_{bh} obtained from Figure 7.3.

Using Equation 7.8a,

$$g_{bh} = jI\left[10\right]^{\frac{-t_{bh}}{f_1}} = 1.3(120.5 - 30)\left[10\right]^{\frac{-22}{24}}$$

$$g_{bh} = 14.2$$

The temperature when the break in the heating curve occurred is T_{bh} and this equals $T_r - g_{bh}$, for example, heat penetration data in Table 7.3, $T_{bh} = 106°C$. The values of g_{bh} and t_{bh} will change when the initial temperature and retort temperature change. Since the physical state transition is most likely to occur at a specific product temperature, that temperature should be used to calculate the g_{bh} and t_{bh}. Thus, a new g_{bh} is calculated first at the new retort temperature and that value is used to calculate a new t_{bh}. A new time–temperature table can then be constructed using Equations 7.8a and b. Table 7.4 shows a new time–temperature table for heat penetration into a canned product at a retort temperature of 122°C and an initial temperature of 40°C on a product exhibiting the heating curve parameters shown in Figure 7.3. Formula methods for calculating a scheduled process for products exhibiting a broken heating curve are more complicated than for products exhibiting simple heating curves. Readers are referred to Toledo (2006). However, if a time–temperature table is created using data from a heat penetration run, it will be possible to calculate the F_o value using the general method as represented by Equation 7.3. The cooling curve parameters are determined the same way as for simple heating curves. For brevity, a new curve will not be shown but the calculated values are $+f_c = 29.3\,min$ and the $j_c = 1.44$. The time–temperature values calculated for the heating and cooling curves based on the latter heating and cooling curve parameters are shown in Table 7.3. Table 7.3 shows

TABLE 7.3

Heating and Cooling Data for Canned Food Exhibiting a Broken Heating Curve Processed at 120.5°C, Come-Up Time 3 min, Cooling Water Temperature 15°C

Heating Time (min)	Temperature (°C)	Cooling Time (min)	Temperature (°C)
0	40	0	119.3
3	42	2	115.9
5	56	4	113
10	80	6	109
15	96	8	95
20	105	10	84
22	108	15	62
25	111	20	46
30	114	25	36
35	116		
40	117		
45	118		
50	118.8		
55	119.3		

calculation for a retort temperature of 122°C and an initial temperature of 40°C. The new value of $g_{bh} = 122 - 106 = 16$°C. The new t_{bh} is calculated using Equation 7.8a.

$$t_{bh} = f_1 \log\left[\frac{jI}{g_{bh}}\right] = 24 \log\left[\frac{1.3(122-40)}{16}\right] = 20 \text{ min}$$

Equations 7.8a and b are used to construct Table 7.4 with values of $t_{bh} = 20$ and $g_{bh} = 16$.

Column 3 in Table 7.3 shows rounded off values of the calculated temperatures. Note that in the heating curve, the calculated temperature value is lower than the initial temperature. This region is in the curved portion of the heating curve and the initial temperature is substituted for the calculated value. Furthermore, in the cooling curve, the calculated temperature values for the first 4 min of cooling are higher than the initial value, while that calculated at 6 min is lower than the initial value. This point at 6 min of cooling is the start of the linear portion of the cooling curve. The curved portion of the cooling curve is estimated by assuming a linear temperature change from the initiation of cooling to the point where the curved segment of the curve joins the linear segment. Thus, if the temperature drops from 120.9°C to 110°C after 6 min, each min of cooling will result in a drop of 1.8°C. This relationship was used to estimate the curved portion of the cooling curve. The lethal values of each time increment in the heating and cooling process are calculated in Table 7.4 and the sum of the lethal values is calculated to obtain the F_o value. The values in Table 7.4 show an F_o value for the 55 min process at 122°C from an initial temperature of 40°C equal to 10.8 min. The cooling curve for Table 7.4 is shown in Figure 7.4.

TABLE 7.4

F_o Value Calculation for Canned Product with Heating and Cooling Data Shown in Table 7.3 When Process Was Carried Out at 122°C from an Initial Temperature of 40°C

Heating	Time (min)	Temp. (°C)	Temp. Adj. (°C)	L	F_{oi}	F_o Cum.
	0	40.00000	40	7.76247×10^{-9}		
	2	34.01191	40	7.76247×10^{-9}		
	4	49.37427	49	6.16595×10^{-8}		
	6	62.05441	62	1.23027×10^{-6}		
	8	72.52066	73	1.54882×10^{-5}		
	10	81.15955	81	9.77237×10^{-5}		
	15	96.72116	97	0.003890451		
	20	106.3533	106	0.030902594	0.086984	
	25	111.0993	111	0.097723722	0.321567	0.408550
	30	114.5735	115	0.245470892	0.857987	1.179553
	35	116.9404	117	0.389045145	1.586290	2.444277
	40	118.5529	119	0.616595002	2.514100	4.100390
	45	119.6515	119.7	0.724435960	3.352577	5.866678
	50	120.4000	120.4	0.851138038	3.938935	7.291512
	55	120.9099	121	0.977237221	4.570938	8.509873
Cooling						
	0	120.9000	120.9	0.954992586		
	2	145.3164	117.2	0.407380278	1.362373	9.872246
	4	126.3627	114.0	0.194984460	0.602365	10.47461
	6	110.1657	110.0	0.077624712	0.272609	10.74722
	8	96.32441	6.000	3.0903×10^{-12}	0.077625	10.82484
	10	84.49627	84.00			
	15	61.91514	62.00			
	20	46.67121	47.00			
					Total F_o	10.8

Note: F_{oi} is the F_o value at any given point.

7.2.6 MULTIPLE HEAT PENETRATION CURVES FOR DETERMINATION OF A SCHEDULED PROCESS

The heating curve parameters f_h and j_h must be representative of the heating rate of the slowest heating container subjected to the scheduled process. There is an inherent variability in the values of empirically derived heating curve parameters. The heating curve parameters should not be averaged using the set of values from the different containers monitored. The heating curve parameters f_h and j_h used to calculate the scheduled process should be from the container that had the slowest heating rate in the family of experimental heating curves to ensure safety of all the canned foods produced using the scheduled process. The heat penetration measurements should consist of at least three separate retort batches with at least five containers monitored for temperature in each batch.

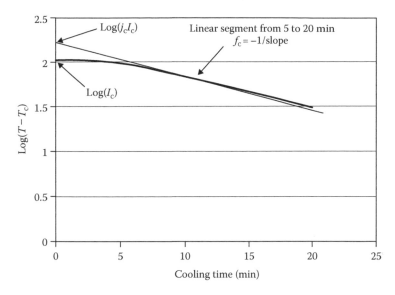

FIGURE 7.4 Cooling curve of product that exhibits a broken heating curve.

7.2.6.1 Retort Temperature Distribution

Heat penetration curves should be determined on a retort similar to that used in production. Prior to conduction of the heat penetration measurements, the temperature distribution should be measured by distributing the temperature sensor at different points in the retort. The retort should be loaded with filled cans during this process and the cans should be positioned the same way they are to be loaded into the retort used in production. The position of the sensors in the retort should be noted to identify what part of the retort is the slowest heating section. The retort temperature vs. time curves should be analyzed to reveal the position that is slow to reach the retort temperature set point and the position where the retort temperature levels off at a value lower than the set point. If the retort temperature shows very large difference between the set point and the slowest heating section, it may be necessary to alter the design of the steam header and the vent. Cans with thermocouples monitored for heat penetration should be positioned in the retort in the slowest heating position and the position that exhibits the largest deviation from the retort temperature set point.

7.2.6.2 Critical Factors Affecting Process Lethality

Process lethality is dependent on the resistance to thermal inactivation of microorganisms in the product inside the container and the actual temperature and time to which the microorganisms have been exposed. Table 7.5 summarizes these critical factors. Resistance of microorganisms to thermal inactivation is used as the basis for target F_o values used in calculating an adequate scheduled process to achieve commercial sterility. Since spore forming microorganisms would not grow at water activity below 0.85 or a pH of 4.6 and below, products with any one of these properties are free from the hazards of botulism and do not require a severe thermal process. However, salt and sugar increases the thermal resistance of microorganisms,

TABLE 7.5
Critical Factors Affecting Process Lethality

Factors Affecting Microbial Resistance to Heat Inactivation	Factors Affecting Heat Transfer
a_w—water activity	Container geometry and dimensions
Salt concentration	Fill weight
Equilibrium pH	Moisture content
Starch and/or sugar concentration	Pressure and temperature (for steam/air or water/air heating media)
	Container placement (random, vertical, or horizontal)
	Spacers between container layers
	Viscosity of fluid contents
	Particulate solids to liquid ratio
	Particulate solids dimensions
	Settling tendency of particulate solids

therefore, once an F_o value is selected as the basis for a scheduled process, the concentration of these components in the product must be maintained in commercially produced products. The critical factors affecting heating rates are listed in Table 7.5. Heating of the contents of containers occurs first by convection between the outside container surface and the heating medium followed by the transfer of heat from the inner container walls to the critical point in the container. The farther heat has to travel to reach the critical point, the slower the rate of heating at the critical point. Therefore, the size of container and the weight of its contents will affect the heat transfer. Product weight will have a large impact on heat transfer particularly for products packaged in flexible containers because expansion of the contents during thermal processing will increase the product thickness. For rigid containers, product weight can be a measure of how tightly the product is packed inside the container and consequently affect circulation inside the container of the fluid component during the heating process. For canned foods consisting of a fluid and particulate solids inside the container, heat is transferred by conduction and convection from the inside container wall to the carrier fluid and eventually to the fluid or the particulate solid located at the critical point. These heat transfer principles suggest that the type of heating medium whether high pressure water, saturated steam, superheated steam, or a mixture of air and steam will affect the heat transfer coefficient between the heating medium and the outside wall of the container. The container contents either solid or fluid will determine whether the predominant heat transfer mechanism between the inside container wall and the contents at the critical point will be by conduction or by a combination of conduction and convection. Thus, viscosity of the fluid in the container will affect the heat transfer coefficient between the inner surface of the container and the fluid and the convective heat transfer between the inner wall and the fluid. Convection heat transfer between the inside fluid and the particulate solids and within the entire container contents will be affected by fluid viscosity. If the content is solid, such as tuna or pumpkin, the amount of water and electrolytes will determine the thermal conductivity and consequently the heating rate by conduction. Thus, salt content will not only affect microbial resistance to thermal inactivation but also the rate of heat transfer. Since the interior of particulate solids must also be commercially sterile, the size and shape of these solids will affect the heating rate. Packing of particulate solids will affect heat transfer by restricting convective currents generated by the fluid contents of the container during the thermal process. Thus, the ratio of particulate solids to liquid must meet conditions specified during the determination of the heating rate parameters.

Solids with a tendency to settle and form a mat at the lower end of each container will affect the rate of heat transfer in the later stages of heating, the most critical point where the majority of the process lethality is delivered. Products such as leafy greens, squash, rice, beans, and pasta products in sauce are examples of container contents having a settling tendency. Products containing components with a tendency to settle need to be processed in an agitating retort. When processed in a stationary retort, cans containing these products should be processed while lying down on their sides to distribute the settled solids over the length of the container and minimize packing of solids when they settle in a small area. The other critical factors in Table 7.1 are concerned with ensuring good distribution of the heating medium through all containers in a retort batch. The importance of defining these critical factors and specifying their values when conducting the heat penetration tests is critical to the safety of a scheduled process.

7.3 QUALITY OF THERMALLY PROCESSED FOODS

It is generally recognized that overprocessing is the primary cause of poor sensory attributes of thermally processed products. The necessity of having the slowest heating point in the container to achieve the target F_o value results in overexposure of the contents of the container near the wall to the high temperature of processing. Since the rate of deteriorative chemical changes in foods during commercial sterilization increases exponentially with increasing temperatures, thermal exposure longer than that required to achieve commercial sterility will lead to excessive deterioration of quality. The solution is not a simple matter of reducing thermal processing time. The safety of thermally processed foods is of paramount concern to the thermal processing authority. Thus, in an argument between a canned food product technologist and the thermal processing authority over the target F_o value set for a scheduled process, the thermal processing authority always wins. High target F_o values are set when there is large variability in the values of the heat penetration parameters and in particular, when the sources of the variability are unknown or difficult to control. Some variables that bear on the selected target F_o value include inconsistency in flow of the heating medium across packages in the retort between batches processed; variability in the number and thermal resistance of the microbial load in the product; large variability in product net weight in the package; and variability in product composition and physical properties that may affect in-container heat transfer and thermal resistance to microbial inactivation. However, thermally processed product quality is not just a function of the target F_o value. The following approaches can be used to produce high-quality products from a thermal process carried out at a designated F_o value.

7.3.1 IMPROVEMENT OF RETORT OPERATION

Rapid come-up of retort to the processing temperature will minimize high-temperature product exposure since come-up time is not accounted for in the implementation of the scheduled process. Short retort come-up time is particularly important in processes carried out at temperatures higher than 121.1°C. Short come-up time also shortens the total time a batch is inside a retort and therefore increases the production rate. Rapid come-up time can be achieved by increasing the rate at which steam is introduced into the retort. Rapid retort cool-down will also be beneficial in a thermal process. Slow cool-down will prolong the residence time of the product at the high temperature and therefore increase the extent of deteriorative quality changes. Rapid cool-down can be achieved by increasing the flow of cooling water into the retort. When containers are made of glass or plastic, care should be taken to avoid sudden changes in temperature but the rate of retort temperature change can be optimized to reduce the time without compromising package integrity. In addition to rapid retort come-up and cool-down, increasing the flow of the heat transfer fluid in the retort will also decrease the temperature variation between different points in the retort. Since the target F_o value is set on the container with the slowest heating rate, reducing retort temperature variability will avoid overprocessing of those containers in the section of the retort with the highest temperature.

7.3.2 CONTAINER DIMENSIONS INCREASING HEAT TRANSFER RATE TO CRITICAL POINT

Rapid heating at the critical point can be achieved by choosing containers with the least dimension from the container wall to the critical point on the side that has the largest surface area. Examples are flat cans which resemble discs rather than tall cylinders and brick-shaped containers with one side having a thickness much smaller than the other two dimensions. The characteristic dimension of an object for heat transfer is the half-thickness of the side with the largest surface area, and the time required for the critical point to achieve commercial sterility is an inverse function of the square of the half-thickness.

7.3.3 PRODUCT FORMULATIONS TO INCREASE IN-CONTAINER CONVECTION

The fluid component of products inside packages serves as the heat transfer medium to transmit heat from the container walls to the critical point. Thus, reduced viscosity of the fluid component of the product will result in rapid heat transfer and thus shorten the processing time. A modified starch that gelatinizes at high temperature and maintains a low-viscosity dispersion at high temperature but increases in viscosity on cooling is an example of an option for increasing heating rates for canned products. Particulate solids with varying shape and dimensions will permit more freedom of movement of the fluid component thus promoting heat transfer. Blanching and cooling of starchy particulate solids before filling will minimize clumping and sloughing off of starchy particles thus maintaining low viscosity of the suspending fluid. Blanching of dry ingredients such as rice and beans before filling will avoid absorption of the suspending fluid thereby maintaining a low viscosity during thermal processing.

7.3.4 CONTAINER ORIENTATION IN STATIONARY RETORTS TO AVOID CLUMPING

Ingredients that have a tendency to clump during thermal processing should be distributed over a large surface area inside the container by orienting containers with the longest dimension perpendicular to the pull of gravity. When these particulate solids ingredients settle, they will be distributed over a large interior surface area and facilitate free movement of the suspending fluid down to the base of the settled solids near the container wall. This technique will minimize the shift in the value of the slope index in the different segments of a broken heating curve thus reducing processing time.

7.3.5 IN-CONTAINER AGITATION DURING THERMAL PROCESSING

When the product is liquid, liquid with small suspended particulate solids, or large particulate solids immersed in a low-viscosity fluid, agitation greatly improves the heat transfer rate. Agitation is imparted by specially designed retorts that may roll the cans on their sides, rotate then end over end, or shake them with rapid acceleration in a reciprocating motion. Agitation prevents clumping of particulate solids in the container and generally reduces the shift in the heat transfer rate between segments of a broken heating curve. Agitation will be of no advantage in a solid pack where the mode of heating is purely conduction.

7.3.6 HIGH TEMPERATURE–SHORT TIME PROCESSING

A quality improvement may result on thermal processes conducted at high temperature because rate of heat inactivation of microorganisms increases more rapidly than rate of deteriorative chemical and physical changes with increasing temperature. This principle is best illustrated by sterilization of a liquid by subjected to instantaneous heating to the processing temperature within a heat exchanger and holding for a specified time in a holding tube at constant temperature to achieve the target F_o value. To obtain an F_o value of 10 min, at a holding tube temperature of 130°C, a hold time of $10 * 10^{(121.1 - 130)/10}$ or 0.13 min will be obtained compared to 10 min at a hold temperature of 121.1°C. The rate of color formation through nonenzymatic browning in milk has been reported by Burton (1954) to have a D_o value of 12.5 min and a z value of 26°C. At 130°C, the D value for nonenzymatic browning will be $12.5 * 10^{(121.1 - 130)/26} = 5.45$ min. For a chemical change where the concentration of the reaction product increases with time, and the initial concentration is C_o, $\log(C/C_o) = t/D$. Therefore, for a 0.13 min process at 130°C, $\log(C/C_o) = 0.13/5.45 = 0.0238$ and $C/C_o = 1.055$ indicating that the concentration of the browning reaction product has increased only by 5.04%. In contrast, processing at 121.1°C, for 10 min, where the D value is 12.5 min, $\log(C/C_o) = 10/12.5 = 0.8$ and $C/C_o = 6.3$. Thus, the concentration of the products of the browning reaction will be sixfold higher at 121.1°C compared to 130°C. For canned foods sterilized within a container, however, the prolonged heating to achieve commercial sterility at the critical point will subject the layers of food near the container wall to excessive heating well past the designated F_o values. Thus, high temperature–short time heating will only be beneficial if the limiting quality factor has relatively low rate of formation (high D value) and the mass of food near the container wall is not much larger than that a short distance away from the critical point. Products with low f_h values can benefit from high temperature–short time sterilization. In contrast, if the rate of formation of the degraded quality factor is high, and the z value for the change is lower than that for microbial inactivation, a high temperature–short time process will not have an advantage.

7.3.7 VARIABLE RETORT TEMPERATURE PROCESSES

The concept for variable temperature thermal processes was reviewed by Durance (1997) who argued that the concept of quality improvement from high temperature–short time processes as discussed in the preceding section would not be valid when applied to product thermally processed inside containers where the container contents are not subjected to agitation. The theory behind variable retort temperature processing is that the buildup of heat-degraded components in the section within the container near the wall can be eliminated by minimizing the temperature gradient between the product near the container wall and the critical point. Teixeira et al. (1975) used computer simulation to demonstrate differences in thiamine retention between processes where the retort temperature was increased to the same temperature as a step (the standard retort process), as a ramp, and as sinusoidal. When the same target F_o values were attained, there was no difference in thiamine retention. Other investigators (Sjöström and Dagerskog 1977, Saguy and Karel 1979, Holdsworth 1985, Banga et al. 1991, Balsa-Canto et al. 2002) investigated various variable retort temperature options for different types of containers and showed quality improvement for some in comparison with conventional retorting. However, the approach has not been embraced widely by the industry as the improvement is minor and retort temperature controllers must be installed and retort operators must be properly trained for these processes. In addition, obtaining approval for these processes as satisfying the regulatory requirements for food safety has not been actively pursued by processors. However, there are versions of variable retort processes that can be considered under the umbrella of retort processes acceptable to food safety regulators as suitable for consistent delivery of thermal lethality and easily verifiable using standard thermal evaluation techniques. An example is a preheat process (Kandala 2005) where the containers are subjected to two-step changes in the retort temperature: a first step to a retort temperature near 100°C and a second step to a temperature of 125°C or higher. Since only the second step contributes to the lethality, the first step is used to establish an initial temperature at the critical point used to calculate the scheduled process.

7.3.8 PROCESS SIMULATION AND QUALITY OPTIMIZATION

Process simulation can be used to verify the effects on quality of a proposed retort temperature schedule. For certain shape of container and type of product, an optimum initial temperature and retort temperature may exist where the product quality is better than a conventional process conducted at 121.1°C. There have been several forms of the objective function used for retort process optimization. Mansfield (1962) proposed a cook value, C, defined as

$$C = \int_0^t [10]^{\frac{T-T_{ref}}{z}} \, dt \tag{7.9}$$

The z value used in Equation 7.9 is the quality factor degradation or the appearance of a degradation product. Equation 7.9 is suitable only for processes where the contents are thoroughly mixed such as in aseptic processes carried out with a heat exchanger and a hold tube. A variation of Equation 7.9 was proposed by Ohlsson (1980), which calculated a volume averaged cook value and took into account the temperature distribution with time inside the container. The volume averaged cook value proposed by Ohlsson is

$$C_{avg} = \frac{1}{V} \int_0^V \int_0^t [10]^{\frac{T-T_{ref}}{z}} \, dt \, dV \tag{7.10}$$

Equation 7.10 does not involve a D value and although useful to obtain relative cook values on quality factors with relatively high D values, it does not account for changes in the sections near the container wall that is subject to high temperature for prolonged periods when the quality factor is heat sensitive with low D value. Silva et al. (1992) argued that the D value of the deteriorative chemical reaction is important and should be considered in the simulation calculations. Equation 7.11 is an expression for both microbial inactivation and extent of degradation or a quality factor or the appearance of degradation products

$$\left[\frac{N}{N_o}\right]_{avg} = \frac{1}{V} \int_0^V \left[10^{\frac{-1}{D_{ref}}\int_0^t [10]^{\frac{T-T_{ref}}{z}} \, dt}\right] dV \tag{7.11}$$

Equation 7.11 can be used to calculate a volume averaged microbial inactivation $(N/N_o)_{avg}$ or extent of quality factor deterioration (fraction of original concentration left). If the quality factor considered is the appearance of a degradation product, a positive sign is used on the exponent of 10 before the integral. To solve Equation 7.11, the container is separated into several control volume elements. The temperature at the center of each volume element at a certain process time increment is then calculated using the heat transfer equation. A finite heat transfer coefficient is used on the control element representing the surface of the container. The expression is evaluated at each time increment and the sum of the average values for each time increment represents the total volume averaged microbial inactivation or quality factor degradation. Equation 7.11 shows that to properly evaluate the value of an objective function that changes during thermal sterilization, it will be necessary to know the temperature distribution inside the container at any time and the D and z values of both microbial inactivation and quality factor deterioration. Since measurement of the actual temperature in different locations inside the container is subject to errors in placing the thermocouple in the exact location, this problem is best solved by measuring the temperature at a single point, using the time–temperature measured to calculate the surface heat transfer coefficient based on known values of the thermophysical properties of the product, and then a temperature distribution at any point in the container at any time during the process is calculated. The process is tedious and requires a computer to conduct the simulation but this approach is a lot less time consuming than having to actually carry out the process at different scheduled thermal processes.

REFERENCES

Balsa-Canto, E., Banga, J.R., and Alonso, A.A. 2002. A novel, efficient and reliable method for thermal process design and optimization. Part II: Applications. *Journal of Food Engineering* 52: 235–247.

Banga, J.R., Perez-Martin, R.I., Gallardo, J.M., and Casares, J.J. 1991. Optimization of the thermal processing of conduction-heated canned foods: Study of several objective functions. *Journal of Food Engineering* 14: 25–51.

Burton, H. 1954. Color changes in heated and unheated milk. I. The browning of milk on heating. *Journal of Dairy Research* 21: 194–198.

Bell, C. and Kyriakides, A. 2000. *Clostridium botulinum: A Practical Approach to the Organism and its Control in Foods*. Oxford, U.K.: Blackwell Science.

CFR 2008a. *Code of Federal Regulations*, USA. Thermally processed low acid foods packaged in hermetically sealed containers. Title 21, Part 113.

CFR 2008b. *Code of Federal Regulations*, USA. Acidified foods. Title 31, Part 114.

CFR 2008c. *Code of Federal Regulations*, USA. USDA canning regulations. Canning and canned products. Title 9, Parts 318 (meat) and 391(poultry).

Durance, T.D. 1997. Improving canned food quality with variable retort temperature processes. *Trends in Food Science and Technology* 8: 113–118.

FDA/CFSAN 2008. *U.S. Food and Drug Administration, Center for Food Safety and Applied Nutrition*. Available at http://www.fda.gov/cfsan/subject/canned food recalls. Accessed on Sept. 30, 2008.

FPI 1995. *Canned Foods – Principles of Thermal Process Control, Acidification, and Container Closure Evaluation*, eds. Gavin A. and Weddig L.M. The Food Processors Institute, Washington, DC.

Holdsworth, S.D. 1985. Optimization of thermal processing—A review. *Journal of Food Engineering* 4: 89–116.

Kandala, R. 2005. *Improvement of the quality attributes of commercially sterile egg product by the use of improved formulations and optimized thermal process*. PhD Dissertation, The University of Georgia, Athens, GA.

Mansfield, T. 1962. High-temperature, short-time sterilization. *Proceedings of the 1st International Congress on Food Science and Technology*, Vol. 4, 311–316, London.

Ohlsson, T. 1980. Optimal sterilization temperatures for sensory quality in cylindrical containers. *Journal of Food Science* 45: 1517–1521.

Pflug, I.J. 1987a. Using the straight line semi-logarithmic microbial destruction as an engineering design model for determining F-value for heat processes. *Journal of Food Protection* 50: 342–346.

Pflug, I.J. 1987b. Factors important the heat process value F for low acid canned foods. *Journal of Food Protection* 50: 347–351.

Saguy, I. and Karel, M. 1979. Optimal retort temperature profile in optimizing thiamine retention in conduction-type heating of canned foods. *Journal of Food Science* 44: 1485–1490.

Silva, C., Hendrickx, M., Oliveira, F., and Tobback, P. 1992. Critical evaluation of commonly used objective functions to optimize overall quality and nutrient retention of heat preserved foods. *Journal of Food Engineering* 17: 241–258.

Sjöström, C. and Dagerskog, M. 1977. Optimization of the time/temperature relationship for heat sterilization of canned food by computer simulation. *Proceedings of Mini Symposium—Mathematical Modelling in Food Processing*, Örenäs, Sweden.

Teixeira, A.A., Zinsmeister, G.E., and Zahradnik, J.W. 1975. Computer simulation of variable retort control and container geometry as a possible means of improving thiamine retention in thermally processed foods. *Journal of Food Science* 40: 656–659.

Toledo, R.T. 2006. *Fundamentals of Food Process Engineering*. 3rd edn. New York: Springer.

USFDA 2008. *Acidified and Low Acid Canned Foods: Regulations, Process Filing Forms, Instruction and Guidelines*. Available at http://www.fda.gov/food/cfsan/acidified and low acid foods. Accessed on Sept. 2, 2008.

8 Alternative Heating Technologies

Marc Regier, Matthias Rother,
and Heike P. Schuchmann

CONTENTS

8.1   INTRODUCTION

Starting in the 1960s to 1970s there has been an increasing shift from originally batch thermal operations toward continuous short time processing operations of food (often combined with high temperatures). This is due to the increasing trend of processing food to obtain products with minimum loss of sensorial and nutritional valuable ingredients, while keeping them safe for consumption (minimum processed food). For example, the high temperature short time pasteurization (UHT) of milk at 140°C ensures a safe product, taking only a few seconds so that the product suffers only slight quality deterioration.

For such short time thermal processes, the heating rate becomes more and more important. Heating times can no more be neglected compared to holding times. Thus, short time processes rely on rapid heat transfer mechanisms: For liquid products, this can be often managed by conventional heating techniques using plate or scraped surface heat exchangers; whereas already with particulates of some millimeters, the technique cannot ensure a sufficient heating of the center of the particle to hit the safety criteria of sterility. For solid foods, this problem becomes even more dominant.

Limited internal heat and mass transfer also is often a problem in mild drying of solid food pieces as vegetables or fruits. In this convective process the drying rates may be limited by two resistances: The resistance against heat condition from the heated surface to the center and/or the internal moisture transport resistance hindering the water transport from the center to the surface. In order to increase the drying rates (at constant product size), the only possibility is to increase the temperature of the convective airflow. The consequence is a large temperature gradient within the food, often leading to overheated (and overdried) surface regions and/or texture failures with corresponding quality loss. The often slow internal heat conduction can only be overcome with alternative heating technologies, thus, giving a technological push to these techniques.

As alternative techniques in thermal food processing, three principles have been applied: ohmic heating, microwave heating, and infrared (IR) heating. Without neglecting their special differences, the history of all these processes does show some similarities. The alternative heating technologies are often also called "new" heating technologies. This is somehow misleading, as all of them already have a long history since they have been first described. For example, the IR or heat radiation was already discovered in the nineteenth century by William Herschel. Its radiative and optical characters similar to visible light were already stated in the same time by Fizeau and Foucault, but it took some time to use it consciously to directly heat food. Long wave IR radiation (also called far-infrared, FIR) is, along with heat transfer by convection and heat conduction, one of the main heat transfer mechanisms in conventional ovens. Starting in the 1950s Lykow studied the IR heating by theoretical and experimental means. Further investigators in those early times were Jubitz, Déribéré, Leconte, and Pavlov (Skjöldebrand 2001).

Shortwave IR (or near-infrared, NIR) radiations were applied to foods starting only in the 1970s, after it had already been used in other industries (paper and textile industry). During these times, the IR equipment as radiators and reflectors were developed and first important optical properties of food were gained. Nevertheless, up to now only a few successful applications of IR heating in the food area can be stated.

The history of microwave heating of food sounds similar. In the 1950s the legendary Percy Spencer, a British radar engineer, observed that a chocolate bar turned warmer when passing a radar horn. In any case, he recognized the chance to apply this type of energy to heat food in a "microwave" oven and filed several patents concerning this method (Regier and Schubert 2001). However, its first application started approximately a decade later. This was caused by the need for high power microwave sources to be developed and the knowledge of important dielectric properties of food. The first major applications were finish drying of potato chips, precooking of poultry and bacon, tempering of frozen food, and drying of pasta (Decareau 1985). Often, the first applications were—if at all—only temporarily successful. More knowledge had to be gathered so that the technique could survive and be successful in industrial application although still being a niche application.

Also similar is the history of the technique of ohmic heating. Although electric conduction and the heat generation by ohmic resistances are known since 1841 (by Joule), it has taken a long time until electric conduction through foods has been used as an alternative food heating technology. First patents had already been filed in the nineteenth century (Ruan et al. 2001). In the early twentieth century, industrial pasteurization of milk was achieved by ohmic heating and several other heating and blanching applications could be found. Nevertheless, these processes disappeared again due to the lack of inert electrode materials and the lack of knowledge of food and process parameters. It was not before 1970 that the interest in these processes grew again due to the availability of new materials, process designs, and pushed by the industry's interest in short time techniques combined with aseptic processes.

As shown, all the alternative heating technologies have a long history, an initial more or less successful application, an interim market-break due to the lack of knowledge or equipment, and a kind of "rebirth" in niches with developing potential.

8.2 PHYSICAL PRINCIPLES OF ALTERNATIVE HEATING TECHNOLOGIES

The alternative heating technologies described here are all based on electromagnetism. The main difference between ohmic heating, microwave heating, and IR heating is the frequency used by the electromagnetic field to generate the heat by coupling with the matter and thus the interactions between the electromagnetic fields and the matter. In Figure 8.1, a schematic view of the electromagnetic spectrum is given.

Starting at very low frequencies (down to static electric fields at frequency 0) the electric fields may be used for heating by applying a voltage to a more or less conductive sample and thus passing an electric current through the sample. At higher frequencies, the electromagnetic fields are able to dissolve from the electrodes and

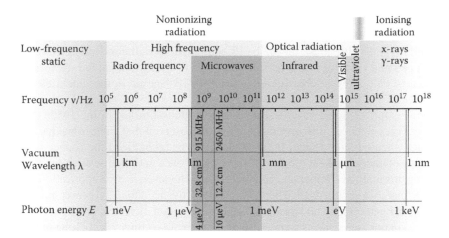

FIGURE 8.1 Electromagnetic spectrum. (Adapted from Regier, M. and Schubert, H., *The Microwave Processing of Foods*, eds., Schubert, H. and Regier, M., Woodhead Publishing, Cambridge, U.K., 2005.)

traveling through the sample, thus being called electromagnetic waves. Both microwaves and IR radiation belong to these electromagnetic waves, only the frequency and thus the wavelength of the waves are different. Whereas waves of frequencies between 300 MHz and 300 GHz are called microwaves (the corresponding wavelength in vacuum or air ranges from 1 m to 1 mm), the term IR radiation belongs to waves of next smaller wavelengths (in vacuum) between 1 mm and 700 nm (300 GHz and 0.47×10^{15} Hz). The wavelength of 700 nm limits the IR waves from visible light of the red color. Moving on in the spectrum toward smaller wavelengths (and thus higher frequency and photon energy) ultraviolet rays, x-rays, and gamma rays follow.

All these waves have frequencies smaller than ultraviolet rays and thus all three alternative heating technologies presented in this chapter are called nonionizing radiation saying that the waves do not have enough energy to directly produce ions from neutral atoms or molecules. The latter is important for the chemical/physical impact of the radiation.

8.2.1 OHMIC HEATING

As mentioned above, ohmic heating means the generation of heat from electromagnetic (or here even electric) energy at very low-electric frequencies from 0 Hz up to several hundreds of hertz. Whereas direct current (DC), corresponding to a frequency of 0 Hz, is disadvantageous due to the "fouling" or dissolution of electrodes, most often the frequency of the normal electrical alternating current (AC) power lines of 50 or 60 Hz, depending on the country, is used due to its high availability.

As expressed by its name, ohmic heating is based on the physical principle of generation of heat in an electrical ohmic resistance by passing an electric current through it. In the case of ohmic heating of food, the ohmic resistance is the food itself, thus showing the prerequisite of food to be suited for ohmic heating: a finite

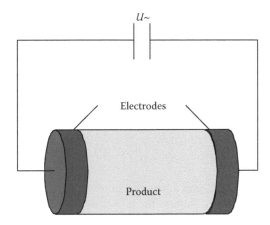

FIGURE 8.2 Schematic view of a body to be heated in a plate capacitor as part of an electrical circuit.

electrical conductivity $\sigma > 0$ or a specific resistivity: $\rho = 1/\sigma < \infty$. Thus, the product body is part of an electric circuit as shown schematically in Figure 8.2.

If a voltage U is applied to the plate electrodes, an electric current I flows through the body having a resistance R. According to Ohm's law, the current is proportional to the applied voltage divided by the resistance R:

$$I = \frac{U}{R} \tag{8.1}$$

The resistance itself is proportional to its length and inversely proportional to its cross-sectional area A, where the constant of proportionality is the specific resistivity (being the inverse conductivity) $\rho = 1/\sigma$:

$$R = \frac{\rho \cdot l}{A} \tag{8.2}$$

The electric power P that generates the heat can be expressed as

$$P = U \cdot I \tag{8.3}$$

Using Equations 8.1 and 8.2 to express the power in terms of the voltage, the resistivity and geometric dimensions become

$$P = \frac{U^2}{R} = \frac{U^2}{\rho \cdot l} \cdot A = \frac{U^2 \cdot \sigma}{l} \cdot A \tag{8.4}$$

Taking into account that the electric field strength E is defined as voltage U divided by the plate distance l and the food volume V the product of area A times the plate distance l, the power can be expressed as

$$P = \frac{\sigma \cdot E^2 \cdot l^2}{l} \cdot A = \sigma \cdot E^2 \cdot l \cdot A = \sigma \cdot E^2 \cdot V \tag{8.5}$$

So the heat is generated within the volume and not only at the surface of the product, as in conventional heating, where there is always a temperature gradient from the surface that is heated by conduction or convection to the center. These temperature gradients in conventional heating are mainly dependent on the food's conductivity (for solid foods) and convective heat transport properties (for liquid foods).

Equation 8.5 leads to the power or heat generation density P/V that is directly proportional to the square of the electric field strength and the conductivity:

$$p_v = \frac{P}{V} = \sigma \cdot E^2 \tag{8.6}$$

Ideally, a plate capacity generates a homogeneous electric field and the food has also locally constant conductivity so that this power density is constant within the volume yielding a so-called volumetric heating with practically no or only small local temperature differences.

Nevertheless, this equation shows that the heat generation is strongly dependent on the electric field strength and proportional to the conductivity so that products with high conductivity have a high heating rate.

One very simple example should be presented here: Let us assume a food consisting of two basic substances of volume ratio V_1/V and V_2/V with different electrical conductivities σ_1 and σ_2. The food is placed in the ohmic heating device—a parallel plate capacitor—so that the two phases are separated in the manner shown in Figure 8.3. According to the volume ratios, the plate areas are occupied correspondingly A_1 and A_2. The voltage U for both substances is the same. According to Equation 8.4, the heat generation within the separated volumes is

$$P_1 = \frac{U^2 \cdot A_1}{l} \cdot \sigma_1; \quad P_2 = \frac{U^2 \cdot A_2}{l} \cdot \sigma_2; \quad P = P_1 + P_2 \tag{8.7}$$

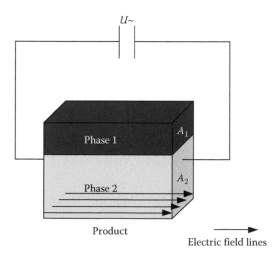

FIGURE 8.3 Sketch of a food consisting of two separated phases within a parallel plate capacitor.

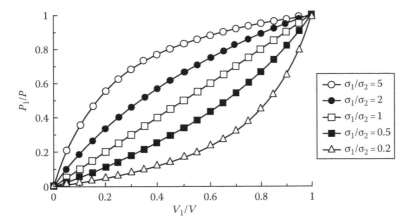

FIGURE 8.4 Ratio of the power dissipated in substance 1 to the absolute dissipated power being dependent on the conductivity ratio of the substances (for the model of separated substances).

Thus, the ratio of the power that is dissipated in substance 1 can be calculated by

$$\frac{P_1}{P} = \frac{A_1 \cdot \sigma_1}{(A_1 \cdot \sigma_1 + A_2 \cdot \sigma_2)} = \frac{(V_1/V) \cdot \sigma_1}{\left[(V_1/V) \cdot \sigma_1 + (V_2/V) \cdot \sigma_2\right]} \tag{8.8}$$

and is plotted in Figure 8.4 for some examples of conductivities. As shown, the part of power dissipated in substance 1 can be much larger or smaller than the corresponding volume ratio depending on the conductivity ratio being large or small.

The temperature increase with time dT/dt within a food is not only dependent on the conductivity but also on its density ρ_m and its specific heat capacity c_p:

$$P = \frac{c_p \cdot m \cdot dT}{dt} \rightarrow \frac{dT}{dt} = \frac{P}{(c_p \cdot m)} = \frac{P/V}{[c_p \cdot (m/V)]} = \frac{\sigma \cdot E^2}{(c_p \cdot \rho_m)} \tag{8.9}$$

On the other hand, the electric field strength may also vary within the food volume due to electrode imperfections or dielectric property or conductivity differences between constituents of foods.

The DC conductivity in foods is mainly based on ions, being solved in water. Thus, the conductivity is a monotonously increasing function of the ion concentration. For low concentrated solutions, this function becomes linear and may be expressed as

$$\sigma = \Sigma c_i \cdot \mu_i \cdot |z_i \cdot e| \tag{8.10}$$

where
  $z_i \cdot e$ is the electrical charge
  μ_i is the mobility of the ion
  c_i is the concentration of the ion (number per volume)

TABLE 8.1
DC Conductivities of Foods at a Temperature of 20°C

Substance	DC Conductivity, $\sigma/(1/\Omega\ m)$
Potable Water	
Typical	0.01–0.05
Upper limit by law (Germany)	0.25
Carrot	1.5–2
Pork	
Leg lean	0.7
Shoulder lean	0.6
Sensorically dry	<0.1
Sensorically watery	>0.8

Source: Data from Rosén, C. *Food Technology,* 26: 36, 1972. With permission.

In the mobility of the ion, its diffusion coefficient is incorporated that is also dependent on the viscosity of the fluid. Typical DC conductivities of foods are presented in Table 8.1.

Since the viscosity and diffusion coefficients and thus mobilities are strongly dependent on the temperature, that is, the higher the temperature, the higher the mobility, there is also an increase of the conductivity with rising temperature. This leads to the effect that at positions of high temperature even more heat is generated, resulting in a so-called runaway heating effect. At high voltages also a rupture of the cell walls leading to cell lysis may occur, giving an additional rise of the conductivity (Bhale 2004).

The electric field distribution within a food consisting of several phases may be very complex (the field lines are not parallel anymore) but may be described theoretically by Maxwell's equations. These equations which will be shown later are coupled in the case of ohmic heating with the laws describing heat and mass transfer (e.g., heat conduction, radiation, and convection). Up to now, this coupled problem of differential equations can only be solved numerically, when some preconditions and approximations are assumed (Sastry and Palaniappan 1992, Zhang and Fryer 1993).

In any case, the temperature change will not be completely even (but mostly more even than in conventional processes). For pure liquid foods, this lack of homogeneity is generally not a problem since a strong mixing due to turbulences is very simple to achieve. For solid foods and particulate mixtures of solid foods with liquids, however, this may eventually cause quality changes. Thus, the evaluation of locally and temporally resolved temperature distributions is essential for an optimal and safe process. Whereas in pure fluids the temperatures may be monitored at exemplary places with temperature probes as resistance thermometers or thermocouples, this task is much more difficult but even more important for solid foods or even solid particles moving within a fluid. For this task there may be only one sophisticated measuring tool: temperature mapping using magnetic resonance imaging. For the principles of this technique the reader is referred to Knoerzer et al. (2005) but, briefly, is a noninvasive technique with high local, temporal, and an acceptable temperature resolution which can give more information about chemical concentrations, structures, diffusion coefficients, or flow velocities in model and real foods. Due to its high costs, it is only applied in research where it can give very useful information for modeling and model evaluation. Some impressive results on temperature distributions during ohmic heating of food models consisting of several phases have been reported (Ruan et al. 2001, Ye et al. 2004).

Concerning the use of ohmic heating for microorganism inactivation processes in pasteurization or sterilization, it has to be stated that the inactivation mechanism is mainly due to the thermal component. A very mild electroporation may occur additionally, when high voltages (electric field strength >10–20 kV/cm) (Raso and Heinz 2006) and low frequencies are applied, which allows electric charges to build up at cell membranes forming pores.

8.2.2 MICROWAVE HEATING

Microwaves are a kind of electromagnetic waves within a frequency band of 300 MHz–300 GHz. In the electromagnetic spectrum (Figure 8.1), the microwaves are embedded between the radio frequency range at lower frequencies and IR and visible light at higher frequencies. Thus, microwaves belong to the nonionizing radiations.

Electromagnetic waves can be basically described with Maxwell's equations, as follows:

$$\nabla \cdot \vec{D} = \rho \tag{8.11}$$

$$\nabla \cdot \vec{E} = -\frac{\partial \vec{B}}{\partial t} \tag{8.12}$$

$$\nabla \cdot \vec{B} = 0 \tag{8.13}$$

$$\nabla \cdot \vec{H} = \vec{j} + \frac{\partial \vec{D}}{\partial t} \tag{8.14}$$

Equations 8.11 and 8.13 describe the charge density ρ as a source of an electric field and that no magnetic monopole as source for the magnetic field exists. On the other hand, Equations 8.12 and 8.14 show the coupling between time-dependent electric and magnetic fields.

The interaction of electromagnetism with matter is included by the material equations also called constitutive relations, that is, Equations 8.15 through 8.17, where the permittivity or dielectric constant ε (interaction of nonconducting matter with an electric field), the electrical conductivity σ, and the permeability μ (interaction with a magnetic field) describe their behavior. The zero-indexed values belong to vacuum values, so that ε and μ are relative values.

$$\vec{D} = \varepsilon_0 \varepsilon \cdot \vec{E} \tag{8.15}$$

$$\vec{B} = \mu_0 \mu \cdot \vec{H} \tag{8.16}$$

$$\vec{j} = \sigma \cdot \vec{E} \tag{8.17}$$

Equation 8.17 is equivalent to Ohm's law (Equation 8.1), when the definitions of the current density j and of the electric field (within a homogeneously filled plate capacitor with plate distance l and plate area A) are taken into account:

$$\frac{I}{A} = \sigma \cdot \frac{U}{l} \Rightarrow I = \sigma \cdot \frac{A}{l} \cdot U = \frac{U}{(1/\sigma)(l/A)} = \frac{U}{\rho \cdot (l/A)} = \frac{U}{R} \tag{8.17a}$$

In general the material parameters μ, ε, and σ are complex tensors with direction-dependent behavior. For food substances, some simplifications are effective for most practical uses: since food behaves nonmagnetically, the relative permeability can be set to $\mu = 1$ and the permittivity tensor can be reduced to a complex constant with real (ε') and imaginary part (ε''), which may include the conductivity σ, when an isotropic material is assumed.

Maxwell's equations cover all aspects of electromagnetism. In order to describe the more specific theme of electromagnetic waves, the corresponding wave equations for the electric or the magnetic field can be easily derived, starting from Maxwell's equations (Regier and Schubert 2005):

$$\Delta \vec{E} - (\mu_0 \cdot \mu \cdot \varepsilon_0 \cdot \varepsilon) \left[\frac{\partial^2 \vec{E}}{\partial t^2} \right] = 0 \tag{8.18}$$

The corresponding wave equation for the magnetic component yields the same equation, when E is replaced by B. By comparing the wave equation with the standard one, it can be inferred that in this case the wave velocity is defined by

$$c = \frac{1}{\sqrt{\mu_0 \cdot \varepsilon_0 \cdot \mu \cdot \varepsilon}} = \frac{c_0}{\sqrt{\mu \cdot \varepsilon}} \tag{8.19}$$

The differential wave equation can be solved, for example, by the so-called linearly polarized plane wave. Linearly polarized means that, for example, the electric field consists of only one component, for example, in the z-direction E_z while other components do not exist. If this component only depends on one local coordinate, for example, x (and the time), the wave is called plane wave. Equation 8.18 then reduces to

$$\frac{\partial^2 E_z}{\partial x^2} - \left[\frac{1}{c^2} \cdot \frac{\partial^2 E_z}{\partial t^2} \right] = 0 \tag{8.20}$$

Often used as solutions also for more complex cases are time harmonic functions of the form

$$\vec{E} = \vec{E}_0 \cos(\vec{k} \cdot \vec{x} - \omega t); \ \vec{E} = \vec{E}_0 \sin(\vec{k} \cdot \vec{x} - \omega t); \ \vec{E} = \Re \left[\vec{E}_0 \cdot e^{\left[i(\vec{k} \cdot \vec{x} - \omega t) \right]} \right] \tag{8.21}$$

where the wave vector pointing to the direction of propagation with its absolute value is defined by

$$\vec{k}^2 = \frac{\omega^2}{c^2} \tag{8.22}$$

where $\omega = 2\pi f$ is the circular frequency of the wave.

It should be noted that the separate wave equations for the electric and the magnetic fields are not sufficient to completely replace Maxwell's equations. Instead, further conditions must describe the dependency between the magnetic and the electric fields.

The dispersion (the dependence of the velocity of light on the frequency in materials) is included in the behavior of k. To include the absorption within matter, a complex permittivity and with this a complex wave vector k have to be introduced.

When, additionally, a finite conductivity σ in Equation 8.17 is allowed so that a current occurs, instead of the simple wave, that is, Equation 8.18, the expanded Equation 8.23 is valid:

$$\Delta\vec{E} - (\mu_0 \cdot \mu \cdot \sigma)\left[\frac{\partial\vec{E}}{\partial t}\right] - (\mu_0 \cdot \mu \cdot \varepsilon_0 \cdot \varepsilon)\left[\frac{\partial^2\vec{E}}{\partial t^2}\right] = 0 \tag{8.23}$$

Again, time harmonic functions for the electric field are solutions of Equation 8.23, which have been reduced to

$$\Delta\vec{E} + (\omega^2 \cdot \mu_0 \cdot \mu \cdot \varepsilon_0)\left[\varepsilon - i \cdot \frac{\sigma}{\varepsilon_0\omega}\right]\vec{E} = 0 \tag{8.24}$$

This shows that an imaginary term of the permittivity ε may also express a finite conductivity σ.

An exemplary (plane wave) solution of the absorbing problem where the permittivity ε has an imaginary part $\varepsilon = (\varepsilon' - i\varepsilon'_{total})$ with

$$\varepsilon''_{total} = \varepsilon'' + \frac{\sigma}{\varepsilon_0\omega} \tag{8.25}$$

is also a solution of

$$\frac{\partial^2 E_z}{\partial x^2} + (\omega^2 \cdot \mu_0 \cdot \mu \cdot \varepsilon_0)(\varepsilon' - \varepsilon''_{total})E_z = 0 \tag{8.26}$$

and can be written as

$$E_z = g \cdot e^{[(ik+k)x]} + h \cdot e^{-[(ik+k)x]} \tag{8.27}$$

This describes an exponentially damped wave, with (real) wave number k and damping constant κ, both dependent on the permittivity ε. A comparison of coefficients yields

$$k = \omega \sqrt{\frac{\mu_0 \cdot \mu \cdot \varepsilon_0 \cdot \varepsilon}{2} \left[\sqrt{1 + \left(\frac{\varepsilon''^{*2}}{\varepsilon'^2} \right)} + 1 \right]} \tag{8.28}$$

and

$$k = \omega \sqrt{\frac{\mu_0 \cdot \mu \cdot \varepsilon_0 \cdot \varepsilon'}{2} \left[\sqrt{1 + \left(\frac{\varepsilon''^{*2}}{\varepsilon'^2} \right)} - 1 \right]} \tag{8.29}$$

The corresponding electric field penetration depth, that is, the distance in which the electric field is reduced to the fraction $1/e$, can thus be calculated by

$$\delta_E = \frac{1}{k} = \frac{1}{\omega} \sqrt{\frac{2}{(\mu_0 \cdot \mu \cdot \varepsilon_0 \cdot \varepsilon') \left(\sqrt{1 + \left(\frac{\varepsilon''^{*2}}{\varepsilon'^2} \right)} - 1 \right)}} \tag{8.30}$$

An important consequence of the frequency dependency of κ (and δ_E) is that microwaves of 915 MHz do have an approximately 2.5 times larger penetration depth than waves of 2450 MHz, if similar permittivities of both frequencies are assumed. This larger penetration depth helps to heat larger (industrial) pieces more homogeneously.

With the assumption of the excitation and the propagation of a plane wave first, estimations of the field configurations within the food are possible. For example, the laws of the geometric optics can be inferred, which are also valid for microwaves when the object size is much larger than the wavelength. Thus, the particular center heating of objects of dimensions in centimeter range with convex surfaces (like eggs) can be easily understood, since at the convex surface the microwave rays are refracted and focused to the center. For objects about the same size as the wavelength or smaller, direct field modeling by numerical solutions of Maxwell's equations becomes quite important.

In order to calculate temperature changes within an object by microwave heating, it is important to determine the power density starting from the electromagnetic field configuration. In most cases (of nonmagnetic material), the knowledge of the electric field is enough to calculate the heat production by power dissipation. This power dissipation (per unit volume) p_v is determined by ohmic losses, which are calculable by

$$p_v = \tfrac{1}{2} \sigma_{\text{total}} \cdot (\vec{E})^2 = \tfrac{1}{2} (\omega \cdot \varepsilon_0 \cdot \varepsilon_{\text{total}}) \cdot (\vec{E})^2 \tag{8.31}$$

where, according to Equation 8.25, the total conductivity may be expressed as

$$\sigma_{total} = \sigma + (\omega \cdot \varepsilon_0 \cdot \varepsilon'') \tag{8.32}$$

Equation 8.31 is equivalent to Equation 8.6 in ohmic heating apart from the factor ½, which comes from the integration of the time harmonic electric field over one wave period. It also shows the increase of the power dissipation with increased microwave frequency (when the permittivity is constant). The dependency on the squared electric field magnitude results in the power dissipation penetration depth δ_p being only half the value of the electric field penetration depth δ_E, that is,

$$\delta_p = \frac{1}{\omega} \sqrt{\frac{1}{2(\mu_0 \cdot \mu \cdot \varepsilon_0 \cdot \varepsilon') \left(\sqrt{1 + \left(\dfrac{\varepsilon''^{*2}}{\varepsilon'^2} \right)} - 1 \right)}} \tag{8.33}$$

Typical values for the penetration depth are listed in Table 8.2 (Tang 2005).

8.2.3 INFRARED HEATING

At higher frequencies or shorter wavelength than microwaves, the electromagnetic waves are called IR radiation. As in the case of microwaves, the food to be heated is treated by IR waves originating from a radiator (Ginzburg 1969).

In the case of IR radiators, the most convenient way to produce high spectral intensities is to use heated surfaces that emit heat radiation according to the spectral laws of Planck. The thermal radiation is not a discrete frequency or wavelength but has a continuous spectrum. The IR spectrum can be subdivided into three parts: the long waves (also called FIR) with wavelength in vacuum between 1 mm and 4 μm, the medium waves (or medium infrared, MIR) with wavelengths in vacuum between 3 and 1.4 μm, and the shortwaves (called NIR) with wavelengths in vacuum between 1.4 and 0.7 μm. The wavelength λ_{max}, where the maximum spectral heat flow is emitted at a certain absolute temperature T, follows Wien's law:

$$\lambda_{max} = 2900 \frac{\mu m}{(T/K)} \tag{8.34}$$

Radiation with smaller and larger wavelength is emitted but with less spectral intensity, and so, 25% of the radiated energy has a wavelength smaller than λ_{max}.

The total amount of heat emitted by a so-called blackbody (a perfect radiator) can be expressed by the Stefan–Boltzmann equation where A is the radiator surface and $\sigma_{SB} = 5.7 \times 10^{-8}$ W/(m²·K⁻⁴), the Stefan–Boltzmann constant:

$$\dot{Q} = A \cdot \sigma_{SB} \cdot T^4 \tag{8.35}$$

TABLE 8.2
Penetration Depth of Microwaves in Food Materials

Material	Temperature (°C)	Penetration Depth (δ_p/mm at 915 MHz)	Penetration Depth (δ_p/mm at 2450 MHz)
Water			
Distilled/deionized	20	122.4	16.8
0.5% Salt	23	22.2	10.9
Ice	−12	—	11,615
Corn oil	25	467	220
Fresh fruits and vegetables			
Apple ("Red Delicious")	22	42.6	12.3
Potato	25	21.3	9.0
Asparagus	21	21.5	10.3
Dehydrated apple ("Red Delicious") % moisture (wet basis)			
87.5	22	48.9	12.9
30.3		33.7	11.9
9.2		387	289
68.7	60	33.1	14.5
34.6		36.8	13.2
11.0		71.5	29.9
High protein products			
Yogurt (premixed)	22	21.2	9.0
Whey protein gel	22	22.2	9.6
Cooked ham	25	5.1	3.8
	50	3.7	2.8
Cooked beef	25	13.0	9.9
	50	9.5	8.9

Source: Adapted from Tang, J., *The Microwave Processing of Foods*, eds., Schubert, H. and Regier, M., Woodhead Publishing, Cambridge, U.K., 2005.

Some examples of the radiator temperature and the corresponding wavelength with maximum heat flow as well as the heat flow density for blackbody radiators are shown in Table 8.3.

Radiators are not perfect blackbodies, but they do emit only a fraction (the so-called emissivity ε) of the theoretical maximum heat flow possible. This emissivity of a substance ranges from 0 to 1 and is in general dependent on the wavelength and the temperature, so the radiator is called selective. The emitted electromagnetic spectrum is characteristic for the substance and may be used for

TABLE 8.3

Examples of Radiator Temperature-Dependent Wavelength with Maximum Spectral Intensity and Total Heat Flow Density

Radiator Temperature (°C)	λ_{max} (μm)	λ_{max} (W/m²)	Remark
−270	1000	4.3×10^{-6}	Upper wavelength limit of FIR
690	3	5.0×10^{4}	Upper wavelength limit of MIR
1800	1.4	6.0×10^{5}	Upper wavelength limit of NIR
3870	0.7	1.7×10^{7}	Upper wavelength limit of visible light

chemical analysis by the so-called IR spectroscopy. The emissivity of a radiator is equal to its absorptivity α that is the fraction of radiation (meeting a body) that is absorbed by it:

$$\varepsilon(\lambda, T) = \alpha(\lambda, T) \qquad (8.36)$$

In certain wavelength ranges, the emissivity of a substance is rather constant (or is assumed to be so). Then the radiator emits a constant fraction of the radiation of a blackbody and it is called to be a gray body. Foods may also be described as gray bodies, with absorptivity $\alpha_F = \varepsilon_F$. Thus, Equation 8.35 has to be modified to

$$\dot{Q} = A \cdot \varepsilon \cdot \sigma_{SB} \cdot T^4 \qquad (8.37)$$

Typical values for food substances range between 0.7 and 1, whereas the absorptivity may strongly decrease for (polished) metals. Examples of emissivities are given in Table 8.4.

Radiation that is not absorbed may be transmitted or be reflected by the food, so the sum of fractions that are absorbed ε, reflected r, and transmitted τ equals 1, that is,

$$\varepsilon + r + \tau = 1 \qquad (8.38)$$

Due to the small penetration depth of IR radiation in food in the order of millimeters (as shown later), the reflectivity r may often be expressed as

$$r = 1 - \alpha = 1 - \varepsilon \qquad (8.39)$$

Regular reflection takes place at the surface of a material and produces the gloss or the shine of polished surfaces. For organic materials, the reflectivity is rather small

TABLE 8.4

Examples of Emissivities in the IR Wavelength Range

Material	Remarks	Emissivity ε
Metal	Polished	<0.05
	Unpolished	$0.25 < \varepsilon < 0.7$
	Painted	0.9
Wood	Painted	0.9
Paper	White	0.9
Beef	Lean	0.74
	Fat	0.78
Water	Fluid	0.955
	Solid (ice)	0.97
Dough		0.85
Toast	Burnt	1.00

Source: Adapted from Fellows, P.J., *Food Processing Technology*, Woodhead Publishing, Cambridge, U.K., 1996.

(approximately 4%) (Skjöldebrand 2001). Body reflection means the entering of the light in the material and a diffuse reflection after scattering and absorption. This body reflection is responsible for the optical color and pattern of the material. The amount reflected is dependent on the wavelength. Skjöldebrand (2001) states food values of 50% for wavelength shorter than 1.25 μm and 10% above as rules of thumb. For each food and state of processing this has to be measured to choose the right IR radiator. In any case, the food also absorbs heat radiation from the radiator of temperature T_R, according to Equation 8.37 but with the corresponding food area A_F, absorptivity (= emissivity ε_F) and emits heat radiation with the same parameters but of the food temperature T_F. By measuring this radiation, it is even possible to determine the equivalent surface temperature of the food, for example, by IR thermometers or cameras. The net heat transfer from the emitter to the food is accordingly

$$\dot{Q}_{net} = \sigma \cdot \varepsilon_F \cdot A_F (T_R^4 - T_F^4) \qquad (8.40)$$

The waves may, due to the low-molecular density in gases, penetrate air practically without any absorption loss and hit the molecules in the food. At these IR frequencies even the smallest molecular dipoles, as for example the water molecule and the smallest ions, are too inert to follow the alternating electromagnetic field. However, the parts of the molecules (atoms) are able to absorb the energy by vibrational motions as bending or stretching of electronic bonding lengths within a molecule. Whereas the energy of the vibrations is characteristic for the molecule, the amount of motions is characteristic for the temperature. Due to the continuous spectrum of IR radiators, a lot of molecules can take part in this absorption process with different vibrations.

This absorption within the material is mostly described by a Lambert–Beer law stating an exponential decrease of radiative heat flow with the penetration in a material:

$$\dot{Q} = \dot{Q}_0 \cdot e^{(-\delta_p x)} \tag{8.41}$$

where
    $\dot{Q}_0$ is the radiative heat flow at the material's surface
    x is the distance from the surface
    δ_p is the so-called penetration depth

This penetration depth is, of course, not only dependent on the material but also on the wavelength so that a calculation of heat distribution by IR radiation with its continuous wavelength spectrum is complex. For food substances a rule of thumb is that penetration depths are in the range of 0.1–8 mm. The larger the IR wavelength the smaller the resulting penetration depth will be. Some exemplary values are given in Table 8.5.

The larger penetration depth of NIR radiation makes it possible to increase the heat flux without overheating or burning effects at the surface layers. Nevertheless, also in this wavelength range the penetration depth is rather small compared to microwaves, radio waves, or even ohmic heating. In case of small food layers to be heated, for example, in drying of pasta, spices, teas, cocoa, flours, grains of aromatic fluids, or when surface heating is of particular interest as for example in baking or surface pasteurization, IR may be an interesting alternative, often combined with other heating and drying techniques.

8.3 ALTERNATIVE HEATING EQUIPMENT

8.3.1 Ohmic Heating

In most cases an ohmic heating device is a part of larger equipment mainly for inactivation processes of solid/fluid food mixtures as soups, crèmes, etc. Thus, additionally to the ohmic heating device it may consist of feeding equipment, for example, a pump for fluids, holding equipment (holding the temperature at a constant level for a defined time), cooling equipment (for fast reduction of the temperature), and a tank for storage or direct aseptic filling equipment. Whereas the other techniques are well established, the absence of inert electrode materials as well as of accurate and robust control equipments kept ohmic heating of foods from being successful at the beginning. Thus, the electrodes and the control equipment are critical factors for its successful application. The electrodes must withstand the food material and should not show electrolytic effects that may cause dissolution of the electrode material into the food. This may be achieved using food-compatible materials (e.g., metal-coated titanium electrodes in some commercial equipment) and a suitable electrical current density with an electric filed strength E under 100 V/cm (Lebovka et al. 2005), sometimes combined with high AC frequencies up to hundreds of kilohertz (then also stainless steel electrodes may be used). The inside part of the tubes for food transportation has to be nonconducting, for example, high temperature resistant plastics.

TABLE 8.5

Penetration Depth of IR Radiation in Foods Data

Product	λ_{max} (µm)	Spectral Range (µm)	δ_p (mm)
Bread	1.12	$\lambda < 1.25$	6.25
		$1.25 < \lambda < 1.51$	1.52
Bread crumbs	1.30	$0.8 < \lambda < 1.25$	3.8
		$1.25 < \lambda < 2.50$	1.4
		$0.8 < \lambda < 2.50$	1.9
	1.32	$0.8 < \lambda < 1.25$	3.8
		$1.25 < \lambda < 2.50$	1.4
		$0.8 < \lambda < 2.50$	1.9
	1.41	$0.8 < \lambda < 1.25$	3.8
		$1.25 < \lambda < 2.50$	1.4
		$0.8 < \lambda < 2.50$	1.8
Bread crust	1.30	$0.8 < \lambda < 1.25$	2.5
		$1.25 < \lambda < 2.50$	0.6
		$0.8 < \lambda < 2.50$	1.2
	1.32	$0.8 < \lambda < 1.25$	2.5
		$1.25 < \lambda < 2.50$	0.6
		$0.8 < \lambda < 2.50$	1.1
	1.41	$0.8 < \lambda < 1.25$	2.5
		$1.25 < \lambda < 2.50$	0.6
		$0.8 < \lambda < 2.50$	1.1
Potato	1.12	$\lambda < 1.25$	4.76
		$1.25 < \lambda < 1.51$	0.48
		$\lambda > 1.51$	0.33
	1.24	$\lambda < 1.25$	4.17
		$1.25 < \lambda < 1.51$	0.47
		$\lambda > 1.51$	0.31
Pork	1.12	$\lambda < 1.25$	2.38
		$1.25 < \lambda < 1.51$	0.28

Source: Adapted from Skjöldebrand, C., *Thermal Technologies in Food Processing*, eds., Richardson, P., Woodhead Publishing, Cambridge, U.K., 2001.

The electrodes can be located along the flow direction and thus the electric field lines are parallel to the flow (in-line field) or located perpendicular to it (cross-field). Optimum control of the process can be achieved by using not only one pair of electrodes but also multiple sets of electrodes serially along the flow direction of the food. The advantage of this arrangement is to react more flexibly on changes in electrical conductivity along the flow path with increasing temperature. Nevertheless, feed-forward control processes are recommended for better control (Biss et al. 1987). Two heating systems are well known and in commercial application: one system by RAZTEK is used for liquid egg pasteurization. The second offered by APV, and mainly distributed in Japan, is used for liquid/particulate fruit products of particles up to 2 cm as for example strawberries for yoghurt (Ohlsson and Bengtsson 2002). Other applications are blanching of potatoes and vegetables as well as processing low-acid convenient, ready-to-eat meals. A consortium evaluated these processes and concluded the viability and efficiency of the process (Zoltai and Swearingen 1996).

8.3.2 Microwave Heating

As shown in Figure 8.1, the frequency range of microwaves adjoins to the range of radio frequencies that is strongly used for broadcasting. However, the microwave frequency range is used for telecommunication purposes like mobile phones, radar, or in modern times' bluetooth connections, too. In order to prevent possible interference problems, special frequency bands have been assigned for industrial, scientific, and medical (so-called ISM) applications where a certain radiation level has to be tolerated by other applications like bluetooth communication devices. In the range of microwaves, the ISM bands are located at 433, 915, and 2450 MHz whereby the first frequency is not commonly used, the second is not generally permitted all over the world. Outside of the permitted frequency range, leakage is highly restricted.

Whereas 915 MHz has some considerable advantages for industrial applications (mainly the larger penetration depth as stated in Section 8.2), for microwave ovens in households the only used frequency is 2450 MHz.

Apart from the regulations concerning interference, two types of safety regulations exist:

- Regulation concerning the maximum exposure or absorption of a human, working in a microwave environment
- Regulation concerning the maximum emission or leakage of the microwave equipment

The exposure limits for humans are based on the estimation of thermal effects caused by microwaves in the human body. Especially sensible organs like the eye, with a reduced thermal balancing possibility and geometric focusing effects, are taken into account. Thus, the limit for human exposure is generally considered to be safe is at a level of 1 mW/cm^2 body surface in most countries. As for ionizing radiation, for microwaves it becomes common to express the exposure or absorption by humans using the value of the specific absorption rate (SAR), which is defined as the quotient between the incident power and the body weight. For microwaves, the International Commission on Non-Ionizing Radiation Protection (ICNIRP 1998) recommends a value for the SAR to be set to 0.4 W/kg.

The maximum emission of a microwave equipment is limited to a value of 5 mW/cm^2 measured at a distance of 5 cm from the point where the leakage has the maximum level. Thus, the permissible leakage level is higher than the maximum exposure limit. The power density of nonfocused radiation, which is normally the case for leakage, decreases proportional to the inversed square of the distance from the source. Thus, a leakage which just manages to stay in the limits of 5 mW/cm² at a distance of 5 cm is already below the maximum exposure limit of 1 mW/cm² at a distance of 11.2 cm. Each microwave system consists normally of three basic parts: a microwave source, a waveguide, and an applicator. By far the mostly used microwave source for industrial and domestic applications is the magnetron tube (Metaxas 1983). A description of the magnetron will, therefore, follow. More detailed descriptions can be found elsewhere (Metaxas and Meredith 1983, Püschner 1966).

A magnetron consists of a vacuum tube with a central electron emitting cathode of highly negative potential. This cathode is surrounded by a structured anode that forms cavities which are coupled by the fringing fields and have the intended microwave resonant frequency. Due to the high electric DC field between cathode and anode, the emitted electrons are accelerated radially, but an orthogonal magnetic DC field deflects the electrons, yielding a spiral motion. The electric and the magnetic field strengths are chosen appropriately so that the resonant cavities take energy from the electrons. This phenomenon can be compared to the excitation of the resonance by whistling on a bottle. The stored electromagnetic energy can be coupled out by a circular loop antenna in one of the cavities into a waveguide or a coaxial line.

The power output of a magnetron can be controlled by the tube current or the magnetic field strength. Its maximum power is generally limited by the anode temperature, which has to be prevented from melting. Practical limits at 2.45 GHz are approximately 1.5 and 25 kW for air and water cooled anodes, respectively (Roussy and Pearce 1995). The 915 MHz magnetrons have larger cavities (lower resonant frequency means larger wavelength) and thus can achieve higher powers per unit. The efficiencies of modern 2.45 GHz magnetrons range at approximately 70% mostly limited by the magnetic flux of the used economic ferrite magnets (Yokoyama and Yamada 1996), whereas the total efficiency of microwave heating applications often is lower due to unmatched loads.

Transmission lines (e.g., coaxial lines) and waveguides are used for guiding the microwaves with minimum losses to the applicator. As losses of waveguides at higher frequencies, like microwaves, are generally lower than those of coaxial lines waveguides are usually used in microwave power applications. Waveguides are simply hollow structures of conducting materials (metals) normally of constant cross section, whereby rectangular and circular forms are of most practical use. The internal size defines a minimum frequency f_c (the so-called cut-off frequency) by the solution of the wave equations and appropriate boundary conditions. Below this frequency, waves do not propagate without being strongly damped. For rectangular waveguides with width a and height b the cut-off frequency is

$$f \geq \frac{\sqrt{(m/a)^2 + (n/b)^2}}{2\sqrt{\mu \cdot \mu_0 \cdot \varepsilon \cdot \varepsilon_0}} = \left\{ \frac{1}{2a\sqrt{\mu\mu_0\varepsilon\varepsilon_0}}, a \geq b; \quad \frac{1}{2b\sqrt{\mu\mu_0\varepsilon\varepsilon_0}}, b \geq a \right\} = f_c \quad (8.42)$$

Within the waveguide the wave may spread out in so-called modes, which define the electromagnetic field distribution within the waveguide. The most commonly waveguide is of rectangular cross section with a width a equal to double the height b and is used in transversal electric (TE10) mode.

The waveguide itself can be already used as applicator for microwave heating when the material to be heated is introduced by wall slots and the waveguide is terminated by a matched load. This configuration is then called traveling wave device since the location of the field maxima changes with time. A radiation through the slots only occurs if wall current lines are cut and the slots exceed a certain dimension, which can be avoided (Roussy and Pearce 1995). More common in the food industry and domestic field are standing wave devices described in the next section, where the microwaves irradiate by slot arrays or horn antennas.

For receiving a high power absorption and little back-reflection of microwaves from the applicator to the source, the impedance of the load containing applicator has to be matched with the corresponding impedance of the source and the waveguide. In order to achieve such a situation, tuners are introduced. Tuners are waveguide components used to match the load impedance to the impedance of the waveguide, so that tuners minimize the amount of reflected power, which results in the most efficient coupling of power to the load. Due to change in the load during processes, this matching has to be controlled continuously or optimized for a mean load. An often used tuner is the so-called three-stub-tuner with three stubs entering the waveguide section being adjusted. The rest of the reflected power has to be prevented from coming back to and overheating the microwave source. Therefore circulators—directional-dependent microwave traveling devices are used—let the incident wave pass and guide the reflected wave into an additional load. As side effect, by heating this load, the reflected power can also be determined.

Common applicators can be classified into three types by the type of field configurations: near-field applicators, single mode applicators, and multimode applicators.

In the case of near-field applicators, the microwaves originating from a horn antenna or slot arrays directly "hit" the product to be heated. The power should be set to level that can be practically completely absorbed by the product so that only minor parts of the power are transmitted and transformed into heat in dielectric loads (usually water) behind the product. Similarly to the traveling wave device, in near-field applicators standing waves do not exist. Consequently, a relatively homogeneous electrical field distribution (depending on the mode irradiated from the waveguide) within a plane orthogonal to the direction of propagation of the wave can be achieved. The near-field applicators as well as the traveling wave devices work best with materials with high losses. In order to heat substances with low dielectric losses efficiently by microwaves, applicators with resonant modes which enhance the electric field at certain positions are better suited. The material to be heated should be located at positions where the electric field is concentrated.

Single mode applicators consist generally of one feeding waveguide and a tuning aperture and a relatively small microwave resonator with dimensions in the range of the wavelength. The standing wave yields a defined electric field pattern, which can then be used to heat the product. It has to be noted that this type of applicators has to be well matched to the load since the insertion of dielectric material naturally shifts the resonant modes. Small dimensions of the applicator are needed to avoid modes that differ from the desired mode, since the number of modes per frequency range grows very fast with the cavity's dimensions.

For multimode applicators, by increasing the dimensions of the cavity a fast transition from the single mode to the multimode applicator occurs due to the strongly increasing mode density with applicator size. Additionally, it has to be taken into account that common microwave power generators like magnetrons do not emit a single frequency but rather a frequency band.

In industrial, as well as in domestic applications, the multimode applicators play by far the most important role since most of the conveyor-belt-tunnel applicators and the microwave ovens at home are of the multimode type due to their typical dimensions. Despite the high number of stimulated modes, often a nonhomogeneous field distribution which is constant in time will develop. This field distribution mainly depends on the cavity and the product geometry and the dielectric properties of the material to be processed.

In opposition to the case of the single mode application, usually this inhomogeneous field distribution which would result in an inhomogeneous heating pattern is not desired since it is hard to control. An undesired inhomogeneous heating pattern can be prevented by changing the field configuration by varying cavity geometries (e.g., stirrer mode) or by moving of the product (conveyor belt, turntable), which also influences the field distribution.

Industrial applications mostly need continuous processing, due to the desired high throughputs. Today's industrial ovens may have either one high-power magnetron or several low-power magnetron devices. Whereas common industrial high-power magnetrons have longer operating lifetimes, the low-power magnetrons have the advantage of very low prices due to the high production numbers for the domestic market.

As mentioned above, an important hurdle for all microwave ovens, especially continuously operated systems, is the avoidance of leakage radiation through the product in- and outlet. For fluids or granular products with small dimensions (centimeter range), the legislative limits can be guaranteed by the small in- and outlet sizes together with the absorption of the entering product sometimes coupled with additional dielectric loads just in front of the openings. Especially in the case of larger product pieces, in- and outlet gates which completely close the microwave application device have to be used. A scheme of a conveyor belt oven, with its alternative power sources and openings, is shown in Figure 8.5.

8.3.3 INFRARED HEATING

As in the case of the other alternative heating technologies, the IR heating equipment is just one part of the process equipment, so that several process steps are combined with IR heating to form a whole process. As previously stated, apart from reheating and frying applications in catering IR heating is mainly used in drying processes and baking or roasting ovens as additional heat source to convective or conductive heat transfer mechanisms. In all these cases, water is removed from the product by evaporation and has to be taken away by convective air or vacuum equipments. In continuous processes, a transportation of the product through the zones of IR heating, for example, by conveyor belts, has to be guaranteed.

Due to the small penetration depth of IR radiation, products of larger dimensions have to be prepared for the radiative heat transfer, for example, by suitable size reduction if homogeneous heating is wished. On the other hand, a product movement

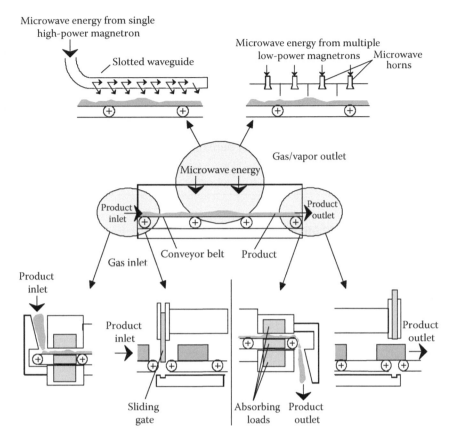

FIGURE 8.5 Typical industrial conveyor belt microwave oven. (Adapted from Regier, M. and Schubert, H., *Thermal Technologies in Food Processing*, ed., Richardson, P., Woodhead Publishing, Cambridge, U.K., 2001.)

as a fluidization or at least a turning of the product may be helpful to treat it from all sides with similar radiative heat, apart from the case of baking, where only surface crust formation is the goal. The IR equipment consists of a radiator that basically radiates in all directions and a reflector that is responsible for the heat radiation to be directed to the place where it is needed.

In general, industrial IR radiators for heating purposes may be divided into two groups: gas-heated radiators where a gas burns and produces the heat by this chemical reaction and electrically heated radiators where an ohmic heating of a metal within a gas atmosphere, a ceramic body or a quartz tube, produces the IR radiation. To produce long waves, gas-heated radiators as well as tubular/flat metallic heaters and ceramic heaters are used. The corresponding radiator temperature is rather low between 700°C and 800°C, so that in an air atmosphere approximately only 50% of the heat is transferred directly by radiation, while the rest is transferred by convection (Fellows 1996). For medium waves, practically exclusively quartz tube radiators at a temperature of approximately 950°C are used. This type of radiator may also be heated up to 2200°C, yielding shortwave IR radiation. Quartz tubes also heat lamps as halogen heaters and "IR guns" are used. Some more information on exemplary radiators is summarized in Table 8.6.

Reflectors consist of polished metallic surfaces with its low absorption but high reflection potential (see Table 8.4), which are individually formed to reflect the IR radiation to the surface to be heated. As geometric arrangements, individual metallic/gold reflectors (one for each radiator), individual gilt twin quartz tube reflectors (one for two quartz tubes), or flat metallic/ceramic cassette reflectors (one for a few radiators) are used commercially (Figure 8.6). The degree of heating is controlled by measurement of temperature and mostly pure on/off control mechanisms with short

TABLE 8.6
Characteristics of IR Radiators

Type of Waves	Radiator Type	Maximum Radiator Temperature (°C)	Maximum Process Temperature (°C)	Fraction of Radiative Heat Transfer	Heating/ Cooling Time (s)	Expected Lifetime
Short waves	Heat lamp (halogen lamp)	2200	300	75	1	5000 h
	IR gun	2300	1600	98	1	
	Quartz tube	2200	600	80	1	5000 h
Medium waves	Quartz tube	950	500	55	30	Years
Long waves	Element	800	500	50	<120	
	Ceramic	700	400	50	<120	Years

Source: Adapted from Fellows, P.J., *Food Processing Technology*, Woodhead Publishing, Cambridge, U.K., 1996.

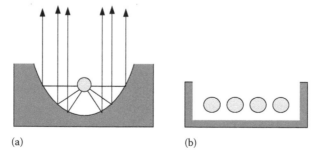

(a)   (b)

FIGURE 8.6 Arrangement of radiators in different type of reflectors: (a) individual reflector and (b) flat metallic/ceramic cassette reflector. (Adapted from Skjöldebrand, C., *Thermal Technologies in Food Processing*, ed., Richardson, P., Woodhead Publishing, Cambridge, U.K., 2001.)

cycle times of seconds to minutes at maximum of individual heaters. The heaters serve as heating source arranged similarly as in the case of microwave applications (Figure 8.5) whereby their number and position as well as the direction of radiation, which is important for nonflat products as breads, together with the speed of transportation through the heating zone, give a strong flexibility within the heating process. This flexibility can be enhanced if the on/off control is converted into a control of the radiator temperature and the heat flow from the radiator.

8.4 SPECIAL SAFETY AND QUALITY ASPECTS FOR FOODS PROCESSED BY ALTERNATIVE HEATING TECHNOLOGIES

8.4.1 OHMIC HEATING

8.4.1.1 General Considerations

Ohmic heating has the potential for rapid and uniform heating of a carrier medium containing particulates. For optimal quality of the processed product, an even heating pattern of the product flow is desired. This highly depends on the physical properties of the treated product, namely density, viscosity, conductivity, and heat capacity. Processed particulates should have a density similar to the carrier medium, otherwise sinking or floating of particulates may occur. This can lead to inconsistent processing. One way to soften this effect is to raise the viscosity of the carrier medium in order to lower sinking and floating velocity. It has to be kept in mind that processing can change rheological properties, when particulates release moisture or thickening agents undergo a phase change. Similar demands are made on the electric conductivities of carrier phase and particulates. Particulates with high conductivity compared to the medium tend to heat faster than the carrier medium, while those with lower conductivity leads to an underheating of particulates compared to the carrier medium (Zoltai and Swearingen 1996, Salengke and Sastry 2007). This can be easily illustrated when particulates and medium are regarded as parallel-connected ohmic resistances. A high conductivity in a volume element corresponds to a relatively higher current and thus more dissipated power. Nonconductive materials should be minimized as they can only be heated via heat conduction. When particulates and carrier medium have similar conductivities, specific heat capacity has an influence on which phase heats faster. Particulates tend to have lower heat capacity due to reduced moisture content (Zoltai and Swearingen 1996). From a controlling point of view, it has to be kept in mind that temperature sensors monitor the temperature of the continuous phase. Thus it is difficult to detect under- or overprocessing of particulates online (Zoltai and Swearingen 1996).

If the process is well adjusted and controlled, the risk of fouling and product burning can be significantly reduced compared to conventional processes. However, depending on product, temperature, electrode size, and voltage, a critical current density can be exceeded where arcing is most likely to occur. This happens particularly above the boiling point of the carrier medium when gas bubbles are formed (Reznick 1996). When a product formulation is adjusted for changes in market and ingredient cost or availability, the heating characteristic may change and, thus, additional process data might be necessary (Larkin and Spinak 1996). One important question concerning the safety of ohmic processes is the disintegration of electrode material. Only few results have been published in this field of research, yet. However, Amatore et al. (1998) pointed out that this topic deserves some attention. They provide the crucial steps that have to be mastered choosing an electrode material and state that an increase in frequency of the AC impressed reduces electrolytic reactions and, thus, contamination of the product. Wu et al. (1998) reported that corrosion of electrode material significantly diminished at a current frequency above 500 Hz and was no longer visible at frequencies exceeding 5 kHz. Reznick (1996) observed that stainless steel electrodes operated at a frequency above 100 kHz had been working for more than 3 years, with no marks of any metal dissolution. It was also pointed out that insoluble pure carbon electrodes are available, which allow for the safe use of electrical power at the common frequency of 50–60 Hz. Literature reports from possible applications of ohmic heating in the fields of food preservation, cooking, blanching, thawing, and as pretreatment of drying and extraction processes. However, most publications on application of ohmic heating are in the sector of food preservation, closely connected to the keywords sterilization, pasteurization, and aseptic processing.

8.4.1.2 Pasteurization and Sterilization—Aseptic Processing of Solid–Liquid Mixtures

Ohmic heating can be used to produce a variety of high quality shelf stable low- and high-acid products containing particulates. This requires for the combination of ohmic with aseptic processing. The size of the particulates in such foods is limited to 2.5 cm for three reasons. Firstly, it has to be ensured that enough space is maintained to allow flowing past the electrodes. Secondly, at that size (aseptic) filling is easily possible. Lastly, particulates larger than 2.5 cm affect convenience as they require cutting prior to consumption. The particle concentration typically ranges from 20% to 70% (Zoltai and Swearingen 1996). To optimize the quality of ohmically processed foodstuffs, some general comments on the used raw materials can be made. Generally, fresh raw materials will result in the highest quality products. This is crucial for vegetables and fruits. Eliot et al. (1999) proposed a precooking of cauliflower in saline solution at low temperatures (40°C–50°C) in order to obtain high quality products. Sarang et al. (2007) suggested a similar pretreatment at 100°C for solid components of a ready-to-eat soup. If meat is introduced in the formulation, it has to be taken into account that processing is significantly shortened compared to conventional sterilization processes in which meat is thoroughly cooked and tenderized. Thus, ohmically treated meat can lack tenderness and might require a pretreatment in order to meet consumer demands. Bones and fat should be minimized due to lacking electric conductivity.

Fresh pasta was found to produce higher quality products than dried pasta. Precooking pasta in conductive solutions can help to adjust its conductivity. The quality of sauces can be improved if they are homogenized prior to ohmic processing in order to stabilize fat and thermally fragile protein components within the food matrix. This measure can also be used to adjust viscosity. It should be noted that a possible addition of hydrocolloids not only affects viscosity of the carrier fluid but also significantly raises its conductivity. According to Marcotte et al.

(1998), this effect highly depends on the ash content of the hydrocolloid, being extremely high for carrageenan and xanthan and comparably lower for gelatin, pectin, and starch. Agitation of the carrier medium should be minimized to avoid air enclosures that can possibly lead to uneven heating (Zoltai and Swearingen 1996, Eliot et al. 1999).

Processed food samples were found to have texture, color, flavor, and nutrient retention that were comparable to or exceeded that of traditional processing methods (Zoltai and Swearingen 1996). Similar findings were made by Leizerson and Shimoni (2005) for pasteurization of orange juice. Higher retention of flavor compounds, such as limonene and myrcene, was observed when compared to conventionally processed juice. In a series of sensorial triangle tests, persons could not differ between fresh an ohmically processed juice, meanwhile, conventionally processed juice was filtered out.

One important issue concerning consumer safety, but also regulatory requirements, is to validate the efficiency of inactivation. It has to be stated that the significant effect that causes chemical changes or the inactivation of microorganisms in ohmic treatments is of thermal nature (Leizerson and Shimoni 2005). However, it has to be mentioned that there are reports on special electrical effects at moderate voltage gradients that will be dealt with later in this chapter.

Kim et al. (1996) used the markers M1 and M2 that will be further explained in Section 8.4.2 and alginate beads inoculated with *Bacillus stearothermophilus* spores to validate an ohmic sterilization process. In the range of pasteurization temperatures, the inactivation kinetics of enzymes as *B. licheniformis* amylase or *B. amyloliqufaciens* amylase can be used as time–temperature integrators in order to show at which process parameters a pasteurization process becomes marginal relating to food safety (Tucker 2004).

8.4.1.3 Cooking of Meat Products

One promising application of ohmic heating is the processing of meat. In an ohmic heating/plate heating combination process for hamburger patties, Ozkan et al. (2004) could not find a significant difference in moisture and oil content or mechanical properties compared to conventionally prepared samples. Although Shirsat et al. (2004) were able to analytically detect slight differences in color and springiness between ohmically cooked and steam cooked meat emulsion batters, these differences were not perceived in sensory evaluation. Hence, the differences found would probably not have an impact on consumer acceptance of ohmically processed meat. The only quality problem found by Piette et al. (2004), while studying the ohmic preparation of sausages, was also related to texture as products appeared too soft. It was, therefore, proposed to overcome this problem by the addition of binders.

8.4.1.4 Blanching

Sensoy and Sastry (2004) suggested applying ohmic treatment for blanching of mushrooms prior to canning. It was observed that, due to volumetric heating, there was no dicing of the mushrooms necessary and processing times could be significantly reduced from 5 to 7 min to less than 1 min. Icier et al. (2006) showed that due to more rapid heating, the color of pea puree could be better preserved compared to a conventional process. However, results were better on applying lower electric fields. Thus, some limitations of applied voltage gradients might be advisable for ohmic blanching.

8.4.1.5   Thawing

The main problem of ohmic thawing is because the electric conductivity of frozen food is about two orders of magnitude lower than that of thawed food. Some possible advantages of ohmic thawing using a prototype unit were reported (Roberts et al. 1998) as follows: (a) there was no water used or wastewater generated, (b) thawing time could be significantly reduced, (c) the process was easily controlled, (d) volumetric heating allowed for relatively uniform thawing, and (e) ohmic thawing improved the quality of thawed shrimp eliminating the leaching of water-soluble proteins and reducing microbial contamination. The principle of ohmic thawing has basically been proofed (Li and Sun 2002) but limited research on ohmic thawing has been reported in the literature.

8.4.1.6   Pretreatment for Drying and Extraction

Literature reports on the influence of ohmic pretreatments of fruits and vegetables on drying rates and extraction yields. Zhong and Lima (2003) showed that the drying rate of sweet potato tissue could be significantly improved when it had been ohmically pretreated with ACs at a frequency of 60 Hz and a field strength of 50–90 V/cm, far below the voltage gradients applied in pulsed electric field treatments. Similar observations were made by Lebovka et al. (2005, 2006) for air-drying of potato tissue. It was concluded that electroporation effects can occur at relatively low voltage gradients below 100 V/cm. In another publication of the same group (Praporscic et al. 2006), it was shown that this treatment allows a high level of mechanical softening of apple and potato tissues even at moderate temperatures below 50°C. Thus, juice yields could be significantly increased, particularly for the processing of apples. Wang and Sastry (2002) pointed out that significant thermal effects should not be observable at temperatures below 60°C and, thus, concluded from observed results on improved juice yields, that cell walls must be susceptible to some extent to an electrical breakdown at low temperatures and moderate fields. However, these effects have not yet been proofed and the quality of such pretreated foods needs to be examined.

8.4.2   Microwave Heating

8.4.2.1   Effects on Microorganisms and Chemical Compounds

In the past, literature reported on experimental results that cannot be explained solely due to a thermal effect of microwaves on chemical structures and microorganisms. Thus, assumptions have been made frequently on how a possible nonthermal effect of microwaves on cells, spores, or chemical compounds could be explained. Koutchma et al. (2001) compared a continuous conventional heating process with a continuously working microwave system and calculated D-values for *Escherichia coli* at multiple temperatures. They found that the resulting D-values of the microwave driven process at the temperature were far below those of the conventional system.

Sinell (1986) examined the influence of 2450 MHz microwaves on *E. coli*, *Staphylococcus aureus*, and *Salmonella typhimurium*. The derived *D*-values varied significantly although not as strong as in the results reported by Koutchma et al. (2001). It was concluded that heating in a microwave field exerts effects on microorganisms that significantly differ from a conventional treatment in a water bath. However, it is generally believed that inactivation due to dielectric heating in the range of microwave frequency is solely caused by thermal effects (Mudgett 1985).

On a theoretical basis, it has been reported that electromagnetic energy can only be absorbed in quantums (Rosén 1972). According to this author, for a nonthermal effect of microwaves the quantum energy of microwaves had to exceed the bond energy of chemical bonds, but in reality lay four to five orders of magnitude below (Table 8.7). Reuter (1979) came to the same conclusion, and additionally showed that not even hydrogen bonds can be broken by microwave quantums. Elevated inactivation levels observed sometimes in microwave applications could be explained considering that before a temperature equilibration can take place, temperatures in hot spots reach levels that falsify integrated measured values. An elevation of the temperature of a few degrees in the critical temperature range can potentiate a success of inactivation (Rosén 1972).

An important contribution to the influence of microwaves on microorganisms on an experimental level was made by Goldblith and Wang (1967). They studied the inactivation of *E. coli* cells and *B. subtilis* spores heated in a 2450 MHz microwave

TABLE 8.7
Quantum Energies Compared to Bond Energies

Irradiation	Quantum Energy (eV)
Gamma rays	1,240,000
X-rays	124,000
UV rays	4.1
Heat radiation (25°C)	0.026
Microwaves	0.00001

Bond Type	Bond Energy (eV)
Covalent bonds	
H–OH	5.20
H–CH$_3$	4.50
H–NHCH$_3$	4.00
H$_3$C–CH$_3$	3.80
Hydrogen Bonds	
O–H...O (H$_2$O)	0.20
O–H...O (CH$_3$COOH)	0.32
O–H...O (peptides)	0.07

Sources: Adapted from Rosén, C., *Food Technol.*, 26, 36, 1972; Reuter, H., *ZFL*, 30, 242, 1979.

field and heated conventionally. To do so, inactivation was first conducted in a microwave and the resulting temperature characteristic monitored. This temperature characteristic was then modeled conventionally with an open flame. From his results, nonthermal effect of microwaves on microorganisms could not be substantiated. In a second experiment, ice was added to the *E. coli* suspension to keep temperatures low during irradiation with microwaves. This setup was chosen to separate a possible nonthermal effect of the microwaves from an overlaying thermal effect that exists without any doubt. After 100 s of exposure to microwaves no inactivation could be detected. It was thus concluded that nonthermal effects on microorganisms at a frequency of 2450 MHz were not validated (Goldblith and Wang 1967). Goldblith et al. (1968) extended these experiments on the destruction of thiamine. They concluded that the entire destruction of vitamin B1 observed was solely due to heat. Irradiating samples kept at 0°C with microwaves did not result in any destruction of thiamine.

Another interesting approach was pursued by Vela and Wu (1979). As a first step they exposed soil with different water contents previously contaminated with bacteria, to microwave irradiation at a frequency of 2450 MHz and observed a positive correlation between moisture content and inactivation (in dry soil, no inactivation took place). To eliminate the influence of the soil on the inactivation, experiments were adjusted, and freeze-dried and rehydrated samples of several bacterial strains were used. While in rehydrated samples an inactivation could be observed, the freeze-dried samples survived long exposure to microwaves. Vela and Wu (1979) did not find any evidence of a nonthermal effect in their results. They concluded that a possible nonthermal effect, if it existed, must strongly depend on the presence of water.

A series of publications of a particular research group (Kozempel et al. 1997, 1998, 2000) have been looking into microwave pasteurization at low temperatures. They started developing a chilled microwave batch process in 1997 to examine the influence of microwaves on microorganisms in liquid media. Over the years the process was improved so that at the end a completely continuous process with an exactly adjustable residence time distribution was implemented. In order to avoid local heat peaks, the process was modified to enforce turbulent flow. While the first results on "cold pasteurization" in 1997 seemed quite promising and it was predicted that nonthermal effects existed without doubt, in 2000 it was finally stated that there was no convincing evidence that microwaves have the ability to kill bacteria without the influence of heat.

An FDA report from 2000 dealt in detail with the problem of possible nonthermal effects of microwaves that could be used to increase food safety (U.S. Food and Drug Administration 2000). As possible nonthermal effects could support the inactivation of microorganisms, the report mentions selective heating, electroporation, rupture of cell walls, and cell lysis. Despite frequent reports on the observation of nonthermal effects, the authors recommend to assume the presence of thermal effects solely. Also, no evidence was found to consider elevated heat resistance of pathogenic bacteria exposed to microwaves.

A complete literature overview on this still controversially discussed topic can be found in the literature (Fung and Cunningham 1980, Rosenberg and Bögl 1984, 1987b, U.S. Food and Drug Administration 2000).

It is to be stated that, as opposed to the cell phone and communication sector, in the food industry, possible nonthermal effects do not endanger the consumer. If nonthermal effects had an influence on food safety, it would increase it. No chemical changes can be induced by microwaves that do not occur in conventionally heated foodstuffs. However, the temperature distribution in microwave-treated foodstuffs develops completely different compared to conventional processes. So observed unexpected changes in quality (positive or negative) can be explained by uncommon heating patterns that can only occur applying volumetric heating. In principle, it can be stated where microwaves increase heating uniformity an enhancement in quality can be expected, meanwhile, in processes where extreme uneven heating patterns occur, quality levels fall below the ones of conventional processes.

8.4.2.2   Uneven Heating Problem

Depending on the compounds and the geometry of a food sample, uneven heating can occur in a microwave field. This is caused by a series of effects that interact with, amplify, and cancel each other. The temperature and frequency dependence of the dielectric properties of the heated body is to be mentioned. Also, dielectric properties vary in space caused by an inhomogeneous composition of the sample. The electric field decays exponentially entering a body that causes dielectric losses. The higher the frequency the more will the electric field strength be already damped in outer regions of the load (Sinell 1986). In addition to that, it must be mentioned that the electric field independently of the load does not develop homogeneously in the applicator. In household ovens, depending on the geometry of the applicator standing wave patterns will cause the load to be heated more intensely in some parts of the oven than in others (Ringle and Donaldson 1975). The geometry of the heated sample plays an important role. Edges and corners generally overheat faster than round or arced surfaces. Small bodies with a diameter below 2 mm are barely heated, while bodies with a diameter below 5 cm tend to overheat in the center due to focusing effects (Kessler 1996). On the surface of microwave-heated foodstuffs, the effect of evaporative cooling can be observed. The effect is the greater the larger is the difference between the temperature of the surface and the surrounding air (Giese 1992). Heat conduction within the load and heat transfer at the surfaces can have a big influence on the resulting heating pattern (Sinell 1986). Sometimes "runaway heating" effects can occur if the absorbed energy exceeds by far the energy that can be transported by heat conduction. This happens, for example, if small areas of food melt, meanwhile, most of the sample stays frozen.

8.4.2.3   Nutritional Value of Microwave-Processed Food

Jonker and Til (1995) conducted a study feeding microwave-cooked and conventionally cooked diets to rats over a period of 13 weeks in order to test whether microwave-cooked diets had any adverse effects on mammal health. Their findings did not indicate any difference between the tested diets. As stated before, the deleterious effect of microwaves on chemical compounds and thus on nutrients is based on thermal effects. Erbersdobler and Meissner (1994) analyzed the formation of hydroxymethylfurfural while holding milk at a constant temperature of 80°C and 90°C using microwave or conventional heating, without finding a significant difference. After comparing different raw, pressure-cooked, and microwave-cooked legumes, Khatoon and Prakash (2004) concluded that microwave cooking did not exert any deleterious effect on the nutrient content except for heat labile nutrients like thiamine or ascorbic acid. A literature overview on the nutritional value of microwave-processed food is given by Cross and Fung (1982) stating that no significant nutritional differences exist between foods prepared by conventional and microwave methods and any differences reported are minimal.

8.4.2.4 Use of Microwaves in the Industry

Albeit microwaves are commonly used in private households, in industrial processes microwave technology is still underrepresented compared to its potential benefits. They can be used for (re)heating, baking, (pre)cooking, tempering, blanching, sterilization, pasteurization, and drying (Regier and Schubert 2001). However, these processes will always compete with known and well-established standard processes (Mohr and Hanne 1981). At first, microwaves were only used in batch operation to avoid microwave leakage. In the meantime, the development of continuous microwave systems has lost a lot of its challenge. This was necessary as industrial applications often demand for continuous processes. In addition to that, these systems tend to be more efficient compared to batch systems. This is obvious as in continuous processes inhomogeneous field patterns carry less weight (Rosenberg and Bögl 1984).

However, there are problems connected to the use of microwaves that have to be taken into account. One problem is the choice of the packaging material. Because of high temperature peaks that can occur in microwave applicators, any packaging material must have a sufficiently high melting point (U.S. Food and Drug Administration 2000). According to Reuter (1980), such material should be transparent to microwaves. Metal components just as metal-coated foils are not suited for the use in microwaves as they reflect the waves (Reuter 1980). In contrast, George (1993) reported that metal shielding can have positive effects if used to avoid the overheating of edges and corners. The so-called susceptors allow for additional heat dissipation at selected parts of the product permitting, for example, the browning of surfaces. They usually have a laminate structure built of a structural layer, mostly paper, and a microwave interacting layer as aluminum-coated poly(ethylene-terephthalate) (PET) foil. A second paper layer can shield the food from direct contact to the active layer (Mountfort et al. 1996). Nevertheless, due to extreme temperatures that can develop in these processes, it cannot be excluded that chemical compounds migrate from the susceptor material to the food as shown by Begley et al. (1991) for the adhesive di-glycidyl ether of bisphenol A (DGEBA). A tenfold increase in migration of PET-oligomers from susceptor materials used in microwaves compared to food heated in a PET-tray in a conventional oven has been reported (Castle et al. 1989). Risch (1993) found that a second layer can help to prevent PET-oligomers migration serving as a functional barrier. So, if susceptors are used, it has to be carefully proofed that legal requirements concerning migration of chemicals are met and above all the safety of the consumer is ensured. The general feeling is that there is no safety concern with susceptors (Nemitz 1962).

8.4.2.5 Pasteurization and Sterilization

Focusing on pasteurization and sterilization processes, over the decades, a consequent shift toward elevated temperatures and shorter process times could be observed. This is caused by the fact that the inactivation kinetics of microorganisms are a lot more dependent on a shift in temperature than the degradation of quality attributes as vitamins and texture. Due to limited heat conductivity, this advantage cannot be used for the processing of solid and highly viscous foods. Here, the overall savings in process time cannot compensate for the quality degradation in surface regions (Ohlsson 1980, Mohr and Hanne 1981). There lies one chance for the use of microwaves as they can lead to an acceleration of a process due to the dissipation of electromagnetic energy into heat inside of the product. This leads to improvements in texture, taste, and nutritional value joint with an increased productivity (Giese 1992).

Harlfinger (1992) suggested a process for microwave sterilization consisting on four phases: heating, temperature equilibration, holding, and cooling. Because of inhomogeneous heating patterns in microwave fields, a special relevance lies on the second point. This is also valid for microwave pasteurization. The effect of evaporative cooling on the surface as source of an inhomogeneous temperature distribution cannot be overcome by simply prolonging the times for equilibration and holding. It was proposed to avoid this problem by treating the product in waterproof packaging or by using external heating sources to avoid the phenomena (Mudgett 1985).

The most critical point in pasteurization and sterilization processes is to assure that any point in the treated body has reached the working temperature of the process. As uneven heating is commonly observed in microwave processes and temperature distributions cannot be easily measured in a microwave field, alternative approaches are necessary for the validation of the sterile criteria. One interesting approach is the use of time–temperature integrators. In many foods under the influence of heat, chemical reactions between sugars and proteins take place. The products of these reactions can be used as temperature monitors for the experienced time–temperature effect. By analyzing the formation of such chemical markers (M1: product of glucose and proteins; M2: product of ribose and proteins) the homogeneity of the temperature distribution can be evaluated and thus, it can be concluded which parts of the sample match the sterile criteria (Prakash et al. 1997). However, to ensure the effectiveness temperatures must exceed 116°C. Hence, the markers M1 and M2 are suited only for sterilization processes (Kim and Taub 1993). The marker M1 can be used to evaluate sterilization processes of packed foods on temperature homogeneity (Wang et al. 2004b). For pasteurization processes, an enzyme-based time–temperature integrator has been developed. For this, *Aspergillus oryzae* α-amylase is immobilized in a polyacrylamide gel. Its dielectric properties can be easily matched to a variety of foods by simply adjusting its salt concentration (Raviyan et al. 2003). In addition to that, enzyme-based time–temperature integrators on the α-amylase of *B. amyloliquefaciens* have been used to track and evaluate pasteurization processes (Van Loey et al. 1997).

8.4.2.6 Drying

According to Zhang et al. (2006), the possible advantages of microwaves in drying processes are as follows. Firstly, the correlation between dielectric loss and moisture content, the so-called leveling effect, makes the electromagnetic energy rather dissipate in wet regions than in regions of the product that have already been dried. So the microwaves know where the water is. Secondly, focusing effects can lead to a preferred heating of the product center that cannot be as easily reached by heat conduction from the surface. Thirdly, volumetric heating allows for rapid energy dissipation throughout the product. Fourthly, a minor migration of water-soluble constituents can be observed compared to conventional processes. Fifthly, in combination with vacuum the process temperature can be lowered. At a pressure of 50 mbar, the evaporating temperature of water falls to 33°C. Lastly microwave-aided drying processes are more efficient in the classic falling rate period. Actually, in the beginning the falling rate period is delayed observing an extended constant rate period. So, as soon as the evaporation temperature has been reached, the drying rate is proportional to the applied microwave power and thus constant as long as the microwave power does not have to be adjusted. The latter is necessary when the dielectric loss of the dried product falls at levels where the field strength exceeds the values leading to a dielectric breakdown. This dependence between field strength and dielectric loss can be easily seen in Equation 8.43:

$$p_V = \tfrac{1}{2}(\omega \cdot \varepsilon_0 \cdot \varepsilon_r'')E^2 \tag{8.43}$$

where
  ε_0 is a constant
  ω is the circular working frequency of the used magnetron and thus constant

Assuming that the process starts at an uncritical power level where all the power is delivered directly to the load. The moisture content and thus ε_r'' will fall continuously in any part of the product giving rise to an increasing maximal field strength in the applicator. This field strength will, without adjustment of the power level, finally reach the critical breakdown value that strongly depends on the geometry of workload and applicator and its configuration toward each other. According to Metaxas and Meredith (1983), under atmospheric pressure this field strength can be as low as 1 kV/cm presuming unfavorable conditions as sharp metallic objects. Under vacuum conditions this high frequency breakdown electric field strength is considerably lower, for the frequency 2.45 GHz reaching its minimum at a pressure of approximately 2.5 mbar. The breakdown then commonly appears as localized glow discharge. If a discharge appears next to the product surface the surface scorches and thus, the quality of the product is reduced.

Khraisheh et al. (2004) dried potato cylinders conventionally at 30°C–60°C and in a microwave oven at power levels of 10.5–38 W. Microwave-dried samples showed improved rehydratability and retained at least twice the vitamin C content of convective-dried samples. Alibas (2007) used microwave drying, microwave-assisted air-drying, and conventional air-drying to dry 50 g portions of pumpkin slices and observed that the best color retention was reached using a process applying 350 W microwaves combined with a mild heat treatment of 50°C. Baysal et al. (2003) pointed out that quality was not significantly affected using microwave or conventional drying for the drying of garlic and carrot but drying time could be reduced considerably. Sumnu et al. (2005) noted that it was not only the drying rate but also the achievable final water content that changed significantly using microwave for the drying of carrots. Using microwaves, moisture contents of 0.12 kg/kg dm were reached easily. If IR radiation was applied in addition, this value dropped to

0.018 kg/kg dm. Regier et al. (2005) showed that the drying time of carrots could be reduced by a factor of 2–4 using microwave vacuum drying compared to a conventional drying process (50°C–70°C) while providing the same carotene stability. In another study on microwave vacuum drying of carrots, Lin et al. (1998) reported that microwave vacuum-dried carrots had a higher retention of carotenes and vitamin C showed a softer texture and lower density combined with a high rehydration potential. In sensory panel tests, microwave vacuum-dried carrots were evaluated equal to or better than freeze-dried samples. Other results on quality aspects on microwave-dried carrots can be found (Cui et al. 2005, Suvarnakuta et al. 2005, Wang and Xi 2005). Other product-related influences of microwave-assisted drying on quality aspects have been reported for potato (Wang et al. 2004a), tapioca starch pearls (Fu et al. 2005), asparagus (Nindo et al. 2003), pumpkin (Wang et al. 2007), mushrooms (Giri and Prasad 2007), and banana (Drouzas and Schubert 1996).

Generally speaking, the use of microwaves in (combination) drying processes allows for an enhancement in terms of efficiency, drying time meanwhile obtaining a comparable or enhanced quality. Possible process configurations according to Zhang et al. (2006) are

- Microwave-assisted air-drying
- Microwave vacuum drying
- Microwave-enhanced spouted bed drying
- Microwave finish drying following osmotic dehydration
- Microwave-assisted freeze drying
- Three drying stage combinations including microwave drying

Major challenges in future research include the determination of combination order, type, and conversion point of microwave-related drying, as well as its modeling and optimization.

8.4.2.7 Tempering and Thawing

Li and Sun (2002) summarize the possible benefits of microwave thawing naming shorter thawing time, smaller space for processing, reduced drip loss, less microbial problems, and chemical deterioration. However, the difference of the dielectric loss between frozen water and liquid water of three orders of magnitude can easily lead to so-called runaway heating as soon as a part of the product is completely thawed (Rosenberg and Bögl 1987a). To avoid these problems three measures can be applied. Firstly, instead of fully thawing the food often the food is tempered to just below the latent heat of fusion. Secondly, to provide uniform heating and allow for the tempering of larger products usually the frequency of 915 MHz is used due to its larger penetration depth. Lastly, surface cooling at temperatures just below the freezing point can minimize drip loss and quality deterioration (Decareau 1985). An intelligent way to fully thaw frozen food has been developed by Ferrite Inc, combining microwave heating with the mixing effect of a massager. Virtanen et al. (1997) developed a process to thaw frozen food using a feedback temperature control. They found that thawing time could be reduced as much as the factor of 7 compared to conventional thawing. Surface cooling below the freezing point improved the performance of the process.

8.4.2.8 Special Applications

Gerard and Roberts (2004) used microwaves to increase the yield of apple cider production. A heat treatment of apple mash at 70°C increased the juice yield up to 11% compared to untreated samples. In addition to that he could show that this treatment increased the extraction of polyphenolic compounds up to 61%. The potential use of microwaves for frying purposes has been mentioned by Oztop et al. (2007). It has been showed that microwave-heated cumin seeds had a better retention of characteristic flavor components of cumin (i.e., total aldehydes) compared to conventionally roasted seeds, meanwhile observing the same yields of volatile oils (Behera et al. 2004). However, Megahed (2001) concluded from his trials on microwave roasting of peanuts that even short heating times had significant effects on oil quality, lowering its stability and helping to accelerate oil rancidity.

8.4.3   Infrared Heating of Foodstuffs

8.4.3.1 General Considerations

IR radiation is omnipresent in the environment and directly coupled to the human reception of heat. Chemical changes in foodstuffs that could be troublesome are solely attributed to excessive heating and can thus be controlled as in conventional heating processes by limiting the power applied. Typical features of IR applications are as follows. Firstly, efficient heat transfer reduces processing time and energy costs. Secondly, the ambient air is not necessarily heated and can be kept at mild levels. Thirdly, there is a good controllability and safety due to easy-to-implement on–off controls that have an immediate impact on the heating process. Nevertheless, due to rapid heating rates there is a danger of overheating accompanied by possibly problematic chemical changes in the product. Thus, temperatures especially at the surface have to be carefully observed and controlled. Lastly, heating is usually more uniform compared to conventional processes as surface irregularities have a smaller effect on the rate of heat transfer (Sakai 1994).

Possible applications of IR radiation in the food processing industry have to be broken down to the three bands of IR radiation: FIR radiation with wavelengths exceeding 3 μm, mid range IR radiation ranging from 1.4 to 3.0 μm and NIR radiation with wavelengths below 1.4 μm. According to Sheridan, the selection of 3.0 μm as boundary is relevant to cooking effects—in his study of meat—as beyond these wavelengths no crust formation takes place. In the mid-IR band as produced by conventional gas powered plaque heaters, the surface temperature of 1100 K can char the product. The resulting crust is porous and can act as an insulator against the surrounding medium (Sheridan and Shilton 1999). In contrast it was reported that for the baking of bread, FIR radiation formed dryer crusts and had a higher rate of color development compared to NIR-baking (Sakai and Hanzawa 1994). Sheridan and Shilton (1999) cooked meat using a FIR (767 K) and alternatively a mid-IR (1098 K) source. Although the shorter wavelength mid-IR radiation penetrated the surface of the sample more efficiently, the temperature at the surface fast reached the boiling temperature of water and thus led to drying-out of the surface layer. Some energy was lost due to vaporization and product yield declined. So the overall balance between the two compared processes was stated to be neutral. They followed that for sole heating applications FIR is a valid option allowing for energy savings and small cookout losses. Usually IR-sources radiate over a whole frequency band and foodstuffs have a certain absorption spectrum varying with frequency. Jun showed how this could be used to selectively heat foodstuffs by introducing optical band-pass filters in his process. Heating soy protein and glucose for 5 min, he managed to attain a relative temperature difference of 12 K between the samples compared to IR heating without filter (Jun and Irudayaraj 2003). This finding can give rise to possible quality enhancement where selective heating is wanted as for example in the processing of several component ready-to-eat meals.

Sakai summarizes industrial applications of FIR in Japan. Possible operations that can be performed or enhanced using FIR radiation are cooking, baking, drying, thawing, and pasteurization (Sakai and Hanzawa 1994). However, the large majority of published work on IR applications in the food sector deals with the drying of foodstuffs.

8.4.3.2 Drying

The advantages of IR over conventional drying are a higher drying rate, energy savings, and a uniform temperature distribution. However, according to most publications dealing with IR drying at atmospheric pressure, energy and time considerations rather turn the balance toward an IR-process than enhanced quality measures (Hebbar et al. 2004, Wang and Sheng 2004, Wang and Sheng 2006). If there are quality relevant IR typical observations, these can be attributed to sole temperature effects. Nowak and Lewicki (2005) dried apple slices using NIR radiation at a power of 7.9 kW/m^2, and compared the results with samples dried conventionally at 65°C and 75°C in terms of quality parameters like color, water adsorption, and texture. They proposed that the observed differences could be explained by the two parameters: drying rate and final material temperature. While high drying rates lead to tissue damage and a fragile product, a high final temperature of the product leads to chemical changes, as for example browning reactions. It was, hence, concluded that the way how the heat was delivered was not decisive.

Baysal et al. (2003) compared the drying characteristic of carrots and garlic for microwave-, FIR-, and hot air dried samples. While the microwave and IR-dried carrots showed a better rehydration characteristic, hot air dried carrots and garlic showed better color retention. They concluded that the choice of a convenient drying process highly depends on the wanted product characteristic. Fu and Lien (1998) constructed a pilot far-IR drier for the dehydration of shrimp. They optimized the process by adjusting plate distance, plate temperature (122°C–358°C), and the temperature of the surrounding air (43°C–117°C). The best quality as indicated by low levels of formed thiobarbituric acid was achieved at a plate temperature of 286°C, an air temperature of 81°C, and a plate distance of 7.9 cm. However, taking into account drying time and energy efficiency in addition to quality the recommendation would be significantly different (357°C, 43°C, and 12.5 cm). Therefore, the decision for a certain process will always be a trade-off between cost and quality measures.

As IR radiation allows for the transfer of energy without contact, the load drying processes can be carried out at significantly reduced pressures and thus reduced temperatures. For banana, it has been demonstrated that this process can be optimized further when low pressure oversaturated steam is applied (Mongpraneet et al. 2004, Nimmol et al. 2007).

8.4.3.3 Special Applications

Some special applications of IR radiation in the food industry are referred to in the literature, and have a direct impact on food quality. The quality of frozen oysters can be enhanced when the product is irradiated preliminary to freezing. A thin surface layer with higher density then acts as moisture barrier reducing drip loss. The "boiling" of eggs can be achieved using FIR radiation. The product stands out due to a brighter color, a reduced risk of cracks as eggs are not in contact with each other and a reduced risk of microbial contamination as the product is not in contact with water (Sakai and Hanzawa 1994).

FIR heating has been used for the blanching of carrot slices. As a pretreatment for freezing, it was sufficient to inactivate enzymes at the surface where the slicing operation has disrupted the cells. Better texture preservation and less cell damage compared to a conventional hot water treatment were observed (Galindo et al. 2005). For the baking of bread, Olsson (2005) reported that NIR was a helpful tool to design baking processes, easing the control of crumb thickness, texture, and color. In a series of publications, Hashimoto and Sawai have proposed a pasteurization process based on FIR radiation that is supposed to be milder than a conventional heat treatment (Hashimoto et al. 1992, Sawai et al. 2000, 2003). However, they have not proofed yet that their findings do not affect enzymes and chemical reactions in a similar manner. Other authors reported on findings that there was no special characteristic of damaged cells exposed either to FIR or conductive heating. Thus, there is no evidence of effects of high intensity FIR light on microorganisms or biological substances that go beyond temperature effects. However, under specific conditions, especially for surface pasteurization, FIR radiation can be very effective (Sakai and Hanzawa 1994).

REFERENCES

Alibas, I. 2007. Microwave, air and combined microwave-air-drying parameters of pumpkin slices. *Lebensmittel-Wissenschaft und-Technologie—Food Science and Technology* 40: 1445–1451.

Amatore, C., Berthou, M., and Hebert, S. 1998. Fundamental principles of electrochemical ohmic heating of solutions. *Journal of Electroanalytical Chemistry* 457: 191–203.

Baysal, T. et al. 2003. Effects of microwave and infrared drying on the quality of carrot and garlic. *European Food Research and Technology* 218: 68–73.

Begley, T. H., Biles, J. E., and Hollifield, H. C. 1991. Migration of an epoxy adhesive compound into a food-simulating liquid and food from microwave susceptor packaging. *Journal of Agricultural and Food Chemistry* 39: 1944–1945.

Behera, S., Nagarajan, S., and Rao, L. J. M. 2004. Microwave heating and conventional roasting of cumin seeds (*Cuminum eyminum* L.) and effect on chemical composition of volatiles. *Food Chemistry* 87: 25–29.

Bhale, S. D. 2004. Effect of ohmic heating on color, rehydration and textural characteristics of fresh carrot cubes. M.Sc. Thesis. Baton Rouge, LA: Louisiana State University.

Biss, C. H., Coombes, S. A., and Skudder, P. J. 1987. The development and application of ohmic heating for continuous heating of particulate foodstuffs. In *Engineering Innovation in the Food Industry*, pp. 11–20. Rugby, U.K.: The Institution of Chemical Engineers.

Castle, L. et al. 1989. Migration of poly(ethylene-terephthalate) (PET) oligomers from PET plastics into foods during microwave and conventional cooking and into bottled beverages. *Journal of Food Protection* 52: 337–342.

Cross, G. A. and Fung, D. Y. C. 1982. The effect of microwaves on nutrient value of foods. *CRC Critical Reviews in Food Science and Nutrition* 16: 355–381.

Cui, Z. W. et al. 2005. Temperature changes during microwave-vacuum drying of sliced carrots. *Drying Technology* 23: 1057–1074.

Decareau, R. V. 1985. *Microwaves in the Food Processing Industry*. Orlando, FL: Academic Press.

Drouzas, A. E. and Schubert, H. 1996. Microwave application in vacuum drying of fruits. *Journal of Food Engineering* 28: 203–209.

Eliot, S. C., Goullieux, A., and Pain, J. P. 1999. Processing of cauliflower by ohmic heating: Influence of precooking on firmness. *Journal of the Science of Food and Agriculture* 79: 1406–1412.

Erbersdobler, H. F. and Meissner, K. 1994. Influence of microwave cooking on food quality. *Ernahrungs-Umschau* 41: 148.

Fellows, P. J. 1996. *Food Processing Technology*. Cambridge, U.K.: Woodhead Publishing.

Fu, W. R. and Lien, W. R. 1998. Optimization of far infrared heat dehydration of shrimp using RSM. *Journal of Food Science* 63: 80–83.

Fu, Y. C., Dai, L., and Yang, B. B. 2005. Microwave finish drying of (tapioca) starch pearls. *International Journal of Food Science and Technology* 40: 119–132.

Fung, D. Y. C. and Cunningham, F. E. 1980. Effect of microwaves on microorganisms in food. *Journal of Food Protection* 43: 641–650.

Galindo, F. G., Toledo, R. T., and Sjoholm, I. 2005. Tissue damage in heated carrot slices. Comparing mild hot water blanching and infrared heating. *Journal of Food Engineering* 67: 381–385.

George, R. M. 1993. Recent progress in product, package and process design for microwaveable foods. *Trends in Food Science and Technology* 4: 390–394.

Gerard, K. A. and Roberts, J. S. 2004. Microwave heating of apple mash to improve juice yield and quality. *Lebensmittel-Wissenschaft und-Technologie—Food Science and Technology* 37: 551–557.

Giese, J. 1992. Advances in microwave food processing. *Food Technology* 46: 118–123.

Ginzburg, A. S. 1969. *Application of Infrared Radiation in Food Processing*. London: Leonard Hill Books.

Giri, S. K. and Prasad, S. 2007. Drying kinetics and rehydration characteristics of microwave-vacuum and convective hot-air dried mushrooms. *Journal of Food Engineering* 78: 512–521.

Goldblith, S. and Wang, D. 1967. Effect of microwaves on *Escherichia coli* and *Bacillus subtilis*. *Applied Microbiology* 15: 1371–1375.

Goldblith, S., Tannenbaum, S. R., and Wang, D. 1968. Thermal and 2450 MHz microwave energy effect on the destruction of thiamine. *Food Technology* 22: 1266–1268.

Harlfinger, L. 1992. Microwave sterilization. *Food Technology* 46: 57–61.

Hashimoto, A. et al. 1992. Effect of far-infrared irradiation on pasteurization of bacteria suspended in liquid-medium below lethal temperature. *Journal of Chemical Engineering of Japan* 25: 275–281.

Hebbar, H. U., Vishwanathan, K. H., and Ramesh, M. N. 2004. Development of combined infrared and hot air dryer for vegetables. *Journal of Food Engineering* 65: 557–563.

Icier, F., Yildiz, H., and Baysal, T. 2006. Peroxidase inactivation and colour changes during ohmic blanching of pea puree. *Journal of Food Engineering* 74: 424–429.

ICNIRP. 1998. International commission on non-ionizing radiation protection. Guidelines for limiting exposure to time-varying electric, magnetic, and electromagnetic fields (up to 300 GHz), *Health Physics*, 74: 494–522.

Jonker, D. and Til, H. P. 1995. Human diets cooked by microwave or conventionally: Comparative subchronic (13-wk) toxicity study in rats. *Food and Chemical Toxicology* 33: 245–256.

Jun, S. and Irudayaraj, J. 2003. Selective far infrared heating system—Design and evaluation. I. *Drying Technology* 21: 51–67.

Kessler, H. G. 1996. *Lebensmittel-und Bioverfahrenstechnik: Molkereitechnology*. München: Kessler.

Khatoon, N. and Prakash, J. 2004. Nutritional quality of microwave-cooked and pressure-cooked legumes. *International Journal of Food Sciences and Nutrition* 55: 441–448.

Khraisheh, M. A. M., McMinn, W. A. M., and Magee, T. R. A. 2004. Quality and structural changes in starchy foods during microwave and convective drying. *Food Research International* 37: 497–503.

Kim, H. and Taub, I. 1993. Intrinsic chemical markers for aseptic processing of particulate foods. *Food Technology* 47: 91–99.

Kim, H. J. et al. 1996. Validation of ohmic heating for quality enhancement of food products. *Food Technology* 50: 253–256.

Knoerzer, K., Regier, M., and Schubert, H. 2005. Measuring temperature distributions during microwave processing. In *The microwave Processing of Foods*, eds. H. Schubert and M. Regier, pp. 22–40. Cambridge, U.K.: Woodhead Publishing.

Koutchma, T., Le Bail, A., and Ramaswamy, H. S. 2001. Comparative experimental evaluation of microbial destruction in continuous-flow microwave and conventional heating systems. *Canadian Biosystems Engineering* 43: 31–38.

Kozempel, M. F. et al. 1997. Preliminary investigation using a batch flow process to determine bacteria destruction by microwave energy at low temperature. *Lebensmittel-Wissenschaft und-Technologie—Food Science and Technology* 30: 691–696.

Kozempel, M. F. et al. 1998. Inactivation of microorganisms with microwaves at reduced temperatures. *Journal of Food Protection* 61: 582–585.

Kozempel, M. F. et al. 2000. Development of a process for detecting nonthermal effects of microwave energy on microorganisms at low temperature. *Journal of Food Processing and Preservation* 24: 287–301.

Larkin, J. W. and Spinak, S. H. 1996. Safety considerations for ohmically heated, aseptically processed, multiphase low-acid food products. *Food Technology* 50: 242–245.

Lebovka, N. I. et al. 2005. Does electroporation occur during the ohmic heating of food? *Journal of Food Science* 70: E308–E311.

Lebovka, N. I., Shynkaryk, M. V., and Vorobiev, E. 2006. Drying of potato tissue pretreated by ohmic heating. *Drying Technology* 24: 601–608.

Leizerson, S. and Shimoni, E. 2005. Effect of ultrahigh-temperature continuous ohmic heating treatment on fresh orange juice. *Journal of Agricultural and Food Chemistry* 53: 3519–3524.

Li, B. and Sun, D. W. 2002. Novel methods for rapid freezing and thawing of foods—A review. *Journal of Food Engineering* 54: 175–182.

Lin, T. M., Durance, T. D., and Scaman, C. H. 1998. Characterization of vacuum microwave, air and freeze dried carrot slices. *Food Research International* 31: 111–117.

Marcotte, M., Ramaswamy, H. S., and Piette, J. P. G. 1998. Ohmic heating behavior of hydrocolloid solutions. *Food Research International* 31: 493–502.

Megahed, M. G. 2001. Microwave roasting of peanuts: Effects on oil characteristics and composition. *Nahrung-Food* 45: 255–257.

Metaxas, A. C. and Meredith, R. J. 1983. *Industrial Microwave Heating*. London, U.K.: Peter Peregrinus.

Metaxas, A. C. and Meredith, R. J. 1983. *Industrial Microwave Heating*. In IEE Power Engineering Series. Peregrinus, London 1983, ISBN 0-906048-89-3.

Mohr, E. and Hanne, H. 1981. Möglichkeiten der anwendung von ultrahochfrequenz-verfahren auf dem lebensmittelsektor. *ZFL* 32: 217–219.

Mongpraneet, S., Abe, T., and Tsurusaki, T. 2004. Kinematic model for a far infrared vacuum dryer. *Drying Technology* 22: 1675–1693.

Mountfort, K. et al. 1996. A critical comparison of four test methods for determining overall and specific migration from microwave susceptor packaging. *Journal of Food Protection* 59: 534–540.

Mudgett, R. 1985. Dielectric properties of foods. In *Microwaves in the Food Processing Industry*, ed. R. V. Decareau, pp. 15–37. Orlando, FL: Academic Press.

Nemitz, G. 1962. Vergleichende untersuchungen über die wasserwiederaufnahme gefriergetrockneter und warmluftgetrockneter gemüse. *Die industrielle Obst-und Gemüseverwertung* 47: 409–412.

Nimmol, C. et al. 2007. Drying of banana slices using combined low-pressure superheated steam and far-infrared radiation. *Journal of Food Engineering* 81: 624–633.

Nindo, C. I. et al. 2003. Evaluation of drying technologies for retention of physical quality and antioxidants in asparagus (*Asparagus officinalis*, L.). *Lebensmittel-Wissenschaft und-Technologie—Food Science and Technology* 36: 507–516.

Nowak, D. and Lewicki, P. P. 2005. Quality of infrared dried apple slices. *Drying Technology* 23: 831–846.

Ohlsson, T. 1980. Optimal sterilization temperatures for flat containers. *Journal of Food Science* 45: 848–852.

Ohlsson, T. and Bengtsson, N. 2002. Minimal processing of foods with thermal methods. In *Minimal Processing Technologies in the Food Industry*, eds. T. Ohlsson and N. Bengtsson, pp. 4–33. Cambridge, U.K.: Woodhead Publishing.

Olsson, E. E. M., Tragardh, A. C., and Ahrne, L. M. 2005. Effect of near-infrared radiation and jet impingement heat transfer on crust formation of bread. *Journal of Food Science* 70: E484–E491.

Ozkan, N., Ho, I., and Farid, M. 2004. Combined ohmic and plate heating of hamburger patties: Quality of cooked patties. *Journal of Food Engineering* 63: 141–145.

Oztop, M. H., Sahin, S., and Sumnu, G. 2007. Optimization of microwave frying of potato slices by using Taguchi technique. *Journal of Food Engineering* 79: 83–91.

Piette, G. et al. 2004. Ohmic cooking of processed meats and its effects on product quality. *Journal of Food Science* 69: E71–E78.

Prakash, A., Kim, H. J., and Taub, I. A. 1997. Assessment of microwave sterilization of foods using intrinsic chemical markers. *Journal of Microwave Power and Electromagnetic Energy* 32: 50–57.

Praporscic, I. et al. 2006. Ohmically heated, enhanced expression of juice from apple and potato tissues. *Biosystems Engineering* 93: 199–204.

Püschner, H. A. 1996. *Heating with Microwaves*. Berlin: Philips Technical Library.

Raso, J. and Heinz, V. 2006. *Pulsed Electric Field Technology for the Food Industry, Fundamentals and Applications*. Berlin: Springer.

Raviyan, P. et al. 2003. Physicochemical properties of a time-temperature indicator based on immobilization of *Aspergillus oryzae* alpha-amylase in polyacrylamide gel as effected by degree of cross-linking agent and salt content. *Journal of Food Science* 68: 2302–2308.

Regier, M. and Schubert, H. 2001. Microwave processing. In *Thermal Technologies in Food Processing*, ed. P. Richardson, pp. 178–207. Cambridge, U.K.: Woodhead Publishing.

Regier, M. and Schubert, H. 2005. Introducing microwave processing of food: Principles and technologies. In *The Microwave Processing of Foods*, eds. H. Schubert and M. Regier, pp. 22–40. Cambridge, U.K.: Woodhead Publishing.

Regier, M. et al. 2005. Influences of drying and storage of lycopene rich carrots on the carotenoid content. *Drying Technology* 23: 989–998.

Reuter, H. 1979. Das dielektrische erwärmen von lebensmitteln teil 1. grundlagen. *ZFL* 30: 242–246.

Reuter, H. 1980. Das dielektrische erwärmen von lebensmitteln teil 2. neuere anwendungen in der industriellen verarbeitung. *ZFL* 31: 7–12.

Reznick, D. 1996. Ohmic heating of fluid foods. *Food Technology* 50: 250–251.

Ringle, E. C. and Donaldson D. B. 1975. Measuring electric field distribution in a microwave oven. *Food Technology* 29: 46–54.

Risch, S. 1993. Safety assessment of microwave susceptors and other high-temperature packaging materials. *Food Additives and Contaminants* 10: 655–661.

Roberts, J. S. et al. 1998. Design and testing of a prototype ohmic thawing unit. *Computers and Electronics in Agriculture* 19: 211–222.

Rosén, C. 1972. Effects of microwaves on food and related materials. *Food Technology* 26: 36–40, 55.

Rosenberg, U. and Bögl, W. 1984. Mikrowellen in der lebensmittelindustrie. *ISH-Heft* 44: 120–135.

Rosenberg, U. and Bögl, W. 1987a. Microwave thawing, drying, and baking in the food-industry. *Food Technology* 41: 85–91.

Rosenberg, U. and Bögl, W. 1987b. Microwave pasteurization, sterilization, blanching, and pest control in the food industry. *Food Technology* 41: 92–99.

Roussy, G. and Pearce, J. A. 1995. *Foundations and Industrial Applications of Microwaves and Radio Frequency Fields*. Chichester, U.K.: John Wiley & Sons.

Ruan, R., Ye, X., Chen, P., and Doona, C. J. 2001. Ohmic heating. In *Thermal Technologies in Food Processing*, ed. P. Richardson, pp. 241–265. Cambridge, U.K.: Woodhead Publishing.

Sakai, N. and Hanzawa, T. 1994. Applications and advances in far-infrared heating in Japan. *Trends in Food Science and Technology* 5: 357–362.

Salengke, S. and Sastry, S. K. 2007. Experimental investigation of ohmic heating of solid–liquid mixtures under worst-case heating scenarios. *Journal of Food Engineering* 83: 324–336.

Sarang, S. et al. 2007. Product formulation for ohmic heating: Blanching as a pretreatment method to improve uniformity in of solid–liquid food heating mixtures. *Journal of Food Science* 72: E227–E234.

Sastry, S. K. and Palaniappan, S. 1992. Mathematical modeling and experimental studies on ohmic heating of liquid-particle mixtures in a static heater. *Journal of Food Process Engineering* 15: 241–261.

Sawai, J. et al. 2000. Far-infrared irradiation-induced injuries to *Escherichia coli* at below the lethal temperature. *Journal of Industrial Microbiology and Biotechnology* 24: 19–24.

Sawai, J. et al. 2003. Inactivation characteristics shown by enzymes and bacteria treated with far-infrared radiative heating. *International Journal of Food Science and Technology* 38: 661–667.

Sensoy, I. and Sastry, S. K. 2004. Ohmic blanching of mushrooms. *Journal of Food Process Engineering* 27: 1–15.

Sheridan, P. and Shilton, N. 1999. Application of far infra-red radiation to cooking of meat products. *Journal of Food Engineering* 41: 203–208.

Shirsat, N. et al. 2004. Texture, colour and sensory evaluation of a conventionally and ohmically cooked meat emulsion batter. *Journal of the Science of Food and Agriculture* 84: 1861–1870.

Sinell, H. J. 1986. Der einfluß der mikrowellenbehandlung auf mikroorganismen im vergleich zur konventionellen hitzebehandlung. *DFG-Abschlußbericht Si/55* 24: 1–4.

Skjöldebrand, C. 2001. Infrared heating. In *Thermal Technologies in Food Processing*, ed. P. Richardson, pp. 208–228. Cambridge, U.K.: Woodhead Publishing.

Sumnu, G., Turabi, E., and Oztop, M. 2005. Drying of carrots in microwave and halogen lamp-microwave combination ovens. *Lebensmittel-Wissenschaft und-Technologie— Food Science and Technology* 38: 549–553.

Suvarnakuta, P., Devahastin, S., and Mujumdar, A. S. 2005. Drying kinetics and beta-carotene degradation in carrot undergoing different drying processes. *Journal of Food Science* 70: S520–S526.

Tang, J. 2005. Dielectric properties of foods. In *The Microwave Processing of Foods*, eds. H. Schubert and M. Regier, pp. 22–40. Cambridge, U.K.: Woodhead Publishing.

Tucker, G. S. 2004. Food waste management and value-added products-using the process to add value to heat-treated products. *Journal of Food Science* 69: R102–R104.

U.S. Food and Drug Administration. 2000. *Kinetics of Microbial Inactivation for Alternative Food Processing Technologies: Microwave and Radio Frequency Processing*. Washington, DC: Food and Drug Administration.

Van Loey, A. et al. 1997. The development and use of an alpha-amylase-based time–temperature integrator to evaluate in-pack pasteurization processes. *Lebensmittel-Wissenschaft und-Technologie—Food Science and Technology* 30: 94–100.

Vela, G. R. and Wu, J. F. 1979. Mechanism of lethal action of 2450 MHz radiation on micro-organisms. *Applied and Environmental Microbiology* 37: 550–553.

Virtanen, A. J., Goedeken, D. L., and Tong, C. H. 1997. Microwave assisted thawing of model frozen foods using feed-back temperature control and surface cooling. *Journal of Food Science* 62: 150 154.

Wang, W. C. and Sastry, K. 2002. Effects of moderate electrothermal treatments on juice yield from cellular tissue. *Innovative Food Science and Emerging Technologies* 3: 371–374.

Wang, J. and Sheng, K. C. 2004. Modeling of muti-layer far-infrared dryer. *Drying Technology* 22: 809–820.

Wang, J. and Sheng, K. C. 2006. Far-infrared and microwave drying of peach. *Lebensmittel-Wissenschaft und-Technologie—Food Science and Technology* 39: 247–255.

Wang, J. and Xi, Y. S. 2005. Drying characteristics and drying quality of carrot using a two-stage microwave process. *Journal of Food Engineering* 68: 505–511.

Wang, J., Xiong, Y. S., and Yu, Y. 2004a. Microwave drying characteristics of potato and the effect of different microwave powers on the dried quality of potato. *European Food Research and Technology* 219: 500–506.

Wang, J. et al. 2004b. Kinetics of chemical marker M-1 formation in whey protein gels for developing sterilization processes based on dielectric heating. *Journal of Food Engineering* 64: 111–118.

Wang, J., Wang, J. S., and Yu, Y. 2007. Microwave drying characteristics and dried quality of pumpkin. *International Journal of Food Science and Technology* 42: 148–156.

Wu, H. et al. 1998. Electrical properties of fish mince during multi-frequency ohmic heating. *Journal of Food Science* 63: 1028–1032.

Ye, X. F. et al. 2004. Simulation and verification of ohmic heating in static heater using MRI temperature mapping. *Lebensmittel-Wissenschaft und-Technologie—Food Science and Technology* 37: 49–58.

Yokoyama, R. and Yamada, A. 1996. Development status of magnetrons for microwave ovens. *Proceedings of 31st Microwave Power Symposium*, pp. 132–135. Boston, MA, Nov. 1996.

Zhang, L. and Fryer, P. J. 1993. Models for the electrical heating of solid–liquid food mixtures. *Chemical Engineering Science* 48: 633–643.

Zhang, M. et al. 2006. Trends in microwave-related drying of fruits and vegetables. *Trends in Food Science and Technology* 17: 524–534.

Zhong, T. X. and Lima, M. 2003. The effect of ohmic heating on vacuum drying rate of sweet potato tissue. *Bioresource Technology* 87: 215–220.

Zoltai, P. and Swearingen, P. 1996. Product development considerations for ohmic processing. *Food Technology* 50: 263–266.

9 Acrylamide Formation and Reduction in Fried Potatoes

Franco Pedreschi

CONTENTS

9.1 INTRODUCTION

In April 2002, Swedish researchers shocked the food safety world when they presented preliminary findings of acrylamide in some fried and baked foods, most notably potato chips and French fries, at levels of 30–2300 µm/kg. Reports of the presence of acrylamide in a range of fried and oven-cooked foods have caused worldwide concern because this compound has been classified as probably carcinogenic in humans with significant toxicological effects namely neurotoxic and mutagenic effects (Rosen and Hellenäs 2002, Tareke et al. 2002).

Before its discovery in foods, acrylamide was known as an industrial chemical and a component of cigarette smoke. Acrylamide is a known carcinogen substance in experimental animals which occurs in carbohydrate-rich foods as a result of cooking methods at high temperatures. As acrylamide has not been detected in unheated or boiled foods, it was considered to be formed during heating at high temperatures. They attributed this fact to the higher temperatures reached in Maillard nonenzymatic browning reactions required for desirable color, flavor, and aroma production, specially in the surface of fried potatoes (Coughlin 2003). For instance, Tareke et al. (2002) showed that acrylamide was formed by heating above 120°C certain starch-based foods, such as potato chips, French fries, bread, and processed cereals. French fries and potato crisps exhibit relatively high values of acrylamide at 424 and 1739 µg/kg, respectively. Recent epidemiological studies by the University of Maastricht supported by the Dutch Food Safety Agency indicate a positive association between dietary acrylamide and the risk of certain types of cancer (Hogervorst et al. 2007). These researchers observed increased risks of endometrial and ovarian cancer with increasing dietary acrylamide intake, particularly among never smokers. Risk of breast cancer was not associated with acrylamide intake. Choosing a balanced and varied diet, and avoiding overcooking of certain starchy foods, will contribute to reducing dietary intake of acrylamide.

Acrylamide concentration determination nowadays appears to be a necessity since very high concentrations of this potentially toxic molecule were detected in amylaceous fried foodstuffs and were found to cause cancer in rats (Rosen and Hellenäs 2002). The analytical methods for acrylamide determination rely on using (a) gas chromatography and mass spectrometry—GC–MS (Tareke et al. 2000) or (b) liquid chromatography and tandem mass spectrometry—LC–MS–MS (Rosen and Hellenäs 2002). Granby and Fagt (2004) validated an analytical method for analyzing acrylamide in coffee. Recently, an LC–MS analytical methodology for simultaneous analysis of acrylamide and their precursors, such as asparagine and glucose, was implemented with a detection limit for acrylamide of 20 µg/kg, for French fries analysis (Nielsen et al. 2006).

Acrylamide formed in potatoes during frying is highly related to the color of potato chips (Mottram et al. 2002, Rosen and Hellenäs 2002, Stadler et al. 2002, Pedreschi et al. 2005). Some international research groups have separately confirmed a major Maillard reaction pathway for acrylamide formation (Coughlin 2003). For instance, Mottram et al. (2002) showed how acrylamide could be formed from food components during heat treatment as a result of the Maillard reaction between amino acids and reducing sugars. On the other hand, Stadler et al. (2002) have shown also that acrylamide can be released by the thermal treatment of certain amino acids, such as asparagine, particularly in combination with reducing sugars, and of early Maillard reaction products (*N*-glycosides).

Wide variations of acrylamide concentration in foods are, at least partially, caused by different levels of precursors of acrylamide in various batches of raw materials (levels of asparagine and sugars fluctuate widely in raw potato tubers). For example, Rydberg et al. (2003) showed that both addition of glucose and asparagine to the levels naturally occurring in potatoes would increase the acrylamide levels in fried potatoes. Both potato variety and field site (fertilization and storage conditions) had a noticeable influence upon acrylamide formation since they affect acrylamide precursor concentrations in the tubers. In addition to potato tuber composition, other factors involved in acrylamide formation are the processing conditions (pretreatments, temperatures and times of frying, type of frying, and postfrying treatments).

The potential capability of different potato varieties to form acrylamide during heat treatment correlated well with the concentration in the tubers of reducing sugars (especially glucose and fructose) and asparagine. The potato cultivars show large differences in their potential to form acrylamide which was primarily linked to their sugar contents (Amrein et al. 2003). Since potato tubers are especially high in the aminoacid asparagine, it is now thought that Maillard reaction is most likely to be responsible for the majority of the acrylamide found in potato chips and French fries (Friedman 2003). Martin and Ames (2001) found that asparagine was the free amino acid present in the highest amount in potatoes (93.9 mg/100 g). Davies (1977) and Hippe (1988) reported that asparagine is present in potatoes in varying, relative high amounts of 0.5%–3% of dry matter. On the other hand, potential of acrylamide formation is also related to the sugar content, such as glucose and fructose (Biedermann et al. 2002). Sugars accumulate in potato tubers, when there is an imbalance between starch degradation, starch synthesis, and respiration of carbohydrates. Storage temperature and physiological age of the tubers are the most important factors that affect this process of sweetening. Potatoes aimed for processing are stored at relatively high temperature (e.g., 8°C). In practice, a limit of 1.5–2.0 mg/g of fresh weight of reducing sugars in potato tubers is used as an indicator for suitability for processing (Burton 1969). Beside diversity in storability between cultivars, large variation is often found between different potato lots/fields of the same cultivar within and between years and hence they should be managed accordingly (Olsson et al. 2004). The amount of sugars and aminoacids in potato tubers is influenced by different factors, such as potato cultivar, fertilizers, climate, storage conditions, etc.

Acrylamide is formed during frying mainly due to the reaction of asparagine and reducing sugars and this reaction pathway is clearly correlated to Maillard reaction (Mottram et al. 2002, Stadler et al. 2002, Pedreschi et al. 2005). For this reason, it has been proposed by several authors to perform quick and easy measurement by way of image analysis of chips browning, rather than using painstaking chromatographic methods (Pedreschi et al. 2007). However, standard procedures for acrylamide determination involve slow and expensive methods of chromatography and mass spectroscopy and thus cannot be used for routine analysis. It was therefore a logical step to develop alternative methods based on image analysis of chips browning to measure acrylamide concentration. Currently, a substantial body of research has been carried out worldwide to build greater understanding of acrylamide, how it is formed in foods, what the risks are for consumers, and how to reduce occurrence levels. This chapter shows not only the principal mechanism(s) of acrylamide formation but also several acrylamide mitigation procedures reported in the literature that could be very useful for fried potato manufacturers in order to reduce to reasonable levels the acrylamide generation in their processing lines.

9.2   ACRYLAMIDE FORMATION MECHANISMS

The main pathway of acrylamide formation in fried potato products is the reaction of free asparagine and reducing sugars. Therefore the contents of these precursors in fried products are important and have to be controlled. During the last decades, the contents of reducing sugars have been in potatoes used for the manufacturing of fried products to avoid excessive browning. Maillard reaction has been suggested as the major pathway for acrylamide formation in foodstuffs, and asparagine in particular is tightly linked to acrylamide generation (Motram et al. 2002, Stadler et al. 2005). The link of acrylamide to asparagine, which directly provides the backbone of the acrylamide molecule, has been established by labeling experiments (Zyzak et al. 2003, Stadler et al. 2004). Studies to date clearly show that asparagine is mainly responsible for acrylamide formation in heated foods after condensation with reducing sugars or a carbonyl source (Gokmen and Palakzaglu 2008). Moreover, the sugar asparaginase adduct, N-glycosylasparagine, generates high amounts of acrylamide, suggesting the early Maillard reaction as a major source of acrylamide (Stadler et al. 2005). Additionally, decarboxylated asparagine (3-aminopropionamide), when heated, can generate acrylamide in the absence of reducing sugars (Zyzak et al. 2003).

Good evidence supporting the early Maillard reaction as the main reaction pathway involving early decarboxylation of the Schiff base, rearrangement to the resulting Amadori product, and subsequent β-elimination to release acrylamide has been presented by Yaylayan et al. (2003). Maillard reaction is a cascade of consecutive and parallel reaction steps which need to be studied more from a kinetic point of view. Unfortunately, many of the kinetic aspects during the complex Maillard reaction are still unclear. Besides, more research is needed to understand the role of the water and the physical state of food matrix (amorphous vs. crystalline) in acrylamide formation (Blank 2005).

There are two major hypotheses published thus far pertaining to the formation of acrylamide from asparagine by Maillard-type reactions (Blank et al. 2005). Mottram et al. (2002) have suggested that α-dicarbonyls are necessary coreactants in the Strecker degradation reaction affording the Strecker aldehyde as precursor of acrylamide. Glycoconjugates, such as N-glycosides and related compounds formed in the early phase of the Maillard reaction, have been proposed as key intermediates leading to acrylamide (Stadler et al. 2002). The key mechanistic step is decarboxylation of the Schiff base leading to Maillard intermediates that can directly release acrylamide. However, as the key intermediates were not or only partially characterized, the chemical reactions leading to acrylamide remained largely hypothetical.

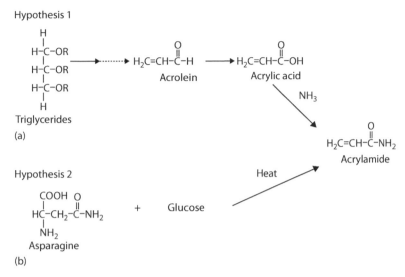

FIGURE 9.1 Possible routes of formation of acrylamide according to Becalski et al. 2003. (Reprinted from Becalski, A. et al., *J. Agric. Food Chem.*, 51, 802, 2003. With permission.)

At the same time, the Maillard reaction has played an important role in improving the appearance and taste of various fried foods (Zhang and Zhang 2007). These authors provide an excellent summary of studies in the formation of acrylamide and contributions from various laboratories. The cooking process per se—baking, frying, and microwaving—as well as the temperature itself seem to be of limited influence in acrylamide generation. It is the thermal input that is crucial, i.e., the temperature and heating time to which the product is subjected.

Finally, two possible routes of acrylamide formation have been reported as shown in Figure 9.1 (Becalski et al. 2003): (a) Acrylamide could be produced from oils and nitrogen-containing compounds present in foods. The most plausible scheme would include the formation of acrolein from the thermal degradation of glycerol (Umano et al. 1987), oxidation of acrolein to acrylic acid, and finally reaction of acrylic acid with ammonia—which potentially could be generated by pyrolysis of nitrogen-containing compounds—leading to the formation of acrylamide. (b) Acrylamide could be formed, by rearrangement, from nitrogen-containing compounds already present in foods.

Fried potato color is the result of Maillard, nonenzymatic browning reactions that depend on the superficial reducing sugar content, and the temperature and frying period (Marquez and Añon 1986). A linear correlation between parameter total color difference ΔE and acrylamide formation in French fries suggesting a relationship between the acrylamide formation and the degree of nonenzymatic browning developed during frying according to the pretreatments employed was detected (Pedreschi et al. 2007). In Figure 9.2, it can be seen that acrylamide concentration showed a good linear correlation ($r^2 = 0.854$) with the color of the fried potato strips represented by the total color difference ΔE pretreated in different ways (final moisture

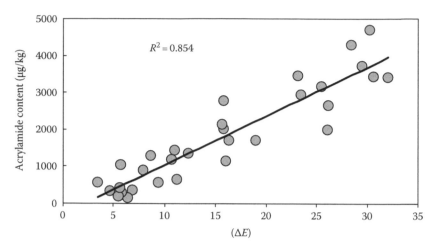

FIGURE 9.2 Acrylamide content vs. color parameters DL^* for potato strips fried at 150°C, 170°C, and 190°C for all the pretreatments tested. (Reprinted from Pedreschi, F. et al., *J. Food Eng.*, 79, 1287, 2007. With permission.)

content ~40/100 g total basis). As the samples get darker during frying, the L^* value diminished and a^* value increased (results not shown) which results in a general increase of the total color difference ΔE. As the frying temperature increases from 150°C to 190°C, the resultant chips get more red and darker as a result of nonenzymatic browning reactions that are highly dependent on oil temperature and frying time. Blanching reduces the ΔE values of French fries probably due to the leaching out of reducing sugars previous to frying inhibiting in this way nonenzymatic browning reactions and leading to lighter and less red French fries (Pedreschi et al. 2005). Lightness of French fries decreased as the acrylamide formation increased since the pieces get darker as a result of Maillard reactions. Besides, the redness chromatic component of French fries increased with the acrylamide formation since the pieces tend to get a more intense red color as the Maillard nonenzymatic reaction develops.

9.3 ACRYLAMIDE MITIGATION METHODS

Recent findings of acrylamide in foods have focused research on the possible mechanisms of formation. Some authors presented a mechanism for the formation of acrylamide from the reaction of the amino acid asparagine and a carbonyl-containing compound at typical cooking temperatures (Zyzak et al. 2003). Among the variables that can affect acrylamide formation in fried potato are: glucose and asparagine concentrations, cut potato surface area and shape, cooking temperature and time, and other processing conditions. Mestdagh et al. (2005) have shown determined that the origin of the deep-fat frying oils did not seem to affect the acrylamide formation in fried potatoes. So, frying oil type (canola, cotton, seed, olive, peanut, shortening, soybean, and sunflower) per se is not an important variable for acrylamide formation, with the exception of olive oil which increases significantly acrylamide generation.

Generally, there are two basic ways to reduce acrylamide in potato crisps (Matissek and Raters 2005). Firstly, through the raw material itself, potatoes have a considerable potential to form acrylamide and fried potato is the food category which has probably chalked up the highest concentrations of acrylamide reported so far (Friedman 2003). The level varies depending on the type of potato, several other agronomical factors (e.g., fertilization, climate, location, etc.), storage conditions, or mainly due to the asparagine levels. Another crucial way to reduce acrylamide level in the end of the product is by influencing the technology used in the production of potato chips. Some of the measures introduced by the industry are (Matissek and Raters 2005) (a) to optimize the temperature/profile during the deep-frying process; (b) to increase end-product moistness; and (c) to use optoelectronic sorting process to remove dark potato crisps.

Reducing acrylamide levels while maintaining product quality (flavor, color, texture, safety, etc.) in commercial fried potato production is a crucial challenge. Some strategies at different levels of production have been reported by Hanley et al. (2005) to mitigate acrylamide formation in fried potatoes: (a) prevention of acrylamide formation by acrylamide removal of the essential precursors (asparagine and a source of a carbonyl moiety—generally a reducing sugar); (b) interruption of the reaction by the addition of chemically reactive compounds that are able to react with intermediates in the Maillard reaction; (c) removal of acrylamide after it has been formed; and (d) minimization of acrylamide formation by changing frying conditions (frying, pressure, time, etc.).

Many additives have been found to have the inhibitory effect of acrylamide formation in the Maillard reaction (Zhang and Zhang 2007). Acrylamide formed in the Maillard reaction may also be reduced via the addition of exogenous chemical additives, which should comply with the following conditions: (a) the addition level should be properly controlled according to corresponding criteria of food additives; (b) the selected additives should not be toxic; and (c) the use of additives should not affect the sensorial properties of the fried potatoes. On the other hand, several approaches have been identified for reducing acrylamide formation in home-prepared food made from potato. They include proper storage of raw potato tubers (e.g., <10°C) and soaking potato slices in water or acid solutions for at least 15 min before frying (Jackson and Al-Taher 2005). More research is needed to identify methods for consumers to reduce acrylamide formation in home-prepared foods. Additionally, the consumer acceptability of foods cooked under conditions that prevent or reduce acrylamide levels needs to be determined.

Details of several relevant issues to reduce acrylamide formation, at different stages of commercial production, are presented in the following sections.

9.3.1 RAW MATERIAL

Because of the economical importance of potatoes, there is a need to determine the factors which influence the formation of acrylamide in fried products. These factors can be of intrinsic or extrinsic nature (De Wilde et al. 2004). Intrinsic factors vary with the composition of the raw material, while extrinsic factors are influenced by the agricultural treatment of the crop. For the intrinsic factors, special attention will be paid to the aminoacids in particular asparagine and the reducing sugars. Some relevant extrinsic factors are storage conditions, fertilization, type of soil, reconditioning, and different types of varieties.

At this level, tubers that in general have principally low-reducing sugar and asparagine content are preferred for frying operation. Fiselier and Grob (2005) found that a suitable limit for the reducing sugars in prefabricates for French fries is a simple and efficient measure to reduce the exposure to acrylamide from the predominant source for many consumers. Potato tubers contain considerable amounts of acrylamide precursors, i.e., free asparagine, glucose, and fructose, which may explain the high concentrations of acrylamide in certain potato products (Amrein et al. 2003). Currently, served French fries are one of the primary sources of acrylamide and it is very important to reduce acrylamide concentrations. Any modification performed on the potato constituents will inevitably impact on the Maillard reaction and its products and, concomitantly, the organoleptic properties including taste and color of the fried potatoes.

Acrylamide levels in potato products are also considerably affected by asparagine and reducing sugars content in raw potatoes. Asparagine represents typically 40% of the total aminoacid content and its level is about 10^3 mg/kg. Compared to this aminoacid, concentrations of glucose and fructose vary largely ranging between 10^1 and 10^3 mg/kg (Dunovska et al. 2004). Asparagine is considered as the main aminoacid responsible for acrylamide formation, whereas the presence of reducing sugars is necessary to affect the conversion of this aminoacid into acrylamide (Becalski et al. 2003). Bearing in mind that in potatoes free asparagine is present in considerable excess compared to the amount of reducing sugars, the reducing sugars will be the limiting factor in the acrylamide formation in fried potatoes, and thus they will particularly determine the formation of acrylamide (Amrein et al. 2004). For this reason, it is a major issue to control the precursors in potato tubers by using the following techniques (Zhang and Zhang 2007): (a) choice of plant cultivars low in reducing sugars; (b) avoidance of storage temperatures below 8°C; (c) optimization of blanching process to extract the maximum amount of sugar and asparagine; and (d) avoidance of adding reducing sugars in further treatments.

9.3.1.1 Selection of Proper Cultivars

Another important factor which affects acrylamide level is the potato cultivar. It is feasible to select potato varieties with low levels of reducing sugars. For example, a maximum of 1 g/kg reducing sugars has been suggested as a way to diminish significantly the formation of acrylamide. Selecting cultivars for food use which contain low levels of asparagine and/or devising conditions to hydrolyze asparagine to aspartic acid chemically or enzymatically with asparaginase or other amidases prior to food processing may result in low-acrylamide foods (Friedman 2003).

9.3.1.2 Development of New Cultivars

Both the storage conditions and nutritional properties of potato can be improved by using methods in genetic engineering (Rommens et al. 2006). The genetic modification was accomplished without inserting any foreign DNA into the plant genome. Interestingly, French fries derived from the genetically modified tubers displayed a strongly enhanced visual appearance and improved aroma accumulating much lower levels of acrylamide. Breeding programs through the world have focused on developing potatoes that accumulate low levels of sugar in potato in storage (Silva and Simon 2005). Changes in tuber reducing sugar and sucrose concentrations in storage are genetically determined, in combination with significant environmental effects. Several genes influencing sugar accumulation have been identified in both cultivated potatoes and their wild relatives and significant progress has been realized in new cultivars which can be stored at below 10°C. On the other hand, the variation in asparagine concentration in tubers allows for the possibility of selecting for low-asparagine genotypes. Screening of wild potato germoplasm will offer another wealth of genetic diversity to help attain goal.

9.3.1.3   Influence on Cultivar and Storage Conditions

Little is known about the influence of fertilization in potatoes over acrylamide formation after frying; some found a correlation between the nitrogen content of the soil and the aminoacid content; however, some experiments have shown that nitrogen fertilizers have no effect on the levels of asparagine. The level of nitrogen fertilization in potatoes appears to have an influence on the formation of acrylamide after frying, when no fertilizers are added to soil (De Wilde et al. 2004). Storage of potatoes below 8°C is known to cause increased levels of reducing sugars. Storage at 8°C or above will therefore reduce the potential for the formation of acrylamide in the potato product upon frying. Avoiding cold conditions can be readily achieved for short-term storage. There is a practical problem for longer agricultural storage. To maintain supplies of potatoes throughout the year it is necessary for producers to store potatoes for periods of several months. This can only be achieved successfully if the storage conditions prevent the potatoes from sprouting. Cool temperatures are often used to prevent sprouting. Alternatively, chemical treatments of potatoes without suppressing agents can be used, although the use of such chemical is not always desired by the consumer or may not be permitted. Optimal long-term storage regimes should be identified to limit the formation of sugars and consequently reduce the potential of acrylamide formation, while avoiding the problems associated with sprouting or spoilage.

9.3.2   Pretreatment Procedures

The principal objective of pretreatments of potato pieces is either to minimize the concentration of acrylamide precursors in potato tubers or to minimize or to inhibit Maillard reaction during frying.

9.3.2.1   Soaking

Levels of reducing sugars can be lowered by soaking the cut potatoes in water at room temperature before they are fried. Pedreschi et al. (2004) showed that glucose content in potato slices decreased slightly as the soaking time in water increased due to the water extraction of this component (Figure 9.3). On the other hand, asparagine content tended to remain constant even for 90 min of soaking time (Figure 9.4). When comparing the control (no soaking in water) with samples soaked in water

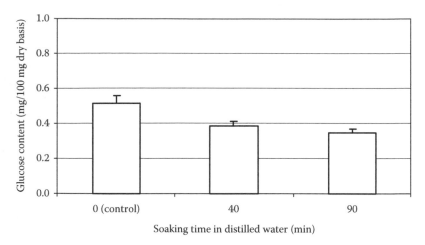

FIGURE 9.3 Glucose content of potato slices soaked 0 (control), 40, and 90 min in distilled water before frying. (Reprinted from Pedreschi, F. et al., *Lebensmittel-Wissenschaft und-Technologie—Food Sci. Technol.*, 37, 679, 2004. With permission.)

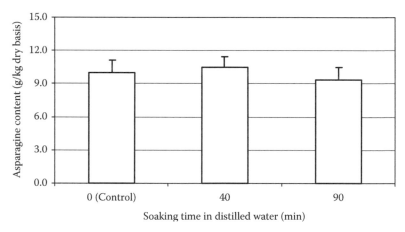

FIGURE 9.4 Asparagine content of potato slices soaked 0 (control), 40, and 90 min in distilled water before frying. (Reprinted from Pedreschi, F. et al., *Lebensmittel-Wissenschaft und-Technologie—Food Sci. Technol.*, 37, 679, 2004. With permission.)

for 40 and 90 min, the decrease in glucose content was 25% and 32%, respectively, while asparagine content remained almost constant in 9.95 ± 0.99 g/kg dry basis. Both the fried control and soaked samples showed a marked increase in acrylamide formation as the frying temperature increased from 150°C to 190°C (Figure 9.5). For the three temperatures tested, acrylamide formation was higher in the control than in soaked samples suggesting that the soaking process leads to a higher leaching of one important acrylamide precursor, such as glucose, that finally results in lower acrylamide formation.

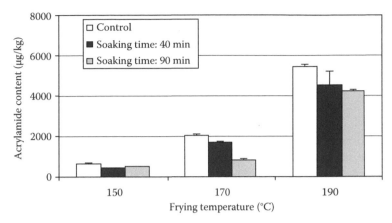

FIGURE 9.5 Acrylamide content of potato slices soaked 0 (control), 40, and 90 min in distilled water after being fried at 150°C, 170°C, and 190°C. (Reprinted from Pedreschi, F. et al., *Lebensmittel-Wissenschaft und-Technologie—Food Sci. Technol.*, 37, 679, 2004. With permission.)

9.3.2.2 Blanching

Levels of reducing sugars and asparagine and asparagine can be lowered by blanching the cut potatoes in hot or warm water before they are fried. Haase et al. (2003) reported that a reduction of the sugar content by blanching could reduce the acrylamide concentration by about 60% according to the raw material (potato variety and field site) and the production process variables (e.g., blanching conditions and frying temperatures). Blanching led to a significant reduction of acrylamide formation in potato strips after frying at the three frying temperatures tested. The longer the blanching time the lower acrylamide formation after frying, and the lower glucose and asparagine content in potato strips before frying. Not only glucose but also asparagine content (Figures 9.6 and 9.7, respectively) decreased drastically as the temperature and time of blanching increased leading to French fries with less acrylamide content after frying. Long-time blanching treatments, such as that of 50°C for 80 min and 70°C for 45 min, resulted in the lowest levels of acrylamide formation (342 and 538 μg/kg as average values for the three frying temperatures tested). These two blanching treatments, after frying at 190°C, led to the lowest acrylamide contents (564 and 883 μg/kg, respectively). Blanched samples resulted in prepared potato strips lighter in color than those of the control or the samples immersed in water at ambient temperature (visual observations). Blanching led to a significant reduction of acrylamide formation in potato strips after frying at the three frying temperatures tested (Figure 9.8).

9.3.2.3 Asparaginase

Acrylamide elimination could also be achieved via lactic acid fermentation (Baardseth et al. 2006) and use of asparaginase (Hendriksen et al. 2005, Cieserova et al. 2006, Pedreschi et al. 2007). Asparagine is the most important aminoacid in potatoes and it is not clear whether control of asparagine levels would be practicable. Assessment

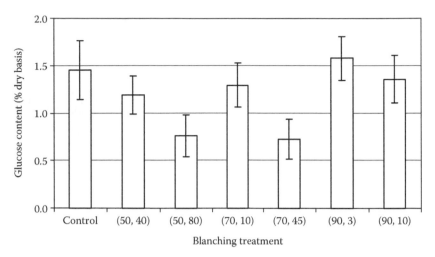

FIGURE 9.6 Glucose content of potato strips blanched in hot water at different temperature–time combinations before frying. Control corresponds to unblanched potato strips. First numbers inside parenthesis indicate the blanching temperature (°C); second numbers indicate the blanching time (min). (Reprinted from Pedreschi, F. et al., *J. Food Eng.*, 79, 1287, 2007. With permission.)

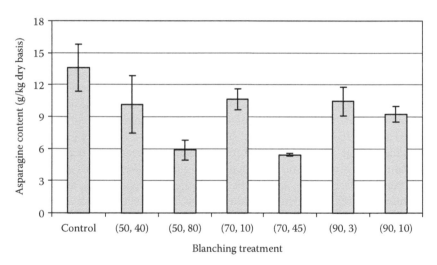

FIGURE 9.7 Asparagine content of potato strips blanched in hot water at different temperature–time combinations before frying. First numbers inside parenthesis indicate the blanching temperature (°C); second numbers indicate the blanching time (min). (Reprinted from Pedreschi, F. et al., *J. Food Eng.*, 79, 1287, 2007. With permission.)

of asparagine level change would need long-term assessment of potato varieties and feasibility studies. The use of enzyme asparaginase is a possible approach to interrupt the interaction of asparagine with reducing sugars, but further investigation is required. Ciesarová et al. (2006) evaluated the impact of L-asparaginase on the

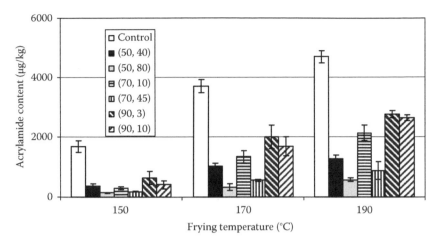

FIGURE 9.8 Acrylamide content of potato strips blanched at different temperature–time combinations after being fried at 150°C, 170°C, and 190°C. First numbers inside parenthesis indicate the blanching temperature (°C); second numbers indicate the blanching time (min). Control corresponds to unblanched potato strips. (Reprinted from Pedreschi, F. et al., *J. Food Eng.*, 79, 1287, 2007. With permission.)

acrylamide content reduction after high heat treatment in a model system as well as in potato-based material. They found that an important mitigation of acrylamide content (90%–97%) was achieved also in products prepared from dried potato powder treated by L-asparaginase.

Novozymes A/S have developed a production of asparaginase based on cloning of *Aspergillus oryzae*. The *A. oryzae* asparaginase has been cloned and expressed in commercial relevant yields in *A. oryzae*. The *A. oryzae* asparaginase has pH optimum at pH 6–7 with good activity between pH 5 and 8, which may be the pH range for production of potato products like French fries and crisps (tested by Novozymes A/S). The optimum temperature of the asparaginase activity measured at pH 7 is 60°C (Novozymes A/S).

Novozymes A/S have introduced an enzyme solution of asparaginase, which can reduce acrylamide levels by up to 90% by converting asparagine into another common amino acid, aspartic acid, without altering the appearance or taste of the final product. Acrylaway®, the commercial name of the enzyme solution, has also received generally recognition as safe (GRAS) status from the United States, approval from the Danish authorities, and a positive evaluation from the Joint FAO/WHO Expert Committee on Food Additives (Vang Hendriksen et al. 2006).

On another front, Baarsdseth et al. (2006) concluded that acrylamide formation during production of French fries can be lowered effectively by lactic acid fermentation of potato pieces before frying. The reduction is due to reduced levels of reducing sugars rather than reduction of available asparagine.

9.3.2.4 Salt Solutions

Water activity at the surface of the potato piece influences the mechanism of acrylamide formation during frying. One possibility to decrease water activity was to increase the local salt concentration. This seems to be a very useful way to lower acrylamide content because the fried potatoes are often salted before consumption. Pedreschi et al. (2008) study the effect of NaCl soaking after blanching over the acrylamide formation in potato chips after frying. These authors determined soaking of blanched potato slices in the 3 g/100 g NaCl solution for 5 min at 25°C, reduces by 11% acrylamide formation in potato chips after frying. However, when the slices are blanched directly in the 3 g/100 g of the NaCl solution at 60°C for 30 min, their acrylamide formation increased surprisingly by ~90%. Soaking of blanched potato slices in the NaCl solution reduced potato chip acrylamide formation by ~36% after frying at 190°C. However, when the blanching of the slices was made directly in the NaCl solution at 60°C for 30 min, the acrylamide content of the potato chips fried at 190°C increased by 80%. Earlier studies have shown that soaking in a NaCl solution of blanched potato slices before frying diminished drastically acrylamide formation in potato chips (Pedreschi et al. 2007).

9.3.2.5 Antioxidants

Many Maillard reaction products have antioxidant effects and are well known to be health protective concerning diseases associated with oxidative damage and stress, such as cancer, diabetes, inflammation, arthritis, immune deficiency, aging, etc. The relationship between the antioxidants effects and acrylamide reduction has not elucidated yet. Zhang et al. (2007) demonstrated that the addition of antioxidant of bamboo leaves could effectively reduce acrylamide in various heat treated foods. However, it is difficult to obtain affirmative conclusions on either positive or negative relationship between the addition of antioxidants and the reduction of acrylamide. The positive or negative mechanisms of action of added antioxidants over acrylamide formation have not been found so far (Zhang and Zhang 2007).

9.3.2.6 Aminoacids and Proteins

Some authors diminished acrylamide formation in fried snacks products by adding amino acids, such as lysine, glycine, and cysteine (Tae Kim et al. 2005). The addition of glycine or glutamine during blanching of potato chips reduced the amount of acrylamide by almost 30% compared to no addition (Claeys et al. 2005). Acrylamide formation can be reduced significantly as well by introducing other aminoacids, such as cysteine, lysine, or glycine, which would compete with asparagine for the carbonyl compounds in the Maillard reaction and/or enhance acrylamide elimination (Claeys et al. 2005). On the other hand, soy protein hydrolysate can also be used to reduce acrylamide due to the fact that soy protein hydrolysate is believed to reduce acrylamide by introducing additional aminoacids to compete with asparagine for key reaction intermediates (Wedzicha et al. 2005).

9.3.2.7 Organic Acids

Lowering the pH using organic acids of the food system to reduce acrylamide generation may attribute to protonating the α-amino group of asparagine, which subsequently cannot engage in nucleophilic addition reactions with carbonyl sources

(Jung et al. 2003). Lowering the pH of the cut potatoes (e.g., with citric acid 0.5%–1.0% <20 min) has been shown to lower the levels of acrylamide formed. However, this approach can cause souring of flavor if a precise procedure is not followed and also the frying oil can become rancid. The correlation between pH decrease and acrylamide reduction varies among products due to multiple factors of different starting pH values of the products (Zhang and Zhang 2007).

Jung et al. (2003) showed that lowering the pH with citric acid before frying was an efficient way to considerably diminish acrylamide formation in French fries. These authors also explain the mechanism by which lowering the pH of the potatoes reduces acrylamide formation after frying. Pedreschi et al. (2007) confirmed the previously reported citric acid immersion effect in acrylamide reduction in French fries. These authors detected that there were no significant differences ($P > 0.05$) between the glucose (Figure 9.9) and asparagine (Figure 9.10) content of the control sample and that immersed in citric acid solutions. Strip immersion in citric acid solution of 10 g/L reduced significantly acrylamide formation in 86%, 47%, and 28% with respect to the control, for the frying temperatures of 150°C, 170°C, and 190°C, respectively (Figure 9.11). The percentage of acrylamide reduction (from 86% to 28%), by the immersion in the citric acid solution previous to frying, diminished as the oil temperature increased. This reduction of acrylamide formation in French fries by dipping the potato strips in citric acid solutions was attributed to both pH lowering and leaching out of free asparagine and the reducing sugars from the surface layer of potato cuts to the solutions (Jung et al. 2003). Since in this case there is almost no removal of acrylamide precursors during soaking in citric acid solution (glucose and asparagine), the reduction acrylamide could be attributed only to the pH decrease.

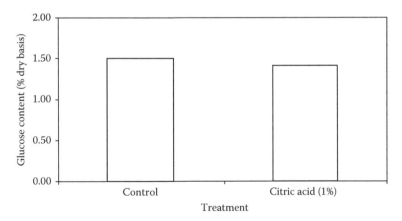

FIGURE 9.9 Glucose content of control and potato strips dipped in a citric acid solution of 10 g/L for 60 min. Control corresponds to potato strips dipped in distilled water for 60 min. (Reprinted from Pedreschi, F. et al., *J. Food Eng.*, 79, 1287, 2007. With permission.)

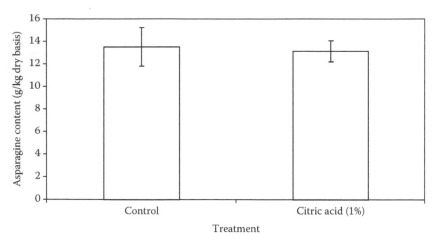

FIGURE 9.10 Asparagine content of control and potato strips dipped a citric acid solution of 10 g/L for 60 min. Control corresponds to potato strips dipped in distilled water for 60 min. (Reprinted from Pedreschi, F. et al., *J. Food Eng.*, 79, 1287, 2007. With permission.)

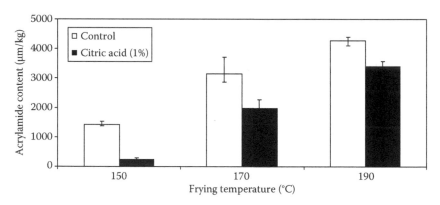

FIGURE 9.11 Acrylamide content of control and potato strips dipped in a citric acid solution of 10 g/L for 60 min after being fried at 150°C, 170°C, and 190°C. Control corresponds to potato strips dipped in distilled water for 60 min. (Reprinted from Pedreschi, F. et al., *J. Food Eng.*, 79, 1287, 2007. With permission.)

9.3.3 PROCESS OPTIMIZATION

The principal objective at this level is to control critical parameters related with acrylamide formation and Maillard reaction. The control of important processing parameters could be regarded as the most efficient way to reduce acrylamide. These parameters include heating temperature, heating time, temperature profile, oil type, pressure, etc. (Zhang and Zhang 2007). The type of frying oil used appears to have little influence on the formation of acrylamide.

9.3.3.1 Temperature Profile

For conventional fried potato products, frying temperatures above 175°C have been shown to increase levels of acrylamide significantly. The initial frying temperature should not exceed 175°C and possibly should not exceed 170°C. The addition of potato to the hot oil and the subsequent evaporation of water cool the oil. Further reduction in the formation of acrylamide can be achieved by further lowering the temperature to 150°C toward the end of cooking. However, use of lower temperatures for extended cooking times can affect quality, such as increased moisture, which results in loss of consistency/crispness. Also, low-frying temperatures can result in increased fat uptake which can have health implications. To enable better control over frying temperatures, it is necessary to improve the reliability and accuracy of temperature controls on frying equipment. Acrylamide reduction may be achieved by low temperature heating, such as low temperature vacuum frying and short frying times (Granda et al. 2004).

9.3.3.2 Vacuum Frying

Among several deep-fat frying technologies, vacuum frying has a significant strategic importance for future fried manufacturing and in reducing acrylamide formation (Garayo and Moreira 2002, Granda et al. 2004). Some authors reported that by lowering frying temperature at atmospheric pressure of potato chips from 185°C to 165°C, it was possible to reduce the acrylamide formation to a half (Haase et al. 2003, Pedreschi et al. 2004, 2006). Granda et al. (2004) applied vacuum frying for producing potato chips and they could reduce acrylamide formation by 94%.

9.3.3.3 Frying Time

There is not a simple linear increase between the acrylamide level and heating temperature. However, acrylamide contents increased with frying time. Among the oils commonly used in industries, oil type has low impact over acrylamide formation in fried products.

9.3.3.4 Moisture Profile in the Product

As the potato pieces fry, their moisture content decreases. However, color development and acrylamide formation only begin when sufficient drying has reached.

9.3.4 POSTFRYING OPERATIONS

To limit acrylamide formation during relatively long frying times, Kita et al. (2004) introduced a new step in potato chip processing: a postdrying step after frying. These authors found that shorter frying times followed by postdrying step resulted in low-moisture potato chips. Additionally, the pots-drying treatment gave significant decreases in acrylamide formation in potato chips.

9.3.4.1 Sorting Out of Overheated Items

One could expect good correlation between nonenzymatic browning development and acrylamide formation, since several studies reported excellent linear relationships between browning and acrylamide accumulation in chips and in model systems (Mottram et al. 2002, Stadler et al. 2002, Pedreschi et al. 2005). The "Centre pour l'Agronomie et l'Agro-Industrie de la Province de Hainaut" (CARAH) and a Belgian industrial partner (Rovi-Tech s.a.) developed a high-speed imaging system coupled with artificial neural networks (ANN) computing, which consists in a fully automatic device that takes a snapshot of every chip tested, then issues result for both color category and acrylamide concentration. The system is intended for analysis of incoming potato batches as well as for checking prefried chips for quality control in food distribution (Pedreschi et al. 2008).

9.3.4.2   Controlled Degradation/Transformation/Evaporation of Acrylamide

Methods to eliminate acrylamide once it has been formed in fried potatoes are not so advisable since they generally required the destruction or affect strongly negative the sensorial properties of final products. Prevention of acrylamide formation is the best way to control its content in fried potatoes.

9.4   CONCLUSIONS

Researchers worldwide have found more evidences related to toxicity of acrylamide in humans. Since Maillard reaction has been confirmed by many groups as the principal mechanism of acrylamide formation in fried potatoes, the ultimate challenge in frying of potatoes will be to achieve a substantial reduction of acrylamide while keeping desirable product attributes, such as flavor and color (which are generated by similar Maillard reaction pathways). Although many possible ways to reduce acrylamide content have been confirmed, the corresponding effects of sensory attributes in most of the reduction studies have not been clearly reported yet. Among the pretreatments mentioned to reduce acrylamide content, the previously reported citric acid immersion effect in acrylamide reduction after frying seems to be due to pH lowering and not to acrylamide precursor leaching out. During frying of potato pieces, acrylamide formation decreased dramatically as the frying temperature decreased. Control temperature profile and the pressure of frying are important technologic process issues related to reduce acrylamide formation in fried potatoes.

ACKNOWLEDGMENTS

The author acknowledges financial support from FONDECYT Project No. 1070031 and the Danish Ministry for Food, Agriculture, and Fisheries (Project: Reduction of the Formation and Occurrence of Acrylamide in Food). Gratitude is also expressed to Novozymes A/S for supplying asparaginase and providing guidelines for its application.

REFERENCES

Amrein, T., Bachman, S., Noti, A. et al. 2003. Potential of acrylamide formation, sugars and free asparagine in potatoes: A comparison of cultivars and farming systems. *Journal of Agriculture and Food Chemistry* 51: 556–560.

Amrein, T., Schobachler, B., Escher, F., and Amadó, R. 2004. Acrylamide in gingerbread: critical factors for formation and possible ways for reduction. *Journal of Agriculture and Food Chemistry* 52: 4282–4288.

Baardseth, P., Blom, H., Mydland, L., Skrede, A., and Slinde, E. 2006. Lactic acid fermentation reduces acrylamide formation and other Maillard reactions in French fries. *Journal of Food science* 71: C28–C33.

Becalski, A., Lau, B.P., Lewis, D., and Seaman, S.W. 2003. Acrylamide in foods: Occurrence, sources, and modeling. *Journal of Agricultural and Food Chemistry* 51: 802–808.

Biedermann, M., Biedermann-Brem, S., Noti, A., and Grob, K. 2002. Methods for determining the potential of acrylamide formation and its elimination in raw materials for food preparation, such as potatoes. *Mitteilungen aus Lebensmitteluntersuchung und Hygiene* 93: 653–667.

Blank, I., Robert, F., Goldmann, T. et al. 2005. Mechanisms of acrylamide formation: Maillard induced transformation of asparagine. In *Chemistry and Safety of Acrylamide in Food*, eds. M. Friedman and D.S. Mottram, pp. 171–189. New York, NY: Springer.

Blank, L. 2005. Current status of acrylamide research in food: Measurement, safety assessment and formation. *Annual New York Academic Science* 1043: 30–40.

Burton, W. 1969. The sugar balance in some British potato varieties during storage. II. The effects of tuber age, previous storage temperature, and intermittent refrigeration upon low-temperature sweetening. *European Potato Journal* 12: 81–95.

Ciesarova, Z., Kiss, E., and Boegl, P. 2006. Impact of L-asparaginase on acrylamide content in potato products. *Journal of Food and Nutrition Research* 45: 141–146.

Claeys, W., De Vleeschouwe, K., and Hendrickx, M. 2005. Effect of aminoacids on acrylamide formation and elimination kinetics. *Biotechnology Progress* 21: 1525–1530.

Coughlin, J.R. 2003. Acrylamide: What we have learned so far. *Food Technology* 57: 100–104.

Davies, A.M.C. 1977. The free amino acids of tuber varieties grown in England and Ireland. *Potato Research* 20: 9–21.

De Wilde, T., De Meulanaer, B., Mestdagh, E. et al. 2004. Acrylamide formation during frying of potatoes: Through investigation of the influence of crop and process variables. *Czech Journal of Food Science* 22: 15–18.

Dunovska, L., Hajlová, J., Aajka, T., Holadova, K., and Hajkova, K. 2004. Changes of acrylamide levels in food products during technological processing. *Czech Journal of Food Science* 22: 283–286.

Fiselier, K. and Grob, K. 2005. Legal limit for reducing sugars in pre-fabricating targeting 50 µg/kg acrylamide in French fries. *European Food Research and Technology* 220: 451–458.

Friedman, M. 2003. Chemistry, biochemistry, and safety of acrylamide. A review. *Journal of Agricultural and Food Chemistry* 51: 4504–4526.

Garayo, J. and Moreira, R. 2002. Vacuum frying of potato chips. *Journal of Food Engineering* 55: 181–191.

Gokmen, V. and Palakzaglu, T. 2008. Acrylamide formation in foods during thermal processing with a focus on frying. *Food and Bioprocess Technology* 1: 35–42.

Granby, K. and Fagt, S. 2004. Analysis of acrylamide in coffee and dietary exposure to acrylamide from coffee. *Analytica Chimica Acta* 520: 177–182.

Granda, C., Moreira, R.G., and Tichy, S.E. 2004. Reduction of acrylamide formation in potato chips by low-temperature vacuum frying. *Journal of Food Science* 69: 405–441.

Haase, N.U., Matthäus, B., and Vosmann, K. 2003. Minimierungsansätze zur Acrylamid-Bildung in pflanzlichen Lebensmitteln-aufgezeigt am Beispiel von Kartoffelchips. *Deutsche Lebensmittel-Rundschau* 99: 87–90.

Hanley, A.B., Offen, C., Clarke, M., Ing, B., Roberts, M., and Burch, R. 2005. Acrylamide reduction in processed foods. In *Chemistry and Safety of Acrylamide in Food*, eds. M. Friedman and D.S. Mottram, pp. 387–392. New York, NY: Springer.

Hendriksen, H., Kornbrust, B., Ernst, S., Stringer, M., and Heldt-Hansen, H. 2005. Asparaginase mediate reduction of acrylamide formation in baked, fried and roasted products. *Journal of Biotechnology* 118: S1–S135.

Hippe, J. 1988. HPLC-analysis of the concentrations of free asparagine and glutamine in potato tubers grown in varying amounts of nitrogen. *Potato Research* 31: 535–540.

Hogervorst, J., Schouten, E., Konings, E., Goldbohm, A., and van den Brandt, P. 2007. A prospective study on dietary acrylamide intake and the risk of endometrial, ovarian and cancer breast. *Cancer Epidemiology Biomarkers and Prevention* 16: 2304–23113.

Jackson, L. and Al-Taher, F. 2005. Effect of consumer food preparation on acrylamide formation. In *Chemistry and Safety of Acrylamide in Food*, eds. M. Friedman and D.S. Mottram, pp. 447–466. New York, NY: Springer.

Jung, M.Y., Choi, D.S., and Ju, J.W. 2003. A novel technique for limitation of acrylamide formation in fried and baked corn chips and in French fries. *Journal of Food Science* 68: 1287–1290.

Kita, A., Brathen, E., Knutsen, S., and Wicklund, T. 2004. Effective ways of decreasing acrylamide content in potato crisps during processing. *Journal of Agriculture and Food Chemistry* 52: 7011–7016.

Márquez, G. and Añón, M.C. 1986. Influence of reducing sugars and amino acids in the color development of fried potatoes. *Journal of Food Science* 51: 157–160.

Martin, F.L. and Ames, J.M. 2001. Formation of Strecker aldehydes and pyrazines in a fried potato model system. *Journal of Agricultural and Food Chemistry* 49: 3885–3892.

Matissek, R. and Raters, M. 2005. Analysis of acrylamide in food. In *Chemistry and Safety of Acrylamide in Food*, eds. M. Friedman and D.S. Mottram, pp. 293–302. New York, NY: Springer.

Mestdagh, F., De Melauner, B., van Poucke, C., Detavernier, C., Cromphout, C., and van Peteghem, C. 2005. Influence of oil type on the amount of acrylamide generated in a model system and in French fries. *Journal of Agriculture and Food Chemistry* 53: 6170–6174.

Mottram, D., Wedzicha, B., and Dodson, A. 2002. Acrylamide is formed in the Maillard reaction. *Nature* 419: 448–449.

Nielsen, N., Granby, K., Hedegaard, R., and Skibstead, L. 2006. A liquid chromatography–tandem mass spectrometry method for simultaneous analysis of acrylamide and the precursors, asparagine and reducing sugars in bread. *Analytica Chimica Acta* 557: 211–220.

Olsson, K., Svensson, R., and Roslund, C.A. 2004. Tuber components affecting acrylamide formation and color in fried potato: Variation by variety, year, storage temperature and time. *Journal of the Science of Food and Agriculture* 84: 447–458.

Pedreschi, F., Kaack, K., and Granby, K. 2004. Reduction of acrylamide formation in fried potato slices. *Lebensmittel-Wissenschaft und-Technologie-Food Science and Technology* 37: 679–685.

Pedreschi, F., Moyano, P., Kaack, K., and Granby, K. 2005. Color changes and acrylamide formation in fried potato slices. *Food Research International* 38: 1–9.

Pedreschi, F., Kaack, K., and Granby, K. 2006. Acrylamide content and color development in fried potato strips. *Food Research International* 39: 40–46.

Pedreschi, F., Kaack, K., Granby, K., and Troncoso E. 2007. Acrylamide reduction under different pre-treatments in French fries. *Journal of Food Engineering* 79: 1287–1294.

Pedreschi, F., Mery, D., and Marique, T. 2008. Grading of potatoes. In *Computer Vision Technology for Food Quality Evaluation*, ed. D.-W. Sun, pp. 305–318. Burlington, MA: Academic Press.

Rommens, C., Ye, J., Richael, C., and Swords, K. 2006. Improving potato and processing characteristics through all-native DNA transformation. *Journal of Agriculture and Food Chemistry* 54: 9882–9887.

Rosen, J. and Hellenäs, K.E. 2002. Analysis of acrylamide in cooked foods by liquid chromatography tandem mass spectrometry. *Analyst* 127: 880–882.

Rydberg, P., Eriiksson, S., Tanake, E., Karlsson, P., Ehrenberg, L., and Törnquist, M. 2003. Investigations of factors that influence the acrylamide content of heated foodstuffs. *Journal of Agricultural and Food Chemistry* 51: 7012–7018.

Silva, E.M. and Simon, P.W. 2005. Genetic, physiological, and environmental factors affecting acrylamide concentration in fried potato products. In *Chemistry and Safety of Acrylamide in Food*, eds. M. Friedman and D.S. Mottram, pp. 371–386. New York, NY: Springer.

Stadler, R.H., Blank, I., Varga, N. et al. 2002. Acrylamide from Maillard reaction products. *Nature* 419: 449–450.

Stadler, R.H., Robert, F., Riediker, M.C. et al. 2005. In depth-mechanistic study on the formation and potential strategies of control. *Nutritional Reviews* 62: 449–467.

Tae Kim, C., Hwang, E., and Joo Lee, H. 2005. Reducing acrylamide in fried snack products by adding aminoacids. *Journal of Food Science* 70: 354–358.

Tareke, E., Rydberg, P., Karlsson, P., Eriksson, S., and Tornqvist, M. 2002. Analysis of acrylamide, a carcinogen formed in heated foodstuffs. *Journal of Agricultural and Food Chemistry* 50: 4998–5006.

Umano, K. and Shibamoto, K. 1987. Analysis of acrolein from heated cooking oils and beef fat. *Journal of Agricultural and Food Chemistry* 35: 909–912.

Van Hendriksen, H., Stringer, M.A., Ernst, S., Held-Hansen, P., Schafermayer, R., and Corrigan, P. 2006. Novozymes A/S; Procter & Gamble, Inc. Patent N° WOO6053563.

Wedzicha, B.L., Mottram, D.S., Elmore, J.S., Koutsidis, G., and Dodson, A.T. 2005. Kinetic models as a route to control acrylamide formation in food. In *Chemistry and Safety of Acrylamide in Food*, eds. M. Friedman and D.S. Mottram, pp. 235–253. New York, NY: Springer.

Yaylayan, V.A., Wnorowski, A., and Perez-Locas, C. 2003. Why asparagine needs carbohydrates to generate acrylamide. *Journal of Agricultural and Food Chemistry* 51: 1753–1757.

Zhang, Y. and Zhang, Y. 2007. Formation and reduction of acrylamide in Maillard reaction: a review based on the current state of knowledge. *Critical Reviews in Food Science and Nutrition* 47: 521–542.

Zhang, Y., Chen, J., Zhang, X., Wu, X., and Zhang, Y. 2007. Addition of antioxidant of bamboo leaves (AOB) effectively reduces acrylamide formation in potato crisps and French fries. *Journal of Agriculture and Food Chemistry* 55: 523–528.

Zyzak, D., Sanders, R.A., Stojanovic, M. et al. 2003. Acrylamide formation mechanism in heated foods. *Journal of Agricultural and Food Chemistry* 51: 4782–4787.

Juliana Morales-Castro and
L. Araceli Ochoa-Martínez

CONTENTS

10.1 INTRODUCTION

As demand for natural, minimally processed, and fresh foods is increasing, food preservation methods aimed at maintaining quality and extending storage life, such as modified atmosphere packaging (MAP) and controlled atmosphere packaging (CAP), have been subject of study and research worldwide. The term modified atmosphere technology includes controlled atmosphere storage, ultra low oxygen storage, superatmospheric oxygen storage, gas packaging, vacuum packaging (VP), passive MAP, and active packaging (Hertog and Banks 2003). All these techniques share a common principle: the manipulation or the control of the atmosphere composition that surrounds a product in order to maintain quality during the shelf life.

Although MAP requires strict control and adherence to quality assurance programs in order to be successful, the technology has shown rapid growth with MAP market accounting for 30–35 billions of lb per year, an amount that exceeds other food processing categories such as aseptic, retort pouch/tray, and canned or frozen foods (Forcinio 1997). The meat industry relies heavily on MAP in several processes, so the market share of MAP in western European countries as a percentage of the total retail meat market is 10%–40% while in Norway it is still higher, approximately 60% (Sorheim and Nissen 2002).

In a modified atmosphere, the normal proportions of gases in air are changed intentionally. Although modified atmosphere and controlled atmosphere have been used indistinctly, they are not the same. Controlled atmosphere implies a precise control of the atmosphere in specific concentrations while in modified atmosphere such control is not applied. MAP can be defined as "the enclosure of food products in gas barrier materials, in which the gaseous environment has been changed" (Young et al. 1988) in order to inhibit spoilage agents and therefore either maintain a higher quality within a perishable food during its natural life or actually extend the shelf life (Church and Parsons 1995).

In controlled atmosphere storage the atmosphere composition is controlled throughout the storage period, while in MAP the composition is modified at the beginning and stand additional modification due to food characteristics (respiration rate (RR) and biochemical changes), package properties (permeability), and storage conditions (temperature), which implies that a gas control is difficult to exert.

This chapter is aimed at presenting the most significant contributions regarding the effect of MAP on food quality and some of the work on modeling of modified atmospheres reported recently.

10.1.1 Concepts and Definitions

MAP is defined as a form of packaging that involves the removal of air from the pack and its replacement with a single gas of mixture of gases (Blakistone 1999). Depending on the type of product, a specific mixture of gas will be used. Two types of MAP have been defined: VP and gas, and gas flush or gas exchange packaging (Church and Parsons 1995). The modification of the atmosphere can be achieved using different methods: VP, gas packaging, compensated vacuum, passive atmosphere modification, active packaging; oxygen, CO_2, and ethylene absorbers (Blakistone 2000). VP is used to inhibit aerobic bacteria, oxidative reactions, and spoilage agents, by means of making oxygen unavailable, in order to achieve longer shelf life. Oxygen is reduced to less than 1%, but some anaerobic/microaerophilic organisms can survive and nonoxidative reactions can occur, contributing to some sort of deterioration. Deformation through product compression occurs, which limits VP to soft products such as bakery products. Gas packaging, on the other hand, pretends to overcome the problems related with the use of vacuum. The modification of the headspace atmosphere inside a package can be obtained by mechanical replacement of air with a gas mixture or by generating the atmosphere in two ways: actively or passively. The mechanical replacement of the atmosphere is obtained through gas flushing and compensated vacuum.

The atmosphere generated by the passive method involves the packaging of the product under ambient conditions using a permeable film (Kader et al. 2003) and is commonly used in packaging of fruits and vegetables. During the storage, the RR of the product consumes the surrounding oxygen producing CO_2, which changes the atmosphere. Passive atmosphere modification is a complex process where the interplay of many elements conjugates to modify the atmosphere, until a steady state is reached (Al-Ati and Hotchkiss 2002). The active method involves the use of certain additives into the packaging film or inside the packaging containers, to modify the headspace atmosphere. Some of the additives used include oxygen absorbents, carbon dioxide absorbents/emitters, and ethylene absorbents (Kader et al. 1998).

Another term that has been introduced is equilibrium modified atmosphere packages (EMAP), which describes a systematic approach consisting of alteration of the air around the commodity to a gas combination of 1%–5% O_2 and 3%–10% CO_2 (balance N_2). By knowing the RR of the produce at the desired storage temperature, a packaging configuration can be designed to keep the gas concentrations constant during the storage time of the produce. Packages are designed by selecting the suitable film permeability at the storage temperature, the amount of MPV (minimally processed vegetables) in the package, and the package dimensions.

A key aspect of MAP technology is "temperature," since this is the single most important element for this technology, being mentioned as "... more important than the modified atmosphere" (Brody 1996). Temperature control, in conjunction with good sanitation practices and high initial product quality, assure the success of MAP. Programs like Good Manufacturing Practices (GMP) and Hazard Analysis Critical

Control Points (HACCP) programs should be maintained when processing foods under MAP. A common misconception is that MAP can overcome temperature and microbiological abuse, which is wrong, since MAP can help extend shelf life and it will not improve quality but can slow loss of quality (Day 1991, Brody 1999, Zagory 2000).

10.1.2 BASIS OF MAP

Four mechanisms have been identified as responsible for the bacteriostatic effect of gases in MAP (Sivertsvik et al. 2002):

- Alteration of cell membrane function including effects on nutrient uptake and absorption
- Direct inhibition of enzymes or decreases in the rate of enzyme reactions
- Penetration of bacterial membranes, leading to intracellular pH changes
- Direct changes in the physical–chemical properties of proteins

In order for CO_2 to be effective, a certain amount (depends on the food) has to be dissolved into the product to inhibit bacterial growth. A factor to consider when CO_2 is used is its high solubility in water and fat, which increases substantially as temperature decreases. A ratio between the volume of gas and volume of food product (G/P ratio) of 2:1 or 3:1 has been established, as necessary for CO_2 to exert its action. The main factor that controls antimicrobial activity by MAP is dissolved CO_2 concentration (Devlieghere et al. 1998).

10.2 MAP APPLICATIONS

MAP is used extensively in most food categories: fresh meat, cured and processed meats, fruits and vegetables, fresh cut, bakery, dairy and seafood. More recent applications that can be mentioned are in precooked pastas, prepared salads, or cottage cheese. MAP is part of the food industry around the world and new applications are daily evaluated so MAP has been the subject of numerous publications. Entire books devoted to this topic that covers reports of thousands of publications on MAP research have been published (Brody 1989). Book chapters offer concise information on MAP. Several publications dealing with MAP and intelligent and active packaging also offer recent advances on these technologies. Research on this topic has been carried out for many years (Blakistone 1999). MAP has been around for over a century, starting from the very simple concept of changing air composition (oxygen 21%, nitrogen 78%, and carbon dioxide around 1%) to the up-to-date technology where smart packaging, active devices, novel gases, and sensors have been incorporated along with atmosphere prediction through mathematical modeling, to mention just a few the advances in this area. MAP and controlled atmosphere storage have been the subject of numerous reviews and analysis from different points of view: microbiological safety (Hintlian and Hotchkiss 1986, Brody 1989, Farber 1991, Phillips 1996, Farber et al. 2003), processing of commodities (Young et al. 1988, Jayas and Jeyamkondan 2002), packaging of fresh produce (Zagory and Kader 1988,

Phillips 1996, Rai et al. 2002, Farber et al. 2003, Soliva-Fortuny and Martin-Belloso 2003), processing of fish and fishery products (Skura 1991, Sivertsvik et al. 2002), general approach (Day 1989, Church and Parsons 1995), quality and shelf life (Rai et al. 2002), grains processing (Jayas and Jeyamkondan 2002), and mathematical modeling (Al-Ati and Hotchkiss 2002, Fonseca et al. 2002).

10.2.1 MEAT INDUSTRY

10.2.1.1 Fresh Meats

The meat industry has been using MAP and CAP technologies as well as VP for some time. In the United States and Canada more than 80% of beef is distributed as vacuum-packaged primal cuts from meat packets to retailers and institutional markets, while in the United Kingdom, about 38% of red meat is marketed under MAP (Brody 1999). Fresh poultry is also distributed in master pack in bulk under MAP to retail grocery and institutional establishments. Precooked and marinated poultry is packaged under MAP and vacuum, respectively.

Meat is a highly perishable product for being rich in nutrients and available water and so it can sustain any type of microorganisms, such as bacteria, yeast, and mold. Its shelf life is established by sensory (color deterioration, odor, and exudate retention) and microbial factors (Church and Parsons 1995).

Meat quality is determined by its color and microbiological shelf life at the time of purchase and by its juiciness and tenderness at the time of consumption. Particularly, visual appearance seems to be the quality factor most influencing in consumer's purchase decision. Color detrimental changes affect meat consumer acceptability, being a limiting factor to marketability, since red color is associated to its freshness while brown color to bacterial spoilage or meat from mature animals (Jayas and Jeyamkondan 2002). Microorganisms involved in fresh meat spoilage are *Pseudomonas* or *Acinetobacter* among others (Brody 1989). Shelf life extension of meat is dependent on both the product characteristics (type, dimensions, and initial microbial load) and processing variables (atmosphere, storage temperature, packaging materials, and design) (Church and Parsons 1995). The meat chemical composition (unsaturated fat content, pH) for each type of animal, product composition (fat, protein content), headspace volume in packaged products, and microbial initial load (inversely proportional to shelf life) determine shelf life for each type of product. Shelf life of meat can be established on the basis of the "practical storage shelf life" (PSL) that has been defined by the International Institute of Refrigeration as "The period of storage during which the product retains its characteristics properties and remains suitable for consumption or the intended process." Chilled meat has a PSL between 10 days and 4 weeks depending on the type of meat, when stored between 0°C and 2°C. Primal cut-vacuum packed under 0°C has a PSL on the range of 3–12 weeks, while meat gas flushed with 100% CO_2 and 0°C has a PSL up to 16 weeks. For frozen beef, the PSL is suggested between 12 and 24 months, depending on storage temperature (−12°C to −24°C), and for beef retail cuts, the retail display life, in days, for over-wrapped packages, varies between 1 and 3 days, after storage in vacuum packs at 0°C. When a modified atmosphere (MA) of 80% O_2, 20% CO_2, is applied, the retail display life increases to over 7 days. Oxygen, carbon dioxide, and nitrogen are the most common gases used in MAP of meat; however, the most successful method to extend shelf life of meat has been the use of enriched CO_2 atmospheres. CO_2 increases its solubility at lower temperatures so carbonic acid is formed and a slight drop of pH occurs in meat, which favors microbial inhibition. The mechanism of action of CO_2 resides in its lipid solubility, so it accumulates in lipid bilayer of the microbial cell membrane and thus affecting cell permeability conducing to intracellular pH changes. In addition, it affects enzyme and protein functionality, and thus, inhibiting metabolic activities (Farber 1991).

Fresh meat under CO_2 atmospheres can be preserved using two methods: vacuum or gas method in addition to refrigeration. VP uses oxygen limiting packaging materials to inhibit the growth of psychrophilic bacteria involved in microbial spoilage of refrigerated meat. For vacuum packages, the applied suction leaves a residual O_2 content of 1%–2% reducing thus, significantly, oxidative rancidity. The percentage that is consumed by meat respiration and bacterial activity produces CO_2 increasing its levels in the range of 10%–40% (Jayas and Jeyamkondan 2002). This method is recognized as the most used for distribution of primal and subprimal cuts to the retail stores in United States. Gas addition, the other method used, involves the elimination of the air from the package and its replacement with CO_2, which implies a higher cost. Recommended concentrations for different meats vary widely. For red meat, concentrations suggested are: 60%–85% O_2/15%–40% CO_2 and 70%–80% O_2/20%–30% (Chruch and Parsons 1995).

Vacuum skin packaging (VSP) system was developed as a response to overcome some of the disadvantages of VP. The systems involves two stages: (a) instantaneous heating of the lower surface upper film of the package before the film descends over the meat surface and (b) tight disposition of the plastic film on the meat surface in order to avoid wrinkles and purges. This system showed higher quality (microbiological and physicochemical analysis) over VP with lower counts of aerobic mesophiles, anaerobes, and lactic acid bacteria (LAB). Lower pH values correlated with higher LAB counts, higher meat firmness, slower met tenderization, and higher luminosity, and redness. All these factors improved meat color (Barros-Velázquez et al. 2003). A most recent MAP technology reported as an alternative to MAP for meat consists in modified atmosphere storage under subatmospheric pressure, developed in an effort to decrease food packaging waste (Smulders et al. 2006). Its effect on the microbial load and sensory effects on meat quality has been reported (Smulders et al. 2006), at a given pressure between 20 and 30 kPa at 0°C–1°C and 3°C–4°C, showing results similar to storage under vacuum and with pure CO_2.

10.2.1.2 Effects of MAP on Microbial Population

The European Commission regulation on microbiological criteria for foodstuffs states that *Listeria monocytogenes* should be below 10^2 cfu/g during the shelf life of ready-to-eat foods (European Commission 2005). MAP inhibits better than LAB groups of *Pseudomonas*, *Enterobacteriaceae*, lypolytic, and proteolitic bacteria. *Pseudomonas* growth has been found to be inhibited in about 50% in the presence of 10% CO_2, a value that did not increase at higher CO_2 concentrations. Once *Pseudomonas* growth is restricted, *Lactobacillus* can grow more rapidly on meat under anaerobic conditions and are able to inhibit the growth of the competing flora.

It has been suggested that in designing a packaging system the focus should be in creating proper conditions where LAB predominate since their growth is slow and spoilage changes are lower (Rao and Sachindra 2002). When meat is stored under CO_2 shelf life is extended up to 15 times of that obtained under air. The effect is attributed to a shift in microbial flora from aerobic to anaerobic bacteria (i.e., LAB), which causes a slight drop in pH favoring a negative environment for most pathogens and Gram negative bacteria. Some of the pathogens most important for food safety are *L. monocytogenes* and *Salmonella typhimurium* and it has been found that modified atmospheres of CO_2/N_2 can be effective growth inhibitors alone or combined with potassium lactate and sodium diacetate as antimicrobials at 10°C in pork chops and cured ham (Michaelsen et al. 2006). Skandamis and Nychas (2002) reported a study where they applied MAP alone and in combination with oregano essential oil (OEO), storing at 0°C, 5°C, 10°C, and 15°C with atmospheres applied of 40% $CO_2/30\%$ $N_2/30\%$ O_2, 100% CO_2, 80% $CO_2/20\%$ air, vacuum packed and air. Shelf life depended on the packaging conditions and increased in the following order: air < vacuum pack < 40% $CO_2/30\%$ $N_2/30\%$ O_2 < 80% $CO_2/20\%$ air < 100% air while OEO increases even further shelf life: at 0°C, 32–65 days, 5°C, 14–27 days, 10°C, 8–22 days, and 15°C, 5–15 days.

VSP versus VP has been evaluated on several cuts from different anatomical parts of beef and on three spoilage bacterial groups (aerobic mesophiles, anaerobes, and coliforms) at 4°C for 40 days. It was observed that for all three groups VSP showed the highest inhibition on microbial growth. However, it seems that VSP has a slightly negative effect on the aging process, since tenderness is one of the most important quality attributes of meat (Vázquez et al. 2004). The effect of MAP on beef quality can be influenced by breed as it was reported by Insausti et al. (2008) who found an interaction between breed and storage time in beef steaks from four Spanish cattle breeds packaged under modified atmospheres (60% O_2, 30% CO_2, and 10% N_2). It is interesting that shelf life was limited by color and odor degradation between days 5 and 10 of the study, probably due to lipid oxidation. A similar study using various beef steaks from different USDA quality grades and muscles showed that MAP was more effective with low choice quality grade showing higher lean color scores and less discoloration at 7 and 10 days (Behrends et al. 2003).

10.2.1.3 Use of Carbon Monoxide in Meat Packaging

High oxygen atmospheres are common in meat packaging but sustain spoilage bacteria growth, accelerate lipid oxidation, off-flavors, bone darkening of bone-cuts, and premature browning during cooking (Cornforth and Hunt 2008). Observed shelf life is around 7 days (Sorheim et al. 1997), limiting distribution. These mentioned difficulties can be overcome by the addition of carbon monoxide (CO) to the atmosphere, given that myoglobin has an affinity 30–50 times greater for CO than for O_2. A CO enriched atmosphere leads to formation of carboxymyoglobin (bright red), so the presence of O_2 does not affect myoglobin reduction. CO addition (processing aid) has been approved in New Zealand and Australia, and a similar system has been used in Norway for several years, because CO increases color stability and shelf life compared with high oxygen systems (70% O_2 and 30% CO_2). From 1985 the CO system has been used in Norway for case-ready modified atmosphere packaged meats with 0.4% CO, 60% CO_2, and 39.6% N_2, where the product remains under CO during display. In the United States this packaging system was approved in 2002 as GRAS for master packaging, although it requires removal of packages from the CO containing atmosphere before display and sale; besides, in 2004, the FDA concluded that MAP system containing 0.4% CO, 20%–100% CO_2, and 0%–80% N_2 was safe for primary packaging systems (retail packages). However, the use of CO_2 was discontinued in Norway, in the same year, after 19 years of successful applications to comply with regulations of European trading partners (Wilkinson et al. 2006, Cornforth and Hunt 2008).

Low oxygen MAP, based on anaerobic atmosphere, uses a mixture of carbon dioxide (20%–30%), CO (0.4%), and nitrogen (balance gas), to keep a consumer's appealing red color. This technology presents considerable advantages over its predecessor, high oxygen MAP, due to the hurdle effect of anaerobic conditions, refrigeration and elevated CO_2, with a considerable shelf life extension of 23 days, similar to the shelf life of vacuum products.

A big debate is going on in the United States regarding the use of CO for meat packaging (Riley and Huffman 2006) due to a media campaign launched by meat processors to persuade Congress to ban this technology following a petition filed to the Food and Drugs Administration (FDA) on November 2005. This prompted a response from the American Meat Science Association who commissioned a report with a detailed analysis of meat packaging technologies, in order to "bring clarity to the discussion currently taking place within the meat industry and with policy makers and consumers concerning the use of CO as component of fresh meat Packaging system" (Cornforth and Hunt 2008). This report establishes the advantages of the low oxygen CO-MAP technology over its predecessors, high oxygen MAP, vacuum and PVC packages and also about the negligible risk of CO toxicity due to CO treated meat consumption or the process itself. It also cautions about the need for "adherence to label instructions for product shelf life and the use of odor and overall appearance as spoilage indicators."

The controversy is mainly focused on two issues: associated risk to the consumer, due to the toxic nature of the gas and the potential for masking spoilage, which is considered a safety risk for consumers, due to the fact that although discoloration, offensive odors, and flavors indicate meat spoilage, consumers rely on color and odor when making purchase decisions. However, under CO conditions, discoloration is unlikely to occur, so it is necessary to be specially careful in processing meat under CO, since CO does not affect pathogen growth. This system has been recommended for master pack to be transported longer distances, where those times are needed to reach distant markets (Wilkinson et al. 2006).

Numerous studies have been conducted on CO use that confirm higher redness values in CO-MAP packaged than in aerobic packages for Ground meat (Luño et al. 1998), pork sausages (Martínez et al. 2005), and pork (Sorheim et al. 1997). For pork chops under CO-MAP, red color persisted for a longer time than package chops under high O_2 atmospheres (Wicklund et al. 2006). A similar result was observed in top sirloin steaks under CO-MAP versus those packaged in high O_2 MAP and under vacuum.

During storage in MAP using CO_2 atmospheres, LAB are the predominant microflora, which results in sour and acidic off-odors. By using CO, carboximyoglobin is formed and it did not mask product spoilage, which is one of the drawbacks mentioned for CO_2 atmospheres. CO atmospheres have also been used to avoid the phenomenon of premature browning presented in meat stored in high O_2 atmospheres. Positive results have also been confirmed with fresh pork sausages (Laury and Sebranek 2006) and shelf life of pork chops under CO atmospheres has been established in 8 weeks under refrigerated storage for a master-packaging system (Wilkinson et al. 2006).

After analyzing numerous studies conducted using CO, the evidence supports the safety and beneficial effects of using CO in fresh meats through improving red color stability and flavor, inhibition of premature browning, and decrease of spoilage and pathogenic microorganisms, when compared with PVC packaging. A negligible risk of CO toxicity is also reported, but emphasis is placed on adherence to label instructions for product shelf life and to the use of odor and overall appearance as spoilage indicators. MAP added with CO offers benefits and some drawbacks that should be considered on designing packaging systems that bring added value to consumers (Cornforth and Hunt 2008).

10.2.1.4 Residual Oxygen and Other Applications

Even with modified atmospheres of 60% N_2 and 40% CO_2 and vacuum packages, small amount of oxygen is left on the packages and care should be taken because residual oxygen inside the package can promote adverse reactions such as lipid oxidation. On MAP and VP, raw and cooked, beef O_2 levels between 1.15% and 1.26% were detected (Smiddy et al. 2002a). Similar findings were reported for cooked chicken patties, with concentrations of 0.9%–1.1% and .11%–0.15% for MAP and VP, respectively, with the most oxidized samples for MAP products (Smiddy et al. 2002b). This can be explained by the fact that gaseous environment within MAP is not static since product respiration, microbial metabolism, and gas exchange act continuously to change the composition of the atmosphere. Concentrations of residual oxygen in MAP packs can be attributed to a number of factors such as oxygen permeability and poor sealing ability of packaging film (package leaks), food ability to trap air, and ineffective gas flushing. A study for processed cooked meats concluded that in 88% of packs oxygen was present after only 24 h of package with concentration up to 1.2% O_2 at day 1 and after 21 days of storage, which indicates that MAP is dynamic changing constantly due to the factors mentioned above with the consequent quality deterioration (Smiddy et al. 2002c). It has been estimated that meat losses in the order of 2%–20% are due to discoloration and off-odors, since at high oxygen pressure the surface oxymyoglobin layer is thicker and it looks red, but high oxygen concentration in MAP for meats induces lipid oxidation, especially membrane phospholipids, which cause meat rancidity.

Chicken is another product where MAP has effectively proved to increase shelf life. Packaging methods, including vacuum package and MAP, have been effective in extended shelf life of poultry and inhibit growth of spoilage and pathogenic microorganisms. When poultry and its products are stored aerobically, *Pseudomonas* and yeast are the main microorganisms dominating while under MAP storage, LAB prevails. Refrigerated shelf life for fresh processed poultry and cooked poultry without MAP has been estimated in 3–10 and 5–16 days, while under MAP conditions, shelf life reaches 12–18 and 21–30 days, respectively (Forcinio 1997). Gas concentrations used for MAP-processed poultry have been reported between 70% and 80% CO_2, finding better results with 80% CO_2 (17 days, compared to 14 days) (Sarantópoulos et al. 1998). A chemical indicator to assess meat quality is the presence of Biogenic amine (BA) product of the enzymic decarboxilation of specific amino acids due to microbial enzyme activity. The main BAs are polyamines such as spermine and spermidine that occur naturally in pork and beef meat but during storage, histamine, putrescine, tyramine tryptoamine, and cadaverine may be formed. The presence of these products has been documented in different studies and tyramine, putrescine, and cadaverine were proposed as quality indicators or MA-stored broiler chicken (Rokka et al. 2004). Fresh pork quality can be judged on color lightness, microbial growth, and lipid oxidation. Atmospheres containing 45% oxygen and 20% CO_2 have been shown to be effective in preserving shelf life up to 8 days of storage, based on color acceptability, microbial load reduction, and total volatile basic nitrogen (TVBN) reduction properties at 4°C (Zhang and Sundar 2005). Shelf life of pork meat is about 7 weeks stored at −1.5°C to 0.5°C under vacuum, being considered close to the maximum shelf life that can be attained using this technology.

10.2.2 Dairy Products

The main food products market (meat, poultry, and fresh produce) has moved toward MAP. A similar trend, but to a lesser extent, has been observed in the dairy and cheese markets, due to the need for satisfying consumer demands for smaller portions of product, adequately packaged and ability to see and inspect the product before it is purchased. Protection is given by the special atmosphere inside the package (with additional shelf life while maintaining freshness), where the product is protected against contamination and visibility of the product is possible due to the clarity of the film barrier. MAP application in dairy products focuses mainly on cheese packaging, with different MAP requirements for hard cheeses and soft cheeses. Stabilized cheeses (which contain no active LAB and include processed cheeses, cream cheeses, and feta) should be packaged in low O_2 and high CO_2 levels and MAP and active packaging should be used. Active cheeses contain live LAB and present a challenge since they produce varying amounts of CO_2 through respiration and so an appropriate packaging material must be selected to permit CO_2 to permeate and avoid expansion or even explosion of the package and CO_2 absorbers can be considered and so the use of AP is highly recommended for control of CO_2 and O_2 besides considering antimicrobial films. Mold ripened cheese represents the biggest challenge when using MAP, since the mold should have optimum conditions to grow and keep it alive. A careful control of CO_2 and O_2 through package should, therefore, be controlled so extensive research is required for active packaging for this type of cheeses (Floros et al. 2000).

The main limiting factor of shelf life is mold growth, for both spoilage and safety problems (Floros et al. 2000). Filamentous fungi such as *P. commune*, *P. roqueforti*, *A. versicolor*, and *P. nalgiovense* can be controlled by elimination of oxygen from the package headspace, so VP is a suitable method commonly used for cheese. However, two disadvantages of VP are the difficulty to open the package and the low quality plastic image. MAP has, thus, been used to overcome these limitations, facilitating opening and extending shelf life (Subramaniam 1989). Cheddar cheese is packaged under vacuum and under MAP, being suggested that the most suitable gas mix consists of 75% CO_2/25% N_2. Sliced, grated cheese is also packaged using MAP with mixtures of 70% N_2/30% CO_2. MAP has also been used in bulk packaging and some reports mention that cheese blocks use a resealable system introduced in the U.K. market in 1992 with an innovation in 1995. Italian cheeses used MAP for packs of 25 and 50 lb for food food service and industrial users (Subramaniam 1989).

Floros et al. (2000) concluded, from experiments conducted at very low O_2 levels, that fungi growth is much slower and has a different appearance when oxygen levels are reduced to 1% or below which suggests the application of very low O_2 levels to significantly extend the shelf life of some cheeses.

MAP at the retail level has been applied in a Greek whey cheese called "Myzithra Kalathaki," which was packaged in four different modified atmospheres, attaining 40 and 33 days of shelf life under 40% and 60% CO_2. Inhibition of *Enterobacteriaceae* and mold and yeast during the storage period was achieved, although LAB predominate toward the end of the storage period with good sensory characteristics during 30 days of storage while control samples were not acceptable after 10–12 days of storage (Dermiki et al. 2007). Another Greek fresh whey cheese, "Anthotryros," got shelf life extension of 10 and 20 days (from 7 days for fresh whey cheese under aerobic conditions), at atmospheres of 30%/70% (CO_2/N_2) and 70%/30% (CO_2/N_2), respectively, by means of inhibit growth of mesophilic bacteria (Papaioannou et al. 2006). Mozzarella cheese has been the subject to different studies. CO_2 enriched atmospheres with 100% CO_2 increased shelf life by 385%. In studies conducted with mozzarella cheese slices, CO_2 (100% and 50%) proved to be effective since shelf life increased 385% and 246% traduced in 63 days, and 45 days for 100% CO_2 (Alves et al. 1996). Sometimes CO_2 can promote light induced oxidation. Shredded Cheddar cheese, for example, showed higher concentrations of aldehydes and fatty acids and lower concentrations of alcohols and esters than cheeses stored under nitrogen when it was stored under 100% CO_2 with fluorescent light (Colchin et al. 2001). Also, CO_2 affects sensory attributes such as in the case of goat cheese, whose texture and appearance were negatively affected at the end of the storage period of 28 days, although cheese packaged in air atmosphere was unacceptable after 14 days of storage (Olarte et al. 2002). A softer taste and more cohesive and friable structure was obtained under MAP of Parmigiano Reggiano chesse (50% CO_2/50% N_2 and 30% CO_2/70% N_2) stored at 4°C (Romani et al. 2002). The safety aspect of MAP regarding pathogen growth was studied with fresh pasteurized goat cheese (Cameros cheese) finding that *L. monocytogenes* can survive and grow under MAP conditions of 100% CO_2 (Olarte et al. 2002), although MAP is effective in inhibiting spoilage organism and extending shelf life (27 days compared to 7 days under air refrigeration).

MAP is not always the most suitable method for shelf life extension. Specifically for dairy products, MAP cannot provide sufficient shelf life or maintain quality, so a similar approach is the use of direct addition of CO_2. Such technology was developed by Chen and Hotchkiss (1991a, b) and consists of the direct injection of CO_2

into the products joined with a high barrier packaging. The technique was named "direct addition of carbon dioxide" to differentiate it from MAP. The principle of its application resides, as in MAP, in the addition of gas to the product in order to extend shelf life by inhibiting microbial activity (Hotchkiss et al. 2006). Direct addition of carbon dioxide technology has been successfully applied to refrigerated cottage cheese, getting a shelf life extension of 200%–400% (Chen and Hotchkiss 1991a, b) and maintaining natural characteristics of the product. Mannheim and Soffert (1996) also attained a shelf life extension of 150% by flushing the headspace with carbon dioxide at 8°C. Cottage cheese deterioration is attributed to the microbial growth of Gram-negative psychrotrophic bacteria such as *Pseudomonas*, *Proteus*, *Aeromonas*, or *Alcaligens* sp. resulting in undesirable off-flavors, pigment formation, or slimy curd. Yeast and molds cause damage on flavor, texture, and appearance (Chen and Hotchkiss 1991a).

Addition of carbon dioxide to milk has been successful for delaying or inhibiting native microflora. Hotchkiss et al. (2006) experimented at high scale with milk stored in a stainless steel liquid bulk tanks for rail shipment, and the effectiveness of CO_2 was again confirmed when milk reached a shelf life of 14 days, 4 days longer than control milk. Hotchkiss et al. (2006) concluded that direct CO_2 addition to raw bulk milk substantially improves and extends shelf life of the products, increasing product safety and quality.

As with other products a safety issue emerges from the presence of *Listeria* spp. in raw milk, due to the risk that it represents. There is a possibility that temperature abuse occurs and shelf life extension obtained by inhibition of spoilage microorganisms using dissolved CO_2 can result in toxicity preceding sensory spoilage. By dissolving CO_2 in packages of cottage cheese (maintaining a headspace concentration of 35% v/v), the growth of *Clostridium sporogenes* and *Listeria monocytogenes* was inhibited up to 63 days at 4°C. Some other dairy products, such as yogurt and dried milk powder, are also packaged using modified atmospheres. A general practice for packaging of dried milk powders is the use of modified atmospheres in cans or drums with mixes of CO_2 and N_2. The addition of CO_2 for yogurt shelf life extension has been reported, focusing mainly on sensory attributes effects.

10.2.3 MAP IN SEAFOOD

Shelf life extension attained in MAP using high CO_2 concentrations (25%–100%) and VP for meat products is in the order of weeks and months, while for fish, shelf life increases in days or up to 1 week, depending on species and temperature. This broad difference resides in different spoilage microflora and pH values for meat and fish. As with other foods the most important factor to preserve fish quality is temperature, which combined with MAP can give shelf life extension that varies from 50% to 100% for fresh fish and 100% to 200% for heat treated fish (Dalgaard 2000). Fish and shellfish are highly perishable products due to high a_w, neutral pH, and presence of autocatalytic enzymes. Spoilage occurs mainly for microbial activity but also important are biochemical reactions, such as auto-oxidation or enzymatic hydrolysis of the lipid fraction resulting in off-odors and off-flavors and fish softening due to tissue enzyme activity.

Torrieri et al. (2006) pointed out that MAP will affect shelf life of fish depending on the fish species, fat content, initial microbiological contamination, background of the fish, treatment the fish undergoes after slaughtering (handling and storage condition), ratio of gas/product (g/p), and most importantly, packaging method and temperature of storage (Bøknæs et al. 2001, Sivertsvik et al. 2002). Based on this, it is difficult to generalize about the relationship between shelf life extension and fish acceptability under MAP, but combination with chilled storage offers benefits such as shelf life extension, longer transport distances of the product, and reduced financial losses (Farber 1991, Bøknæs et al. 2001). It is convenient to distinguish two categories of MA package products: those eaten without any heat treatment previous to consumption (ready-to-eat products, sushi, smoked salmon, cooked shell fish) and those products subjected to heat treatment to inactivate vegetative pathogens before serving (most fresh fish) (Sivertsvik et al. 2002).

Fish deteriorative indicators such as TVBN and trimethylamine nitrogen (TMAN) are used as indices for spoilage of fish in refrigerated storage. Rejection limits of 40 and 12 mg/100 g have been set for TVBN (Directive 95/149/EEC) and TMAN-N (Directive 91/493/EEC), respectively. Thus, shelf life has been determined on the basis of those parameters. One additional indicator for shelf life under MAP is growth of *Photobacterium phosphoreum*, which has resulted to correlate much better with remaining shelf life than total viable counts (TVC) (Dalgaard 2000). Fresh fish products packed under MAP are a trend of the European retail market due to the convenience of selling these products in the same chill cabinets as with other foods. One disadvantage of this system is the shorter shelf life in comparison with other meat products and off-flavors detected in seafood in retail. Off-flavors in seafood are caused by high numbers of spoilage bacteria due to inappropriate time–temperature storage conditions, insufficient degree of freshness of the fish raw material before packaging, or insufficient hygienic practice prior to packaging (Bøknæs et al. 2001). Some of these problems can be overcome by MAP products prepared from previously frozen fish that allow prolonging shelf life of thawed fish compared to the corresponding fresh MAP. Baltic Sea cod shelf life was extended from 11–12 days at 2°C, for fresh MAP fillets, compared to more than 20 days for the thawed MAP product, due to inhibition of *P. phosphoreum* and the reduction in trimethylamine (TMA) production.

MAP has been proved beneficial to different sea products, being cod one of the most widely studies. Under MA packaging cod increases acceptable shelf life by 50% depending on storage temperature, raw material quality, handling, gas mixture, gas to product (g/p) volume ratio, and packaging material (Sivertsvik 2007). The main spoilage microorganisms identified for wild cod stored under aerobic conditions in ice are *Shewanella putrefaciens* (intense and unpleasant off-odors), which reduces Trimetilamine Oxide (TMAO) to TMA and produces hydrogen sulfide (H_2S). For packaged cod stored under vacuum and MA the key microorganism is *P. phosphoreum* whose growth is directly related to the shelf life of packaged fresh cod (Dalgaard 2000, Sivertsvik 2007). Shelf life of cod under MA conditions has been reported from 10 to about 20 days at 0°C–3°C and around 1 week at 4°C–5°C. The use of MAP during the frozen storage has been evaluated, but no positive effect was observed when it was evaluated against the application of MAP during the chilling period of thawed fish (at the retail display) since a significant drip loss occurred. However, when MAP was applied during chilling, quality parameters, namely *P. Phosphoreum* growth and trymethylamine production, were inhibited which supported the use of MAP after the fish is thawed (Bøknæs et al. 2001). Fresh garfish fillets were maintained in good conditions under MAP for 20 days at 0°C (20 days in air) and 9 days at 5°C under air and MAP and when a previous frozen storage was applied, shelf life reached 17 days, contrasted with 10 days under air (Dalgaard 2000). Similar results to those for cod were observed for hub Mackerel (*Scomber japonicus*) at 20–21 days of shelf life under MAP (70% CO_2 and 30% N_2) due to lower rate of fish spoilage manifested by lower TBA values, drip loss, TMAN values, and odor score higher than the lower acceptability limit of 6 after 20–21 days (Goulas and Kontominas 2007). Garfish is a fish that has been identified to cause histamine fish poisoning (HFP), at concentrations between 750 and 1200 ppm of histamine, so a procedure to avoid this problem is to package it after frozen storage under modified atmospheres (40% CO_2 and 60% N_2) since freezing storage causes inhibition or *P. phosphoreum* responsible for histamine formation (Dalgaard 2000).

MAP in shrimp was effective at 2°C, inhibiting *L. monocytogenes* growth, but not at 5°C–8°C under 50% CO_2, 30% N_2, and 20% O_2 (Mejlholm et al. 2005), which gave a sensory shelf life of 25–26 days. Cold-smoked salmon under VP and MAP (60% CO_2/40% N_2) presented a shelf life of 4 and 5 and 1/2 weeks before it presented detriment in sensory characteristics such as sourness, rancidity, bitterness, and soft texture (Paludan-Müller 1998). Regarding safety, growth of *L. monocytogenes* can be prevented by use of diacetate and MAP to comply with the EC Regulation, which establishes 100 cfu/g limits during the period of shelf life (EC2073/2005), in the classification of ready-to-eat foods. In spite of MAP application, temperature continues to be the most important factor in maintaining quality of salmon as evidenced by data gathered by Sivertsvik (2003), establishing a shelf life of 10, 17, and >24 days at temperatures 4.4°C, 2.0°C, and 1.0° C. Such values contrasted with those attained when salmon was stored under air, with shelf lives of 7, 11, and 21 days, respectively (Randell et al. 1999, Sivertsvik et al. 2003). Temperature is the single most important factor influencing shelf life of fresh and lightly preserved seafood (Dalgaard 2000). Fish can get benefit from modified atmosphere when stored under bulk using controlled atmosphere storage. Hake fish extended its shelf life to 24 days when it was stored under controlled atmosphere storage for 12 days (60% CO_2/15% O_2/25% N_2), followed by in the case of packaging retail using MAP with the same atmosphere, showing values for TVBN and TMA-N of 40 and 12 mg/100 g, respectively (Ruiz-Capillas et al. 2001). The use of near-infrared (NIR) spectroscopy has been proposed as a means to evaluate fish quality under MAP, as is reported by Bøknæs et al. (2002), who evaluated quality attributes, such as drip loss, water holding capacity, and dimethylamine content. Process parameters were also analyzed through NIR spectroscopy, such as frozen storage temperature, frozen storage period, and chill storage period of thawed–chilled MAP on Barents Sea cod fillets. The atmosphere evaluated for cod was composed of 40% CO_2, 40% N_2, and 20% O_2.

10.2.3.1 Pathogens in Fish and MAP

Different studies have reported that storage of products under MA may not increase the risks above those presented at storage under air, from *Salmonella*, *Staphylococcus*, *C. perfringens*, *Yersinia*, *Campilobacter*, *Vibrio parahaemolyticus*, and *Enterococcus* (Silliker and Wolfe 1980, Reddy et al. 1996). However, *C. botulinum* type E, nonproteolytic type B and *Listeria monocytogenes*, are the pathogens that remain under the scrutiny in packed fish under anaerobic conditions (Sivertsvik et al. 2002). The most important factor to control *C. botulinum* is temperature and under abuse conditions most fish, despite packaging method, can become toxic (Sivertsvik et al. 2002). *Listeria monocytogens* growth in ready-to-eat shrimp is inhibited by 100% CO_2 atmospheres at 3°C, while other methods, such as air and vacuum, allow it to grow. In addition, CO_2 atmosphere is also effective to retard growth of psychrotophic bacteria resulting in better scores for odor and appearance after 15 days of storage (Rutherford et al. 2007).

10.2.4 FRUITS AND VEGETABLES

MAP is a common technology in postharvest handling of fruits and vegetables. The application of MAP goes back to the pioneer work of modified atmospheres on apples and pears in New York State that gave rise to controlled atmosphere storage (CAS) technology. MAP uses in horticultural products (whole) have been restricted to few commodities due to unavailability of appropriate films that can provide safe modified atmospheres (MAs) at different temperatures, higher cost of MAP technology (film cost and packaging line modification) that will impact profit margins negatively, and maintenance of package integrity during marketing operations (Kader and Watkins 2000). However, for fresh cut products (minimally processed products), storage time is shorter, temperature control is stringent, and the products represent a higher value which offers a potential for MAP application. According to the International Fresh-cut Produce Association (IFPA), fresh-cut products are fruits and vegetables that have been trimmed and/or peeled and/or cut into 100% usable product that is bagged or prepackaged to offer consumers high nutrition, convenience, and flavor while still maintaining its freshness (Lamikanra 2002). Fruits and vegetables are complex systems characterized by respiration (product and temperature dependent), different storage temperatures, water absorption, and short shelf life, so its physiology is different from other food products since, essentially, they are living tissues until they are processed or consumed. The benefits that the use of MAs conveys to these products can be summarized in lowering RR, delaying senescence (a natural form of deterioration), inhibiting ethylene production, greater ascorbic acid retention, better pigment retention and texture (less softening and lignification), limited growth of spoilage organisms, and maintenance of sensory attributes. The use of MAP to horticultural products brings additional benefits, which include excellent branding options and product differentiation channel, waste reduction throughout distribution, semicentralized manufacturing options, reduction of labor and waste at retail level, expanded radii of distribution systems, and quality advantages transferred to the consumer (Day 1991). Usually, the atmospheres used in fruits and vegetables involve low levels of O_2 (1%–10%) and higher levels of CO_2 (20%). CO_2 is the only gas with a direct antimicrobial effect and a recommendation regarding O_2 levels, for both safety and quality, falls between 1% and 5% (Farber et al. 2003). Under modified atmosphere storage, the concentration of O_2, CO_2, and C_2H_4 within the plant tissue determines the physiological or biochemical response of that tissue (Rai et al. 2002). As previously mentioned, the modification of the atmosphere can be reached actively (the package is flushed with a gas mixture before it is closed) or passively (package is sealed under normal air conditions).

When atmosphere modification is chosen as the preservation method for fruits and vegetables, considerations that should be taken into account include

- Lowest O_2 level to reduce the RR and ripening or senesce but safe enough to prevent low O_2 injury
- Highest safe CO_2 level to attenuate the RR and ripening and to prevent high CO_2 injury
- Highest relative humidity (RH) levels to reduce moisture loss and preserve freshness
- Lowest temperature to reduce RR but safe enough to prevent chilling injury
- Lowest ethylene levels to suppress ripening

For the specific case of fruits of vegetables, besides the additional shelf life, the benefits of film packaging to create MA conditions on fruits and vegetables can include (Kader and Watkins 2000)

- Maintenance of high relative humidity and reduction of water loss
- Improving of sanitation, by reducing contamination during handling
- Minimization of surface abrasions by avoiding contact between the commodity and the material of the shipping container
- Reducing of spread of decay from one unit to another
- Possibility of exclusion of light, when needed
- Using of the films as carriers of fungicides, sprout inhibitors, scald inhibitors, or other chemicals
- Ease of brand identification

Overall, MA packaged fruits and vegetables offer quality advantages such as color, moisture, flavor and maturity retention, and less artificial preservatives.

To develop MAP systems for fruits and vegetables, it is essential to have basic knowledge on film characteristics as well as produce characteristics since there are many variables and elements that affect the atmosphere modification. MAP technology allows the interaction between the physiological parameters of the commodity and the film characteristics when four key processes are taking place simultaneously: respiration of the produce, transpiration of the produce, permeation of gases through the packaging material, and heat transfer (Al-Ati and Hotchkiss 2002). These processes are affected by temperature and gas concentrations, mainly, although physiological parameters such as maturity, permeability properties such as film thickness, chemical composition, also affect the final atmosphere reached inside a package.

For a successful MAP design it is necessary to know RR and respiratory quotient (RCO_2/RO_2) of the produce, film barrier properties suitable for respiration rate, film area, produce weight, and headspace volume (Al-Ati and Hotchkiss 2002). The safety of MAP in fresh produce has been under scrutiny for long time. Numerous publications have revised and evaluated the effect of modified atmospheres on the microbial growth of pathogens (Hintlian and Hotchkiss 1986, Farber 1991, Phillips 1996, Farber et al. 2003) and concluded that MAP is still a technology vulnerable for pathogen growth.

10.2.4.1 MAP in Fresh Whole Products

Bananas are a commodity where the use of modified atmosphere for marine shipment is already a reality. Different studies have demonstrated the usefulness of MA for long-term storage of bananas (Hewage et al. 1995, Ketsa 2000). Other factors that shorten shelf life of bananas are senescent spotting that can be alleviated using MA even at higher temperatures such as 12°C–18°C or with a heat treatment of 42°C, since aging and senesce spotting have been established that depend on cultivar and oxygen level. Shelf life could be extended up to 28 and 42 days, respectively, by delaying compositional changes in texture, pulp to peel ratio, and total soluble solids (Kudachikar et al. 2007). For a nonhead Chinese cabbage, Bok Choy, weight loss, and chlorophyll retention had higher values under MAP, with a shelf life determined by sensorial evaluation (overall preference) of 10 days, compared to 4 and 4 days for polyethylene (PE) and bianially oriented polypropylene (BOPP) (Lu 2007). Bartlett pears, in studies carried out by Drake et al. (2004), could be stored under MAP and then transferred to regular storage in very good conditions. MAP has been useful for alleviating chilling injury of fruits, a common physiological disorder presented when susceptible fruits and vegetables are stored at refrigeration temperatures. Chilling injury symptoms were prevented by using MA packaging of "Solo" papaya packaged under low density polyethylene (LDPE) or Pebax-C film, at 7°C and 13°C for 30 days, but at 7°C, the fruit failed to ripen normally and show skin scald and internal injury, so the optimal temperature was 13°C, coupled with MAP that maintained acceptable levels of dietary antioxidants such as ascorbic acid, total carotenoids, and lycopene (Sing and Rao 2005). A similar effect has been reported for citrus fruit, where shelf life is limited by rind disorders (rind breakdown, stem-end rind breakdown [SERB] and shriveling and collapse of the button tissue) and is also susceptible to chilling injury. Microperforated films that create a modified atmosphere with elevated CO_2 and lowered O_2 levels (2%–3% CO_2 and 17%–18% O_2) were effective in reducing rind disorders (Porat et al. 2004).

Few studies have been conducted on MAP of table grapes. MAP combined with natural fungicides is a good alternative to limit decay of table grapes, since they are very sensitive to water loss and fungal infection (mainly by *Botrytis cinerea*). SO_2 or (E)-2-hexenal, natural fungicide, were evaluated in "Superior seedless" table grapes, in combination with passive MAP obtained by using two polypropylene (PP) films. Treatments were stored for 7 days at 0°C followed by 4 days at 8°C and 2 days at 20°C. Control clusters showed the highest weight losses and decay while almost no losses occurred under MAP treatments. After shelf life, only a slight decrease on total polyphenolics was noticeable for all treatments (Artés-Hernández et al. 2006).

Studies show that MAP can enhance the effect of other treatments for postharvest decay in fruits and vegetables. This has been reported for peaches where three treatments alone, and in combination, were evaluated for their potential to control postharvest diseases: hot water, yeast antagonist (*Candida oleophila* I-182), and MAP. The combination of yeast antagonist and MAP had higher effectiveness than yeast antagonist or MAP alone. Previous studies reported that MAP combined with cold storage enhanced the control activity of yeast antagonists, due to elevated levels of CO_2 that can suppress pathogen growth and stimulate yeast growth, environment where yeast antagonist can compete advantageously with pathogens under those conditions, since its mode of action involves competition with the pathogen for the nutrients and space (Karabulut and Baykal 2004). MAP combined with the application of citokinin was used in green asparagus. A dipping treatment in 20 ppm 6-benzylaminopurine (6-BA) for 10 min in combination with a passive MA developed using LDPE gags or an active MA with a gas mixture of 10 kPa O_2 + 5 kPa CO_2 was flushed and stored for 24 days at 2°C. The combination of 6-BA with MAP was effective in delaying RR and maintain a better color, firmness, and overall appearance, besides higher retention chlorophyll and ascorbic acid and less fiber (Zhang et al. 2008).

10.2.4.2 Fresh Cut Products

The greater expansion on the application of MAP came in the form of MAP of lettuce and salad preparation rather than for fruit-ripening control. In 2003, in the United States, total sales for lightly processed (fresh cut) produce increased from near 0 to 3.9 billion (Beaudry 2007) and by 2005, the sale value for just fresh-cut salads was estimated around $3 billion in the United States (Beaudry 2007), most of which employ modified atmospheres, so MAP is crucial for this kind of products. Quality attributes that define the acceptability of minimally processed fruits and vegetables are visual appearance, texture, flavor, nutritional value, and safety (Lin and Zhao 2007). MAP has proved to be a valuable tool to preserve these products from quality losses and excellent overviews have documented that information (Day 1991, Raj et al. 2002, Jayas and Jeyamkondan 2002, Farber et al. 2003, Soliva-Fortuny and Martin-Belloso 2003, Rico et al. 2007).

The physiology of fresh cut products is that of a wounded tissue and the response of each fruit or vegetable is dependent on variety, surrounding gas concentrations, water vapor pressure, and any other inhibitor present (Lin and Zhao 2007). The limiting factors to shelf life of these products are browning and tissue softening, so special attention should be given to those in order to extend shelf life. These products are characterized by an increase in RR which brings a very perishable product with shelf life of days, so the use of MAP is aimed, mainly, at extension of shelf life. The mechanical operations to which fresh cut products and other minimal processing procedures are subjected can cause a wide spectrum of responses of the wounded tissue: ethylene production, increase in respiration, membrane deterioration, water loss, susceptibility to microbiological spoilage, loss of chlorophyll, formation of pigments, loss of acidity, increase in sweetness, formation of flavor volatiles, tissue softening, enzymatic browning, and lipolysis and lipid oxidation (Rico et al. 2007). An issue of concern with MAP in fruits and vegetables is the microbiological safety of these products, since it has been demonstrated that MAP produce is vulnerable to atmospheres with low O_2 levels that inhibit the growth of most aerobic spoilage organisms (Farber et al. 2003, Soliva-Fortuny and Martin-Belloso 2003). The potential risk is represented by psychrotropic foodborne pathogens such as *L. monocytogenes*, *Yersinia enterolitica*, *Aeromonas hydrophila*, nonproteolytic *C. botulinum*, *Salmonella* spp., *E. coli* O157:H7 and *Shigella* spp. Products characteristics and a strict low temperature control are the key elements for successful and microbiological safety of MAP produce.

MAP has been used as a barrier to preserve the initial color of fresh cut products. Enzymatic browning in fruits causes considerable losses and the intensity of browning is influenced by the amount of active forms of the enzyme and by the phenolic content in the fruit tissue. Enzymatic reactions are catalyzed by polyphenol oxidases (PPO). Research reports have concluded that MAP alone cannot prevent effectively browning of fresh cut products, so it has been combined with other substances or techniques, such as dips using ascorbic acid, 4-hexylresorcinol (4-HR), or cysteine. The rate of softening of fresh cut fruits has been affected by the atmosphere packaging composition, as in the case of Golden Delicious Apples and Conference Pears (Soliva-Fortuny and Martin-Belloso 2003). Ready-to-eat carrots (Ayhan et al. 2007), celery sticks (Gomez and Artes 2005), and asparagus (Villanueva et al. 2005) are fresh cut products benefited from MAP. Specifically for these products, minimally processed fruits and vegetables or ready-to-eat products, the application of a multiapproach technology or hurdle effect is very effective, given their perishability. On this basis, MAP in combination with other technologies has been studied for a great number of products. Ozone and MAP for asparagus were effective in delaying the rate of increase of cell wall constituents (An et al. 2006). Hot water treatments with MAP were applied with peaches and nectarines, and the results showed that the combination of both treatments could maintain good quality fruit postharvest handling, for just 1 week (Malakou and Nanos 2005), and a mild heat pretreatment (MHOT) was applied on peach to induce a firming effect and stored under passive and active MAP (Steiner et al. 2006).

10.3 MAP COMBINED WITH OTHER TECHNOLOGIES: MULTIAPPROACH STRATEGY TO PROLONG SHELF LIFE

Hurdle technology is the combination of different preservation techniques as a preservation strategy, being the most commonly used hurdles: temperature control, water activity, acidity, redox potencial, use of preservatives, MAP, and competitive microorganisms (Rico et al. 2007).

10.3.1 ACTIVE PACKAGING

There has been an attempt to break with the traditional MAP based on use of gas mixtures and films, by an alternative incorporating intrinsic and extrinsic elements to assure food safety proactively. This approach is denominated "Novel MAP," and the intrinsic elements are intelligent or active packaging, while the extrinsic elements refer to the processing environment (Yuan 2003). Intelligent or active packaging technologies are denominated "Interactive packaging," a concept used to describe those technologies incorporated into the package or the packaging film to maintain or monitor the quality or safety of the product.

The mechanisms of interaction include

- Interaction with the food product (e.g., antimicrobial effects)
- Alteration of the packaging atmosphere (scavengers, moisture control and freshness indicator, time temperature indicators, antimicrobial films, edible coatings and films, aroma enhancements)
- Response to environmental changes
- Communication with consumers

In terms of the extrinsic elements, quality of the product and the storage temperature are considered the most important. Active packaging refers to those technologies intended to interact with the internal gas environment and/or directly with the product, in order to get a benefit in quality or shelf life. A modification of the gas environment occurs by removing (absorb or scavenge) gases from or adding gases or vapors (emit) to the package. Some of the additives used include oxygen absorbents, carbon dioxide scavengers or emitters, humidity absorbers or controllers, enzymatically active systems, and antimicrobial agents (Blakistone 1999, Kader 2003, Lopez-Rubio et al. 2004). There are numerous studies carried out in active packaging. For meats, Perry et al. (2006) presented an overview on the utilization of active and intelligent packaging systems for meat and muscle-based products. Gill (2003) also presented some of the applications of MAP Technology on meats. The main route of gas exchange in MAP is through gas permeable films, perforations in films, or both. When referring to active or intelligent packaging, gas can be modified by flushing it with a specific mixture to get a precise initial atmosphere composition. Also, gases may be actively released or scavenged in the package, a partial vacuum can be imposed, biological active materials can be incorporated in the package, or sensors may be used to respond to the product or package conditions (Beaudry 2007). A description of the additives used in active packaging for a variety of foods is presented by Lopez-Rubio et al. (2005).

10.3.2 ANTIMICROBIAL COMPOUNDS

Natural antimicrobial compounds, such as spices, have been evaluated to act in conjunction with MAP. Eugenol, thymol, and carvacrol as antimicrobial agents, in combination with MAP, have been added to packages containing table grapes. The essential oils did not modify the gaseous atmosphere, but decreased, drastically, microbial counts (molds and yeasts and mesophiles) and extending shelf life considerably. The mechanism responsible for these results could be the well-known antioxidant activity reported for these compounds that could delay the common oxidative processes involved in fruit ripening and senescence (Guillén et al. 2005). A previous study applied eugenol, menthol, or thymol combined with MAP on grapes (Valverde et al. 2005). Some studies did the same on beef patties adding rosemary extract and ascorbate/citrate. A similar approach was studied by addition to packages with either 100% N_2 or 80% O_2/20% N_2. Inhibition of lipid oxidation but not protein oxidation occurred for both antioxidants, while an antioxidant effect on meat color was found for both substances at high O_2 atmospheres (Lund et al. 2007).

MAP combined with lauric acid to extend shelf life of minced cod of shrimp tails (MAP:50% CO_2:50% N_2) produced higher scores on sensory properties and acceptability commercially after 1 month of storage (Pastoriza et al. 2004). MAP combined with oregano can give additional shelf life, as in the case of sea bream (*Sparus aurata*) that was stored under MAP (40% CO_2/30% O_2/30% N_2) and oregano essential oil under refrigerated conditions. Salted fish had acceptable sensory scores during the first 15–16 days of storage, while MAP provided 27–28 days of storage, but MAP with oregano essential oil maintained sensory acceptance after 33 days of storage. Biochemical parameters evaluated were TVBN and TMAN (Goulas and Kontominas 2005). Rosemary and ascorbate/citrate (1:1) in combination with MAP demonstrated their efficiency to inhibit lipid oxidation but not protein oxidation in minced beef patties (Lund et al. 2007). Some atmospheres exert an additional protection when they are combined with antioxidants on functional ingredients, such as the case where MAP and VP prevented lipid oxidation in dry fermented sausages added with linseed oil an antioxidant helping to maintain nutritional benefits after 5 months of storage (Valencia et al. 2006). Some atmospheres exert an additional protection with which they are combined with antioxidants or functional ingredients such as linseed oil and antioxidants to dry fermented sausages and packaged under aerobic, vacuum and modified atmospheres and stored during 5 months. MAP and vacuum had a protective effect on the nutritional characteristics of linseed oil and antioxidants, and lipid oxidation was undetectable by thiobarbituric acid reactive substances (TBARs) and peroxides assays (Valencia et al. 2006). Broccoli packaged in LDPE with an ethylene absorber gave the highest shelf life in terms of chlorophyll retention, weight loss, and texture (Jacobson et al. 2004).

10.3.3 MAP and Other Antimicrobial Factors

Ultraviolet type C (UV-C) treatments have been used in combination with MAP. Lettuce was packed in individual bags of bioriented polypropylene (PPB), sealed, and placed at 5°C during 10 days. UV-C was effective in reducing growth of psychrotrophic and coliform bacteria as well as yeast growth, but LAB growth was not affected or even stimulated and sensory quality was not adversely affected (Allende and Artes 2003). Meat products have been added with lactate and diacetate to control spoilage organisms and pathogens such as *S. typhimurium* and *L. monocytogens*, being a common practice in the meat industry in products such as moisture enhanced pork and processed ready-to-eat products such as frankfurters and hams. Given the possibility of pathogen growth under MAP, studies conducted combined lactate and diacetate with MAP to enhance the effectiveness of either treatment alone (100% CO_2 for sliced ham and 99.5% CO_2 plus 0.5 CO for porch chops) and vacuum. Results showed that high CO_2 concentrations or MAP and antimicrobials are effective to inhibit *S. typhimurium* and *L. monocytogenes* but the combination did not act synergically on the action of either treatment (Michaelsen et al. 2006). MAP and VP have been combined with films incorporated with natural antioxidant (alpha-tocopherol) or synthetic antioxidant, in order to decrease oxidative rancidity on mechanically deboned turkey meat (MDTM), where the lowest values corresponded to meat packaged under these treatments (Pettersen et al. 2004). MAP in combination with lactic acid strains can inhibit spoilage microorganisms growth as in the case of frankfurter-type sausages and sliced cooked cured pork, inoculated with *Leuconostoc mesenteroides* L124 and *Lactobacillus curvatus* L442, used as antagonistic cultures and their bacteriocins (Metaxopoulos et al. 2002). MAP combined with superchilling, called "partial freezing," can give positive results on shelf life extension, typical of 7 days, and it consists in lowering the temperature to 1°C–2°C below the initial freezing point so some ice is formed inside the product. Salmon (bulk whole and fillets) maintained high sensory characteristics for 21 days in 90% CO_2 atmosphere at 0°C (Sivertsvik et al. 2003). In addition, the combination of MAP with additives (citric acid and potassium sorbate) has been used to decrease the production of 3-methyl-1-butanol (microbial spoilage index) in ready-to-eat desalted cod (Fernández-Segovia et al. 2006).

10.4 EDIBLE FILMS AND MAP

Edible coatings are another alternative for atmosphere modification in fruits and vegetables, given their potential activity to regulate transfer of moisture, oxygen, carbon dioxide, aroma, and taste compounds in a food system in order to improve quality and extend shelf life of fresh produce (Lin and Zhao 2007). Edible films can act as moisture barriers (alleviate moisture loss), gas barriers (decrease respiration and delay deterioration, retard enzymatic oxidation and browning discoloration, and texture softening), controllers on the exchange of volatile compounds (prevents loss natural volatile flavor and color components), and carriers of functional ingredients (antimicrobial, antioxidants, nutraceuticals, color and flavor ingredients) that could reduce microbial loads, delaying oxidation and discoloration and improving quality (Lin and Zhao 2007). The use of edible coatings in fruits and vegetables provides benefits such as moisture loss reduction, restriction of oxygen entrance, decrease of respiration and ethylene production, seal in flavor volatiles and through the incorporation of additives, retardation of discoloration and microbial growth (Ahvenainen 1996), which in turn will cause stability, quality, functionality, and safety.

In order for an edible coating to meet the needs for preservation of fruits and vegetables for a specific produce application, three aspects should be considered: barrier property to moisture, oxygen and carbon dioxide, produce characteristics, surrounding atmosphere and temperature. The functionality of the coating will depend on the material used (chemical composition) and its shape. Materials for edible coatings are based on their physical and chemical characteristics, which will determine its application. Basically, proteins, lipids, carbohydrates, and resins are the basic materials for coating formation and in a similar manner to the synthetic film formation, the addition of plasticizers, cross-linking agents, antimicrobial, antioxidants, and texture agents is needed to produce a wide range of edible coatings. Polyethylene glycol, shellac, and wood rosin are common materials for laminates and emulsions. The expected performance of an edible coating for fruits and vegetables depends on its functional properties: permeation, absorption and diffusion into water, oxygen, and carbon dioxide. Physical properties important for material selection include water solubility, hydrophilic and hydrophobic nature, ease of formation on coatings, and sensory properties. In order to fulfill the requirements for shelf life extension of produce, edible coatings create a modified atmosphere (or controlled atmosphere) inside the fruits and vegetables that delay ripening and senesce likewise as the use of modified atmosphere storage. The use of only one edible material is insufficient to meet the specific conditions for each fruit and vegetable, so the combination of more than one material is the common practice for producing edible coatings (Farber et al. 2003). An overview of the materials for edible coatings for fruits and vegetables can be revised for more detailed information (Lin and Zhao 2007).

Recent examples of the application of edible films to modify atmosphere applications can be mentioned. Chitosan films were compared with low-density polyethylene (LDPE) in the packaging of tomato (*Licopersicon esculentum*) and bell pepper (*Capsicum annuum*). Chitosan packages seemed to maintain quality desirable attributes of the fruits, after 30 days of storage. Oxygen concentrations were in the range of 1%–2%, while CO_2 was in average 3%. CO_2 concentrations preserve the green color but also produced calyx decoloration. Profile components analysis (PCA) for sensory cores showed variance in firmness, surface gloss, sharpness of edge, and green colors (Srinivasa et al. 2006). Quitosan was combined with $CaCl_2$ to coat mushrooms (whole and sliced) and package them with PVC wraps or two polyolefins films. The modified atmosphere generated inside the package was most important for quality of whole and sliced mushrooms than the coating treatments (Kim et al. 2006). A coating based on calcium caseinate and whey protein isolates was used to cover peeled minicarrots irradiated at 0.5 or 1 kGy and packaged under air or MAP to be stored at 4°C for 21 days. The coating protected carrots from dehydration during storage in air and in conjunction with irradiation at 1 kGy maintained firmness of carrots at the same condition. MA was beneficial in retarding whiteness of uncoated carrots but was not effective in preventing loss of firmness or inhibiting, significantly, microbial growth. As a follow up to this study, another investigation in minicarrots was aimed at following the effect of antimicrobial treatments, applied *trans*-cinnamaldehyde, and irradiation with a coating of calcium caseinate and combined with MAP. The best combination treatment was MA with irradiation at 0.5 kGy independently if the product was coated or uncoated, during the 21 days of storage (Caillet et al. 2006). Polylactic acid (PLA) with higher water vapor permeability can be used to maintain the quality and sanitary conditions of freshly harvested green peppers in MAP at 10°C (Koide and Shi 2007). Edible coatings for a large number of foods have been developed and studied and recent publications offer very valuable information on all those developments (Cagri et al. 2004, Cha and Chinnan 2004, Cutter 2006, Lin and Zhao 2007).

10.5 RISKS AND NEGATIVE EFFECTS OF MAP

The efficiency of MAP to extend shelf life of different food products has been well documented. As already described the improvement on quality is well established. However, an issue of concern is related to the risks associated to the inhibition of spoilage microorganisms that allow consumers to reject spoiled foods, since MAP may allow growth of psychrotropic pathogenic bacteria such as *Clostridium botulinum* and *L. monocytogenes.* A study conducted in the United Kingdom, during September to November of 2003, analyzed 2981 VP-MAP cooked ready-to-eat meat samples. From these, 66% had satisfactory or acceptable microbiological quality and 33% were unacceptable from the microbial point of view, containing *L. monocytogenes* at 100 cfu/g or higher, and *Campylobacter jejuni*. Regarding other microorganisms, acceptable levels ($<10^2$ cfu/g) or satisfactory (<20 cfu/g) for *Staphylococcus aureus* and *Clostridium perfringens* were maintained until the end of shelf life. Chicken, beef, and turkey had the higher levels of unsatisfactory or unacceptable quality than ham or pork (Sagoo et al. 2007). MAP promotes lipid oxidation, especially from polyunsaturated fatty acids present in membranes giving origin to undesirable off-flavors and odors. Usually, lipid oxidation is coupled with discoloration indicating less oxidative stability. Light is another pro-oxidant, so some natural antioxidants such as rosemary and ascorbic acid have proved effective on pork sausages in retarding discoloration and reaching shelf life of 16 days equaling the shelf life of product maintained darkness (Martínez et al. 2007). For some products VP exerts a deleterious effect on sensory properties such as color intensity and flavor, as reported on unpacked ripened sausages that scored higher by panelists on red intensity and global flavor, compared to vacuum-packed ones. Also a greater hydrolytic degradation of their lipid fraction was detected in samples under vacuum contrasted with those unpacked during 40 days (Summo et al. 2006).

Sometimes, vacuum is more adequate for some products such as the case of dry-cured ham, which packaged under vacuum retained quality attributes over 4 months of storage, with only a slight loss of flavors. On the other hand, the equivalent MAP-treated product presented significant decrease in quality aspects, specifically in rancidity, flavor loss, off-flavor formation, visual appearance, and saltiness. Those changes were attributed to the low pH caused by the CO_2 present (Cilla et al. 2005).

10.6 NOVEL MAP-BASED TECHNOLOGIES

10.6.1 SOLUBLE GAS STABILIZATION AND DIRECT CO_2 INJECTION

The effectiveness of MA packaging is limited by the amount of available CO_2 to dissolve into the food which depends on the partial pressure of the gas inside the package and the gas to product volume (g/p) ratio (Sivertsvik and Birkeland 2006). Additionally, CO_2 dissolves into the aqueous part of the product in MA packaging resulting in volume contraction of a flexible package. The package collapse can be ameliorated by introducing gases with less solubility, such as N_2 or O_2 which decreases CO_2 partial pressure; however, this is not the optimal solution for all packaged foods. For MA packaging of seafood, a g/p ratio between 2 and 3 is recommended to ensure bacteriostatic CO_2 availability and to prevent package collapse (Sivertsvik and Birkeland 2006). Norwegian researchers proposed an alternative through a technique called soluble gas stabilization (SGS) aimed toward using less gas volumes that involve CO_2 dissolution into the product prior to retail packaging, since solubility of CO_2 increases at lower temperatures and at higher partial and total pressures. A sufficient amount of CO_2 can be dissolved into the product for 1 to 2 h in pure CO_2 with the consequent package size reduction and increase in shelf life compared to MAP alone. This technique has the advantage of decreasing package collapse, even at high degree of filling (DF) (volume product vs. volume of package) and also reducing packaging volume by 30%–40%. The technology offers similar or even increased food safety, as the traditional MA packaging.

Fresh Atlantic Salmon (*Salmo salar*) fillets, treated with SGS, have shown promising results in extending shelf life (Sivertsvik et al. 2000, 2003), and also sliced meat products (cold cuts) have been reported as potential for SGS treatments (Sivertsvick and Birkeland 2006). Dairy products, such as cottage cheese and raw milk have also been improved their shelf lives through the direct addition of CO_2 prior to packaging. Hotchkiss et al. (2006) reported experiments, conducted at a high scale, in milk stored in stainless steel liquid bulk tanks for rail shipment. The effectiveness of CO_2 was again confirmed (at 45 mM CO_2 under pressure of 138–345 kPa), reaching a shelf life of 14 days, 4 days longer, than control milk. Chicken breast fillets treated with SGS resulted in a shelf life of 24 days compared with those samples stored simply in air that spoiled after 5 days. The SGS improve microbiological and sensorial characteristics and, at the same time, reduced package collapse (Rotabakk et al. 2006). Shrimp has also benefited from SGS since the sensory qualities or RTE shrimp were enhanced while exudates were reduced when SGS treatment was applied (Sivertsvik and Birkeland 2006). Atlantic halibut showed a shelf life of 28 days based on off-odor analysis when it was subjected to SGS, at two different times (1 and 2 h) and two CO_2 levels (200 and 400 kPa) (Rotabakk et al. 2008).

10.6.2 NOVEL GASES AND MAP

Other nonconventional gases such as noble gases have been evaluated for MAP for their potential benefits. Previous studies reported that argon can inhibit certain microorganisms, suppress enzymatic activities, and control degrading chemical reactions in some perishable food products (Spencer 1995). The mechanism of action suggested is that argon is biochemically active, probably due to their enhanced solubility in water compared to nitrogen and possible interference with enzymatic receptors sites, reducing the RR. Nitrous oxide (N_2O) treatment has shown ripening inhibition effect by extending the lag phase, which precedes the ethylene rise and delayed color change in preclimacteric fruits of tomatoes and avocados and also inhibits ethylene action and synthesis in high plants (Rocculi et al. 2004). Minimally processed apples packed in different atmospheres containing argon and N_2O presented some beneficial effects during the 10 day period of storage compared to a control. Also a nonsulfite dipping for the enzymatic browning inhibition was used. More recently, argon and xenon were evaluated in asparagus. Noble gases form clathrate hydrates when dissolved in water and restrict water molecule, an activity that can prolong the shelf life of fruits and vegetables. MAP was successful in limiting weight loss of asparagus, since there was practically null variation (Zhang et al. 2008). The effect of noble gases on the anaerobic metabolism and quality attributes of mushroom and sliced apples had negative results, showing no difference between treatments with helium, argon, neon and nitrous oxide, compared to nitrogen use (Ozdemir et al. 2004).

Nitrous oxide seems to bind lipids and also proteins such as cytochrome C (Gouble et al. 1995, Day 1996), but there is scarce information on its effects on MAP fruit. Since kiwi fruit rapidly deteriorates due to ethylene action and enzymatic activities, it was considered a good candidate for noble gas application and storage at 4°C for 12 days. N_2O was the most effective atmosphere in terms of firmness, color retention, soluble solid content, lower respiratory activity, and less browning in the pericarp and core surfaces. The treatment showed that N_2O maintained discriminatory quality parameters for N_2O sample versus a quality loss for samples in air and in N_2, and acceptable quality for 8 days for sample under argon (Rocculi et al. 2005).

10.6.3 IRRADIATION

Irradiation in combination with MAP can have a synergistic effect on microbial load as the case with minimally processed Chinese cabbage (*Brassica rapa* L.) that once prepared was packaged under MA (air, 99.9% CO_2, and 25% CO_2 + 75% N_2), stored at 4°C, and irradiated (0, 0.5, 1%, and 2 kGy). Quality attributes could be maintained for 3 weeks and irradiation at 1 kGy or above was useful to enhance microbial safety (Ahn et al. 2005).

10.6.4 NEUTRAL ELECTROLYZED OXIDIZING WATER

A multiapproach strategy to prolong shelf life using decontamination treatments, low temperature storage, and MAP can be achieved by employing neutral electrolyzed oxidizing water (NEW). This practice constitutes a novel decontamination method, which posses increased levels of free chlorine and some newly formed compounds with antimicrobial effects, since it is generated in an electrolytic cell from tap water or diluted sodium chloride solution pumped into it. NEW has been evaluated for shelf life studies with shredded cabbage, observing that the microbial population was acceptable after 14 days at 4°C and 8 days of storage at 7°C. The NEW treatment resulted in a shelf life extension or 5 and 3 days, at 4°C and 7°C, respectively (Gómez-López et al. 2007).

10.6.5 INTENSE LIGHT PULSES

Intense light pulses (ILP) have been studied in conjunction with MAP as a new method directed for decontamination of food surface, using short time high frequency pulses of an intense broad spectrum, rich in IV-C light. Since efficiency of ILP is decreased by the presence of proteins and fat, its application on vegetables seems to have potential. Cabbage and lettuce showed log reduction of 0.54 and 0.46 for aerobic psychrotrophic count (APC) after flashing with ILP. However, under the conditions of the study reported, it seemed that ILP treatment alone did not increase shelf life of cabbage and lettuce although there was a reduction of the initial microbial load (Gómez-López et al. 2005).

10.6.6 SUPERATMOSPHERIC OXYGEN ATMOSPHERES

Some fruits and vegetables have higher RRs, which are even higher when they are prepared as minimally processed, so they represent a different challenge. Thus, they require modified atmospheres with high oxygen concentrations and films and a viable alternative is to use microperforated films. However, there is a food safety concern regarding the possibility of pathogenic bacteria contamination of the product through the perforations. The use of superatmospheric O_2 as an effective treatment for inhibiting microbial growth and enzymatic discoloration has been therefore proposed, as well as preventing anaerobic fermentation reactions, undesirable odor and moisture losses (Day 1991, 1996). When CO_2 is added to superatmospheric O_2, the inhibitory effect is increased, but negative implications have also been reported. Baby spinach is a product where neither traditional MAP nor microperforated films meet the requirements for shelf life. Allende et al. (2004) conducted studies using high O_2 levels (80 and 100 kPa) in two types of films (barrier and permeable films) on baby spinach. High O_2 atmospheres cause a significant reduction in aerobic mesophilics growth and sensory quality when compared with samples stored under passive atmosphere. The super atmospheric O_2 treatment was advantageous for both passive MAP and perforated packages.

Other candidates for superatmospheric oxygen are fresh cut vegetables, where CO_2 levels around 20 kPa or above cannot be used as they cause physiological damage. An alternative treatment proposed by Allende et al. (2004) is to use a combined treatment with high O_2 levels and 10–20 kPa CO_2 that could provide adequate suppression of microbial growth and prolonged shelf life. This combined treatment was evaluated for fresh cut butterhead lettuce (*Lactuca sativa* L.) cv Zendria at 1°C, 5°C, and 9°C, under different controlled atmospheres. The best conditions to avoid fermentation were 80 kPa O_2 in combination with 10%–20% kPa CO_2 (Escalona et al. 2006).

Fresh cut melon is another product that has been tested with high O_2 atmospheres. A treatment using O_2 at 70 kPa prevented fermentation and significantly improved the quality of fresh cut melon (initial color and firmness), while preserving its microbiological stability (Oms-Oliu et al. 2008a). High O_2 atmospheres (70%, 80%, and 95%) compared to EMA (5% O_2) showed better results in inhibiting enzymatic browning and microbial reduction (yeast growth reduction) on three ready-to-eat vegetables (mushrooms slices, grated celeriac, and shredded chicory endive) (Jacxsens et al. 2001). Oms-Oliu et al. (2007) also reported positive effects on fresh cut melon var Piel de Sapo when low and high O_2 and CO_2 atmosphere were studied on the effect on polyphenols and vitamins by delaying deteriorative changes associated with those gas concentrations, by decreasing wounding stress and reduce deteriorative changes related to high peroxidase activity on tissue, since those compounds seem to react as substrates for this enzyme. Microbial stability of Fresh cut pears variety "Flor de Invierno" was assured through oxygen atmosphere although both high and low oxygen atmospheres were effective in reducing microbial growth. The drawback was that after 14 days of storage, browning appearance of the cut surfaces developed offodors (Oms-Oliu et al. 2008b).

10.7 MICROPERFORATED FILMS

Given the need for films with high rates of gas exchange, the use of microporous and perforated films has been generalized. The gas diffusion constant for a perforation is approximately 4–8 million times greater than that of an LDPE film, so just one perforation has the gas exchange capacity of a bigger area and so very small perforations are needed for most products (Ben-Yehoshua et al. 2005), and thus the use of microperforated films to achieve atmosphere modification is a good alternative, which gives high permeability while moisture loss is minimized. Perforations or pinholes on the produce package have been suggested and tried as tools to control the atmosphere and relative humidity of the packages by elevating exchange rates of gas and moisture across the film. The number, length, and cross-sectional area of perforations or pinholes have been reported to affect the attained MA and RH. The most important design criterion is relationship between O_2 and CO_2, so the desired permeability of micropores is achieved. Mathematical models for an oxygen and humidity change in perforated packages of fresh produce have been derived by Fishman et al. (1996) and for prediction of O_2 and CO_2 concentrations in microperforated packs of fresh produce by Renault et al. (1994) and Lee et al. (2000). Perforated films, macro- or microperforated, filled a need for high respiring products with additional benefits such as certainty of aerobic conditions, reduction of package condensation while maintaining a high CO_2 level with a low CO_2 level, increasing of water vapor transfer rate, and, consequently, reduction of spoilage.

10.8   MODELING OF MAP

The design of MAP is a difficult task, especially for fruits and vegetables. Advances in modeling of MAP have been focused mainly on fruits and vegetables, as well as seafood, although other foods have also been considered. MAP modeling has been, therefore, focused predominantly on these two food groups.

10.8.1   MATHEMATICAL MODELING ON FRUITS AND VEGETABLES

Numerous factors influence MAP design: respiration parameters, film characteristics (O_2, CO_2 and N_2 permeabilities, thickness, surface area, water vapor transmission rate), temperature, free volume inside the package, and product weight. Mathematical models are useful tools for defining the package requirements for MAP and the development of models for MAP has been very productive. Those models are based, mainly, in mass balances for O_2 and CO_2 to describe the interactions among the product respiration, the film permeability, and the environment. In many cases, it takes a long time to reach gas equilibrium, or transient period during which the product is exposed to unsuitable gas composition and thus preventing the positive effects of the steady state atmosphere (Charles et al. 2006). In a MAP system, gas exchange depends on food product type, freshness, microbial load, lipid content, package permeability, and temperature and storage time. Gas transfer in MAP systems has been described by different models (Fisman et al. 1995, Emond et al. 1998, Simpson et al. 2004). When running a MAP experiment, it is not practical to determine RR of the product at every time, due to time consuming factor and thus modeling of RR has been the interests and subject of numerous publications. A summary of the major models developed for respiration are presented by Fonseca et al. (2002) where their strong points and limitations are discussed. One of the most suitable models based on enzyme kinetics was proposed by Lee et al. (1991) for predicting RR of fresh produce as a function of O_2 and CO_2 concentrations. The model describes the dependence of respiration on O_2 as assumed to follow a Michaelis–Menten type equation and the effect of CO_2 is considered to be of uncompetitive inhibition. In different studies, experimental data have given good fitting to this proposed model and even, recently, one study carried out with mushrooms confirmed that RR of commodities stored under MAP can be predicted using the enzyme kinetics approach (Deepak and Shashi 2007). A similar finding is reported by Li and Zhang (2008) who studied MAP using film windows with different permeabilities on the storage of *Agrocybe Chaxingu Huang*, obtaining that a model based on enzyme kinetics was successful for respiration prediction (Li and Zhang 2008). Modeling for specific applications such as for microperforated or macroperforated films has been reported (Techavises and Hikida 2008). Some of them include O_2, CO_2, and N_2 exchanges in MAP while some others include atmospheric gas and water vapor exchanges. The model is composed of an empirical equation for effective permeability and is based on Fick's law for prediction of gases and water vapor exchange in MAP films with macroperforations. Another model on microperforated films for determination of gas transmission rates (O_2 and CO_2) agrees with the modified Fick's law (that considers the total diffusive pass length of a perforation as the sum of the perforation length and a correction term ε that was 0.56 day for O_2 and 0.46 day for CO_2). This model allows the estimation of the gas flow through the film by a simple measurement of the perforation size with an ocular microscope and can be used for a wide range of conditions, given the range of the microperforations' dimensions used, a feature that establishes the difference with other models (Gonzalez et al. 2008).

A model that predicts RR under different modified atmospheres including superatmospheric oxygen, different to the Michaelis–Menten type enzyme kinetics approach, is proposed for fresh cut melon "Piel de Sapo" variety, where experimental data showed that the best fit to concentrations inside the package was the Weibell model for O_2 concentration and the logistic model for CO_2 concentration. The respiratory activity of fresh cut melon stored under 70 kPa O_2 atmospheres was adequately described through CO_2 production rates (Oms-Oliu et al. 2008b). RR of bananas has been described by a model based on enzymatic kinetics with Arrhenius type temperature dependence which showed a closer agreement to other model based on linear regression (Bhande et al. 2008). Another example of the advances in modeling of MAP is the model proposed by Charles et al. (2006) consisting on modeling an active package with an oxygen absorber for respiring products that allows prediction of the gas exchange dynamics in an active MAP associated with an oxygen absorber. The model was based on the following parameters: vegetable respiration (tomato as a model), film permeability, and oxygen absorption kinetics of the absorber (Charles et al. 2006).

10.8.2 Mathematical Modeling of MAP with Seafood

MAP is a system suitable to apply mathematical modeling due to the control of the gas atmospheres. Global marketing of seafood and most food promotes the need for methods to evaluate and predict the effects of storage and distribution conditions on shelf life. According to Dalgaard (2000) three approaches have been followed to predict shelf life of seafood:

- Kinetic models for growth of specific spoilage organisms (SSOs) at different environmental conditions
- Empirical models for relative rates of spoilage
- Simple models for linear relations between indices of spoilage and remaining product shelf life at different temperatures as discussed below

Predictive microbiology is an approach to shelf life prediction and several microbial spoilage models have been developed (Dalgaard et al. 1997). Prediction of shelf life by microbial models is a complex task due to the changing nature of seafood as extrinsic and intrinsic parameters determine SSOs and spoilage reactions. An iterative model was developed where the effect of temperature and CO_2 on the maximum specific growth rate of *P. phosphoreum* was modeled by a square root equation and by a polynomial equation. Shelf life was estimated based on initial numbers of *P. phosphoreum*, product temperature profiles, and the level of CO_2 in the MA environment. The concept "predicted relative rates of spoilage" (PRS) has been defined as the keeping time at 0°C divided by the keeping time at a given temperature. Mathematical PRS models allow shelf life to be predicted at different storage temperature when shelf life has been determined at one temperature. It is important to mention that the effect of temperature on shelf life of foods differs among types of seafood, so this fact should be taken into account when PRS models are used for shelf life prediction (Dalgaard 2000). The effect of temperature on shelf life and *L. monocytogenes* growth of seafood under MAP has been predicted using a software called seafood spoilage and safety predictor (SSSP) developed by Dalgaard et al. (2002). Shelf life of fish is mainly determined by microbial activity, due to fish spoilage which results from off-odors and off-flavors for both packed and unpacked fish. Microbial growth models can be used to predict the effects of environmental conditions (storage temperature and package atmosphere) on the microbial activity in agreement to the effects of the same storage conditions on product shelf life (Corbo et al. 2005). However, many of these models have been developed using data from experiments conducted in liquid media, environment different to the conditions found in actual food which brings as a consequence that those models often overestimate the growth occurring into the real food, since factors such as structure and microbial interactions can be present. Thus, Corbo et al. (2005) presented a different approach by studying the growth dynamics of microbial growth naturally contaminating packed cod fillets, where growth data of the total bacterial count and total coliforms were used to model kinetically the shelf life of the samples. Using this approach it was concluded that mathematical model might be useful for the industry, allowing it to rely on a more objective measure to determine the shelf life of the product.

A phenomenon called Jameson effect is known to occur when high concentrations of LAB in lightly preserved seafood (smoke and marinated products) inhibit growth of *L. monocytogenes* (Mejlhom and Dalgaard 2007). Interaction between LAB and *L. monocytogenes* (microbial interactions) was predicted by combining a LAB model and a model on growth boundary for *L. monocytogenes*. Several approaches have been undertaken to determine the optimum gas concentrations for different seafood products. One study reported the use of PCA to clarify the contribution of the atmosphere composition and time on the quality changes of fish (Torrieri et al. 2006). Another approach was to use a simplex centroid mixture design with three components (partial pressures of CO_2, O_2, and N_2) and lattice 3 (9 runs + 3 replicates of center point) using a statistical software to determine the optimum MA for filleted formed cod (Sivertsvik 2007). With regard to SGS modeling, one of the studies developed a model to predict the amount of CO_2 absorbed in MA at equilibrium, that considers the CO_2 dissolved in the product prior to packaging since it affects the equilibrium atmosphere (Rotabakk et al. 2008).

10.9 CONCLUSIONS

Given the actual trends on consumer purchase decisions based on the benefits perceived that include taste, convenience, health and safety, and the emerging of new technologies, MAP should be able to fit to work together with those new technologies and future developments will center on active packaging, safety, modeling, and functional characteristics of foods. In a near future, food preservation and preparation will integrate with packaging so active packaging is a potential area for future research. Some experts agree that urgent MAP research should include the following areas (Farber et al. 2003, McKinna et al. 2003, Brody 2006, Floros 2006, Rico et al. 2007):

Active packaging
- Interaction of active packaging with food components (sensory changes)
- Active, smart, or intelligent packaging, such as oxygen scavenging from packaging atmospheres
- Carbon dioxide control by generation or absorption
- Controlled release of antimicrobial preservatives
- Time–temperature integrators tags
- Different active packaging systems to different food types

Functional foods or ingredients
- Evaluation of the effect of MAP on functional components of foods

Safety aspects
- Study the interactions of the background microflora with foodborne pathogens in various MAs used for produce as well as the effects of different gaseous environments on the survival and growth of bacterial foodborne pathogens on whole and fresh-cut produce.
- Examine the potential for growth of *C. botulinum* in a wide variety of MAP produce stored at mildly abusive temperatures such as 7°C–12°C. In addition, other hurdles besides temperature need to be examined to prevent totulinum toxin production.
- Examine the influence of different atmospheres, background microflora and storage temperatures on the survival and growth of *L. monocytogenes* on MAP fresh-cut produce.

- Investigate the behavior of verotoxin-producing *E. coli* on fresh and fresh-cut product, both under MAP and without MAP.
- Explore the survival of the enteric pathogens *Y. enterolitica and Campylobacter* spp. and the behavior of foodborne viruses and protozoan parasites on MAP produce.

Novel related MAP technologies

- Antimicrobial effect of superatmospheric O_2 in the fresh-cut safety

Hurdle technology on the combinations of novel methods of food treatment and packaging need to be examined, for example, irradiation used with MAP and antimicrobial films used in combination with MAP using natural preservatives such as combination of lactofferrin, organic acids, and oregano extracts with MAP and pulsed electric field technology.

REFERENCES

Ahn, H.-J., Kim, J.-H., Kim, J.-K., Kim, D.-H., Yook, H.-S., and Byun, M.-W. 2005. Combined effects of irradiation and modified atmosphere packaging on minimally processed Chinese cabbage (*Brassica rapa L.*). *Food Chemistry* 89: 589–597.

Ahvenainen, R. 1996. New approaches in improving the shelf life of minimally processed fruits and vegetables. Review. *Trends in Food Science and Technology* 7: 179–187.

Al-Ati, T and Hotchkiss, J.H. 2002. Application of packaging and modified atmosphere to fresh-cut fruits and vegetables. In *Fresh-Cut Fruits and Vegetables: Science, Technology and Market*, ed. O. Lamikanra, pp. 305–338. Boca Raton, FL: CRC Press.

Allende, A. and Artes, F. 2003. Combined ultraviolet-C and modified atmosphere packaging treatments for reducing microbial growth of fresh processed lettuce. *Lebensmittel-Wissenschaft und Technologie* 36: 779–786.

Allende, A., Luo, Y., McEvoy, J.L., Artes, F., and Wang, C. 2004. Microbial and quality changes in minimally processed baby spinach leaves stored under super atmospheric oxygen and modified atmosphere conditions. *Postharvest Biology and Technology* 33: 51–59.

Alves, R.M.V., De Luca Sarantopoulos, C.I.G., Van Dender A.G.F., and De Assis Fonseca Faria J. 1996. Stability of sliced mozzarella cheese in modified atmosphere packaging. *Journal of Food Protection* 59: 838–844.

An, J., Zhang, M., Lu Q., and Zhang, Z. 2006. Effect of a prestorage treatment with 6 benzylaminopurine and modified atmosphere packaging storage on the respiration and quality of green asparagus spears. *Journal of Food Engineering* 77: 951–957.

Artés-Hernández, F., Tomás-Barberán, F.A., and Artés, F. 2006. Modified atmosphere packaging preserves quality of SO_2-free 'Superior seedless' table grapes. *Postharvest Biology and Technology* 39: 146–154.

Ayhan, Z., Esturk, O., and Tas, E., 2007. Effect of modified atmosphere packaging on the quality and shelf life of minimally processed carrots. *Turkish Journal of Agriculture and Food* 31: 1–8.

Barros-Velázquez, J., Carreira, L., Franco, C., and Vázquez, B. 2003. Microbiological and physicochemical properties of fresh retail cuts of beef packaged under and advanced vacuum skin system and stored at 4°C. *Journal of Food Protection* 66: 2085–2092.

Beaudry, R. 2007. MAP as a basis for active packaging. In *Intelligent and Active Packaging for Fruits and Vegetable*, ed. C.L. Wilson, pp. 31–56. Boca Raton, FL: CRC Press Taylor & Francis.

Behrends, J.M., Mikel, W.B., Armstrong, C.L., and Newman, M.C. 2003. Color stability of semitendinosus, semimembranosus, and biceps femoris steaks packaged in a high-oxygen modified atmosphere. *Journal of Animal Science* 81: 2230–2238.

Ben-Yehoshua, S., Beaudry, R.M., Fishman, S., Jayanty, S., and Mir, N. 2005. Modified atmosphere packaging and controlled atmosphere storage. In *Environmentally Friendly Technologies for Agricultural Produce Quality*, ed. Ben-Yehoshua, pp. 61–112. Boca Raton, FL: CRC Press Taylor & Francis.

Bhande, S.D., Ravindra, M.R., and Goswami, T.K. 2008. Respiration rate of banana fruit under aerobic conditions at different storage temperatures. *Journal of Food Engineering* 87: 116–123.

Blakistone, B.A. 1999. Preface. In B.A. Blakistone *Principles and Applications of Modified Atmosphere Packaging of Foods*. Aspen Publications, Gaithersbhurg, MD, pp. 1–4.

Blakistone, B.A. 2000. Preface. In *Principles and Applications of Modified Atmosphere Packaging of Foods*, ed. B.A. Blakistone, pp. 1–4. Gaithersburg, MD: Aspen Publishers Inc.

Bøknæs, N., Østerberg, C., Sørensen, R., Nielsen, J., and Dalgaard, P. 2001. Effects of technological parameters and fishing ground on quality attributes of thawed, chilled cod fillets stored in modified atmosphere packaging. *Lebensmittel-Wissenschaft und-Technologie—Food Science and Technology* 34: 513–520.

Bøknæs, N., Jensen, K.N., Andersen, Ch., and Martens, H. 2002. Freshness assessment of thawed and chilled cod fillets packed in modified atmosphere using near-infrared spectroscopy. *Lebensmittel-Wissenschaft und-Technologie—Food Science and Technology* 35: 628–634.

Brody, A.L. 1989. *Controlled/Modified Atmosphere/Vacuum Packaging of Foods*. Trumbull, CT: Food & Nutrition Press.

Brody, A.L. 1996. Fundamentals of modified atmosphere packaging. Conference at the Society of Manufacturing Engineers (SME). Dearborn, MI. December, 1996.

Brody, A.L. 1999. Markets for MAP foods. In *Principles and Applications of Modified Atmosphere Packaging of Foods*, ed. B.A. Blakistone, pp. 14–38. Gaithersburg, MD: Aspen Publishers Inc.

Brody, A.L. 2006. State of the art of active/intelligent food packaging, presented at the Fifth Research Summit, *Food Packaging Innovations: The science, current research, and future research needs*, organized by Institute of Food Technologists, IFT. Baltimore, MD, May 7–9, 2006.

Cagri, A., Ustenol, Z., and Ryser, E. 2004. Antimicrobial edible films and coatings. Review. *Journal of Food Protection* 67: 833–848.

Caillet, S., Millette, M., Salmiéri, S., and Lacroix. M. 2006. Combined effects of antimicrobial coating, modified atmosphere packaging, and gamma irradiation on *Listeria innocua* present in ready-to-use carrots (*Daucus carota*). *Journal of Food Protection* 69: 80–85.

Cha, D.S. and Chinnan, M.S. 2004. Byopolymer-based antimicrobial packaging: A review. *Critical Reviews in Food Science and Nutrition* 44: 223–237.

Charles, F., Sanchez, J., and Gontard, N. 2005. Modeling of active modified atmosphere packaging of endives exposed to several postharvest temperatures. *Journal of Food Science* 70: E443–E449.

Charles, F., Sanchez, J., and Gontard, N. 2006. Active modified atmosphere packaging of fresh fruit and vegetables: Modeling with tomatoes and oxygen absorber. *Journal of Food Science* 68(5): 1736–1742.

Chen, J.H. and Hotchkiss, J.H. 1991a. Effect of dissolved carbon dioxide on the growth of psychrotropic organisms in cottage cheese. *Journal of Dairy Science* 74: 2941–2494.

Chen, J.H. and Hotchkiss, J.H. 1991b. Long shelf life cottage cheese through dissolved CO_2 and high barrier packaging. *Journal of Dairy Science* 74 (Suppl 1): 125.

Church, I. and Parsons, A. L. 1995. Modified atmosphere packaging technology: A review. *Journal of the Science of Food and Agriculture* 67: 143–152.

Cilla, I., Martinez, L., Beltran, J.A., and Roncalez, P. 2005. Dry cured ham quality and acceptability as affected by the preservation system used for retail sale. *Meat Science* 73:581–589.

Colchin, L.M., Owens, S.L., Lyubachevskaya, G., Boyle-Roden, Russek-Cohen, E., and Rankin, S.A. 2001. Modified atmosphere packaged cheddar cheese shreds: Influence of fluorescent light exposure and gas type on color and production of volatile compounds. *Journal of Agricultural and Food Chemistry* 49: 2277–2282.

Corbo, M.R., Altieri, C., Bevilacqua, A., Campaniello, D., Dámato, D., and Sinigaglila, M. 2005. Estimating packaging atmospheres—temperature effects on the shelf life of cod fillets. *European Food Research and Technology* 220: 509–513.

Cornforth, D.P. and Hunt, M. 2008. *Low Oxygen Packaging of Fresh Meat with Carbon Monoxide: Meat Quality, Microbiology, and Safety*. American Meat Science Association white Paper Series, No. 2. January 2008.

Cutter, C.N. 2006. Opportunities for bio-based packaging technologies to improve the quality and safety of fresh and further processed muscle foods. *Meat Science* 74: 131–142.

Dalgaard, P. 2000. Fresh and lightly preserved seafood. In *Shelf-Life Evaluation of Foods*, eds. C.M.D. Man and A.A. Jones, pp. 110–139. Gaithersburg, MA: Aspen Publishers Inc.

Dalgaard, P., Buch, P., and Silberg, S. 2002. Seafood Spoilage Predictor—development and distribution of a product specific application software. *International Journal of Food Microbiology* 73(2-3):343–349.

Dalgaard, P., Mejlholm, O., and Huss, H.H. 1997. Application of an iterative approach for development of a microbial model predicting the shelf-life of packed fish. *International Journal of Food Microbiology* 38:169–179.

Day, B.P.F. 1989. Modified atmosphere packaging. Principles of production and distribution of chilled foods. *Proceedings of International Conference on Modified Atmosphere Packaging*, 17–20. Campden and Chorleywood Food Research Association (CFDRA), Chipping Campden, October 1989.

Day, B.P.F. 1991. A perspective of modified atmosphere packaging of fresh produce in Western Europe. *Food Science and Technology Today* 4: 215–221.

Day, B.P.F. 1996a. High oxygen modified atmosphere packaging for fresh prepared produce. *Postharvest News Information* 7: 31N–34N.

Day, B.P.F. 1996b. High oxygen modified atmosphere packaging: A novel approach for fresh prepared produce packaging. In *Packaging Yearbook* 1996, Ed. Blakistone, B. February 1997, pp. 55–67.

Deepak, R.R. and Shashi, P. 2007. Transient state in-pack respiration rates of mushroom under modified atmosphere packaging based on enzyme kinetics. *Biosystems Engineering* 98: 319–326.

Dermiki, M., Ntzimani, A., Badeka, A., Savvadis, I.N., and Kontominas, M.G. 2007. Shelf life extension and quality attributes of the whey cheese "Mysithra Kalathaki" using modified atmosphere packaging. *Lebensmittel-Wissenschaft und-Technologie—Food Science and Technology* 41: 284–294.

Drake, S.R., Elfving, D.C., Drake, S.L., and Visser, D.B. 2004. Qulaity of modified atmosphere packaged "Bartlett" pears as influenced by time and type of storage. *Journal of Food Processing and Preservation* 28: 348–358.

European Commission. 2005. Commission Regulation (EC) No 2073/2005 of 15 November 2005 on microbiological criteria for food stuffs. *Official Journal of the European Union* L338: 1–26.

Emond, J., Chau, K., Brethc, J., and Ngadym, M. 1998. Mathematical modeling of gas concentration profiles in modified atmosphere bulk packages. American Society of Agricultural Engineers 41(4): 1075–1082.

Escalona, V.H., Verlinden, B.E., Geysen, S., and Nicolai, B.M. 2006. Changes in respiration of fresh cut butterhead lettuce under controlled atmospheres using low and superatmospheric oxygen conditions with different carbon dioxide levels. *Postharvest Biology and Technology* 39:48–55.

Farber, J.N. 1991. Microbiological aspects of modified atmosphere packaging technology—a review. *Journal of Food Protection* 54: 58–70.

Farber, J.N., Harris, L.J., Parish, M.E. et al. 2003. Microbiological safety of controlled and modified atmosphere packaging of fresh and fresh-cut produce. *Comprehensive Reviews in Food Science and Food Safety* 2: 142–160.

Fernández-Segovia, I., Escriche, I., Gómez-Sintes, M., Fuentes, A., and Serra, J.A. 2006. Influence of different preservation treatments on the volatile fraction of desalted cod. *Food Chemistry* 98: 473–482.

Fishman, S., Rodov, V., and Ben-Yehoshua, S. 1996. Mathematical model for perforation effect on oxygen and water vapor dynamics in modified-atmosphere packages. *Journal of Food Science* 61: 956–961.

Fisman, S., Rodov, V., Peretz, J., Ben-Yehoshua, S. 1995. Model for gas exchange dynamics in modified atmosphere packages of fruits and vegetables. *Journal of Food Science* 60(5): 1078–1083.

Floros, J.D. 2006. Packaging & our food system in the future. Presented at the Fifth Research Summit, *Food Packaging Innovations: The science, current research and future research needs*, organized by Institute of Food Technologists, IFT. Baltimore, MD, May 7–9, 2006.

Floros, J.D., Nielsen, P.V., and Farkas, J.K. 2000. Advances in modified atmosphere and active packaging with applications in the dairy industry. *Bulletin of the International Dairy Federation* 346: 22–28. Brussels: International Dairy Federation.

Fonseca, S.C., Oliveira, F.A.R., and Brecht, J.K. 2002. Modelling respiration rate of fresh fruits and vegetables for modified atmosphere packaging: A review. *Journal of Food Engineering* 52: 99–119.

Forcinio, H. 1997. Mapping out a fresh approach. *Prepared Foods* 166: 44.

Gill, C.O. 2003. Active packaging in practice: Meat. In *Novel Food Packaging Techniques*, ed. R. Ahvenainen, 365–383. Cambridge, U.K.: Woodhead Publishing Ltd.

Gomez, P.A. and Artes, F. 2005. Improved keeping quality of minimally processed celery stick by modified atmosphere packaging. *Lebensmittel-Wissenschaft und-Technologie—Food Science and Technology* 38:323–329.

Gómez-López, V.M., Devlieghere, F., Bonduelle, V., and Debevere, J. 2005. Intense light pulses decontamination of minimally processed vegetables and their shelf-life. *International Journal of Food Microbiology* 103: 79–89.

Gómez-López, V.M., Ragaert, P., Ryckeboer, J., Jeyachchandran, V., Debevere, J., and Devlieghere, F. 2007. Shelf life of minimally processed cabbage treated with neutral electrolysed oxidizing water and stored under equilibrium modified atmosphere. *International Journal of Food Microbiology* 117: 91–98.

Gonzalez, J., Ferrer, A., Oria, R., and Salvador, M.L. 2008. Determination of O_2 and CO_2 transmission rates through microperforated films for modified atmosphere packaging of fresh fruits and vegetables. *Journal of Food Engineering* 86: 194–201.

Gouble, B., Fath, D., and Soudain, P. 1995. Nitrous oxide inhibition of ethylene production in ripening and senescing climateric fruits. *Postharvset Biology and Technology* 5: 311–321.

Goulas, A.E. and Kontominas, M.G. 2007. Effect of modified atmosphere packaging and vacuum packaging on the shelf life of refrigerated chub makerel (*Scomber japonicus*): Biochemical and sensory attributes. *European Food Research and Technology* 224: 545–553.

Guillén, F., Zapata, P.J., Martínez-Romero, D., Castilo, S., Serrano, M., and Valero, D.J. 2005. Improvement of the overall quality of table grapes stored under modified atmosphere packaging in combination with natural antimicrobial compounds. *Journal of Food Science* 72: S185–S195.

Hertog, M.L.A.T.M. and Banks, N.H. 2003. Improving MAP through conceptual models. In *Novel Food Packaging Techniques*, ed. R. Ahvenainen, pp. 337–362. Cambridge, U.K.: Woodhead Publishing Ltd.

Hewage, S.K., Wainwright, H., Wijerathnam, S.W., and Swinburne, T. 1995. The modified atmosphere storage of bananas as affected by different temperatures. In *Postharvest Physiology and Technologies for Horticultural Commodities: Recent Advances*, 172–176. Agadir, Morocco: Institute of Agronomique and Veterinaire Hassan II.

Hintlian, C.B. and Hotchkiss, 1986. The safety of modified atmosphere packaging: A review. *Food Technology* 39: 70–76.

Hotchkiss, J.H., Werner, B.G., and Lee, E.Y.C. 2006. Addition of carbon dioxide to dairy products to improve quality: A comprehensive review. *Comprehensive Reviews in Food Science and Food Safety* 5: 158–168.

Insausti, K., Beriain, M.J., Lizaso, G., Carr, T.R., and Purroy, A. 2008. Multivariate study of different beef quality traits from local Spanish cattle breeds. *Animal* 2: 447–458.

Jacobson, A., Nielsen, T., Sjholm, I., and Wendin, K. 2004. Influence of packaging material and storage conditions on the sensory quality of broccoli. *Food Quality and Preference* 15: 301–310.

Jacxsens, L., Devlieghere, F., Van der Steen, C., Debevere, J. 2001. Effect of high oxygen modified atmosphere packaging on microbial growth and sensorial qualities of fresh-cut produce. *International Journal of Food Microbiology* 71: 197–210.

Jayas, D.S. and Jeyamkondan, S. 2002. Modified atmosphere storage of grains, meats, fruits and vegetables. *Biosystems Engineering* 82: 235–251.

Kader, A.A. and Watkins, C. 2000. Modified atmosphere packaging—toward 2000 and beyond. *Hort-technology* 10(3): 483–486.

Kader, A.A., Singh, R.P., and Mannapperuma, J.D. 1998. Technologies to extend the refrigerated shelf life of fresh fruits. In *Food Storage Stability*, eds. I.A. Taub and R.P. Singh, pp. 419–434. Boca Raton, FL: CRC Press.

Kader, A.A., Singh, R.H., and Mannapperuma, J.D. 2003. Technologies to extend the refrigerated shelf life of fresh fruits. In I.A. Taub and R.P. Sings. Eds. *Food Storage Stability*. Boca Raton, FL: CRC Press, pp. 419–434.

Karabulut, A. and Baykal N. 2004. Integrated control of postharvest diseases of peaches with a yeast antagonist, hot water and modified atmosphere packaging. *Crop Protection* 23: 431–435.

Karagul-Yucceer, Y., Coggins, P.C., Wilson, J.C., and White, C.H. 1999. Carbonated yogurt: sensory properties and consumer acceptance. *Journal of Dairy Science* 82: 1394–1398.

Ketsa, 2000. Development and control of senescent spotting in banana. *Food Preservation Science* 26: 176–178.

Kim, K.M., Ko, J.A., Lee, J.S., Park H.J., and Hanna, M.A. 2006. Effect of modified atmosphere packaging on the shelf-life of coated, whole and sliced mushrooms. *Lebensmittel-Wissenschaft und-Technologie—Food Science and Technology* 39: 364–371.

Koide, S. and Shi, J. 2007. Microbial and quality evaluation of green peppers stored in biodegradable film packaging. *Food Control* 18: 1121–1125.

Kudachikar, V.B., Kulkarni, S.G., Vasantha, M.S., Prasad, B.A., and Aradhya, S.M. 2007. Effect of modified atmosphere packaging on shelf life and fruit quality of banana stored at low temperature. *Journal of Food Science and Technology Mysore* 44: 74–78.

Lamikanra, O. 2002. *Fresh-cut Fruits and Vegetables: Science, Technology and Market*. Boca Raton, FL: CRC Press.

Laury, A. and Sebranek, J. 2006. Use of carbon monoxide combined with carbon dioxide for modified atmosphere packaging of pre- and postrigor fresh pork sausage to improve shelf life. *Journal of Food Protection* 70: 937–942.

Lee, D.S., Haggar, P., Lee, J., and Yam, K.L. 1991. Model for fresh produce respiration in modified atmospheres based on principles of enzyme kinetics. *Journal of Food Science* 56: 1580–1585.

Lee, D.S., Kang, J.S., and Renault, P. 2000. Dynamics of internal atmosphere and humidity in perforated packages of peeled garlic cloves. *International Journal of Food Science and Technology* 35: 455–464.

Li, T. and Zhang, M. 2008. Effects of modified atmosphere packaging with various sizes of silicon gum film window on the storage of *Agrocybe Chaxingu Huan* and the modelling of its respiration rate. *Packaging, Technology and Science* 21: 13–23.

Lin, D. and Zhao, Y., 2007. Innovations in the development and application of edible coatings for fresh and minimally processed fruits and vegetables. *Comprehensive Reviews in Food Science and Food Safety* 6: 60–75.

Lopez-Rubio, A., Almenar, E., Hernández-Muñoz, P., Lagarón, J.M., Catalã, R., and Gavara, R. 2005. Overview of active polymer-based packaging technologies for food applications. *Food Reviews International* 20: 357–387.

Lu, S. 2007. Effect of packaging on shelf life of minimally processed Bok Choy (*Brassica chinensis L.*). *Lebensmittel-Wissenschaft und-Technologie—Food Science and Technology* 40: 460–464.

Lund, M.N., Hviid, M.S., and Skibsted, L.H. 2007. The combined effect of antioxidants and modified atmosphere packaging on protein and lipid oxidation in beef patties during chill storage. *Meat Science* 76: 226–233.

Luño, M., Beltrán, J.A., and Roncalés, P. 1998. Shelf-life extension and colour stabilization of beef packaged in a low O_2 atmosphere containing CO: Loin steaks and ground meat. *Meat Science* 48: 75–84.

Malakou, A. and Nanos G.D. 2005. A combination of hot water treatment and modified atmosphere packaging maintains quality of advanced maturity 'Caldesi 2000' nectarines and 'Royal Glory' peaches. *Postharvest Biology and Technology* 38: 106–114.

Mannheim, C.H. and Soffert, T. 1996. Shelf life extension of cottage cheese by modified atmosphere packaging *Lebensmittel-Wissenschaft und-Technologie—Food Science and Technology* 29: 767–771.

Martínez, L., Djenane, D., Cilla, I, Beltrán, J.A., and Roncalés, P. 2005. Effect of different concentrations of carbon dioxide and low concentration of carbon monoxide on the shelf life of fresh pork sausages package in modified atmosphere. *Meat Science* 71: 563–570.

Martínez, L., Cilla, I., Beltrán, J.A., and Roncalés, P. 2007. Effect of illumination on the display life of fresh pork sausages package in modified atmsphere. Influence of the addition of rosemary, ascorbic acid and black pepper. *Meat Science* 75: 443–450.

McKinna et al., 2003. Packaging in the Australian Red Meat Industry. Meat and Livestock Australia, MLA-0126. 6th October 2003.

McMeeking, T.A., Ross, T., and Olley, J. 1992. Application of predictive microbiology to assure the quality and safety of fish and fish products. *International Journal of Food Microbiology* 15: 13–32.

Mejlholm, O., Bøknæs, N., and N., Dalgaard, P. 2005. Shelf life and safety aspects of chilled cooked and peeled shrimps (*Pandalus borealis*) in modified atmosphere packaging. *Journal of Applied Microbiology* 99: 66–76.

Mejlholm, O. and Dalgaard, P. 2007. Modeling and predicting the growth of Lactic acid lightly preserved seafood and their inhibiting effect on Listeria monocytogenes. *Journal Food Protection* 70(11): 2485–2487.

Metaxopoulos, J., Mataragas, M., and Drosinos, E.H. 2002. Microbial interaction in cooked cured meat products under vacuum or modified atmosphere at 4°C. *Journal of Applied Microbiology* 93: 363–373.

Michaelsen, A.R., Sebranek, J., and Dickson, J. 2006. Effects of microbial inhibitors and modified atmosphere packaging on growth of *Listeria monocytogenes* and *Salmonella enterica Typhimurium* and on quality attributes on infected pork chops and sliced cured ham. *Journal of Food Protection* 69: 2671–2680.

Olarte, C., Gonzalez-Fandos, E., Gimémez, M., Sanz, S., and Portu, J. 2002. The growth of *Listeria monocytogenes* in fresh goat cheese (Cameros cheese) packaged under modified atmospheres. *Food Microbiology* 19: 75–82.

Oms-Oliu, G., Odriozola-Serrano, I., Soliva-Fortuny, R., and Martín-Belloso, O. 2008b. The role of peroxidase on the antioxidant potential of fresh-cut 'Piel de Sapo' melon packaged under different modified atmospheres. *Food Chemistry* 106: 1085–1092.

Oms-Oliu, G., Raybaudi-Massilia, Martinez, R.M., Soliva-Fortuny, R., and Martín-Belloso, O. 2008a. Effect of superatmospheric and low oxygen modified atmospheres on shelf-life extension of fresh-cut melon. *Food Control* 19: 191–199.

Oms-Oliu, G., Soliva Fortuny, R., and Martin-Belloso, O. 2007. Respiratory rate and quality changes in fresh-cut pears as affected by superatmospheric oxygen. *Journal of Food Science* 72(8): E456–E463.

Oms-Oliu, G., Soliva Fortuny, R., and Martin-Belloso, O. 2008c. Modeling changes of headspace gas concentrations to describe the respiration rate of fresh-cut melon under low superatmospheric oxygen atmospheres. *Journal of Food Engineering* 85(3): 401–409.

Ozdemir, I.S., Varoguaux, P.J., Tournemelle, F., and Vildiz, F. 2004. Effect of noble gases, nitrous oxide and nitrogen on the anaerobic metabolism and quality attributes of mushroom (*Agaricus bisporus* L.) and sliced apple (*Malus sylvestris* Mil.). *Sciences des Aliments* 24: 233–245.

Papaioannou, G., Chouliara, I., Karatapanis, A.E., Kontominas, M.G., and Savvaidis, I.N. 2006. Shelf life of a Greek whey cheese under modified atmosphere packaging. *International Dairy Journal* 17: 358–364.

Pastoriza, L., Cabo, Bernanderz, M., Sampedro, G., and Herrera, J.J. 2004. Combined effects of modified atmosphere packaging and lauric acid on the stability of pre-cooked fish products during refrigerated storage. *European Food Research and Technology* 215: 189–193.

Paludan-Muller C. 1998. Evaluation of the role of Carnobacterium piscicola in spoilage of vacuum- and modified-atmosphere-packed cold-smoked salmon stored at 5°C. International *Journal of Food Microbiology* 39: 155–166.

Perry, J.P., Grady, M.N., and Hogan, S.A. 2006. Past, current and potential utilization of active and intelligent packaging systems for meat and muscle based products: A review. *Meat Science* 74: 113–130.

Pettersen, M.K., Mielnik, M.B., Eie, T., A., Skrede, G., and Nilsson, A. 2004. Lipid oxidation in frozen, mechanically deboned turkey meat as affected by packaging parameters and storage conditions. *Poultry Science* 83: 1240–1248.

Phillips, C.A. 1996. Review: Modified atmosphere packaging and its effects on the microbiological quality and safety of produce. *International Journal of Food Science and Technology* 31: 463–479.

Porat, R., Weiss, B., Cohen, L., Daus, A., and Aharoni, N. 2004. Reduction of postharvest disorders in citrus fruit by modified atmosphere packaging. *Postharvest Biology and Technology* 33: 35–43.

Rai, D.R., Oberoi, H.S., and Baboo, B. 2002. Modified atmosphere packaging and its effect on quality and shelf life of fruits and vegetables-an overview. *Journal of Food Science and Technology* 39: 199–207.

Randell, K., Hattula, T., Skytta E., Sivertsvik, M., Bergslien, H., and Ahvenainen, R. 1999. Quality of filleted salmon in various retail packages. *Journal of Food Quality* 22: 483–497.

Reddy, N.R., Paradis, A., Roman, M.G., Solomon, H.M., and Rhodehamel, E.J. 1996. Toxin development by *Clostridium botulinum* in modified atmosphere-packaged fresh tilapia fillets during storage. *Journal of Food Science* 61: 632–635.

Renault, P., Souty, M., and Chambroy, Y. 1994. Gas exchange in modified atmosphere packaging. 1: A new theoretical approach for micro-perforated packs. *International Journal of Food Science and Technology* 29: 365–378.

Rico, D., Martin-Diana, A.B., Barat, J.M., and Barry-Ryan, C. 2007 Extending and measuring the quality of fresh-cut fruit and vegetables: A review. *Trends in Food Science & Technology* 18: 373–386.

Riley, J. and Huffman, R. 2006. Low oxygen CO packaging update. *2006 Meat Industry Research Conference*. AMSA. October 4–5, 2006. Hollywood, FL.

Rocculi, P., Romani, S.M., and Dalla Rosa, M. 2004. Evaluation of physico-chemical parameters of minimally processed apples package in non conventional modified atmosphere. *Food Research International* 37: 329–335.

Rocculi, P., Romani, S., and Dalla Rosa, M. 2005. Effect of MAP with argon and nitrous oxide on quality maintenance of minimally processed kiwifruit. *Postharvest Biology and Technology* 35: 319–328.

Rokka, M., Eerola, S., Smolander, M., Alakomi, H.-L., and Ahvenainen, R. 2004. Monitoring of the quality of modified atmosphere packaged broiler chicken cuts stored in different temperature conditions: Biogenic amines as quality indicating metabolites. *Food Control* 15: 601–607.

Romani, S., Sacchetti, G., Pittia, P., Pinnavaia, G.G., and Dalla Rosa, M. 2002. Physical, chemical, textural and sensorial changes of portioned parmigiano regiano cheese packed under different conditions. *Food Science and Technology International* 8: 203–211.

Rotabakk, B.T., Birkeland, S., Jeksrud, W.K., and Sivertsvik, M. 2006. Effect of modified atmosphere packaging and soluble gas stabilization on the shelf life of skinless chicken breast fillets. *Journal of Food Science* 71: S124–S131.

Rotabakk, B.T., Birkeland, S., Lekand, O.I., and Sivertsvik, M. 2008a. Enhancement of modified atmosphere packaged farmed Atlantic halibut (*Hippoglossus Hipoglossus*) fillet quality by soluble gas stabilization. *Food Science and Technology International* 14(2): 179–186.

Rotabakk, B.T., Willer, J., S., Lekand, O.I., and Sivertsvik, M. 2008b. A Mathematical method for determining equilibrium gas composition in modified atmosphere packaging and soluble gas stabilization systems for non respiring foods. *Journal of Food Engineering* 85: 479–490.

Ruiz-Capillas, C., Morales, J., and Moral, A. 2001. Combination of bulk storage in controlled and modified atmospheres with modified atmosphere packaging system for chilled whole gutted hake. *Journal of the Science of Food and Agriculture* 81: 551–558.

Rutherford, T., Marshall, D.L., Andrews, L.S., Coggins, P.C., Schilling, M.W., and Gerard, P. 2007. Combined effect of packaging atmosphere and storage temperature on growth of *Listeria monocytogenes* on ready-to-eat shrimp. *Food Microbiology* 24: 703–710.

Sagoo, S.K., Lilttle, C.L., Allen, G., Williamson, K., and Grant, K.A. 2007. Microbial safety of retail vacuum-packed and modified-atmosphere-packed cooked meats at end of shelf life. *Journal of Food Protection* 70: 943–951.

Sarantópoulos, C.I.G.L., Vercelino Alves R.M., Contreras, C.J.C., Galvão, M.T.E.L., and Gomes, T.C. 1998. Use of modified atmosphere masterpack for extending the shelf life of chicken cuts. *Packaging Technology and Science* 11: 217–229.

Satomi, K. 1990. Gas exchange packaging. In *Food Packaging*, ed. T. Kadoya, pp. 269–278. London: Academic Press.

Selman, J. 1987. New methods of preservation. *Food Manufacturing* 53: 55–56.

Silliker, J.H. and Wolfe, S.K. 1980. Microbiological safety considerations in controlled atmosphere storage of meats. *Food Technology* 34: 59–63.

Simpson, R., Almonacid, S., Acevedo, C., and Cortes, C. 2004. Simultaneous heat and mass transfer applied to non-respiring foods packed in modified atmosphere. *Journal of Food Engineering* 61(2): 279–286.

Sing, S.P. and Rao, D.V.S. 2005. Effect of modified atmosphere packaging (MAP) on the alleviation of chillin injury and dietary antioxidants levels in "Solo" papaya during low temperature storage. *European Journal of Horticultural Science* 70: 246–252.

Sivertsvik, M. 2003. Novel methods for packaged seafood. *Workshop: The influence of Packaging Methods and Food Preservation Technologies on Human Health*. Krakov, Poland, April 22–23, 2003.

Sivertsvik, M. 2007. The optimized modified atmosphere for packaging of pre-rigor filleted farmed cod (*Gadus morhua*) is 63 ml/100 ml oxygen and 37 ml/100 ml carbon dioxide. *Lebensmittel-Wissenschaft und-Technologie—Food Science and Technology* 40: 430–438.

Sivertsvik, M. and Birkeland, S. 2006. Effect of soluble gas stabilization (SGS), modified atmosphere, gas to product volume ratio and storage on the microbiological and sensory characteristics of ready-to-eat shrimp. *Food Science and Technology International* 12: 445–454.

Sivertsvik, M. and Jensen, J.S. 2005. Solubility and absorption rate of carbon dioxide into non-respiring foods. Part 3: Cooked meat products. *Journal of Food Engineering* 70(5):499–505.

Sivertsvik, M., Jeksrud, W.K., and Rosnes, J.T. 2002. A review of modified atmosphere packaging of fish and fishery products—significance of microbial growth, activities and safety. *International Journal of Food Science and Technology* 37: 107–127.

Sivertsvik, M., Rosnes, J.T., and Kleiberg, G.H. 2003. Effect of modified atmosphere packaging and superchilled storage on the microbial and sensory quality of Atlantic salmon (*Salmo salar*) fillets. *Journal of Food Science* 68: 1467–1472.

Skandamis, P.N. and Nychas, G.J. E. 2002. Preservation of fresh meat with active and modified packaging conditions. *International Journal of Food Microbiology* 79: 35–45.

Skura, B.J. 1991. Modified atmosphere packaging of fish and fish products. In *Modified Atmosphere Packaging of Food*, eds. B. Ooraikul and M.E. Stiles, pp. 148–168. New York: Ellis Horwood.

Smiddy, M., Fitzgerald, M., Kerry, J.P., Papkovsky, D.B., Sullivan, C.K.O., and Guilbault, G.G. 2002a. Use of oxygen sensors to non-destructively measure the oxygen content in modified atmosphere and vacuum packed beef: Impact of oxygen content on lipid oxidation. *Meat Science* 61: 285–290.

Smiddy, M., Papkovskaia, N., Papkovsky, D.B., and Kerry, J.P. 2002b. Use of oxygen sensors for the non-destructive measurement of the oxygen content in modified atmosphere and vacuum packs of cooked chicken parties: Impact of oxygen content on lipid oxidation. *Food Research International* 35: 577–584.

Smiddy, M., Papkovsky, D., and Kerry, J. 2002c. Evaluation of oxygen content in commercial modified atmosphere packs (MAP) of processed cooked meats. *Food Research International* 35: 571–575.

Smulders, F.J.M., Hiesberger, J., Hofbauer, P., Dogl, B., and Dransfield, E. 2006. *Journal of Animal Science* 84: 2456–2462.

Soliva-Fortuny, R.C. and Martin-Belloso, O. 2003. New advances in extending the shelf lie of fresh-cut fruits: A review. *Trends in Food Science and Technology* 14: 341–353.

Sorheim, O. and Nissen, H. 2002. Current technology for modified atmosphere packaging of meat. *FoodInfo Online Features* 25 January 2002. Food Science Central.

Sorheim, O., Aune, T., and Nesbakken, T. 1997. Technological, hygienic and toxicological aspects of carbon monoxide used in modified atmosphere packaging of meat—a review. *Trends in Food Science and Technology* 8: 307–312.

Spencer, K.C. 1995. The use of argon and other noble gases for the MAP of Foods. *International Conference on MAP and Related Technologies*. Campden & Chorleywood Research Association. Chipping Campden U.K. 6–7 September 1995.

Srinivasa, P.C., Harish Prashant, K.V., Susheelamma, N.S., Ravi, R., and Tharanathan, R.N. 2006. Storage studies of tomato and bell pepper using eco-friendly films. *Journal of the Science of Food and Agriculture* 86: 1216–1224.

Steiner, A., Abreu, M., Correia, L., Beirao-da-Costa, M.L., Empis, J., and Moldao-Martins, M. 2006. Metabolic response to combined mild heat pre-treatments and modified atmosphere packaging on fresh-cut peach. *European Food Research and Technology* 222: 217–222.

Subramaniam, P.J. 1989. Dairy Foods, multi-component products, dried foods and beverages. In *Principles and Applications of Modified Atmosphere Packaging of Foods*, ed. B.A. Blakistone, 158–193. Gaithersburg, MD: Aspen Publishers Inc.

Summo, C., Caponio, F., and Pasqualone, A. 2006. Effect of vacuum-packaging storage on the quality level of ripened sausages. *Meat Science* 74: 249–254.

Techavises, N. and Hikida, Y. 2008. Development of a mathematical model for simulating gas on water vapor exchanges ion modified atmosphere packaging with macroscopic perforations. *Journal of Food Engineering* 85: 94–104.

Torrieri, E., Cavella, S., Villani, F., and Masi, P. 2006. Influence of modified atmosphere packaging on the chilled shelf life of gutted farmed bass (*Dicentrarchus labrax*). *Journal of Food Engineering* 77: 1078–1086.

Valencia, I., Ansorena, D., and Astiasarán, I. 2006. Stability of linseed oil and antioxidants containing dry fermented sausages: A study of the lipid fraction during different storage conditions. *Meat Science* 73: 269–277.

Valverde, J.M., Guillen, F., Martinez-Romero, D., Castillo, S., Serrano, M., and Valero, D. 2005. Improvement of table grapes quality and safety by the combination of modified atmosphere packaging (MAP) and eugenol, menthol, or thymol. *Journal of Agricultural and Food Chemistry* 53: 7458–7464.

Vázquez, B. I., Carreira, L., Franco, C., Fente, C., Cepeda, A., and Barros-Velázquez, J. 2004. Shelf life extension of beef retail cuts subjected to an advanced vacuum skin packaging system. *European Food Research and Technology* 218: 118–122.

Vercelino, Alves R.M., De Luca Sarantopoulos, G., Van, Dender A.G.F., and De Assis Fonseca Faria J. 1996. Stability of sliced mozzarella cheese in modified atmosphere packaging. *Journal of Food Protection* 59(8): 838–844.

Villanueva, M.J., Tenorio, M.D., Sagardoy, M., Redondo, A., and Saco, M.D. 2005. Physical, chemical, histological and microbiological changes in fresh green asparagus (*Asparagus officinalis*, L.) stored in modified atmosphere packaging. *Food Chemistry* 91: 609–619.

Wicklund, R.A., Paulson, D.D., Tucker, E.M. et al. 2006. Effect of carbon monoxide and high oxygen modified atmosphere packaging and phosphate enhanced case ready pork chops. *Meat Science* 74: 704–709.

Wilkinson, BG.H.P., Janz, J.A.M., Morel, P.C.H., Purchas, R.W., and Hendriks, W.H. 2006. The effect of modified atmosphere packaging with carbon monoxide on the storage quality of master-packaged fresh pork. *Meat Science* 73: 605–610.

Young, L.L., Reviere, R.D., Cole, A., and Benjamin-Cole, A. 1988. Fresh red meats: A place to apply modified atmospheres. *Food Technology* 41: 65–69.

Yuan, J.T.C., 2003. Modified atmosphere packaging for shelf-life extension. In *Microbial Safety of Minimally Processed Foods*, eds. J.S. Novak, G.M Sapers and V.K. Juneja, pp. 205–220. Boca Raton, FL: CRC Press.

Zagory, D. 2000. What modified atmospheres packaging can and can't do for you. *Sixteenth Annual Postharvest Conference*, Yakima, WA. March 14–15, 2000.

Zagory, D. and Kader, A.A. 1988. Modified atmosphere packaging of fresh produce. *Food Technology* 41: 70–77.

Zhang, M. and Sundar S. 2005. Effect of oxygen concentration on the shelf-life of fresh pork packed in a modified atmosphere. *Packaging Technology and Science* 18: 217–222.

Zhang, M., Zhan, Z.G., Wang, S.J., and Tang, J.M. 2008. Extending the shelf life of asparagus spears with a compressed mix of argon and xenon gases. *Lebensmittel-Wissenschaft und-Technologie—Food Science and Technology* 41: 686–691.

11 Effects of Chilling and Freezing on Safety and Quality of Food Products

Liana Drummond and Da-Wen Sun

CONTENTS

11.1 INTRODUCTION

A great variety of chilled and frozen foods is currently available to the consumer. Raw, partially cooked, as well as ready-to-eat options are generally perceived to be "fresh" and "healthy" alternatives to an increasing group of consumers who are unable or unwilling to cook full meals from scratch, particularly after a busy day at work. For most people, chilling and freezing are conventional and familiar preservation methods as many have some experience cooling or freezing food at home and the procedures are not seen to change food products considerably. Low-temperature storage has grown to be one of the major preservation techniques associated to high-quality foods. Frozen foods account for one of the biggest sectors in the food industry. According to data reported by the British Frozen Food Federation (BFFF 2007), the frozen food market (retail and foodservice) accounted for a total value of 87.5 million pounds in 2006 with the United States, Germany, and the United Kingdom as the three top-consuming countries, in both volume and value. In the United Kingdom, the retail frozen food market showed an annual value growth of 1.7% for the year ending in March 2007, the value of total retail frozen foods for the period amounting to just under 4.5 million pounds (BFFF 2007). For consumers, frozen foods represent a practical and straightforward manner to have access to ready-meals or food ingredients at short notice, providing a year-round supply of seasonal produce. The chilled food market has evolved consistently over the last decades and is still a major growing category of foods (Dennis and Stringer 2000). The sector offers continuing opportunities for new as well as improved products in order to keep up with consumer demand for fresh and convenient foods. Not surprisingly, recent growth has been largely in the areas of healthy eating and added-value products (Dennis and Stringer 2000).

In this chapter, the effects of chilling and freezing on the safety and quality of food products, which have some common aspects, will be explored first. Discussion on specific quality properties and safety concerns as affected by chilling and freezing will then follow. Although such concerns depend on the product and overall process in question, some general considerations are applicable to most products.

11.2   SAFETY AND QUALITY ISSUES AT LOW TEMPERATURES

The overall quality evaluation of any food product must evidently include its safety for consumption. While quality and safety can be considered together, an important distinction is that food product safety should be a primary and indisputable premise during food handling, manufacture, and distribution; while the quality factors and properties can vary considerably according to the target consumer group, their preferences, and expectations. Quite often, product quality will be a compromise between demand, cost, and operational limitations. While consumers might be willing to compromise in certain aspects of product quality, that will be unlikely whenever food safety is concerned. The responsibility is on the manufacturer, not only to supply food that is fit for consumption but also to ensure that safeguards are in place so that products will reach consumers at optimum conditions, and that relevant information is available for consumers on handling, storage, and on final cooking instructions, if applicable. While in the past many food manufactures relied mainly on sampling plans for monitoring food safety and quality, the inherent deficiencies of this system meant that product hold and disposal were a constant possibility. These have now been largely overcome by a wider use of quality and safety assurance programs. Instead of checking for compliance, manufacturers employing safety and quality have control measures built into the process so that a proactive approach is in place. Although product safety cannot be compromised, steps taken to achieve the required safety level during product shelf life may vary according to economical and operational capability. It is often the case that product quality and safety parameters are achieved and maintained using a combination of processing steps rather than relying on one operation alone, and several approaches and techniques might be available to the producer.

Specific safety concerns will depend on the food product and ingredients, as well as on the processing steps involved, which might, for example, favor or bring into prominence a particular group of microorganisms. Nevertheless, the dominant microflora of concern when considering the microbial growth and spoilage in chilled foods will generally consist of psychrotrophs, capable of growing at temperatures down to −5°C (Adams and Moss 1995). While mesophilic organisms will not grow at low temperatures, they may become a problem in situations of temperature abuse. In many cases, the product comprises several ingredients and their respective susceptibilities and characteristics must be considered when designing the process. The microbial flora and initial load will vary considerably between different foods and these will determine the necessary specific measures to be taken during the process. The importance of aforementioned safety assurance programs is further stressed by the fact that chilling or freezing per se cannot be considered as reliable bactericidal steps, especially for pathogens. The main purpose of chilling or freezing is to extend shelf life. The quality and integrity of the raw ingredients and the proper operational conditions and their efficient monitoring and control are therefore vital to minimize quality problems and ensure product safety.

In regard to undesirable chemical and biochemical reactions, oxidation and enzymatic reactions are frequently a source of quality loss for food stored under low-temperature conditions. Although reaction rates are reduced, deterioration can still occur at temperatures well below 0°C, eventually leading to constituents' degradation and development off-flavors and odors that will ultimately limit product shelf life. The effect of storage conditions on product moisture content and distribution is also frequently responsible for a gradual deterioration of product texture, appearance, and palatability. However, the combination of the beneficial effects of low-temperature storage and other preservative methods and agents have made it possible to develop an increasing variety of products which are stable and safe for consumption, for satisfactory lengths of time.

11.3 CHILLING AND FREEZING—LOW-TEMPERATURE PRESERVATION

The principles of low-temperature preservation have been employed for many years and are for the most part due to the advantageous negative effect of reduced temperature on various chemical and biochemical reactions responsible for food spoilage, as well as on microbial growth and spore germination. Additionally, during rapid chilling some microbial injury and death is also likely to occur due to cold-shock. Cold-shock is also possible during freezing, though a slow temperature drop is actually more harmful to organisms due to longer exposure to high solute concentrations (Adams and Moss 1995). Preformed microbial toxins and other undesirable chemical products are not affected by chilled or frozen storage and therefore contaminated products will still be unfit for consumption after thawing. Because the preservation effect is largely (though not entirely) due to the low temperature during storage, once the food product is defrosted or warmed up, the food is as perishable as prior to chilling or freezing. Temperature fluctuations during storage may cause moisture migration and loss, accelerate spoilage reactions, and may even provide sufficient conditions for microbial growth during thawing. The rate of temperature reduction, the control of storage temperature and conditions, and the rate and conditions during thawing are among the relevant aspects for the manufacture of high-quality frozen food products. However, some of these factors might not be under the control of the food manufacturer, as products ultimately have to be transported and prepared for eating by the consumer.

As previously mentioned, very often low-temperature storage is used in combination with other preservation methods to ensure food safety and enhance quality properties, and to expand the range of consumer goods available. Because chilling and freezing cannot be considered in isolation as a reliable bactericidal step, conditions are usually dictated by maximum quality retention rather than on microbial lethal capacity. Typical products may consist of uncooked ingredients, but more often components will receive some prior heat treatment, for example, blanching for some fruits and vegetables or pasteurization for meat and dairy products. For many pasteurized, fermented, or cured foods, and for some products packed under vacuum or modified atmosphere conditions, low-temperature storage is an intrinsic part of the preservation system.

As with most food products, the quality of chilled and frozen foods is primarily dependent on the quality of the raw materials employed, as cold storage cannot compensate for or improve the use of inadequate or poor quality ingredients. Indeed, it is widely recognized that product freshness and overall quality prior to chilling or freezing will, to a substantial degree, determine the final quality of product before eating.

Many of the detrimental effects of chilling or freezing on product quality properties can be greatly reduced or even eliminated by employing appropriate packaging, particularly for cooked products. Lipid oxidation rate, a determinant in many frozen foods shelf life, increases with product exposure to air and the reaction in some cases is catalyzed by light. Unwrapped frozen products are also prone to dehydration and freezer burn, thus altering product appearance and sensory properties.

The desired quality properties sought and expected from a food will be characteristic of the product. Apart from the initial microbial load and overall integrity of the raw materials, time taken to reduce product temperature and storage temperature control are frequently the most likely causes of ordinary quality and safety problems arising during manufacturing. In general (though not always), temperature should be reduced as quickly as possible to minimize quality loss and prevent or minimize microbial growth or spore germination, and always, storage should be maintained at the chosen temperature with minimal fluctuation to prevent undesirable effects on product texture and flavor, as well as to avoid present microorganisms to grow or produce toxins. For frozen foods, the thawing method might also have a significant bearing on the quality of the unfrozen product, particularly in relation to the amount or moisture loss (thaw drip).

11.4 EFFECTS OF CHILLING ON SAFETY AND QUALITY OF FOOD PRODUCTS

Chilling comprises not only the reduction of food product temperature but also the maintenance of the low storage temperature throughout product's shelf life. Separate equipment should be used for each step to ensure that temperature reduction is rapid enough to maintain quality properties and prevent microbial growth and to prevent excessive temperature variation during storage once products are cooled to the desired level. Chilled foods are typically stored at temperatures close to but above their freezing point, ranging between 8°C and −1°C. The definition of the U.K. Chilled Foods Association for chilled food refers to only prepared food and excludes unprepared foods such as raw meat, poultry, and fish as well as some dairy products such as butter and cheese (CFA 1997). In contrast, the definition given by U.K. Institute of Food Science and Technology incorporates all perishable foodstuff stored at refrigerated temperatures (IFST 1990). Different storage ranges can be employed depending on the food product in question. For example, fresh and smoked meats and fish are kept at the lower temperature range (between −1°C and 1°C); most pasteurized foods and ready-to-eat products (such as soups, salads, pasta, and sandwiches) can be kept between 0°C and 5°C; and cooked meats and fish products, cured meats, cheese, and butter may be stored between 0°C and 8°C (Fellows 2000).

Domestic refrigerators should operate between 1°C and 5°C but often temperatures much higher than these are encountered (James et al. 2007). Commercial chillers might also operate at ultrachilled temperatures between −3°C and −7°C (Sun et al. 2005). However, foods stored in the ultra-chilled region are in fact partially frozen.

Chilling can be employed on its own to increase food product shelf life but more often it is combined to other preserving techniques to increase safety and maintain optimum product quality for longer shelf life. When choosing among the available and applicable preservation methods, food manufacturers must evaluate the risks posed by the consumption of the final food product, as a consequence of the raw materials employed and their associated hazards, the preservation steps in place and achievable control levels, and in some cases also in relation to particularly vulnerable consumer groups (infants, infirms, and the elderly), among other factors. Manufacturers may categorize chilled foods into risk classes, according to preestablished criteria that should include the nature of the ingredients; the presence and severity of a control step; particular characteristics of the processing that might make an identified hazard more or less likely to pose a risk to consumers; and the probability of further contamination or hazard development during storage and distribution. For instance, products made up entirely of raw ingredients may be ranked as high risk while in-pack cooked and cooled products may be classed as low risk (Brown 2000).

Chilling storage extends the shelf life of most products and prevents growth of a great proportion of microorganisms but it often changes the nature of spoilage involved. Psychrotrophs will be the dominant microflora and there are a number of pathogens that can still grow and represent a safety risk. Psychrotroph growing rate at refrigerated conditions is reasonably low. However, even small temperature

TABLE 11.1

Pathogenic Microorganisms Capable of Growing at Refrigerated Conditions

Pathogen	Some Common Food Product Associations	MGT (°C)
Aeromonas hydrophila	Water, vegetables, fish, and shellfish	$-0.1–1.2$[a]
Bacillus cereus	Meat, seafood, milk, and vegetables	4.0[a, b]
Nonproteolytic *C. botulinum*	Meat, fish, canned low-acid foods	$3.3–5.0$[a]
Enterotoxigenic *Escherichia coli*	Meat, poultry, fish, cheese, and vegetables	4.0[c, b]
Listeria monocytogenes	Meat, poultry, milk and dairy products, seafood, and vegetables	-0.4[a, b]
Salmonella enteritidis	Meat, poultry, milk, and eggs	5.0[c]
Vibrio parahaemolyticus	Fish and shellfish	5.0[c, b]
Yersinia enterocolitica	Raw and cooked meats, poultry, seafood, and milk	-1.3[a]

Note: MGT, minimum growth temperature.
[a] Walker and Betts 2000 (pp. 168–172).
[b] Fellows 2000 (Table 19.4).
[c] Adams and Moss 1995 (p. 182, 194, 211).

variations can have a marked effect on their growing rates, indicating the importance in maintaining stable storage temperatures. Psychrotrophs can be inactivated at relatively low temperatures, but if allowed to contaminate ready or postcooked products these organisms will be able to grow steadily causing food to spoil and, in some cases, lead to food poisoning incidents. Table 11.1 displays a list of pathogenic microorganisms capable of growing at temperatures below 5°C, and their most commonly associated food products. Psychrotrophic *Clostridium botulinum* is one of the greatest threats for chilled foods due to the extremely lethal and potent toxin it produces. Nonproteolytic *C. botulinum* strains have been reported to grow at temperatures as low as 3.3°C if given sufficient time and appropriate conditions (Brackett 1992).

The need for good manufacturing practices and a proactive approach to safety is therefore essential. Improper storage temperature or temperature abuse during chilled food storage is frequently identified as a significant vulnerability, particularly when refrigeration is the main or only preservation technique employed. Temperature abuse is an even greater concern if the product is not designed to undergo further heating prior to consumption. Cooked and chilled foods are different from raw or minimally processed foods, in which the former is generally free from the most common spoilage microorganisms associated with the food and in the event of postprocess contamination, pathogenic microorganisms will face no competing microflora to grow and multiply. Moreover, mild cooking procedures will not eliminate heat-resistant spore forming pathogens or their toxins.

The most significant nonmicrobiological concern in relation to the deterioration of chilled foods in general is oxidation. Lipid oxidation reduces the shelf life of fat-containing foods changing its color and appearance, developing rancid and off-flavors and odors, destroying nutrients and in some cases generating toxic by-products. For example, in cooked meat products an unpleasant "warmed-over" flavor (WOF) can develop. In fruits and vegetables, oxidation destroys color pigments and vitamins, affecting product appearance, palatability, and nutritional value. In addition, many fruits and vegetables might be susceptible to cold injury. Biochemical reactions catalyzed by specific enzymes are also often a problem. Normally the activity of enzymes is reduced at low temperatures but in the case of cold-water fish, for instance, reaction rate might not be considerably affected. Most enzymes of concern are originally present in the food product via its ingredients, but in some cases they might have been added for a specific purpose, for example, for tenderization or ripening, or derive from microorganisms killed by processing. The main undesirable enzymatic reactions that might occur during chilling and freezing storage of foods are enzymatic browning, lipolysis, glycolysis, and proteolysis. A few examples of food products affected and some of the available pretreatments to reduce enzymatic spoilage are shown in Table 11.2. These will be discussed with more detail in Sections 11.4.1 for meat and poultry and their products; 11.4.2 for fish and fish products; 11.4.3 for fruits and vegetables; and 11.4.4 for dairy products, when they are relevant.

11.4.1 Effects of Chilling on Red Meat and Poultry and Their Products

Changes in the meat tissue start immediately after slaughter, and include some desirable reactions that promote flavor and tenderness. However, microbial growth and enzymatic reactions will eventually render the meat unfit for consumption, hence the need for refrigerated storage. In addition to the beneficial effect of low-temperature

TABLE 11.2

Enzymatic Reactions of Significance in the Spoilage of Chilled Foods

Enzymatic Reaction	Food Products	Main (Negative) Effects	Possible Additional Pretreatments
Browning	Fruits and vegetables	Color change	Vacuum or modified atmosphere packaging Blanching, addition of acid or sugar, protective coatings
Lipid hydrolysis and enzymatic oxidation	Meats, fish, and dairy products	Off-odors and rancidity development	Heat treatment, addition of antioxidants, oxygen exclusion
Glycolysis	Fruits, vegetables, meats, and milk	pH decline	Blanching Addition of salt
Proteolysis	Meats, fish, milk, and dairy products	Loss/deterioration of texture; Bitter flavor development	—

Source: Adapted from Brown, H.M. and Hall, M.N., *Chilled Foods: A Comprehensive Guide*, 2nd ed., Eds., Stringer, M. and Dennis, C., CRC Woodhead Publishing Ltd., Cambridge, U.K., 2000.

storage on the reduction of spoilage reactions rates and microbial growth, moisture lost during chilling also tends to decrease microbial growth as a result of reduced surface water activity (a_w). However, excessive drying can have a detrimental effect on meat juiciness and color, and represent an increased loss in product weight. Up to 2% losses during chilling of beef carcasses have been reported, which prompted many plants to resort to water spraying of carcasses during chilling (Savell et al. 2005, Kinsella et al. 2006). However, the advantages of water spraying must be weighed against the risks of a higher surface water activity, its potential to increase bacterial loads, and its possible effects on carcass appearance.

The tenderness of meat changes dramatically during the first hours after slaughter. During this period, muscle fibers contract and meat becomes relatively tough, due to the onset of "rigor mortis". However, if the carcass is maintained at chilled temperatures (usually 2°C), continuing tenderization may take place while the rate of microbial growth and other degradation reactions is reduced. The rate of carcass chilling has a marked effect on meat tenderness development during rigor mortis onset and dissolution, which also affects drip loss and meat color development. Although in principle it might seem advantageous to increase the rate of carcass chilling as much as it is economically viable, it has been shown that a rapid drop in muscle temperature can lead to meat toughening, also known as cold-shortening. To avoid this effect, a combination of postslaughter time, temperature, and pH is employed. As a general rule, it is recommended that after slaughter the temperature of beef and lamb should not fall below 10°C before the onset of rigor mortis to avoid toughening the meat (Bender 1992, Anonymous 2000). Pork does not seem to be significantly affected by cold-shortening. The rate of rigor mortis advancement depends also on temperature, so after slaughter the carcass can be kept at ambient temperatures to accelerate rigor development, then rapidly chilled in order to condition the meat. Efforts on the development of postslaughter techniques have made it possible to combine increased chilling rates with other tenderizing processes such as electrical stimulation, pelvic suspension, and aging times to prevent cold-shortening and optimize meat quality properties. Electrical stimulation causes a very rapid fall in meat pH and accelerates rigor mortis, allowing cooling to proceed without the development of cold-shortening. Pelvic suspension takes advantage of the fact that stretched muscles do not suffer from cold-shortening during rigor mortis development. It has also been shown that it is possible to reduce the core temperature of beef muscles down to −1°C in 5 h postmortem to produce tender meat, using a technique called very fast chilling (VFC) (Joseph 1996, Van Moeseke et al. 2001). The effect of VFC on beef tenderness has been attributed to the release of bound calcium which seems to stimulate tenderizing enzymes, and to the formation of a frozen crust on the surface of the muscle that prevents shortening. However, the effect of calcium ions on proteolysis enzymes and subsequent meat tenderization has not been conclusively proven (Van Moeseke et al. 2001). The technique has been successfully applied to selected individual beef muscles rather than to a whole carcass, mainly because attainable cooling rate is dependable on meat piece size and shape and shortening seems to depend on the direction of cooling relative to the direction of muscles fibers.

As with red meat, tenderness is also of primary importance for poultry processors. After slaughter, carcasses are usually scalded in hot water to ease feathers' removal.

Chilling before completion of postmortem tenderization should be avoided, even if the process is accelerated by electrical stimulation. Optimum tenderness is usually achieved within 6–8 h for chickens and within 12–24 h for turkeys (Kotrola 2006). Poultry carcasses are usually chilled using either water immersion, water spraying, or air blast. Typically, carcasses can lose between 1% and 3% weight during air chilling, lose or gain weight during water spraying (between –2% and +1.7%), and gain between 4% and 8% weight during immersion chilling (James et al. 2006). Carciofi and Laurindo (2007) developed a mathematical model to describe the amount of water absorbed by poultry carcasses during immersion cooling. The authors showed that improving water stirring conditions can decrease immersion chilling time, yet carcass water uptake is also increased, which in many countries is controlled by regulations. Water uptake was influenced by cooling conditions such as water temperature and hydrostatic pressure. Savell et al. (2005) reported in their review of the effects of chilling systems on meat quality that quickly cooled carcasses were more likely to suffer induced cold-shortening and toughening, while delayed chilling could potentially have a positive effect on postmortem tenderness. Research work to ensure consistent tenderness development in meat, preferably combined with low-temperature storage, to maximize operational conditions, quality properties, and improve safety standards, is a continuing effort.

Among red meat and poultry products, processed meat products can differ considerably in terms of their organoleptic properties, spoilage patterns, and shelf life. Many of the products widely available rely on the essential effect of low-temperature storage as an intrinsic part of the preservation system, to maintain products' optimal characteristics and ensure their safety for consumption for the period of their shelf life. Some common examples are cured, fermented, or cooked meat products, vacuum-packed products as well as those packed under modified atmospheric conditions. Despite the process chosen and its effects on the product microflora, chilled storage will basically prolong product shelf life by slowing down the rate of enzymatic and chemical reactions, and by reducing the overall level of microbial growth. However, another effect is that microorganisms capable of growing under these new conditions will tend to dominate the spoilage microflora, unless previous processing steps are successful in controlling or eliminating their numbers. This might bring into prominence pathogenic microorganisms that would otherwise seldom represent a safety risk in that product. In the case of cooked meat, microorganisms of concern after pasteurization will often be heat-resistant mesophilic spores of *Bacillus* and *Clostridium* species and meat products must therefore be refrigerated or frozen, to prevent spore germination and growth (Silliker et al. 1980). During cooling, the period that the product stays between 50°C and 12°C has been characterized as the most hazardous part of the process during which adequate conditions to avoid germination and growth of heat-resistant spore forming bacteria must be achieved (Gaze et al. 1998). In commercial operations, long cooling times are common especially for large meat pieces or joints (Gaze et al. 1998, Sun and Wang 2000). In recent years, investigations have shown that vacuum cooling can offer the meat industry a rapid and safe cooling alternative (McDonald et al. 2000, Sun and Wang 2001, Zheng and Sun 2004, 2005). Sun and Wang (2001) have illustrated the difference in the average cooling time from 70°C to 10°C for large hams (~7 kg), using air blast, water immersion, and vacuum cooling systems. Approximate times of 7, 5, and 1 h, respectively, were reported. A lower microbial load, as well as rapid cooling times, was reported for vacuum-cooled beef samples (McDonald et al. 2000). Cooling loss tends to increase and a decrease in meat tenderness and juiciness has been associated with such products. However, more recent investigations into a modified system where cooked meat is vacuum cooled together with some of its cooking solution (immersion vacuum cooling) have shown reduced yield losses and improved quality for cooked pork ham (Cheng and Sun 2006a,b).

Lipid oxidation is the main chemical reaction responsible for quality deterioration and shelf life reduction of meat and processed meat products. The susceptibility of different meat species depends on the level of unsaturated fatty acids in the tissue. Fish tissue is most affected, followed by poultry, pork, beef, and lamb (Byrne et al. 2002). A characteristic unpleasant WOF develops after heat treatment of meat. Free iron, a result of myoglobin denaturation, catalyzes the oxidation of lipids, especially phospholipids. Off-flavors accumulate during cooking and the level of WOF is said to increase with cold storage, as a simultaneous loss of the desirable "meaty" flavor takes place. Cooked meat and poultry pies, hamburgers, patties, and coated products such as nuggets, which require cooking and subsequent reheating, are specially affected. Both synthetic and natural antioxidants have been successfully employed to prevent or delay oxidation and the development of WOF components in meat and poultry products (Mielnik et al. 2006, Jayathilakan et al. 2007, Mielnik et al. 2008). The use of vacuum packaging, modified or low oxygen atmosphere packaging has also been reported to successfully decrease the level of WOF development in beef and pork cooked and chilled products (Stapelfeldt et al. 1993, Kerry et al. 1998, Ansorena and Astiasaran 2004). Another possible approach is to cover meat with sauces or vegetable layers to reduce oxygen contact (Olsen and Aaslyng 2007).

11.4.2 Effects of Chilling on Fish and Fish Products

As with red meat and poultry, changes taking place shortly after catch of fish and other seafood affect their quality and stability. Chilling fish immediately after catching is extremely important since fish is recognized as among the most perishable food products. Initially, changes are mainly due to autolytic enzymatic reactions and later microbial enzymes and growth of microorganisms eventually lead to fish spoilage. Adequate chilling can slow down biochemical reactions and microbial growth, reduce mass losses, and maintain the associated quality aspects of fish. Contamination and damage to fish tissue during catching, transport, and processing will increase the rate of spoilage. Gutting fish prior to chilling removes a major reservoir of microbial contamination, yet it exposes internal surfaces to rapid spoilage (Adams and Moss 1995). Filleting also exposes fish to microorganisms, enzymes and oxygen, altering and accelerating spoilage. Nevertheless exports of fish fillets have been on the increase while the numbers on whole fish steadily decrease (Olafsdottir et al. 2006). All these factors make refrigeration of fresh fish critical for maintaining their quality. The rapidly cooling and storing of fish, close 0°C, is however the most important factor for reduction of deterioration rates and extension of quality and shelf life (Whittle et al. 1990). Preservation systems available to the fish industry include the use of ice, chilled seawater, and slurry ice (Anonymous 2000). During storage in ice, and depending on the species, fish can either gain or lose significant weight. Weight lost as drip is accompanied by leaching of proteins and flavor components, while weight gained usually incur in salt absorption, discoloration, and off-odor production (Anonymous 2000).

The color and overall appearance of fish starts changing quite rapidly after catch, the discoloration is a result of myoglobin oxidation. The reaction rate can be reduced with lowering of temperature but some change in color still occurs. The rate and extent of rigor mortis progress and pH decline and myofibril breakdown caused by proteolysis all have an effect on the development of fish meat texture (Haard 1992). As with other meats, lipid oxidation is a significant cause of deterioration and is often determinant of product shelf life. The extent of oxidation varies with the quantity of lipids in the muscle, fatty fish being obviously more susceptible. The kind of lipid present is also relevant, since unsaturated fatty acids are less stable. If lipid oxidation proceeds, off-flavors and odors develop which will render the product unacceptable. The products of lipid oxidation reactions have also been associated to changes in the protein matrix of fish muscles, affecting protein solubility and causing adverse textural changes in the muscles. Unlike the changes occurring in red meat and poultry proteins during tenderization and storage, oxidative reactions may also induce significant nutritional losses in sulfur-containing proteins. For some fish species, it has been established that the breakdown of trimethylamine oxide (TMAO) and the formation of dimethylamine and formaldehyde (FA) are responsible for undesirable texture and organoleptic changes, causing the destabilization and aggregation of proteins (Santos-Yap 1996). Partially dried or salted fish products made from fatty fish are more susceptible than "wet" fish to oxidative rancidity during chilling storage (Anonymous 2000).

The detection of specific spoilage organisms is a reliable method to evaluate the freshness or spoilage level of fish (Olafsdottir et al. 2006). Microorganisms like *Shewanella putrefaciens*, *Pseudomonas* ssp., and *Photobacterium phosphoreum* are frequently responsible for quality changes and development of off-odors in chilled fish, reducing its shelf life. The temperature characteristics of the associated flora will normally reflect the water temperature in which the fish live. Cold-water fish bacterial flora is mainly psychrotrophic while warm-water fish is mainly mesophilic. Therefore, rapid chilling and chilled storage temperatures will be more successful for extending the shelf life of the latter, due to the higher vulnerability of their natural flora to reduced temperatures.

Further processing of fish during the manufacture of fish products such as fillets, patties, and fish sticks (surimi) increases the surface area exposed to oxygen, therefore increasing product susceptibility to deterioration. Processing should whenever possible avoid warm or high-temperature treatments that are unfavorable to fish product stability and quality properties; unless a cook-chill processing is truly intended. Coatings or films are finding increasing applications as protective barriers against contact with air (oxygen) and water, delaying product deterioration during cold storage. Some are reported to also have antimicrobial, binding, and texturizing properties (Lopez-Caballero et al. 2005). Other food-grade additives are usually incorporated to surimi-type fish products to prevent texture hardening during chilled storage (Kok and Park 2007).

11.4.3 Effects of Chilling on Fruits and Vegetables

Fruits and vegetables start deteriorating immediately after they are harvested. Relevant changes that lead to spoilage of fruits and vegetables are oxidation, enzymatic reactions, nonenzymatic browning, color changes due to pigments degradation, and nutritional losses. Harvest triggers many of these detrimental changes, as well as further processing. Peroxidase activity and ascorbic acid content are frequently used as indicators of vegetable quality while produce color is used as an indicator of vegetable or fruit maturity. Pigments can be grouped as anthocyanins, carotenoids, and chlorophylls. Oxidation is mainly responsible for anthocyanin breakdown and the reaction is catalyzed by light. Carotenoid content increases with ripening and such compounds are also easily oxidized. Chlorophyll level decreases with storage and produce ripening. For naturally green produce such as green vegetables (spinach, broccoli, peas) and fruits (avocados, green grapes, kiwi, pears), storage life is often determined by the level of chlorophyll loss after processing. Many studies have focused on the best treatments to ensure prolonged shelf life with minimal quality loss. Refrigeration and the beneficial effects of chilling on reducing detrimental reaction rates are regularly employed to delay produce deterioration, allowing the necessary time for transport and distribution. However, many fruits and vegetables cannot withstand low temperatures without experiencing some signs of chilling injury. Adverse effects depend not only on the fruit or vegetable but regularly also on the particular variety in question. Chilling symptoms may include discoloration, water loss or soaking, development of off-flavors, accelerated or failure to ripe, increase senescence, and decay (Tabil and Sokhansanj 2001, Singh and Anderson 2004). Symptoms may start from 15°C, down to prefreezing temperatures, and in some cases injury signs are not apparent for weeks. Because species may differ significantly in their sensitivity to chilling, much research has been directed into understanding the nature of chilling damage and on the use of pretreatments and different storage conditions so as to avoid or minimize chilling injury (Meir et al. 1995, Purvis 2002, Rivera et al. 2007). In order to extend postharvest storage, produce respiration and transpiration should be reduced as much as possible. This is usually accomplished by a combination of low temperature and increased air relative humidity storage. Usual chilling storage conditions for a few fruits and vegetables and their respective estimated shelf life are shown in Table 11.3.

Frequently, additional treatments are also employed to help prevent undesirable quality changes or reduce their reaction rate in order to preserve the properties that define a high-quality product. Blanching can be used to inactivate enzymes and remove tissue air, as well as to reduce microbial load and maintain texture and color. However, because heat considerably alters fresh quality properties such as turgidity and juiciness, blanching is more widely employed for vegetables than for fruits. For a few types of fruits, particularly those that will undergo further processing, blanching can still be an option. Otherwise, enzymatic reactions may be prevented by chemicals acting as antioxidants. Ascorbic acid (vitamin C) combined with citric acid is frequently used to reduce and delay fruit oxidation. Sulfur dioxide can also be employed to prevent enzymatic as well as nonenzymatic Maillard-type browning and as an antimicrobial. Whole fruits can also receive a protective coating before storage, while pieces or slices can be deaerated, treated using sulfur dioxide (SO_2)

TABLE 11.3
Typical Chilling Storage Conditions for Some Fruits and Vegetables

Product	Storage Temperature (°C)	Relative Humidity (%)	Shelf Life
Apples	−1 to 4	90–95	1–8 months
Bananas	11–16	85–90	5–10 days (ripe)
			10–30 days (green)
Lemons	10–14	90	1–6 months
Oranges	2–7	90	1–4 months
Aubergines	7–12	90–95	1–2 weeks
Broccoli	0–1	95–100	1–2 weeks
Cucumbers	8–15	90–95	1–2 weeks
Peas	0–1	95–100	1–3 weeks
Spinach	0–1	95–100	1–2 weeks

Sources: Adapted from Fellows, P.J., *Food Processing Technology: Principles and Practice*, 2nd ed., CRC Woodhead Publishing Ltd., Cambridge, U.K., 2000; Aked, J., *Fruit and Vegetable Processing—Improving Quality*, Ed., Jongen, W., CRC Woodhead Publishing Ltd., Cambridge, U.K., 2002.

diluted with nitrogen (N_2), then treated with vacuum to eliminate residual SO_2. This procedure has been shown to successfully inactivate catalase and polyphenolixidase (Skrede 1996). The addition of sugars to exclude oxygen and preserve color and appearance is also commonly used. All above treatments are usually combined with cold storage to enhance product shelf life. Ultimately, the choice of treatment will depend on the intended purpose for the product and on its desired shelf life. Long storage periods will eventually have an undesirable effect on quality properties, and a compromise has to be reached. Much research has also been undertaken on the appropriate selection of fruit and vegetable varieties based on their particular properties and resulting in suitability for specific purposes and treatments.

11.4.4 EFFECTS OF CHILLING ON DAIRY PRODUCTS

Specifically dairy products are perishable foods and they benefit greatly from low-temperature preservation. Typically, a few preservation steps are employed to ensure product safety and a longer shelf life. Most dairy products are pasteurized before commercialization, and subsequent chilled storage ensures that remaining microbial flora growth and enzyme activity are significantly delayed. The most relevant safety issues in relation to such products include the microbiological standards of raw milk for processing and the prevention of postpasteurization contamination. Pasteurized milk and dairy products (butter, cream, and cheese) are used both as foods and as ingredients in the manufacture of food products. Pasteurized milk and dairy products still require chilled storage to prolong their shelf life. Possible quality problems include lactic souring, bitty cream, and proteolysis (Early 2000). Proteolysis is caused by the presence of microbial enzymes not inactivated by pasteurization and the reaction is favored by low temperatures. The main adverse effect of chilling on the quality of dairy products is related to the incidence of lipid oxidation. For this reason it is recommended that packaging for butter, cream, and products rich in these ingredients is chosen to exclude light, which promote milk fat auto-oxidation. However, the development of oxidative rancidity is of course dependent on the presence of oxygen. In addition, milk fat is highly susceptible to lipolysis during chilled storage which results in rancidity.

11.4.5 Effects of Chilling on Other Miscellaneous Products

The variety of chilled foods already available is wide ranging and new products are continuously being developed. Although the effect of chilling is largely beneficial for most products in terms of retained quality and prolonged shelf life, for a few products low-temperature storage may actually raise specific quality problems. For instance, moisture migration and retrogradation of starch which lead to staling in bread and other bakery products are promoted at low temperatures, being detrimental to product quality and therefore undesirable. Some gels may shrink and exude their liquid at chilled temperatures (syneresis) and oil/water suspensions may become unstable. Some raw fruits and vegetables prone to chilling injury might be added as ingredients on a ready-to-eat salad, and other likely products may include both raw and cooked ingredients which could bring active enzymes in contact with new and suitable substrates. Similarly, new challenges for product safety assurance arise from chilled storage conditions which bring into prominence groups of microorganisms that might otherwise not be associated with the product. This emphasizes the need for a thorough safety analysis system as well as a comprehensive quality management program able to identify possible hazards and quality problems likely to arise from the product and process characteristics present.

11.5   EFFECTS OF FREEZING ON SAFETY AND QUALITY OF FOOD PRODUCTS

Frozen foods are stored at temperatures ranging between $-18°C$ and $-30°C$. The optimum storage temperature varies among different products and is often a compromise between quality losses and economic factors. Although microbial activity halts and enzymatic reactions are greatly reduced during frozen storage, quality deterioration can still take place at low temperatures and pathogenic microorganisms present will not necessarily be killed. Indeed, Lund (2000) presented a list of outbreaks and illnesses associated with the consumption of contaminated frozen foods.

Even at $-18°C$ foods are not totally frozen, and biochemical and chemical reactions may still occur. Quality problems arise as a result of physical changes during freezing due to ice crystal formation and growth and during storage due to surface dehydration. Temperature fluctuations during storage allow moisture migration and subsequent ice crystal growth, which damages food structure. Even at constant temperature ice crystals can grow due to "Ostwald ripening" (Sahagian and Goff

1996), a result of differences in vapor pressures and relative free energy between small and larger crystals. The low humidity storage environment also induces moisture sublimation in unwrapped products, leading to surface dehydration and freezer burn. Therefore, although in principle frozen foods can be stored for an indefinite period, during long-term storage quality changes inevitably occur, even at controlled conditions.

11.5.1   Freezing Process

For ease of discussion, the freezing process is usually divided into three steps: freezing, frozen storage, and thawing. Process conditions during each of these steps may produce cumulative effects on product final quality as well as independent outcomes. That is to say, failure to control relevant parameters during each step may lead to undesirable quality losses in the final product that cannot be compensated by optimum operation of the subsequent or prior procedures.

11.5.1.1   Effects of Freezing

During the first step, product temperature is lowered, usually from an already chilled condition, until the final frozen storage temperature. Even though most physical and biochemical detrimental changes are significantly reduced at storage temperatures under −18°C, many of these deteriorative reaction rates actually accelerate between 0°C and −15°C, an effect due to solute concentration in the unfrozen water phase. Therefore, a faster freezing rate is sometimes employed first, followed by a slower temperature drop speed until final storage conditions. Freezing rate is of primary importance for product quality. Apart from minimizing the effect of solute concentration on enzymatic and chemical reaction rates, a fast temperature drop promotes formation of several smaller and more uniformly distributed crystals, with less shrinkage of cells as a consequence. In contrast, slow freezing allows an uneven formation of ice crystals, leading to moisture migration and subsequent crystal growth, which causes cell deformation and rupture. Although fast freezing has a damaging effect on microorganisms, killing or injuring a great proportion of the present population, slow freezing negative effects on cell structure are also significantly harmful to microorganisms. The combination of rapid freezing and slow thawing is accepted to have the most injurious effect on microorganisms (Adams and Moss 1995). Nevertheless, freezing is not a reliable killing step for pathogenic microorganisms.

Water provides a medium for undesirable microbial growth and biochemical reactions. During freezing, as ice is formed the remaining liquid phase becomes increasingly concentrated. As a consequence the solution freezing point is progressively depressed and its viscosity increases. The extent of freezing point depression depends on the food composition as the effect of different solutes will vary. Effectively, a great number of foods are not completely frozen, even at −18°C. Heldman (1992) has presented data on the initial freezing temperature of some foods, as well as extensive information on food properties required for calculation of freezing times and equipment requirements. Freezing point depression and unfrozen water content can be calculated by using the model presented by Heldman and Singh (1981) when the initial freezing temperature of the food is known. Frozen water content can also be obtained from enthalpy data, or experimentally from differential scanning calorimetric measurements. If the unfrozen phase becomes so concentrated that its viscosity is increased (usually above 10^2 Pa s), it may form a glass (Slade and Levine 1991). Under this state, water and solute diffusion is minimal, and physical changes as well as chemical and biochemical reactions virtually cease, significantly increasing frozen food stability. Therefore, much effort and research has been directed at determining the glass transition temperature of different food materials and at understanding the effects of storage conditions on the frozen food stability (Goff, 1992, 1994, Reid et al. 1994, Torreggiani et al. 1999, Ohkuma et al. 2008).

The use of power ultrasound to control crystallization processes has been shown to improve both the rate of nucleation and the crystal growth rate. During freezing of fresh foodstuffs, power ultrasound can be used to accelerate the freezing process and it can help control crystal size distribution in the frozen product, improving quality. Freezing rate is enhanced by an increase in ultrasound output power and exposure time; however, the heat produced when ultrasound propagates through the medium limits the power and exposure time applicable during freezing. For instance, Li and Sun (2002b) applied 15.85 W ultrasound power for 2 min during immersion freezing of potatoes and reported that the freezing rate was greatly improved. Additionally, the authors observed that the plant tissue exhibited good cellular structure, with less intercellular void, less cell disruption, and a prevalence of intracellular small ice crystals. This was attributed to the high freezing rate obtained under high ultrasonic level (Sun and Li 2003).

Recently, a great deal of effort and attention has been turned to the application of high pressure processes and on the use of high pressure shift freezing (Li and Sun 2002a). The principle is based on the fact that at certain pressures, water stays unfrozen at temperatures considerably lower than 0°C. Food samples can, for example, be cooled under pressure inside a vessel without freezing, and later rapidly frozen by depressurization. The main advantages of the technique include the homogeneous and instantaneous ice formation due to the high degree of super cooling, resulting in significantly higher quality properties of the frozen product (Hogan et al. 2005).

Although a very fast freezing rate is usually recommended for keeping high quality properties in a frozen product, there are exceptions when fast freezing may actually be detrimental due to excessive stress that may result in physical damage or, in the case of dough freezing, due to negative effects on yeast survival and activity. In some cases, different freezing rates are employed within specific temperature ranges, depending on the effects of temperature on the rate of relevant undesirable reactions. Therefore, procedures for different products are usually established resulting from a compromise between minimum detrimental effects on quality, safety procedures, and cost.

11.5.1.2 Effects of Frozen Storage

During frozen storage, the main concern is in regard to the effects of storage conditions (commercial and domestic) on quality properties of food products (Gormley et al. 2002, Martins et al. 2004). Temperature variation during storage results in moisture migration and recrystallization, forming larger ice crystals which may give products a grainy texture and cause damage to food tissues. These adversely affect product quality by changing its appearance and texture, as well as increasing drip loss during thawing. In fact, improper frozen storage conditions can wipe out all the benefits of a carefully designed freezing step.

Dehydration, or freezer burn, is caused by sublimation of moisture at the surface of foods, exposed to a drier ambient, resulting in an opaque, darker surface appearance, possible toughening and development of off-flavors. Although freezer burns have no safety implications, dehydration causes unattractive changes in product appearance and moisture lost, representing an economic loss to producers.

The packaging of food materials plays an important part on its storage life and quality, and appropriate packaging can prevent or minimize many of the common quality problems associated with frozen foods. Packaging can act as a protective barrier, reducing freezer burn. In addition, it can also avoid or reduce food contact with oxygen, minimizing lipid oxidation, and development of rancidity. As already mentioned, the rate of lipid oxidation very often determines the shelf life of frozen products. A comprehensive handbook on the effects of processing and packaging on several frozen foods has been recently edited by Sun (2006). Examples of the estimated storage life of some frozen food products stored at −18°C are listed in Table 11.4.

11.5.1.3   Effects of Thawing

Thawing of commercial frozen foods may be undertaken at industrial or domestic environments. As such, operating conditions during this step may not be under the control of the same manufacturer responsible for product freezing and storing. Thawing is a significantly longer process than freezing, because water melting at the product surface has a lower thermal conductivity and thermal diffusivity than

TABLE 11.4

Approximated Storage Life of Some Frozen Foods Stored at −18°C

Food Product	Storage Life (Months)
Broccoli	15
Carrots	18
Peas	18
Spinach	18
Whole chicken	18
Minced meat	10
Bacon	2–4
Oily fish	4
White fish	8–10
Shellfish	4–6

Source: Adapted from Anonymous, *Recommendations for the Processing and Handling of Frozen Foods*, 3rd edn., International Institute of Refrigeration, Paris, France, 1986.

ice (Fellows 2000). During thawing, damages to the cell structure deriving from slow freezing rates or improper storage conditions will finally result in the loss of moisture and internal cell constituents (drip loss), which may include important water-soluble vitamins and nutrients. This nutrient-rich phase forms an ideal substrate for enzyme and microbial activity, a significant problem due to the slow nature of the thawing process. During commercial operations, foods are thawed in a manner that minimizes total thawing time while avoiding excessive moisture losses and overheating of already defrosted sections. Microwave or dielectric thawing, where the heat is generated from within the food, is sometimes employed. Other more widely used options are water immersion thawing, moist air, and condensing steam.

11.5.2 Effects of Freezing on Meat, Poultry, and Their Products

In order to minimize microbial growth and reduce the extent of most biochemical reactions, it would be ideal to chill and freeze meat as quickly as possible after slaughter. However, as already discussed, the undesirable effects on meat texture and palatability would be unacceptable for the consumer. Although in principle meat can be frozen pre- or postrigor mortis, it is generally recommended that only postrigor meat should be frozen. Among other effects, thaw drip is greatly increased during thawing of prerigor frozen meat (thaw rigor) (Lawrie 1991).

The main effects of freezing on meat are on its appearance, tenderness, and lipid oxidation. Frozen meats may have a darker or lighter surface color due to dehydration and freeze burn. In frozen poultry, young birds may also exhibit some bone darkening due to hemoglobin leaching. However, cooking before freezing can prevent this latter problem (Sebranek 1996). Oxidation is by far the main adverse change affecting the eating quality of frozen meats. Meat fats are vulnerable to oxidative rancidity and myoglobin oxidation leads to detrimental changes in meat color and overall appearance. As already mentioned, pork and poultry meats, richer in unsaturated fatty acids, are more susceptible to rancidity than other meats. Comminuted meats are also more susceptible to rancidity than whole muscle products, due to the larger surface area exposed to oxygen. Because meat can be frozen without prior heat treatment, little loss of nutrients occurs during the freezing procedure. Although freezing may cause protein denaturation, this appears to have no adverse effect on their nutritional value or digestibility. However, denaturation does have detrimental effects on protein functionality, altering important structural characteristics of the meat product. Product texture and water holding capacity, associated to product moisture retention and drip loss during cooking and thawing, are particularly affected by changes in the protein matrix. Higher nutritional losses usually take place during meat thawing, through thaw drip containing soluble proteins, vitamins, and minerals. The amount of drip depends on prefreezing treatments, as well as on freezing, storage, and thawing conditions. It may vary from 1% to 10% of the meat weight but is usually around 5% (Bender 1992). Research results on the effects of freezing on nutritional properties of meat vary widely, mainly due to difficulties in the consistent analysis of the particular nutrients.

11.5.3 Effects of Freezing on Fish and Fish Products

Fish is a very perishable product for which freezing is particularly suitable as a preservation method. Frozen fish can be efficiently transported and distributed, without substantial losses to their desired quality properties, extending product shelf life and availability. Ideally, fish should be frozen as soon as possible after catch, as undesirable changes in color and overall appearance take place quite rapidly. However, it is often the case that fish has to be chilled and stored before freezing. Research has shown that the preliminary chilling time and method can have a strong influence on frozen fish quality (Losada et al. 2007).

Myoglobin oxidation, which results in discoloration, can still occur during frozen storage. Lipid oxidation also proceeds during prolonged frozen storage and, eventually, off-flavors and odors develop which will make the product unacceptable. The reaction varies with the quantity and nature of lipids present in the muscle, hence fatty fish and fish containing a great proportion of unsaturated fatty acids are more rapidly affected by oxidation. Oxidation products, such as the formation of dimethylamine and FA in some fish species susceptible to TMAO breakdown, have also been associated with the destabilization and aggregation of muscle proteins during frozen storage and in sulfur-containing fish proteins, oxidative reactions may also induce significant nutritional losses (Santos-Yap 1996). Recommendations regarding maximum frozen storage temperatures take into account species' susceptibility to lipid oxidation. Highly susceptible species should be stored at very low temperatures (at least −29°C) while less susceptible species can be stored between −18°C and −23°C. For species prone to TMAO breakdown, storage temperature must be below −30°C. During fish frozen storage, water migration will cause some protein structure damage, while increasing solid concentration in the unfrozen water phase will increase protein denaturation in the frozen muscle. Fish protein denaturation causes a gradual loss of juiciness and texture which will be responsible for excessive drip loss during thawing. Thanks to the presence of the backbone, whole fish is less susceptible to textural problems. The effect on fillets and other processed products depends on fish species, processing conditions, and added ingredients. Accordingly, the most common additives used in fish, seafood, and related products are antimicrobials and antioxidants.

11.5.4 Effects of Freezing on Fruits and Vegetables

Freezing of fruits and vegetables at their peak and as soon as possible after harvest is an alternative capable of retaining produce quality and nutritional characteristics and to prolong shelf life providing year-round availability. Fruits destined to further processing can be continuously available and of consistent quality. Freezing is also often combined with other preserving techniques for longer produce shelf life and superior organoleptic and nutritional properties. Partial dehydration of fresh fruits and vegetables followed by freezing (freeze drying) has the advantages of product weight and volume reduction, and defrosted and reconstituted products have been reported to be of better quality than ordinary dehydrated ones (Dauthy 1995). Freezing of fruits and vegetables has very small detrimental effect on their nutritional content, particularly when compared to other preservation methods. The loss of vitamins and minerals during freezing, storage, and thawing usually occurs as the compounds dissolve in the exuded liquid (drip) but if the juices are consumed with the product, very little loss actually incurs. Vitamins may be susceptible to both enzymatic and chemical degradation, and measures are regularly taken to ensure minimal nutritional loss during processing. When blanching prior to freezing is deemed necessary, conditions are carefully controlled to minimize nutrient loss and damage to heat susceptible produce properties such as texture and flavor.

Vegetable tissue is more susceptible to freezing damage than meat tissue. Gradual solute concentration on the unfrozen water phase and the growth of ice crystals both have a more accentuated negative effect on the integrity and turgidity of plant structure that will not be recovered after thawing. Under slow freezing rates, the formation and growth of ice crystals is uneven, with crystals forming predominantly on extracellular space and disrupting the structure. Solute concentration promotes cellular dehydration and shrinkage, and crystal growth damages cell walls with subsequent intracellular liquid loss after thawing. Contrary to meat, vegetables usually receive a mild heat treatment before freezing, to inactivate endogenous enzymes and reduce microbial load. However, not all produce is suitable for freezing, as many species and particular varieties are prone to low-temperature damage. For some produce, such as various fruits, maintaining the properties that consumers associate with the "fresh" product is crucial, and as a result control over enzyme activity is usually accomplished by methods other than the use of heat inactivation, for instance by oxygen exclusion, acidification, or sulfur dioxide treatments (Fellows 2000).

11.5.5 EFFECTS OF FREEZING ON DAIRY PRODUCTS

Stability problems associated with the freezing of emulsions, lipid oxidation, and precipitation of proteins are some of the difficulties that have held the widespread use of frozen milk and some milk-based products. Since fresh milk has a very high moisture content, the extent of water-to-ice conversion and the considerable changes on physical properties after thawing have so far made freezing unappealing as a preservation method for this commodity, particularly when compared to other available options such as drying and concentration. For other dairy-based products, such as ice creams and frozen yogurts, freezing effects on organoleptic properties are an intrinsic part of their appeal as they are designed to be consumed in the frozen state. The latter products have continuously evolved to establish their place in the frozen goods market and account for the great majority of frozen dairy-based products currently available.

Ice cream ingredients are mixed, pasteurized, homogenized, and aged prior to freezing. Freezing is achieved in two steps. Firstly, extensive ice crystallization in the mixture is promoted while simultaneously incorporating air. Secondly, ice cream is stored at −30°C or under and the remaining water allowed to freeze, hardening the product. As ice cream is an emulsion and a foam (Goff and Sahagian 1996), fast freezing and the formation of a fat network are fundamental to create the desired texture and overall quality properties of the product. Because the effects of freezing on ice creams are highly beneficial and in fact essential, the freezing and storage of the product are of vital importance for its quality. During crystallization, fast freezing rates determine ice nucleation and crystal distribution, and the formation of many small ice crystals is necessary in order to give the final product its characteristic body and texture. During subsequent storage, conditions should promote rapid hardening, preventing crystal growth and maintaining a homogeneous crystal distribution in the product.

11.5.6 EFFECTS OF FREEZING ON OTHER MISCELLANEOUS PRODUCTS

Among other products, bakery products are often frozen. The freezing of bread can be accomplished before (dough freezing), halfway (partially baked bread), or after baking. Each procedure entails different considerations in regard to operation procedures for optimum product quality. The freezing of bread dough (before baking) produces a very practical and popular product that can be bought, stored, and baked as needed, either at home or in catering premises. During baking, two important factors are responsible for optimum bread volume and crumb texture: the production of gas and its retention by the dough. The rising of bread dough gives particular and desired characteristics to the product and it depends on the action of a microorganism, the yeast. Therefore, in contrast with the freezing of most other food products, freezing rate must not be too fast, since this has been shown to be detrimental to yeast activity, yet it should be fast enough to maintain the structural properties of the dough (Inoue and Bushuk 1996). Añón et al. (2004) refer to 7°C/min as an optimal cooling rate to prevent significant damage to yeast cells. Yeast performance depends not only on the freezing rate applied but also on frozen storage temperature and duration. Changes taking place during freezing and the formation and growth of ice crystal can cause significant damage in the gluten structure (Giannou and Tzia 2006), affecting gas retention properties responsible for the desired and necessary rising during later baking. When permitted by local regulations, additives such as potassium bromate and ascorbic acid are employed to increase dough strength and maintain frozen dough stability, while emulsifiers are used to improve bread volume and crumb firmness. Because it is necessary to reduce dough fermentation time before freezing, the resulting baked breads tend to have comparatively less flavor than breads baked from unfrozen dough (Inoue and Bushuk 1996).

In comparison, baked bread can be frozen, stored, and thawed more easily than dough. The main quality problem in relation to baked bread freezing is the effect of low temperatures on staling which may be due to the accelerating effect of moisture loss on the reactions involved, such as the recrystallization and retrogradation of starch (Chen et al. 1997). Accordingly, fluctuations on storage temperature allowing sufficient water migration between regions of high and low moisture have been associated to substantial increases in frozen bread crumb firmness (Inoue and Buhuk 1996). Staling results in loss of bread texture and eating quality thus after baking bread should be rapidly cooled until below frozen temperature to ensure optimum crumb softness and resilience retention. Cooling can then proceed at a slower rate until the desired storage temperature. It is recommended that baked bread is stored at −18°C or lower for maximum quality retention, and preferably wrapped to minimize moisture losses (Inoue and Bushuk 1996).

Frozen ready-meals are also widely available. The quality and safety of such products and their resulting shelf life will of course depend on the employed pretreatments and on storage conditions. However, being a multicomponent product, differences on the effects of freezing on individual ingredients are to be expected and storage conditions will usually determine which of the deteriorative kinetics might prevail (Kennedy 2000). Often, interactions between the particular ingredients may become the imposing factor determining overall product durability. Although the issues discussed here under the various food groups are still relevant, the selection of storage, packaging, and thawing conditions for the resulting ready-meal product will have different effects on deterioration rates of different components. Ingredients can be prepared separately and combined prior to freezing. Moisture migration may be a problem for such products and the manufacturer can opt to separate some of the meal components in different sections of the product container (Kennedy 2000). In some products the meal is presented as separated portions in bags, ready for reheating. Another increasingly common approach is to use the sous-vide technology developed for chilled meals, subsequently freezing and distributing the products in the frozen state (Creed 2006). Because thawing conditions are significant for optimum product quality properties such as texture and drip loss amounts, manufacturers usually recommend that products are cooked from frozen. Detailed information on specific procedures and effects of processing on such products is not so widely available, most often due to manufacturers' reluctance to disclose vital information to potential competitors.

11.6   IMPORTANCE OF THE COLD CHAIN

The success and applicability of effective low-temperature preservation depends strongly on the capacity to integrate distinct operational steps and facilities in order to maintain storage conditions from food production to consumption. In particular, storage temperature should be kept as invariable as possible, since fluctuations can have potential serious consequences for product quality and safety. In frozen products, partial thawing or storage temperature fluctuations can promote moisture migration and ice crystals' recrystallization and growth. These will lead to loss of textural properties, increase in thaw drip, and increase of undesirable deteriorative reaction rates, promoting overall loss of quality in the product. In chilled products, even small storage temperature fluctuations will have a pronounced effect on the growth rates of viable psychrotrophic microorganisms, accelerating spoilage and potentially causing food-related illnesses. Therefore, "cold chain" relates to the maintenance, monitoring, and control of low-temperature storage conditions from initial chilling to product consumption, including product transport and display. It may include a number of processors, as well as transporters, distributors, retailers, and foodservice personnel. There is an obvious negative economical impact to an improperly managed cold chain as significant deviations from recommended storage conditions will lead to earlier product spoilage; furthermore the quality of such products is also very likely to suffer, the benefits of carefully developed processing steps and conditions intended to maximize product attributes would be also drastically reduced. More importantly though, as a result of the increased demand for minimally processed food products and lower additive levels, low temperature storage is the main preservation system of a growing number of chilled products. The effectiveness of the cold-chain system is therefore also fundamental for food safety assurance. An excellent coverage of freezing-related equipment and facilities for cold chain, including storage, transport, and display, is presented by North and Lovatt (2006), Ketteringham and James (2006), Estrada-Flores (2006), and Cortella and D'Agaro (2006). A detailed account of the requirements and equipment for effective monitoring and control of cold chain is presented by Giannakourou et al. (2006).

11.7 CONCLUSIONS

The main effects of chilling and freezing on the safety and on quality properties of various food products were discussed and ongoing developments to improve both the current technological methods and the resulting products' characteristics were presented. Product quality can be improved by manipulating the food material itself, or by developing or adapting the chilling or freezing process. The former procedures may involve different treatments (feeding enrichment or muscle stimulation for animals; pretreatments or variety selection for fruits and vegetables; addition of functional ingredients to formulated products), while the latter may involve the use of novel or combined techniques and approaches (ultrasound freezing; storage at glass transition temperatures; high pressure low temperature processes). Despite all efforts, it must not be forgotten that different forms of thawing and cooking will also have an effect on the quality of the final product before consumption, and these steps are not always performed by the processor. Furthermore, the quality of the raw materials used, adherence to high standards of hygiene, and proper consideration to the control and monitoring of the process are undoubtedly key to consistently safe and high quality products, as well as a sound background for further development and improvements.

Readers are directed to the reference list for comprehensive reviews on particular food commodities and specific chilling and freezing processes, as well as more detailed information on recent development areas.

REFERENCES

Adams, M. R. and Moss, M. O., 1995. *Food Microbiology*. Cambridge, U.K.: The Royal Society of Chemistry.

Aked, J., 2002. Maintaining the post-harvest quality of fruits and vegetables. In *Fruit and Vegetable Processing—Improving Quality*, Ed. W. Jongen, pp. 119–149. Cambridge, U.K.: CRC Woodhead Publishing Ltd.

Añón, M. C., Le Bail, A., and Leon, A. E., 2004. Effect of freezing on dough ingredients. In *Handbook of Frozen Foods*, Eds. Y. H. Hui et al., pp. 571–580. New York: Marcel Dekker, Inc.

Anonymous, 1986. *Recommendations for the Processing and Handling of Frozen Foods*, 3rd edn. Paris, France: International Institute of Refrigeration.

Anonymous, 2000. Recommendations for chilled storage of perishable produce. *Technical Guides of the International Institute of Refrigeration Series*, 4th edn. Paris, France: International Institute of Refrigeration.

Ansorena, D. and Astiasaran, I., 2004. Effect of storage and packaging on fatty acid composition and oxidation in dry fermented sausages made with added olive oil and antioxidants. *Meat Science* 67 (2), 237–244.

Bender, A., 1992. Meat and meat products in human nutrition in developing countries, *FAO Food and Nutrition Papers No. 53*. Rome, Italy, 91 pp.

BFFF, 2007. Retail Frozen Food Market in Lift off March 07. British Frozen Food Federation Reports. Available at http://www.bfff.co.uk.

Brackett, R. E., 1992. Microbiological safety of chilled foods: Current issues. *Trends in Food Science and Technology* 3, 81–85.

Brown, H. M. and Hall, M. N., 2000. Non-microbiological factors affecting quality and safety. In *Chilled Foods: A Comprehensive Guide*, 2nd edn., Eds. M. Stringer and C. Dennis, pp. 225–255. Cambridge, U.K.: CRC Woodhead Publishing Ltd.

Brown, M. H., 2000. Microbiological hazards and safe process design. In *Chilled Foods: A Comprehensive Guide*, 2nd edn., Eds. M. Stringer, and C. Dennis, pp. 287–339. Cambridge, U.K.: CRC Woodhead Publishing Ltd.

Byrne, D. V., Bredie, W. L. P., Mottram, D. S., and Martens, M., 2002. Sensory and chemical investigations on the effect of oven cooking on warmed-over flavour development in chicken meat. *Meat Science* 61 (2), 127–139.

Carciofi, B. A. and Laurindo, J. B., 2007. Water uptake by poultry carcasses during cooling by water immersion. *Chemical Engineering and Processing* 46 (5), 444–450.

CFA—Chilled Food Association, 1997. *Guidelines for Good Hygiene Practice in the Manufacture of Chilled Foods*. London: Chilled Food Association.

Chen, P. L., Long, Z., Ruan, R., and Labuza, T. P., 1997. Nuclear magnetic resonance studies of water mobility in bread during storage. *LWT—Food Science and Technology* 30 (2), 178–183.

Cheng, Q. and Sun, D.-W., 2006a. Feasibility assessment of vacuum cooling of cooked pork ham with water compared to that without water and with air blast cooling. *International Journal of Food Science and Technology* 41(8), 938–945.

Cheng, Q. and Sun, D.-W., 2006b. Improving the quality of pork ham by pulsed vacuum cooling in water. *Journal of Food Process Engineering* 29(2), 119–133.

Cortella, G. and D'Agaro, P., 2006. Retail display equipment and management. In *Handbook of Frozen Food Processing and Packaging*, Ed. D.-W. Sun, pp. 243–258. London: CRC Press/Taylor & Francis.

Creed, P. G., 2006. Quality and safety of frozen ready meals. In *Handbook of Frozen Food Processing and Packaging*, Ed. D.-W. Sun, 459–479. London: CRC Press/Taylor & Francis.

Dauthy, M. E., 1995. Fruit and vegetable processing. *FAO Agricultural Services Bulletin No. 119*. Rome, Italy.

Dennis, C. and Stringer, M., 2000. The chilled foods market. In *Chilled Foods: A Comprehensive Guide*, 2nd edn., Eds. M. Stringer and C. Dennis, pp. 1–16. Cambridge, U.K.: CRC Woodhead Publishing Ltd.

Early, L. R., 2000. Raw material selection: Dairy ingredients. In *Chilled Foods: A Comprehensive Guide*, 2nd edn., Eds. M. Stringer and C. Dennis, pp. 37–62. Cambridge, U.K.: CRC Woodhead Publishing Ltd.

Estrada-Flores, S., 2006. Transportation of frozen foods. In *Handbook of Frozen Food Processing and Packaging*, Ed. D.-W. Sun, pp. 227–242. London: CRC Press/Taylor & Francis.

Fellows, P. J., 2000. *Food Processing Technology: Principles and Practice*, 2nd edn. Cambridge, U.K.: CRC Woodhead Publishing Ltd.

Gaze, J. E., Shaw, R., and Archer, J., 1998. Identification and prevention of hazards associated with slow cooling of hams and other large cooked meats and meat products. In *Review No. 8*. Gloucestershire, U.K.: Campden & Chorleywood Food Research Association.

Giannakourou, M. C., Taoukis, P. S., and Nychas, G. J. E., 2006. Monitoring and control of the cold chain. In *Handbook of Frozen Food Processing and Packaging*, Ed. D.-W. Sun, pp. 279–310. London: CRC Press/Taylor & Francis.

Giannou, V. and Tzia, C., 2006. Quality and safety of frozen bakery products. In *Handbook of Frozen Food Processing and Packaging*, Ed. D.-W. Sun, pp. 481–502. London: CRC Press/Taylor & Francis.

Goff, H. D., 1992. Low-temperature stability and the glassy state in frozen foods. *Food Research International* 25 (4), 317–325.

Goff, H. D., 1994. Measuring and interpreting the glass transition in frozen foods and model systems. *Food Research International* 27 (2), 187–189.

Goff, H. D. and Sahagian, M. E., 1996. Freezing of Dairy Products. In *Freezing Effects on Food Quality*, Ed. L. E. Jeremiah, pp. 299–335. New York: Marcel Dekker, Inc.

Gormley, R., Walshe, T., Hussey, K., and Butler, F., 2002. The effect of fluctuating vs. constant frozen storage temperature regimes on some quality parameters of selected food products. *LWT—Food Science and Technology* 35 (2), 190–200.

Haard, N. F., 1992. Control of chemical composition and food quality attributes of cultured fish. *Food Research International* 25 (4), 289–307.

Heldman, D. R., 1992. Food freezing. In *Handbook of Food Engineering*, Eds. D. R. Heldman and D. B. Lund, pp. 277–315. New York: Marcel Dekker, Inc.

Heldman, D. R. and Singh, R. P., 1981. *Food Process Engineering*, 2nd edn. Westport: AVI Publishing.

Hogan, E., Kelly, A. L., and Sun, D.-W., 2005. High pressure processing of foods: An overview. In *Emerging Technologies for Food Processing*, Ed. D.-W. Sun, pp. 3–32. London: Elsevier.

IFST—Institute of Food Science and Technology (U.K.), 1990. *Guidelines for the Handling of Chilled Foods*, 2nd edn. London: IFST.

Inoue, Y. and Bushuk, W., 1996. Effects of freezing, frozen storage, and thawing on dough and baked goods. In *Freezing Effects on Food Quality*, Ed. L.E. Jeremiah, pp. 183–246. New York: Marcel Dekker Inc.

James, C., Vincent, C., de Andrade Lima, T. I., and James, S. J., 2006. The primary chilling of poultry carcasses—a review. *International Journal of Refrigeration* 29 (6), 847–862.

James, S. J., Evans, J., and James, C., 2007. A review of the performance of domestic refrigerators. *Journal of Food Engineering* 87 (1): 2–10.

Jayathilakan, K., Sharma, G. K., Radhakrishna, K., and Bawa, A. S., 2007. Antioxidant potential of synthetic and natural antioxidants and its effect on warmed-over-flavour in different species of meat. *Food Chemistry* 105 (3), 908–916.

Joseph, R. L., 1996. Very fast chilling of beef and tenderness—a report from an EU concerted action. *Meat Science* 43 (Suppl. 1), 217–227.

Kennedy, C. J., 2000. Freezing processed foods. In *Managing Frozen Foods*, Ed. C. J. Kennedy, pp. 137–158. Cambridge, U.K.: CRC Woodhead Publishing Ltd.

Kerry, J. P., Buckley, D. J., Morrissey, P. A., O'Sullivan, K., and Lynch, P. B., 1998. Endogenous and exogenous [alpha]-tocopherol supplementation: Effects on lipid stability (TBARS) and warmed-over flavour (WOF) in porcine *M. longissimus* dorsi roasts held in aerobic and vacuum packs. *Food Research International* 31 (3), 211–216.

Ketterinham, L. and James, S., 2006. Cold store design and maintenance. In *Handbook of Frozen Food Processing and Packaging*, Ed. D.-W. Sun, pp. 211–226. London: CRC Press/Taylor & Francis.

Kinsella, K. J., Sheridan, J. J., Rowe, T. A. et al. 2006. Impact of a novel spray-chilling system on surface microflora, water activity and weight loss during beef carcass chilling. *Food Microbiology* 23 (5), 483–490.

Kok, T. N. and Park, J. W., 2007. Extending the shelf life of set fish ball. *Journal of Food Quality* 30 (1), 1–27.

Kotrola, N., 2006. Quality and safety of frozen poultry and poultry products. In *Handbook of Frozen Food Processing and Packaging*, Ed. D.-W. Sun, pp. 325–340. London: CRC Press/Taylor & Francis.

Lawrie, R. A. 1997. *Meat Science*, 5th edn. Oxford, U.K.: Pergaman Press.

Li, B. and Sun, D.-W., 2002a. Novel methods for rapid freezing and thawing of foods—a review. *Journal of Food Engineering* 54 (3), 175–182.

Li, B. and Sun, D.-W., 2002b. Effect of power ultrasound on freezing rate during immersion freezing of potatoes. *Journal of Food Engineering* 55 (3), 277–282.

Lopez-Caballero, M. E., Gomez-Guillen, M. C., Perez-Mateos, M., and Montero, P., 2005. A chitosan-gelatin blend as a coating for fish patties. *Food Hydrocolloids* 19 (2), 303–311.

Losada, V., Barros-Velazquez, J., and Aubourg, S. P., 2007. Rancidity development in frozen pelagic fish: Influence of slurry ice as preliminary chilling treatment. *LWT—Food Science and Technology* 40 (6), 991–999.

Lund, B. M., 2000. Freezing. In *The Microbiological Safety and Quality of Food*, vol. I, Eds. B. M. Lund, T. C. Baird Parker, and G. W. Gould, pp. 122–145. Gaithersburg: Aspen Publishers.

McDonald, K., Sun, D.-W., and Kenny, T., 2000. Comparison of the quality of cooked beef products cooled by vacuum cooling and by conventional cooling. *Lebensmittel-Wissenschaft und-Technologie* 33 (1), 21–29.

Martins, R. C., Almeida, M. G., and Silva, C. L. M., 2004. The effect of home storage conditions and packaging materials on the quality of frozen green beans. *International Journal of Refrigeration* 27 (8), 850–861.

Meir, S., Rosenberger, I., Aharon, Z., Grinberg, S., and Fallik, E., 1995. Improvement of the postharvest keeping quality and colour development of bell pepper (cv. 'Maor') by packaging with polyethylene bags at a reduced temperature. *Postharvest Biology and Technology* 5 (4), 303–309.

Mielnik, M. B., Olsen, E., Vogt, G., Adeline, D., and Skrede, G., 2006. Grape seed extract as antioxidant in cooked, cold stored turkey meat. *LWT—Food Science and Technology* 39 (3), 191–198.

Mielnik, M. B., Sem, S., Egelandsdal, B., and Skrede, G., 2008. By-products from herbs essential oil production as ingredient in marinade for turkey thighs. *LWT—Food Science and Technology* 41 (1), 93–100.

North, M. F. and Lovatt, S. J., 2006. Freezing methods and equipment. In *Handbook of Frozen Food Processing and Packaging*, Ed. D.-W. Sun, pp. 199–210. London: CRC Press/Taylor & Francis.

Ohkuma, C., Kawai, K., Viriyarattanasak, C. et al. 2008. Glass transition properties of frozen and freeze-dried surimi products: Effects of sugar and moisture on the glass transition temperature. *Food Hydrocolloids* 22 (2), 255–262.

Olafsdottir, G., Lauzon, H. L., Martinsdottir, E., and Kristbergsson, K., 2006. Influence of storage temperature on microbial spoilage characteristics of haddock fillets (*Melanogrammus aeglefinus*) evaluated by multivariate quality prediction. *International Journal of Food Microbiology* 111 (2), 112–125.

Olsen, J. and Aaslyng, M., 2007. The meal composition approach—a new way of optimising the quality of foodservice products. *Journal of Foodservice* 18 (4), 133–144.

Purvis, A. C., 2002. Diphenylamine reduces chilling injury of green bell pepper fruit. *Postharvest Biology and Technology* 25 (1), 41–48.

Reid, D. S., Kerr, W., and Hsu, J., 1994. The glass transition in the freezing process. *Journal of Food Engineering* 22 (1–4), 483–494.

Rivera, F., Pelayo-Zaldivar, C., Leon, D. E. et al. 2007. Cold-conditioning treatment reduces chilling injury in Mexican limes (*Citrus aurantifolia* S.) stored at different temperatures. *Journal of Food Quality* 30 (1), 121–134.

Sahagian, M. E. and Goff, H. D., 1996. Fundamental aspects of the freezing process. In *Freezing Effects on Food Quality*, Ed. L. E. Jeremiah, pp. 1–50. New York: Marcel Dekker, Inc.

Santos-Yap, E. M., 1996. Fish and Seafood. In *Freezing Effects on Food Quality*, Ed. L. E. Jeremiah, pp. 109–133. New York: Marcel Dekker, Inc.

Savell, J. W., Mueller S. L., and Baird B. E., 2005. The chilling of carcasses. *Meat Science* 70 (3), 449–459.

Sebranek, J. G., 1996. Poultry and poultry products. In *Freezing Effects on Food Quality*, Ed. L. E. Jeremiah, pp. 85–108. New York: Marcel Dekker, Inc.

Silliker, J. H., Elliot, R. P., Baird-Parker, A. C. et al. 1980. *Microbial Ecology of Foods*. London: Academic Press.

Singh, R. P. and Anderson, B. A., 2004. The major types of food spoilage: An overview. In *Understanding and Measuring the Shelf Life of Food*, Ed. R. Steele, pp. 3–23. Cambridge, U.K.: CRC Woodhead Publishing Ltd.

Skrede, G., 1996. Fruits. In *Freezing Effects on Food Quality*, Ed. L. E. Jeremiah, pp. 183–246. New York: Marcel Dekker, Inc.

Slade, L. and Levine, H., 1991. Beyond water activity: Recent advances based on an alternative approach to the assessment of food quality and safety. *Critical Review in Food Science and Nutrition* 30, 115–360.

Stapelfeldt, H., Bjørn, H., Skibsted, L. H., and Bertelsen, G., 1993. Effect of packaging and storage conditions on development of warmed-over flavour in sliced, cooked meat. *Zeitschrift für Lebensmitteluntersuchung und -Forschung A* 196 (2), 131–136.

Sun, D.-W., 2006. *Handbook of Frozen Food Processing and Packaging*, Ed. D.-W. Sun. London: CRC Press/Taylor & Francis.

Sun, D.-W. and Li, B., 2003. Microstructural change of potato tissues frozen by ultrasound-assisted immersion freezing. *Journal of Food Engineering* 57 (4), 337–345.

Sun, D.-W. and Wang, L., 2000. Heat transfer characteristics of cooked meats using different cooling methods. *International Journal of Refrigeration* 23 (7), 508–516.

Sun, D.-W. and Wang, L. J., 2001. Vacuum cooling. In *Advances in Food Refrigeration*, Ed. D.-W. Sun, pp. 264–304. Surrey, U.K.: Leatherhead Publishing.

Sun, S., Singh, R. P., and O'Mahony, M., 2005. Quality of meat products during refrigerated and ultra-chilled storage. *Journal of Food Quality* 28 (1), 30–45.

Tabil, L. G. and Sokhansanj, S., 2001. Mechanical and temperature effects on shelf life stability of fruits and vegetables. In *Food Shelf Life Stability: Chemical, Biochemical, and Microbiological Changes*, Eds. N. A. M. Eskin and D. S. Robinson, pp. 37–86. Boca Raton, FL: CRC Press.

Torreggiani, D., Forni, E., Guercilena, I. et al., 1999. Modification of glass transition temperature through carbohydrates additions: Effect upon colour and anthocyanin pigment stability in frozen strawberry juices. *Food Research International* 32 (6), 441–446.

Van Moeseke, W., De Smet, S., Claeys, E., and Demeyer, D., 2001. Very fast chilling of beef: Effects on meat quality. *Meat Science* 59 (1), 31–37.

Walker, S. J. and Betts, G., 2000. Chilled foods microbiology. In *Chilled Foods: A Comprehensive Guide*, 2nd edn., Eds. M. Stringer and C. Dennis, pp. 153–186. Cambridge, U.K.: CRC Woodhead Publishing Ltd.

Whittle, K. J., Hardy, R., and Hobbs, G., 1990. Chilled fish and fishery products. In *Chilled foods: The State of the Art*, Ed. T. R. Gormley, pp. 87–116. London: Elsevier Applied Science.

Zheng, L. and Sun, D.-W., 2004. Vacuum cooling for the food industry—a review of recent research advances. *Trends in Food Science and Technology*, 15 (12), 555–568.

Zheng, L. and Sun, D.-W., 2005. Vacuum cooling of foods. In *Emerging Technologies for Food Processing*, Ed. D.-W. Sun, pp. 579–602. London: Elsevier.

12 Drying and Dried Food Quality

Xiao Dong Chen

CONTENTS

12.1 INTRODUCTION

Drying is used conventionally and effectively to maintain the quality of foodstuffs to acceptable levels (i.e., edible). Dried food particles are usually used as ingredients (spices or dairy powders, etc.) or bulk cooking materials (rice, flour, etc.). The major motivation for use of drying as a preservation technique is well understood and is illustrated graphically in Figure 12.1. Sun drying is probably the oldest practice and is still popular in some parts of the world today. Water activity (a_w) values are extensively used to predict the stability of foodstuff with respect to microbial growth and enzymatic chemical and physical changes (such as glass transition) that can lead to food degradation during storage (Christian 2000). The water activity of the food system indicates how tightly water is bound, structurally or chemically, in the food matrix (Scott 1957, Labuza 1975). In other words, the water activity of the food describes the energetics of water in a food system, and hence its availability to act as a solvent and participate in chemical or biochemical reactions (Labuza 1977). The quality attributes for dried foods may be appearance related (color in particular), denaturation of proteins leading to solubility problem/textural changes, nutrient deterioration, fat crystallization or recrystallization, mechanical property changes, microbial changes, etc. During the conventional drying at higher temperatures, all the quality attributes are usually deteriorating due to thermal and concentration

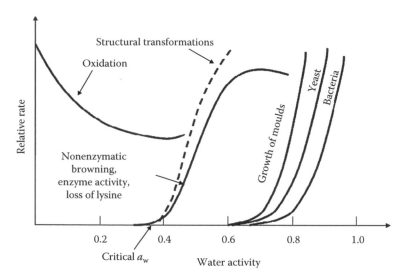

FIGURE 12.1 Stability map for food materials. (Modified from Roos, Y.H., *Lait*, 82, 475, 2002.)

effects. Therefore, the fundamental question is as to what is perceived/measured to be a good quality of dried materials. Around the world, traded food materials are usually stored and transported in large quantities in dried solid form. Hence drying plays a key role in our world economy and indeed in our life.

Products that could be adversely affected by the mixed-in dried materials that can become unstable are

Meat product range
Vegetarian product range
Dairy product range
Poultry product range
Fishery product range
Canned foods (if ingredients were dried first and then mixed in)
Ready meals (if ingredients were dried first and then added in)

Microbial aspects can be separated into two groups: good ones and bad ones. The good ones may be probiotic bacteria that are intentionally added for improving our digestive health. The bad ones would be the spoilage microorganisms (Featherstone 2008), such as

Pathogens
Escherichia coli O157:H7
Salmonella spp.
Listeria monocytogenes
Clostridium perfringens
Staphylococcus aureus
Parasites
Trichinellosis

As shown in Figure 12.1, the minimum water activity of 0.6–0.7 can be seen, which is the limit beyond which a microorganism or group of microorganisms can no longer reproduce for most food materials. Pathogenic bacteria cannot grow below a water activity of 0.85–0.86, whereas yeast and moulds are more tolerant to a smaller water activity of 0.8. Usually there is no growth when water activity is below 0.62. These critical limits may also change somewhat depending on the pH, salt content, antimicrobial additives, heat treatment, and temperature involved (Rahman 2007).

Thermal processing is designed to make foods safe as most of the microorganisms are heat liable. Depending on the natural pH of the product, pasteurization (typically 80°C–100°C) is of the choice for $4.0 \leq pH \leq 4.6$ (foods are classified as acid foods) and sterilization (115°C–125°C) is needed for $pH > 4.6$ (low-acid foods). Those below pH 4.0 are called high-acid foods (Britt 2008). pH is defined by

$$pH = -\log_{10}\left[H^+\right] \tag{12.1}$$

where $[H^+]$ is the concentration of H^+ ions (mol L^{-1}). Most organisms grow best at neutral pH but a few can grow at pH < 4.0. Bacteria are more fastidious about their pH environments than yeast and moulds. Limiting microbial growth using pH alteration is a basic food preservation principle (Figure 12.2).

Drying can be conducted at elevated temperatures (hot-air drying and superheated-steam drying at normal or higher pressures), moderate temperatures (in heat-pump drying, vacuum drying, or low-pressure superheated-steam drying), or low temperatures (freeze drying either at atmospheric level or under vacuum). The "hot" processes can bring the foods to some high temperatures, hence may be viewed as a kind of

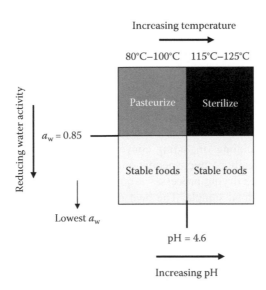

FIGURE 12.2 Thermal processing actions for different pH level foods. (Modified from Britt, I.J., *Food Biodeterioration and Preservation*, ed., Tucker, G., Blackwell Publishing Ltd., Oxford, U.K., 2008.)

thermal processing. The combination of the temperature change and the water content change, and in many ways, their gradient variations during drying can make the quality of the product surface and the product core to be drastically different.

Foodstuffs themselves are complex biological materials, which are main sources of various nutrients, and intended to be used for getting energy and health purposes. When food materials are exposed to drying conditions, the native physical state of the food material is altered leading to changes in microstructure, quality, and safety of food materials. In general, drying processes help maintaining the "acceptable-to-excellent" edible status of various foodstuffs, which have useful life-supporting components such as proteins, vitamins, probiotics, enzymes, etc. Low-temperature drying maintains good qualities of the original products and limits the water activity related growth factors but do not alter the potential for the growth of the bad quality if the food gets into an inappropriate environment.

The shelf life of the food is extended by a drying process that ensures the water activity to be lower than the critical minimum water activity, thus controlling the bioactivity of various useful and harmful biological compounds. Drying involves removal of excess water from the food matrix until a "safe" moisture level is achieved, at which minimum or no physical, chemical, and microbiological reactions occur. The required moisture level (water content or water activity) in dried products for preventing spoilage is different due to dissimilar responses and inactivation mechanisms exhibited by different bioactive compounds.

For thermal deactivation of microorganisms, three stages are considered: pretreatment, heat treatment, and posttreatment. Treating the drying process as a thermal treatment process, it is appropriate to discuss the effects of drying on microorganisms also for three stages: predrying, in-drying, and postdrying. Predrying operations such as osmotic dehydration, evaporation, freeze concentration, membrane separation, and extraction are employed to remove the excess water from the liquid feed prior to the drying operation. The selection of predrying operation depends on the extent of the prior-water removal, the type and physical form of the material, and the expected damage during processing. Fundamental principles of the water removal during predrying processes vary from process to process. During osmotic dehydration of food materials, for instance dipping fruits or vegetables in concentrated syrup, the mutual diffusion of both sugar and water molecules in opposite directions removes the water from the fruits or vegetables. Postdrying processes include cooling, handling (including conveying), packaging, storage, rehydration, etc., which are basically product-specific operations (Chen and Patel 2008).

High-temperature drying processes require a strong heat supply for removal of moisture from the food sample. The heat can be supplied in many ways, such as microwave, radio frequency, hot gas stream including air, superheated steam, etc. The hot gas stream is the most frequently used heat source for large-scale commercial industries due to easier availability, easier heat recovery, and cheaper costs compared to other heat sources. The driers are often named according to how heat is supplied or what is the drying medium, for instance, solar drier, superheated-steam drier, heat-pump drier, and microwave drier. Heat can be supplied through direct contact conduction, for example, belt drying, drum drying, and tunnel drying. Radiation is also frequently used as the way of heating for batch-scale drying operations, for example, infrared drying.

What the drying does to the product in terms of color and microbial status is highlighted here. Of course the mechanical properties of the dried products are completely different from the original and the variability is very large. Shrinkage and density change is also of great importance but these can be found elsewhere in extensive literatures (Aguilera and Stanley 1999).

12.2 COLOR

Color, flavor, taste, and shape (appearance) are the four important factors affecting people's desire toward having foods in the first instance. Texture plays a subsequent and an important role. "Color" forms a first impression about a cooked food or a food product. Vision psychology, science of color, color psychology, the technologies that measure color and the computer imaging have all been developed in order to design foods, quantify color, improving the quality of the foods. These are complex matters and in engineering practice, only the major yet straightforward issues are addressed. The color of the product affects the perception of the goodness or the badness of the product. *Birren*'s statistical analysis is frequently used (Li 1998). The measurement of color is usually conducted with the CIELAB system, or the $L^*-a^*-b^*$ system established in 1976. The parameters L^*, a^*, and b^* are measured through a colorimeter such as the Hunter Lab instrument. A schematic diagram showing the L^*, a^*, and b^* coordinate system is shown in Figure 12.3. $L^* = 0$ corresponds to black; $L^* = 100$ denotes white; therefore, L^* represents the brightness, that is, 0–100 gray scales. As $a^* \rightarrow +a$, the color approaches to the pure red, and as $a^* \rightarrow -a$, the color approaches to pure green, $-b$ toward the pure blue, and $+b$ toward the pure yellow. The product color would be a combination of a^* and b^* at the same

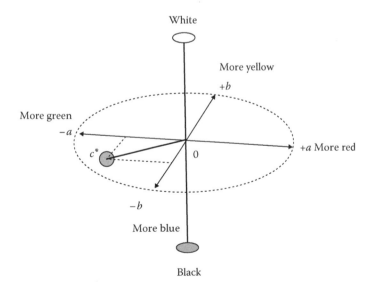

FIGURE 12.3 An illustration of the CIELAB color expression system. (Drawn by Chen, X.D., Moisture diffusivity in food and biological material, plenary keynote at 15th International Drying Symposium (IDS 2006), Budapest, Hungary, August 20–23, 2006.)

level of brightness L^*, based on the basic idea that red, blue, and yellow can make up other colors by combination. The combination value of color is evaluated using the following formula:

$$c^* = \sqrt{\left(a^*\right)^2 + \left(b^*\right)^2} \tag{12.2}$$

This is useful and is called "metric chroma." The greater the c^* the purer the color appears to be. The "metric Hue-angle" is defined as

$$\text{Hue} = \arctan\left(\frac{b^*}{a^*}\right) \tag{12.3}$$

where a^* and b^* may also be called the "color indices." These parameters may be understood (based on Figure 12.3) based on our knowledge of two-dimensional geometry. A three-dimensional parameter is then the following:

$$\Delta E^* = \sqrt{\left(\Delta L^*\right)^2 + \left(\Delta a^*\right)^2 + \left(\Delta b^*\right)^2} \tag{12.4}$$

where ΔL^*, Δa^*, and Δb^* are, respectively, the differences between two points in the coordinate system given in Figure 12.3.

Equation 12.4 may be written as follows to be more practical, that is, by choosing a reference point to benchmark the significance of change:

$$\Delta E^* = \sqrt{\left(L^* - L^*_{\text{ref}}\right)^2 + \left(a^* - a^*_{\text{ref}}\right)^2 + \left(b^* - b^*_{\text{ref}}\right)^2} \tag{12.5}$$

For color "sensory" evaluation, ΔE^* is a very important quantity determining (Li 1998)

Trace level difference, $\Delta E^* = 0 - 0.5$
Slight difference, $\Delta E^* = 0.5 - 1.5$
Noticeable difference, $\Delta E^* = 1.5 - 3.0$
Appreciable difference, $\Delta E^* = 3.0 - 6.0$
Large difference, $\Delta E^* = 6.0 - 12.0$
Very obvious difference, $\Delta E^* = >12.0$

It is not clear yet about what is the relationship between the color perception and water content. Color may be intensified (colorings concentrated) as water is removed. On the other hand, as in hot-air drying, the surface temperature of the product can get very high which promotes heat-sensitive chemical reactions such as mallard reactions, so the product exhibits a "cooked" perception. It is important to emphasize that the color as a quality of the manufactured food is a "surface phenomena," so the surface temperature and surface water content control should be the most important parameters to control.

A fundamental approach may be demonstrated in a study on lycopene degradation (red color change) in a drying process (Goula et al. 2006). Here, experimentally, the temperatures and water contents of the tomato pulp samples were fixed so that the degradation kinetics can be worked out as functions of temperature and water content:

$$\frac{dC}{dt} = -k_d \cdot C \tag{12.6}$$

Here the nutrient (lycopene) is considered to degrade following a first-order reaction mechanism. C is the concentration of lycopene (kg m^{-3}). The rate constant k_d (degradation rate constant) is expressed as

$$k_d = 0.121238 \cdot e^{0.0188 \cdot W} \cdot e^{-\frac{2317}{T}} \ (\text{min}^{-1}) \quad \text{for} \quad W \geq 55 \ (\text{wt\%}) \tag{12.7}$$

$$k_d = 0.275271 \cdot e^{0.00241 \cdot W} \cdot e^{-\frac{2207}{T}} \ (\text{min}^{-1}) \quad \text{for} \quad W \leq 55 \ (\text{wt\%}) \tag{12.8}$$

where
 T is the product temperature (K)
 W is the product moisture content in % (wet basis)

Some combinations of the three color parameters, L^*, a^*, and b^*, may make better sense when correlating to the color changes or the rates of the changes to constituent changes (Kaur et al. 2006) (Figure 12.4):

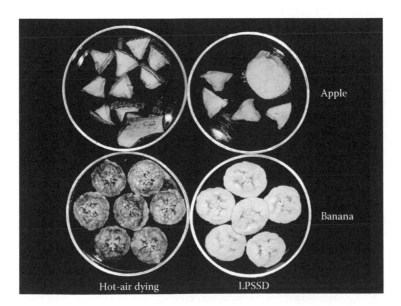

FIGURE 12.4 Color changes depending on drying methods (LPSSD stands for low pressure superheated-steam drying, which employed low temperatures of about 60°C–90°C whilst the air drying was conducted in the same temperature range). (Courtesy of Professor Sakamon Devahastin, KMUT, Thailand.)

(a)   (b)

(c)

FIGURE 12.5 Photographs of shrimp dried using a jet spouted-bed drier at 120°C: (a) dried shrimp before storage, (b) dried shrimp stored under vacuum at 4°C for 16 weeks, and (c) dried shrimp stored under air at 4°C for 16 weeks. (Courtesy of Professor Sakamon Devahastin, KMUT, Thailand.)

$$L^* = 100; \quad L^* = 0$$

No matter what the numerical system is used for quantifying color, visualization by the bare eyes (sensory) cannot be replaced. As mentioned earlier, drying process usually alters the product color (to not as sharp and fresh as the original). Figure 12.5 shows some photos of the products before and after drying. In addition to the temperature effect on reactions, the availability of oxygen also affects color.

12.3 MICROBIAL INACTIVATION

In thermal processing of foods, several important terminologies are used (Li et al. 2006, Patel and Chen 2006).

12.3.1 DECIMAL REDUCTION TIME D

When a living microorganism population, like *E. coli*, is subjected to thermal processing at constant temperature (T), its population will reduce. A plot of the microbial population over time (N vs. t) usually shows an "exponential-like" trend. A semi-log plot of N versus t may be correlated using a linear fit (one kind of mechanism), yielding a straight line with a negative slope ($-D$):

$$D = \frac{t}{\log N_0 - \log N} \tag{12.9}$$

where D is the "decimal reduction time" (s)

The microbial population reduction may then be expressed as

$$\frac{N}{N_0} = 10^{-\frac{t}{D}} \tag{12.10}$$

Obviously, at different T, D should be different. The higher the temperature, the smaller the D value.

12.3.2 THERMAL RESISTANCE CONSTANT Z

The thermal resistance constant Z is a parameter representing the microorganism's resistance to temperature rise:

$$Z = \frac{T_2 - T_1}{\log \dfrac{D_{T_1}}{D_{T_2}}} \tag{12.11}$$

It seems never to have been mentioned, another dimension that influences Z value should be the moisture content, such that D may be labeled as $D_{T,X}$.

12.3.3 THERMAL DEATH TIME F

The thermal death time F is the time required to cause a stated reduction in the population of microorganisms or spores. This time may be expressed as a multiple of D values. For instance, a 99.99% reduction in microbial population is equivalent to $4D$. Usually, F is expressed as F_T^z for a specific temperature T and a thermal resistance constant Z.

12.3.4 RELATIONSHIPS BETWEEN CHEMICAL KINETICS AND THERMAL PROCESSING PARAMETERS

It is generally accepted that at a constant temperature, the microbial population or number concentration (microbe m^{-3}) (N) reduces following a first-order reaction:

$$\frac{dN}{dt} = -k_d \cdot N \tag{12.12}$$

Therefore, the solution for this at constant temperature is

$$\ln \frac{N}{N_0} = -k_d \cdot t \tag{12.13}$$

This is similar to Equation 12.10, except the use of natural logarithms here.

Comparing the Equation 12.13 with Equation 12.10, it is not difficult to arrive at

$$k_d = \ln 10/D \qquad (12.14)$$

During thermal processing, the microbial population is not uniformly distributed within the processing fluid and also the extent of the deactivation is different at different locations and different timing.

Equation 12.12 can also be directly expressed as the following (mathematically the same as Equation 12.6):

$$\frac{dc}{dt} = -k_d \cdot c \qquad (12.15)$$

where c is the mass concentration of the live microbes ($c = m_{microbe} \cdot N$). $m_{microbe}$ is the mass of one average microbe (kg).

The spoilage rate for the specific food system can be influenced by various physicochemical parameters. The following physicochemical parameters are known to be very important:

Temperature
Water activity
pH
Oxygen availability
Nutrients availability
Microstructure
Chemical inhibitors

The first four are probably the most obvious in-drying operations.

Drying rate is an important parameter for the inactivation of biological substances. Different drying rates may have other physical/biological transformations which may be responsible for overall degradation of the food quality. The first group of drying processes which deal with high drying rates might be the high-temperature processes such as spray drying, hot-air tunnel drying, drum drying, superheated-steam drying, fluidized-bed drying, spouted-bed drying, and heat-pump drying. Another group of drying processes, which have low-drying rates and higher processing times, involves relatively lower temperatures compared to the first group of drying processes. Freeze drying, vacuum drying, atmospheric freeze drying, solar drying, and microwave drying can be placed in a group with low-drying rates. In order to achieve higher efficiencies, improved product quality, lower processing times, and reduced production costs, a combined drying approach is often being used. Microwave-vacuum drying, microwave-convection drying, and spray-freeze drying are examples of combined drying processes. Many times, two or more drying techniques are operated in a series to minimize the overall drying time and degradation of useful biological substances.

Low-temperature drying processes are favorable for keeping the high bioactivity of desired biocomponents in the final product. A minimal change in nutritional values is targeted during low-temperature processes. The unwanted microbial activity could also be preserved in the food material. Drying of food using hot air does not necessarily involve high temperatures due to the evaporative cooling effect during water evaporation. Evaporation of water from the food surface to the surrounding air minimizes the temperature rise of the food system. Superheated-steam drying can however lead to high food temperatures during processing. The severity of inactivation of microorganisms during drying is lower than the severity during other thermal treatments where no evaporation is involved and where the temperature of the moist food sample is maintained at as high as 121°C for certain time. High-temperature thermal treatments, although highly effective in deactivating the unwanted bacterial activity, also severely degrade other essential food ingredients.

For the removal of water from the given food geometry, the moisture (liquid and vapor) contained in the solid material must migrate from the interior to the surface. The transport of moisture in the moist food materials can be considered as the moisture transport in porous media, and thus the recognized principles of heat and mass transfer in porous medium can readily be used (Welti-Chanes et al. 2005, Aversa et al. 2007). A common knowledge for migration of moisture from the internal structure to the surface is by assuming the liquid water flow or water vapor diffusion. Many mechanisms have been proposed for describing the internal transport of water, which is generally believed to be a major rate-limiting step. Diffusion phenomena are considered to be extremely complex due to the wide diversity of chemical composition and physical structures of food systems and hence reliable data are limited. The transport of water has been widely modeled using the effective liquid diffusion model (Yusheng and Poulsen 1988, Chen 2006). Figure 12.6 shows a scenario where the heat coming from surrounding hot air is transferred to the product–air interface and subsequently to the moist food product. In order to overcome the latent heat of vaporization, heat must be supplied to the food surface and also into the porous food

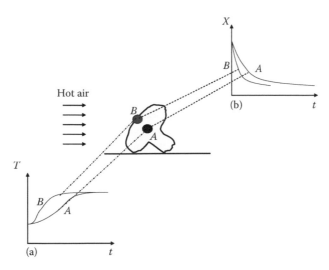

FIGURE 12.6 Qualitative illustration of (a) hot-air drying of a moist food (mushroom for instance) and (b) temperature–time and moisture content–time profiles during drying.

structure. For hot-air drying, the temperature history at the surface and the center of the moist porous material, presented by point A and point B, respectively, are different. The local evaporation rate within the moist food material should be determined by the local driving force, that is, the vapor pressure difference in the solid structure and the local "headspace" such as pores or channel space at the same location. It is shown in Figure 12.6 that the surface water content would drop more rapidly than the water content at the center point. The spatial distribution of water content over the material thickness in the moist food slab would have implications on the deactivation rate of living cells or spores within the matrix. The microbial population near the heated boundary and low water content experiences different history as that in the core region where temperature is lower and water content is high.

Changes in pH away from the optimal value reduce the thermal stability of microorganisms or enzymes within a food system. Thermal inactivation of bacteria has been extensively studied for the corresponding kinetics such as the studies published by Chiruta et al. (1997) and Khoo et al. (2003). In their work on E. coli, they have shown the following first-order inactivation kinetics model to be useful for constant treatment conditions (temperature T and pH) (see Equation 12.13).

$$\ln k_{\mathrm{d}} = C_0 + \frac{C_1}{T} + \frac{C_2}{T^2} + C_3 \mathrm{pH} + C_4 \mathrm{pH}^2 \tag{12.16}$$

where
   T is the absolute exposure temperature (K)
   Equation coefficients C_0, C_1, C_2, C_3, and C_4 for inactivation of E. coli in Carbopol liquid were reported to be −3613, 2.44 × 10⁶, −4.11 × 10⁸, −1.523, and 0.124, respectively

Experiments on thermal treatments of microorganisms at different relative humidity or water activity (a_w) values have revealed a general trend for microorganisms. The thermal stability (survival of the microorganisms in the pre-equilibrated food systems) is increased and gives a peak as the water activity is reduced at the beginning of heat treatment, but the thermal stability is then rapidly reduced as the material being dried to very low water activity levels. This optimal water activity for highest thermal stability is between 0.2 and 0.5.

Drying stage may alter the pH of the food system and cause osmotic imbalance during processing due to water removal and subsequent increment in the concentration of solutes. Acidic foods tend to get more acidic within the food matrix due to higher concentration of solutes. Drying would "fix" the structures of long-chain molecules causing irreversible changes in their preferred three-dimensional positions. When water activity is reduced, the water associated with the cell constituents that are considered to be protective to cell functions is less available, thus reducing the thermal stability of heat-sensitive components. This is why osmotic process by infusing sugar may help preserve the bioactivity. The combination of temperature and water activity is important in determining the extent of deactivation of microorganisms or denaturation of proteins and enzymes. The rate of water removal and the temperature rise are more rapid at the surface of a moist material compared to other locations within the food material when hot-air drying is considered. Typical water content profiles with respect to drying time for surface and center locations are shown qualitatively in Figure 12.6. It is expected that the water content distribution and temperature distribution inside the material being dried are both important phenomena for influencing the local rate of microorganism deactivation. The deactivation of the microorganisms located at the surface or near the surface would be different from the deactivation rate profiles inside the material. The status of mineral components and the possibility of generating insoluble metal compounds may also have an impact on the availability of bioactivity (Watzke 1998).

The inactivation of the microbial population in food during drying is dependent on how microorganisms are distributed within the food matrix. Living cells or active enzymes may be totally encapsulated and stay in the core region of a food material or stay near the boundary region of a food material. The thermal damage to the living cells or active enzymes is also influenced by the food ingredients or additives, which may have protective effects during drying (Leslie et al. 1995, Oldenhof et al. 2005, Zhao and Zhang 2005). For instance, addition of trehalose to yeast suspension significant reduced the damage to the active yeast cells (Berny and Hennebert 1991, Bayrock and Ingledew 1997a,b). The protective effect of a specific food additive may vary for different microorganisms, depending on the type and mode of drying, drying conditions, type and concentration of microbes, and the type of food material itself.

Lievense et al. (1992) proposed an inactivation kinetics model for degradation of *Lactobacillus plantarum* during drying by considering thermal and dehydration inactivation as two separate influences but operating simultaneously. The model had 10 parameters to be obtained from the experimental work. Measurements of the drying parameters were obtained from the fluidized-bed drying with drying temperatures lower than 50°C. The effective diffusivity concept was used to take into account the spatial moisture distribution. The inactivation parameters were measured from "nondrying," heating experiments in which approximately 1 mm thick *L. plantarum*-starch granulate was placed in a Petri dish and stored at $5.0°C \pm 0.5°C$ in a vacuum desiccator for 48 h. After 48 h, the glucose-fermenting activity and moisture concentrations of the sample were measured. They illustrated in their work that thermal inactivation is insignificant at drying temperatures lower than 50°C. Furthermore, they stated that the dehydration inactivation depends on the reached moisture content of the material only and is independent of the drying rate. However, Lievense et al. (1992) did not show if this observation could be true for high-temperature drying processes such as spray drying, where drying rates are much higher. A similar trend was followed by Yamamoto and Sano (1992), who proposed a five-parameter model for enzyme inactivation during drying using a single suspended droplet drying experiment. A sucrose solution of fixed water content containing different enzymes such as β-galactosidase, glucose oxidase, and alkaline phosphatase was incubated at a constant temperature. The thickness of the material used and air temperatures were not reported in the study. The deactivation energy E_d was described as a function of average water content. Again, a binary (water and dissolved solids) diffusion coefficient was introduced to the drying analysis for taking care of the water distribution inside a droplet. They concluded that air temperatures and droplet size significantly affect the inactivation rate. The effect of initial water content is shown to be insignificant. The enzyme activity was experimentally measured using constant temperature and constant moisture content heating experiments, where no evaporation was involved. Again, this work may be classified into the pre-equilibration experiments. In general, inactivation kinetics of microorganisms during drying has been conventionally expressed using the same first-order reaction equation (Equation 12.12).

It is an unstructured model. A structured model which can deal with the non-loglinear behavior has been developed, especially accounting for the tail ends of the survival curves (Geeraerd et al. 2000). For microorganisms distributed in a moist material, which is not yet dried (i.e., in a saturated medium), the inactivation rate constant k_d is usually considered to be a function of temperature and is described using the Arrhenius equation (Wijlhuizen et al. 1979, Johnson and Etzel 1993, Meerdink and Van't Riet 1995):

$$k_d = k_0 \exp\left(-\frac{E}{RT}\right)$$

(12.17)

where
  R is the universal gas constant ($\approx 8.314\,$J mol^{-1} K^{-1})
  E is the activation energy (J mol^{-1})

When the in-drying process is considered, the inactivation kinetics should include material's temperature and moisture concentration effects. The water content profile could be significantly nonuniform within the moist material being dried due to the nature of the drying process. In the literature, the temperature dependence of the inactivation rate constant k_d during drying was described using an Equation 12.7. The temperature distribution may also have an impact on the local distribution of the inactivation rates. The temperature gradient can be significant if the Biot number is quite large (Bi = hL/k; h is the convective heat transfer coefficient (W m^{-2} K^{-1}), which can be determined similarly as that for h_m; k is the thermal conductivity of the moist material (W m^{-1} K^{-1}). Usually a Bi value less than 0.1 is considered, beyond which temperature nonuniformity cannot be ignored (Incropera and De Wit 2002). However, when evaporation takes place, this nonuniformity is damped and even for some large Bi values such as 0.5 or 1, a small evaporating water droplet can have negligible temperature gradient (Chen and Peng 2005).

Corresponding to Equation 12.17 which is a temperature-dependent only formula, the decimal reduction time D should be expressed as

$$D = \frac{\ln 10}{k_d} = \frac{\ln 10}{k_0} \exp\left(\frac{E}{RT}\right)$$

(12.18)

The traditional inactivation kinetics model, Equations 12.12 and 12.17, contains only two parameters, k_0 (pre-exponential factor) and E (deactivation energy), but this model does not include the moisture content effects. Meerdink and Van't Riet (1995) studied the inactivation of the enzyme α-amylase during droplet drying, describing the deactivation energy parameter (E) to be water content dependent. Meerdink and

Van't Riet (1995) concluded that the inactivation rate is more sensitive to changes in material's temperature compared to the drying rate. An approach considered by Meerdink and Van't Riet (1995) is attractive, as it required only four parameters (a, b, k_0, and E) to be obtained from the experimental work. The inactivation rate constant was expressed using the following formulae:

$$k_d = k_0 \exp\left(aX - \frac{E + bX}{RT} \right)$$ (12.19)

Hence the decimal reduction time (D) can be written as follows:

$$D = \frac{\ln 10}{k_0} \cdot e^{-aX} \exp\left(\frac{E + bX}{RT} \right)$$ (12.20)

where X is the water content on a dry basis. The activation energy (E) for inactivation represents the difficulty in deactivating living cells or denaturing proteins and enzymes. Higher activation energy indicates the greater difficulty for deactivating microorganisms or living cells. The activation energy may change with environment conditions around living cells such as temperature, water activity, pH, and pressure. At higher temperature, the activation energy may be lower to destroy the functionalities of living cells or enzymes. Teixeira et al. (1995) reported that the activation energy for inactivation of *Lactobacillus bulgaricus* at $T > 70°C$ was $33.5 \, kJ \, mol^{-1}$, while it was $85.8 \, kJ \, mol^{-1}$ for $T < 70°C$. The activation energy also depends on the food microstructure and the state of the bioactive constituent (e.g., encapsulated, immobilized, etc.). For instance, the activation energy for deactivation of peroxides present in potato, carrot, and two varieties of tomatoes was different from the activation energy for the same enzyme present in pumpkin (Chen and Patel 2008). As a reference, the parameters of Equation 12.6 for inactivation of α-amylase during drying were reported as $k_0 = 1.2426 \times 10^{32}$ (s^{-1}), $E = 247.3 \times 10^3$ ($J \, mol^{-1}$), $a = 121.8$ ($J \, mol^{-1}$), and $b = 341$, respectively.

12.4 CONCLUDING REMARKS

Drying is a heat and mass transfer coupled process, which may also involve significant biochemical and chemical reactions. The composition of the product and size/shape all have impact on drying behavior and the quality changes that happen on the surface of a product or within the product. More quantitative approach to these problems is needed in order to progress the subject as well as the elevation of the technical level of the industry. It is no longer a water content lowering activity; it is much more to it than ever before.

NOTATIONS

a	fitting parameter ($J \, mol^{-1}$)
a_w	water activity
a^*	color index (red and green)

Bi	Biot number (heat transfer based)
b	fitting parameter
b^*	color index (blue and yellow)
c	microbial concentration (mol m^{-3} or kg m^{-3})
c^*	metric chroma
C	lycopene concentration (kg m^{-3})
C_0	fitting constants for $E.\ coli$ inactivation kinetics in carbopol
C_1	fitting constants for $E.\ coli$ inactivation kinetics in carbopol
C_2	fitting constants for $E.\ coli$ inactivation kinetics in carbopol
C_3	fitting constants for $E.\ coli$ inactivation kinetics in carbopol
C_4	fitting constants for $E.\ coli$ inactivation kinetics in carbopol
D	decimal reduction time at temperature T (min)
E	apparent activation energy (J mol^{-1})
ΔE^*	color sensor evaluation parameter
F	thermal death time (min)
h	heat transfer coefficient (W m^{-1} K^{-1})
Hue	metric hue angle
k	thermal conductivity of food material (W m^{-1} K^{-1})
k_0	pre-exponential factor (s^{-1})
kd	inactivation/degradation rate constant (s^{-1})
L	thickness of the slab or film (m)
L^*	color index (black and white)
m_{microbe}	mass of one average microbe (kg)
N	concentration of the live microorganism (cell m^{-3})
N_0	initial number of the live microorganism (cell m^{-3})
R	universal gas constant (J mol^{-1} K^{-1})
t	time (s)
T	temperature (K, °C)
W	percent moisture content on a wet basis
X	water content on a dry basis (kg water/kg dry solids)
X_0	center water content of the food slab (kg water/kg dry solids)
$\overline{X}$	spatial average moisture content on a dry basis (kg water/kg dry solids)
Z	thermal resistance constant

REFERENCES

Aguilera, J.M. and Stanley, D.W. 1999. *Microstructural Principles of Food Processing and Engineering*. Gaithersburg, MD: Aspen Publishers Inc.

Aversa, M., Curcio, S., Calabrò, V., and Iorio, G. 2007. An analysis of the transport phenomena occurring during food drying process. *Journal of Food Engineering* 78: 922–932.

Bayrock, D. and Ingledew, W.M. 1997a. Fluidized bed drying of baker's yeast: Moisture levels, drying rates and viability changes during drying. *Food Research International* 30: 407–415.

Bayrock, D. and Ingledew, W.M. 1997b. Mechanism of viability loss during fluidized bed drying of baker's yeast. *Food Research International* 30: 417–425.

Berny, J.F. and Hennebert, G.L. 1991. Viability and stability of yeast cells and filamentous fungus spores during freeze drying: Effects of protectants and cooling rates. *Mycologia* 83: 805–815.

Britt, I.J. 2008. Thermal processing. In *Food Biodeterioration and Preservation*, ed. G. Tucker, pp. 63–79. Oxford, U.K.: Blackwell Publishing Ltd.

Chen, X.D. 2006. Moisture diffusivity in food and biological materials, Plenary Keynote at *15th International Drying Symposium (IDS 2006)*, Budapest, Hungary, August 20–23, 2006.

Chen, X.D. and Patel, K.C. 2008. Biological changes during food drying processes. In *Drying Technologies in Food Processing*, eds. X.D. Chen and A.S. Mujumdar, pp. 90–109. Oxford, U.K.: Blackwell Publishing Ltd.

Chen, X.D. and Peng, X. 2005. Modified *Biot* number in the context of air-drying of small moist porous objects. *Drying Technology* 23: 83–103.

Chiruta, J., Davey, K.R., and Thomas, C.J. 1997. Thermal inactivation kinetics of three vegetative bacteria as influenced by combined temperature and pH in a liquid medium. *Food and Bioproducts Processing* 75C: 174–180.

Christian, J.H.B. 2000. Drying and reduction in water activity. In *The Microbiological Safety and Quality of Food*, eds. B.M. Lund, A.C. Baird-Parker, and G.W. Gauld, pp. 146–174. Gaithersburg, MD: Aspen Publishers Inc.

Featherstone, S. 2008. Control of biodeterioration in food. In *Food Biodeterioration and Preservation*, ed. G. Tucker, pp. 1–61. Oxford, U.K.: Blackwell Publishing Ltd.

Geeraerd, A.H., Herremans, C.H., and Van Impe, J.F. 2000. Structural model requirements to describe microbial inactivation during a mild heat treatment. *International Journal of Food Microbiology* 59: 185–209.

Goula, A.M., Adamopoulos, K.G., Chatzitakis, P.C., and Nikas, V.A. 2006. Prediction of lycopene degradation during a drying process of tomato pulp. *Journal of Food Engineering* 74: 37–46.

Incropera, F.P. and DeWitt, D.P. 2002. *Fundamentals of Heat and Mass Transfer*. New York: John Wiley & Sons.

Johnson, J.A.C. and Etzel, M.R. 1993. Inactivation of lactic acid bacteria during spray drying. *AIChE Symposium Series* 297: 98–107.

Kaur, D., Sogi, D.S., and Wani, A.A. 2006. Degradation kinetics of lycopene and visual color in tomato peel isolated from pomace. *International Journal of Food Properties* 9: 781–789.

Khoo, K.Y., Davey, K.R., and Thomas, C.J. 2003. Assessment of four model forms for predicting thermal inactivation kinetics of *Escherichia coli* in liquid as affected by combined exposure time, liquid temperature and pH. *Food and Bioproducts Processing* 81C: 129–137.

Labuza, T.P. 1975. Oxidative changes in foods at low and intermediate moisture levels. In *Water Relations of Foods*, ed. R.B. Duckworth, pp. 455–474. New York: Academic Press.

Labuza, T.P. 1977. The properties of water in relationship to water binding in foods: A review. *Journal of Food Processing and Preservation* 1: 167–190.

Leslie, S.B., Israeli, E., Lighthart, B., Crowe, J.H., and Crowe, L.M. 1995. Trehalose and sucrose protect both membranes and proteins in intact bacteria during drying. *Applied and Environmental Microbiology* 61: 3592–3597.

Li, L.T. 1998. *Food Properties*. Beijing, China: China Agriculture Publisher (in Chinese).

Lievense, L.C., Verbeek, M.A.M., Taekema, T., Meerdink, G., and Van't Riet, K. 1992. Modelling the inactivation of *Lactobacillus plantarum* during a drying process. *Chemical Engineering Science* 47: 87–97.

Meerdink, G. and Van't Riet, K. 1995. Prediction of product quality during spray drying. *Food and Bioproduct Processing* 73C: 165–170.

Oldenhof, H., Wolkers, W.F., Fonseca, F., Passot, S.P., and Marin, M. 2005. Effect of sucrose and maltodextrin on the physical properties and survival of air-dried *Lactobacillus bulgaricus*: An *in situ* Fourier transform infrared spectroscopy study. *Biotechnology Progress* 21: 885–892.

Rahman, M.S. 2007. Food preservation: Overview. In *Preservation of Fresh Food Products: Handbook of Food Preservation*, ed. M.S. Rahman, 3–17. Boca Raton, FL: CRC Press/ Taylor & Francis.

Roos, Y.H. 2002. Importance of glass transition and water activity to spray drying and stability of dairy powders. *Lait* 82: 475–484.

Scott, W.J. 1957. Water relations of food spoilage microorganisms. *Advances in Food Research* 7: 83–127.

Teixeira, P.C., Castro, M.H., and Kirby, R.M. 1995. Death kinetics of *Lactobacillus bulgaricus* in a spray drying process. *Journal of Food Protection* 57: 934–936.

Watzke, H.J. 1998. Impact of processing on bioavailability examples of minerals in foods. *Trends in Food Science and Technology* 9: 320–327.

Welti-Chanes, J., Vergara-Balderas, F., and Bermúdez-Aguirre, D. 2005. Transport phenomena in food engineering: Basic concepts and advances. *Journal of Food Engineering* 67: 113–128.

Wijlhuizen, A.E., Kerkhof, P.J.A.M., and Bruin, S. 1979. Theoretical study of the inactivation of phosphatase during spray drying of skim milk. *Chemical Engineering Science* 34: 651–660.

Yamamoto, S. and Sano, Y. 1992. Drying of enzymes: Enzyme retention during drying of a single droplet. *Chemical Engineering Science* 47: 177–183.

Yusheng, Z. and Poulsen, K.P. 1988. Diffusion in potato drying. *Journal of Food Engineering* 7: 249–262.

Zhao, G. and Zhang, G. 2005. Effect of protective agents, freezing temperature, rehydration media on viability of malolactic bacteria subjected to freeze-drying. *Journal of Applied Microbiology* 99: 333–338.

13 Safety and Quality of Irradiated Food

Eileen M. Stewart

CONTENTS

13.1 INTRODUCTION

Food irradiation is the processing of food products by ionizing radiation in order to, among other things, control foodborne pathogens, reduced microbial load and insect infestation, inhibit the germination of root crops, and extend the durable life of perishable produce. (CAC 2003a)

Other diverse applications of irradiation include curing solvent-free links, coatings and adhesives, cross-linking plastic and rubber materials, extracting nitrogen and sulfur oxides from combustion gases to reduce acid rain, decomposing toxic compounds in waste water, and sterilizing medical devices and pharmaceuticals (Cleland 2006).

The regulatory control of food irradiation should take into consideration the *Codex General Standard for Irradiated Foods* (CAC 2003b) and the *Recommended International Code of Practice for Radiation Processing of Food* (CAC 2003a).

According to the Codex code of practice the purpose of regulatory control of irradiated food products should be to

- Ensure that radiation processing of food products is implemented safely and correctly, in accordance with all relevant Codex standards and codes of hygiene practice
- Establish a system of documentation to accompany irradiated food products, so that the fact of irradiation can be taken into account during subsequent handling, storage, and marketing
- Ensure that irradiated food products that enter into international trade conform to acceptable standards of radiation processing and are correctly labeled (CAC 2003a)

The use of ionizing radiation for the preservation of food has been extensively studied for over 100 years making it one of the most thoroughly researched means by which food can be treated in order to make it safer to eat and last longer (Diehl 1995). The two main drivers for the use of food irradiation are the enhancement of food safety and of trade in agricultural products (Borsa 2004). The process should, however, not be used as a substitute for good manufacturing practices but rather as a means of reducing risk. The chapter will outline the effects of the ionizing radiation on the quality and safety of irradiated foodstuffs but does not attempt to cover the complete breadth of information currently available on this topic. The author therefore suggests that publications such as those by Diehl (1995), Molins (2001a), Komolprasert and Morehouse (2004), and Sommers and Fan (2006) be consulted for more detailed information.

13.1.1 SOURCES OF IONIZING RADIATION

The following types of ionizing radiation are permitted for the treatment of food-stuffs (CAC 2003a):

- Gamma rays (or gamma photons) from the radionuclides ^{60}Co or ^{137}Cs
- X-rays generated from machine sources operated at or below an energy level of 5 MeV
- High-energy electrons generated from machine sources operated at or below an energy level of 10 MeV

These three sources of ionizing radiation transfer their energies to materials by ejecting atomic electrons which can then ionize other atoms in a cascade of collisions. All these energy sources can produce similar effects in any irradiated material (Cleland 2006). The major difference between the sources is their penetrating power. The gamma- and x-rays are highly penetrating and can treat food in bulk whereas the high-energy electrons are only useful for surface irradiation or for the treatment of thin packages. A major advantage, however, of using an irradiation system powered by electricity is that it can be switched on and off like a lightbulb, and is not related to the nuclear industry (Stewart 2001a). The choice of source for a particular application will depend on practical aspects such as the thickness and density of the foodstuff being treated, the minimum absorbed dose, with economics also being taken into account (Cleland 2006).

Irradiation of food with gamma rays, x-rays, and electron beam machines may be expected to induce a low level of radioactivity in foods. There is, however, general agreement that the background radioactivity already present in the food far exceeds that induced by any of these sources of ionizing radiation when used at the energy levels recommended by the Codex (Stewart 2001a).

13.1.2 Absorbed Dose

The amount of energy absorbed per unit mass of an irradiated food product is referred to as the "absorbed dose," or more simply as "dose." The absorbed dose is proportional to the ionizing radiation energy absorbed per unit mass of irradiated material and the effects of the treatment are related to this quantity, which is the most important specification for any irradiation process (Cleland 2006). The international unit of absorbed dose is the gray (Gy) which is equivalent to one joule of energy per kilogram of material. It is often more convenient to refer to dose in terms of thousands of gray or kilogray (kGy), with 1000 Gy being equal to 1 kGy.

Although the process of irradiation can be carefully controlled, it is necessary to confirm the absorbed dose received by food and this is normally achieved using dosimetry systems (Ehlermann 2001) whereby dosimeters (or dose meters) are placed at different points within a selected number of packages. A range of dosimetry systems are available but not all are suitable for measuring the low or high doses of irradiation used to treat some foods. Therefore, it is necessary to select a system appropriate for the purpose for which it is being used (Stevenson 1990).

Ionizing radiation is absorbed as it passes through a food, but not all parts received the same absorbed dose as part of the food will receive a maximum dose (D_{max}) and part a minimum dose (D_{min}). The uniformity of the dose distribution ($D_{max}:D_{min}$), also called the overdose ratio, can be improved by irradiating the product from both sides. Prior to treating a food with ionizing radiation it is essential to establish the dose distribution throughout the food package (Stevenson 1990) in order to take into account the differences in the products being irradiated, including factors such as seasonal crop variations, anomalies in bulk density, uneven density variation, and random packaging of agricultural products (Diehl 1995). The dose distribution may be established by placing a number of dosimeters throughout the package or packages of food being treated. By doing so absorbed dose can be measured and positions defined within the package which have received maximum and minimum irradiation doses (Stevenson 1990). For the irradiation of any food, the minimum absorbed dose should be sufficient to achieve the intended technological purpose and the maximum absorbed dose should be less than that which would compromise consumer safety, wholesomeness, or would adversely affect structural integrity, functional properties, or sensory attributes (CAC 2003b). The maximum dose delivered to a food should not exceed 10 kGy, except when necessary to achieve a legitimate technological purpose (WHO 1999).

Dose and the distribution of dose through a product are determined by product parameters and by source parameters (Diehl 1995). Product parameters are mainly the density of the food itself and the density of the packing the individual food containers within the carrier in which the irradiation takes place. Source parameters will differ depending on the type of irradiator used. For gamma radiators relevant factors include isotope, source strength, and geometry, while for machine sources these include factors such as beam energy and beam power.

13.1.3 INTERACTIONS OF IONIZING WITH MATTER

As noted previously, when food is irradiated, energy is absorbed and it is this absorbed energy that leads to the ionization or excitation of the atoms and molecules of the food which, in turn, results in radiolytic changes. When gamma- and x-rays interact with matter, different types of energy transfer can occur but in the case of food irradiation the Compton effect is predominant. In this case, the incident photon interacts with the absorbing medium in such a way that an orbital electron is ejected. The direction of the incident photon changes after the collision and it loses some of its original energy but may go on to react with other atoms to form secondary electrons. It is these ejected electrons that ionize and excite the components of the system producing, for example, free radicals, which in turn are highly reactive and attack other compounds in the medium before forming stable end products. As the Compton electrons produced from either gamma- or x-rays produce ionizations and excitations in the same manner as high-energy electrons, the chemical changes induced in food by irradiation are similar irrespective of the source of ionizing radiation used. The extent of any change produced by irradiation is dependent on the absorbed dose (Stevenson 1992).

These changes may result from direct or indirect action. In "direct" action, a sensitive target such as DNA of a living organism is damaged directly by an ionizing particle or ray, while "indirect" action is caused mostly by the products of water radiolysis, which disappear quickly by reacting with each other and/or with other food components (Stewart 2001a).

13.1.4 CHEMICAL CHANGES OCCURRING IN FOODS

In the case of food irradiation, the radiolysis of water is important. When pure water is irradiated the following highly reactive entities are formed (Stewart 2001):

$$H_2O \rightarrow \cdot OH + e_{aq}^- + \cdot H + H_2 + H_2O_2 + H_3O^+$$

where
  $\cdot OH$ is a hydroxyl radical
  e_{aq}^- is an aqueous (or solvated or hydrated) electron
  $\cdot H$ is a hydrogen atom
  H_2 is a hydrogen molecule
  H_2O_2 is the hydrogen peroxide compound
  H_3O^+ is a solvated (or hydrated) proton

The hydroxyl radical is a powerful oxidizing agent while the hydrated electron is a strong reducing agent and the hydrogen atoms are slightly weaker reducing agents.

Consequently, both oxidation and reduction reactions take place when water-containing food is irradiated. Hydrogen (H_2) and hydrogen peroxide (H_2O_2) are the only stable end products of water radiolysis, being produced in a low yield even when irradiation doses are high (Diehl 1995).

The presence of oxygen during irradiation can influence the course of radiolysis. Hydrogen atoms can reduce oxygen to the hydroperoxyl radical ($\cdot OOH$) which is a mild oxidizing agent. Another oxidizing agent, the superoxide radical ($\cdot O_2^-$), is formed from the reaction of the solvated electron with oxygen. Both the hydroperoxyl and superoxide radical can give rise to hydrogen peroxide. Oxygen can also add to other radicals produced when foods are irradiated giving rise to peroxy radicals ($\cdot RO_2$). Ozone (O_3), a powerful oxidant, can also be formed from oxygen during irradiation (Stewart 2001a).

Temperature can influence radiolytic changes within a foodstuff. In deep-frozen food the reactive intermediates of water radiolysis are trapped and are thus kept from reacting with each other or with the substrate, thus freezing can have a protective effect. During thawing of frozen food, there is an increase in the yield of radiolytic products. However, during transition from the frozen to thawed state, the reactive intermediates react preferentially with each other rather than with the food components. Consequently, any damage to food components is much less than if the food had been irradiated in the unfrozen state (Diehl 1995).

Radiation can affect all the major components of food, that is, carbohydrates, proteins, and lipids. It should, however, be noted that even at the high doses used for sterilization, the changes are small and similar to those produced by other food processing technologies such as pasteurization. As for any other food processing method, irradiation has advantages and disadvantages. Some foods are sensitive to ionizing radiation and too high a dose may cause organoleptic changes rending the food unacceptable to the consumer. Therefore, the actual dose employed is a balance between what is needed and what can be tolerated by the product without unwanted changes (Farkas 2006).

13.2 EFFECT OF IRRADIATION ON THE QUALITY OF FOODSTUFFS

13.2.1 MUSCLE-BASED FOODS

Due to its bactericidal and antiparasitic properties, ionizing radiation is a highly effective means of enhancing the safety of muscle-based foods and extending shelf life (Farkas 1998, Molins 2001b). Enteric pathogens associated with meat and poultry products such as *Clostridium jejuni*, *Escherichia coli* O157:H7, *Staphylococcus aureus*, *Salmonella* spp., *Listeria monocytogenes*, and *Aeromonas hydrophila* can be significantly decreased or eliminated at irradiation doses of less than 3 kGy. Only enteric viruses and endospores of genera *Clostridium* and *Bacillus* are highly resistant to ionizing radiation but even these are significantly affected (Thayer 1995, Ahn and Lee 2006). The sensitivity of microorganisms to ionizing radiation is affected by factors such as temperature during irradiation, stage of growth, the presence of oxygen, availability of water, and composition of the medium in which they are present (Molins 2001b).

The use of radiation processing to extend the shelf life of highly perishable fresh meat and poultry products under refrigerated storage has been extensively studied. The bacteria which most commonly spoil fresh meats and poultry are *Pseudomonas* and *Achromobacter*. *Pseudomonas* are very sensitive to ionizing radiation and doses used to control Salmonellae will wipe out these normal spoilage bacteria (Grant and Patterson 1991, Stevenson 1992). *Achromobacter* are also radiation sensitive, but they will survive low doses of irradiation and consequently the spoilage which will occur on prolonged storage will arise from the outgrowth of these and other bacteria.

A study by Kiss et al. (2001) demonstrated that a dose of 2 kGy can extend the shelf life of aerobically packed minced beef at 4°C storage temperature by 3–5 times as compared with nonirradiated samples. In the same study strains of *E. coli*, *S. aureus*, and *L. monocytogenes* inoculated into minced beef were reduced by 7, 5, and 5 $\log_{10}$ cycles, respectively, by the same dose of 2 kGy. It was found that during 10 days of chilled storage at 4°C the reduced viable cell counts did not change except for that of *L. monocytogenes* which started to grow after 3 days. According to Farkas (1998) recommended doses for radiation processing of frozen poultry are 3–5 and 1.5–2.5 kGy for chilled poultry. These treatments have been found to be effective in reduction of the most resistant serotype of *Salmonella* by about 3 $\log_{10}$ cycles and *Campylobacter* by a still greater rate (Kamplemacher 1984). Kiss and Farkas (1972) demonstrated that using the latter doses, the shelf life of chilled poultry could be extended two- to threefold compared with nonirradiated samples. It was also shown by Kiss (1984) that irradiation of frozen poultry at 3–5 kGy has no effect on the culinary properties of various dishes prepared from chicken.

Combinations of irradiation with other food processing technologies such as modified atmosphere packaging (MAP), refrigeration, freezing, and cooking have been shown to have great potential for improving the quality and safety of fresh and processed meats and poultry (Molins 2001b, Patterson 2001). One of the most common combination treatments is the use of irradiation with refrigeration. For example, a dose of 2.5–3 kGy is sufficient to significantly reduce vegetative pathogens in poultry meat. Many spoilage organisms will also be killed, thus extending the shelf life. However, the maximum potential of shelf life extension can only be achieved if the meat is kept refrigerated after irradiation (Molins 2001b, Patterson 2001). A combination of irradiation with mild heat has also been shown to be effective in controlling spoilage organisms and increasing the safety of certain foods. For example, an irradiation dose of 5 kGy is sufficient to sensitize clostridial spores to subsequent heating which has been found to be of value in heat-treated spiced meats, where the heat treatment required in the canning process could be reduced if spores in the irradiated spices were heat sensitized.

Despite the recognizable advantages of treating meat and poultry products with ionizing radiation, there are some quality aspects which limit adoption of the technology by the meat industry. Irradiation can produce a characteristic aroma, accelerate lipid oxidation, and change the color of the meat (Ahn and Lee 2006). The characteristic odors have been described as "barbecued corn-like" or "bloody sweet." The odors have also been described as being metallic, sulfide, wet dog, goaty, or burnt. The intensity of the characteristic odors developed in irradiated meat products depends on the dose of irradiation applied.

Sensory attributes of irradiated muscle foods have been extensively examined with the conclusions of many researchers pointing toward major involvement of lipids in the characteristic radiation-induced odors (Molins 2001b). Use of ionizing radiation is known to accelerate lipid oxidation in meat due to the generation of hydroxyl radicals, which, as noted previously, are produced from water and act as powerful initiators of lipid oxidation. It has been demonstrated that irradiation accelerates lipid oxidation only when meat is treated and stored under aerobic conditions, especially in cooked meat (Merritt et al. 1975). When oxygen is excluded the initiation step of the chain reaction of lipid oxidation is blocked and it has been reported that preventing exposure to oxygen after cooking meat was more important than packaging, irradiation, or storage of raw meat (Ahn et al. 1992, 1993). Work by Shahidi and Pegg (1994) showed that aldehydes contributed the most to oxidation flavor and rancidity in cooked meat with hexanal being the predominant volatile aldehyde formed. In addition, if meat is irradiated in the frozen state, lipid oxidation is restricted and this is most likely due to the restricted movement of free radicals when produced in a frozen environment (Nam et al. 2002).

Work by Batzer and Doty (1955) initially identified hydrogen sulfide and methyl mercaptan as compounds responsible for the irradiation odor. Merritt et al. (1959) identified the volatiles methyl mercaptan, acetaldehyde, dimethyl sulfide, acetone, methanol, ethanol, methyl ethyl ketone, dimethyl disulfide, ethyl mercaptan, and isobutyl mercaptan in irradiated meats, with all compounds apart from the last two increasing with irradiation dose. Work by Wick et al. (1967) identified methional as a major contributor to irradiation odor in meat on the basis of its low odor threshold, characteristic color, and presence in irradiated meat. Patterson and Stevenson (1995) found that dimethyl sulfide was the most potent off-odor compound in irradiated chicken meat. More recently Lee and Ahn (2003) concluded that the odor of the sulfur compounds was much stronger and more stringent than that of other compounds with the volatiles from lipids accounting for only a small part of the off-odor in irradiated meat. Thus, it has been concluded that the volatiles responsible for the off-odor are sulfur-containing compounds such as methanethiol, dimethyl sulfide, dimethyl disulfide, and dimethyl trisulfide, all of which are generated from the radiolytic degradation of sulfur-containing amino acids present in meat (Ahn and Lee 2004).

It is now well established that irradiation affects the color of fresh and cured meat and poultry in ways that are dose dependent and influenced by the presence of oxygen during irradiation. Other influencing factors in radiation-induced color changes include animal species, muscle type, and the type packaging. In light colored meats, such as poultry breast and pork loin, irradiation results in an increase in redness while brown or gray discoloration occurs in red meat upon irradiation. The pink color in irradiated light meat is characterized as a carbon monoxide–heme pigment complex (Nam and Ahn 2002a, Ahn and Lee 2004). The mechanisms of color change in irradiated beef are different from those of lighter colored meats with the overall beef color being mainly determined by the reducing potential of meat as the contribution of the carbon monoxide–heme to the color is much smaller than that of light meats (Ahn and Lee 2006).

Work by Lebepe et al. (1990) showed increased Hunter Lab a color values in vacuum-packaged pork loins treated at 2.5–3.0 kGy compared to nonirradiated loins, which is indicative of increased redness. Nanke et al. (1998) also demonstrated that irradiated pork and turkey had increased Hunter Lab *a* values, indicative of a redder color, while the equivalent color values for beef decreased and yellowness (Hunter Lab *b* color) increased in direct proportion to dose and storage time. Irradiation has been demonstrated to increase redness in both aerobically and vacuum-packaged raw chicken and turkey meat (Miller et al. 1995, Nam and Ahn 2002b). Luchsinger et al. (1996) also reported that increased red color in irradiated pork was more intense and stable with vacuum packaging than aerobic conditions during refrigerated storage. Work by Nam and Ahn (2003) indicated that the color of beef was significantly affected by aging time. The redness of ground beef was found to be significantly reduced with irradiation, with the color changing from bright red to greenish brown after irradiation.

Given the changes in quality that can potentially occur upon the irradiation of meat, the control of discoloration and off-odor production of irradiated meat is important for it to be acceptable to the consumer. There are a number of ways whereby such changes can be retarded in irradiated muscle-based products. Oxidative rancidity could be delayed by the addition of antioxidants. For example, Chen et al. (1999) demonstrated the beneficial antioxidant effects of sesamol, quercetin, rutin, BHT, and rosemary in irradiated pork patties. Packaging is another critical factor in the development of off-odor in meat and it has been shown that a combination of antioxidants and double (vacuum/aerobic) packaging is effective in controlling the oxidative quality changes of irradiated and cooked meat (Ahn and Lee 2004). Work by Nam and Ahn (2002b) showed that vacuum packaging of aerobically packaged irradiated turkey meat at 1 or 3 days of storage lowered the amounts of sulfur volatiles and lipid oxidation products compared with vacuum-packaged and aerobically packaged meats, respectively. Irradiation was shown to increase the Hunter Lab *a* values of raw turkey meat but exposing the irradiated meat to aerobic conditions alleviated the intensity of redness.

In their succinct review on the impact of irradiation on the safety and quality of poultry and meat products, O'Bryan et al. (2008) concluded that a combination of irradiation dose and temperature, dietary and direct additives, storage temperature, and atmosphere can be used to produce meats and poultry that the average consumer will find indistinguishable from nonirradiated meats. More information, however, is needed on the interactions of irradiation and packaging, with the development of new types of package materials for the irradiated food industry being necessary. For further reading on the irradiation of muscle-based foods the author suggests referring to publications by the IAEA (1993, 1996), Molins (2001b), Ahn and Lee (2006), and O'Bryan et al. (2008).

13.2.2 Fish and Shellfish

Fresh fish and shellfish are highly perishable products with a limited shelf life; therefore, shelf-life extension for even a few days increases the viability of the industry and marketability of seafood products. The radiation processing of fishery products, as detailed in the review by Venugopal et al. (1999), may be one method of improving quality. Low-dose levels of irradiation of 1–3 kGy can often reduce the initial load of potential spoilage bacteria by 1–3 $\log_{10}$ cycles thereby extending the shelf life of fresh fish and shellfish although in some investigations doses up to 5 kGy have been required. Shelf life can be increased by at least 7 days for most species of fish and shellfish using ≤2 kGy radiation processing at which dose spoilage microorganisms are reduced by 99.9% (Andrews and Grodner 2004). Gram-negative bacteria (GNB) are generally considered to be more sensitive than Gram-positive species; thus many of the typical spoilage bacteria are among the least resistant. Gram-positive bacteria such as *S. aureus*, *Micrococcus*, *Bacillus*, and *Clostridium* are among the most irradiation-resistant genera.

Pathogenic bacteria of concern in seafood are those found naturally in the fresh water or marine environment. These include the genera *Vibrio*, *Aeromonas*, and *C. botulinum* type E. *Vibrio* spp. are relatively sensitive to low-dose gamma irradiation compared with other pathogens (Kilgen 2001, Andrews and Grodner 2004, Foley 2006). Use of ionizing radiation to eliminate the risk of *Vibrio vulnificus* is the major justification of using the irradiation technology while *Vibrio parahaemolyticus* O3:K6 is the most process resistant of all the *Vibrio* pathogens tested to date, but is reduced to nondetectable levels following a dose of 1.5 kGy. It has been shown (Kilgen 2001) that a sublethal dose of 1.5 kGy is optimum for elimination of *V. vulnificus* in oysters and that this dose can also extend shelf life of commercially shucked and packaged oyster meat from 14 to 24 days when stored at 4°C in a commercial cooler. Spores are extremely resistant to ionizing radiation, and the Gram-positive bacteria *Micrococcus* (*Deinococcus*) *radiodurans* and *Micrococcus radiophilus* are extremely resistant due to their highly efficient DNA repair enzymes (Venugopal et al. 1999). Fish parasites such as *Anisakis* are normally controlled by freezing the fish prior to processing. Doses of 5 kGy and above are required to eliminate parasites in fish although it has been shown that such high doses can be unacceptable in appearance and flavor (Van Mameren et al. 1969). The major benefits of the application of radiation to fishery products are in the reduction of postharvest losses and the improvement of their hygienic quality (Venugopal et al. 1999). More details on the doses required for improving the microbiological safety and quality of fish and shellfish are presented in the publications by Venugopal et al. (1999) and Kilgen (2001).

The dose level for shelf life extension must be evaluated for each species of fish or shellfish in order to maintain acceptable flavor, texture, and wholesomeness (Grodner and Andrews 1991). Fat content is a major consideration for the irradiation of fish and many high-fat fish have to be vacuum-packaged or packaged in modified atmosphere to prevent oxidative rancidity and other sensory changes such as bleached color (Kilgen 2001). Fish species such as mackerel and salmon which have a high fat content do not have good sensory qualities when processed with doses of >1.5 kGy due to lipid oxidation (Poole et al. 1994). In addition, salmon lose their pigmentation through bleaching when treated with doses as low as 1–2 kGy (Andrews and Grodner 2004). However, it should be noted that vacuum packaging often induces drip loss and provides conditions for potential outgrowth of *C. botulinum* spores if the product is temperature abused. Lean fish are known to undergo the least amount of radiation-induced rancidity (Venugopal et al. 1999). Work by Chen et al. (1996) compared the microbial and sensory quality of irradiated (2 kGy or less) crab products (white lump, claw, and fingers) through 14 days of storage on ice. Irradiation was found to effectively reduce spoilage bacteria with shelf life being extended by more than 3 days beyond the control samples. During storage, fresh crab odor and flavor were similar for irradiated and control samples, while off-flavors and odors developed more rapidly in the controls. Overall acceptability scores for irradiated crab samples were higher than for control samples throughout storage.

For further reading on the irradiation on fish and shellfish the author would point readers toward the publications of Kilgen (2001), Andrews and Grodner (2004), and Venugopal et al. (1999).

13.2.3 FRUITS AND VEGETABLES

A noteworthy review on the irradiation of fruits and vegetables undertaken by Thomas (2001a) stated that the main reasons for treating fruits and vegetables with ionizing radiation include (1) extension of shelf life by delaying the physiological and biochemical processes leading to maturation and ripening, (2) control of fungal pathogens causing postharvest rot, (3) inactivation of human pathogens to maintain microbiological safety, (4) a quarantine treatment for commodities subject to infestation by insect pests of quarantine importance, and (5) increase of juice recovery from berry fruits.

Radiation at dose levels of less than 2 kGy can extend shelf life by 2–12 days for several fresh-cut produce and it would appear that this shelf life extension is caused by the inactivation of spoilage microorganisms. Effectiveness of irradiation depends on the initial quality of the product (Prakash and Foley 2004). The doses required for effective reduction of microbial load will exert the largest effect on produce quality and shelf life (Farkas et al. 1997). Doses of ionizing radiation at levels optimal for shelf life extension are also effective against pathogens found in fresh produce as vegetative pathogens are destroyed while background flora is reduced but not eliminated (Prakash and Foley 2004). Most fruits are sensitive to irradiation even at low doses and the effect on vegetables is varied. Texture is the quality factor most affected by irradiation with a significant loss in firmness being observed in some cases. These changes in texture have been linked to partial depolymerization of cell wall polysaccharides, cellulose, and pectin with damage to cell membranes, as well as activation of pectinmethylesterase and inhibition of polygalacturonase, specific enzymes in the cell wall that act to solubilize pectic substances (Prakash and Foley 2004).

Work by Farkas et al. (1997) on precut bell pepper and carrots inoculated with *L. monocytogenes* studied the effect of irradiation at 1 kGy on viable cell counts of this pathogenic microorganism and changes postirradiation storage together with the naturally occurring contaminating bacterial flora. In the case of bell pepper, the microbiological shelf life was doubled by radiation treatment, with the visual appearance of the vegetable deteriorating faster than the microbiological quality. For carrots, irradiation drastically reduced the load of spoilage bacteria, thereby extending shelf life along with sensorial keeping quality. Irradiation also reduced the viable counts of *L. monocytogenes* by several log cycles. The 1 kGy dose caused a 12% loss of ascorbic acid content of bell pepper, with further losses being less than 10% during storage of both irradiated and nonirradiated samples. Following 10 days of storage, the β-carotene content of irradiated samples was somewhat higher than that of untreated samples.

One product which has been investigated in some detail regarding the effect of ionizing radiation is lettuce. Prakash et al. (2000a) studied the effect of irradiation on the shelf life and quality characteristics of cut romaine lettuce packaged under modified atmosphere. They found that low doses of gamma irradiation at 0.15 and 0.35 kGy increased microbiological shelf life although the 0.35 kGy dose caused softening with a 10% loss in firmness being observed. However, other sensory attributes such as color, generation of off-flavor, and appearance of visual defects were not adversely affected. Goularte et al. (2004) studied the effect of gamma radiation in combination with minimal processing on shredded iceberg lettuce inoculated with *Salmonella* spp. and *E. coli* O157:H7. These workers found that when the lettuce was given a dose of 0.7 kGy, the population of *Salmonella* spp. was reduced by 4.0 $\log_{10}$ cycles while that of *E. coli* was reduced by 6.8 $\log_{10}$ cycles without impairing the sensory attributes of the lettuce. It was, therefore, suggested that such doses of irradiation along with good manufacturing practice could significantly increase the microbiological quality of minimally processed iceberg lettuce. Zhang et al. (2006) also studied the effect of gamma irradiation on the microbiological and physiological quality of fresh-cut lettuce during storage at 4°C. It was found that a dose of 1 kGy exhibited the best preservation effect as the total bacterial count were reduced by 2.25 $\log_{10}$ cycles and the total coliform group were lowered to less than 30 MPN/100g. The total polyphenoloxidase activity of fresh-cut lettuce was significantly inhibited by the same dose and, in addition, loss of vitamin C was significantly lower than that of the nonirradiated lettuce.

Pathogenic bacteria internalized in leaf tissues are not effectively removed by surface treatments but it has been shown that irradiation can inactivate such bacteria. Niemara (2007) investigated the relative efficacy of a sodium hypochlorite wash versus irradiation to inactivate *E. coli* O157:H7 internalized in leaves of romaine lettuce and baby spinach. A cocktail mixture of three isolates of *E. coli* O157:H7 were drawn into the leaves after which the leaves were then washed with a sodium hypochlorite sanitizing solution or treated with ionizing radiation (0.25–1.5 kGy). Results showed that treatment of the leaves with irradiation (but not the chemical sanitizers) effectively reduced viable *E. coli* O157:H7 cells internalized in the leafy green vegetables in a dose-dependent manner. A more complex response to irradiation was observed in the spinach leaves than in romaine lettuce leaves, with a marked tailing effect in spinach at higher doses as compared with a linear response in the lettuce. The specific doses to be used should be determined for each product based on the patterns of antimicrobical efficacy and specific product sensory responses. In further work undertaken by Niemira (2008), irradiation was compared with chlorination for the elimination of *E. coli* O157:H7 internalized in four varieties of lettuce leaf. Bacterial cells of a cocktail mixture of three isolates of *E. coli* O157:H7 were drawn into the leaves of iceberg, Boston, green leaf, and red leaf lettuce after which they were washed with sodium hypochlorite solution or irradiated (0.25–1.5 kGy). In contrast to chlorination, irradiation effectively reduced *E. coli* O157:H7 on all varieties of lettuce examined, with all doses tested being significantly reduced from the untreated control. The efficacy of irradiation was influenced by the variety of lettuce.

Other studies carried out in the recent past have included work on the effects of ionizing radiation on the quality of products such as frozen corn and peas (Fan and

Sokorai 2007), vegetable juice (Song et al. 2007), ready-to-use vegetables (Lee et al. 2006), tomatoes (Schindler et al. 2005), and diced celery (Praskash et al. 2000b) to name but a few. In the study by Praskash et al. (2000b), it was found that the sensory shelf life of celery treated with 1 kGy was 29 days compared to 22 days for celery that was either nonirradiated (control), chlorinated, or treated with 0.5 kGy, and 15 days for acidified and blanched celery. Fan and Sokorai (2007) found the overall irradiation at doses of 1.8 and 4.5 kGy significantly reduced the microbial load and increased the display life of frozen peas and had minimal detrimental effects on the quality of frozen corn and peas. There was a reduction in ascorbic acid content of both vegetables and a loss of texture in peas but irradiation had no significant effect on color, carotenoid and chlorophyll content, or antioxidant capacity. Schindler et al. (2005) showed that gamma irradiation reduced the content of the phenolic compounds in tomatoes but that changes measured were smaller than the naturally occurring differences as the natural content of the phenolics showed a strong variation between different varieties, harvest, and degrees of maturity. Song et al. (2007) carried out radiation pasteurization of fresh vegetable juice (carrot and kale) in order to improve the microbiological quality. All aerobic and coliform bacteria in the carrot juice were eliminated by a dose of 3 kGy which was not the case for the kale juice, although it was noted that the cells that survived in the kale juice did not grow, whereas those of the nonirradiated juices increased upon storage. Amino acids in the juices were stable up to 5 kGy and there was a dose-dependent reduction in ascorbic acid content. It was observed, however, that total ascorbic acid (including dehydroascorbic acid) was stable up to a dose of 3 kGy. It was also found that at 3 days of storage the sensory evaluation results of the irradiated juices were adequate while the quality of the nonirradiated control had deteriorated. Work by Lee et al. (2007) on three ready-to-use vegetables (cucumber, blanched and seasoned spinach, and seasoned burdock) studied the effects of irradiation treatment for eliminating pathogens (*Salmonella typhimurium*, *E. coli*, *S. aureus*, and *L. ivanovii*). Their findings showed that all the bacterial counts of test pathogens inoculated into samples were reduced to below detectable limits by a 3 kGy dose thereby indicating that low-dose irradiation can improve the microbiological safety of ready-to-use vegetables.

Recent studies on some fruits have indicated that irradiation treatment can improve quality. In work by Moreno et al. (2007) the effect of electron beam irradiation on packaged fresh blueberries was investigated at doses greater than 1 kGy. It was found that irradiation treatment with doses up to 1.6 kGy is a feasible decontamination treatment that maintains the overall fruit quality attributes. Generally in terms of overall quality, texture, and aroma, only fruits given a dose of 3.2 kGy were found unacceptable by sensory panelists. It was also noted that only treatment at 3.2 kGy affected the color of the blueberries by the end of the 14 day storage period. Irradiation at doses used in the study did not affect the density, pH, water activity, moisture content, acidity, and juiciness of the fruits. Reyes and Cisneros-Zevallos (2007) studied the effect of ionizing radiation at doses of 1–3.1 kGy on the antioxidant constituents of mango fruit before and during postharvest storage. They observed no change in the phenolic content of the mangoes after irradiation but found that there was an increase in the flavonol constituents following 18 day storage in samples treated at 3.1 kGy. There were no major changes in carotenoid profiles which were indicative of a delay in ripening of irradiated mangoes compared to nonirradiated fruits. However, it was found that a dose greater than 1.5 kGy induced flesh pitting due to localized tissue death. Hussain et al. (2007) carried out work on peach fruit which after harvest were irradiated in the dose range of 1–2 kGy followed by storage under ambient or refrigerated conditions. Evaluation of the anthocyanin content of the fruits showed that irradiation enhanced the color development under both storage conditions and that doses of 1.2–1.4 kGy effectively maintained higher total soluble solid concentration, reduced weight loss, and delayed decay of the fruit by 6 days under ambient conditions and by 20 days when stored under refrigeration.

Fan et al. (2004) reviewed effect of low-dose ionizing radiation on fruit juices. They found that irradiation at low doses effectively inactivated foodborne pathogens and reduced the levels of the mycotoxin patulin and brownness. However, irradiation did induce undesirable chemical changes such as the accumulation of malondialdehyde, formaldehyde, and tetrahydrofuran. They found literature on the negative flavor changes of irradiated juice to be contradictory although they did state that evidence did exist concerning the involvement of volatile sulfur compounds in the development of off-flavor. It was suggested that as for other products, many of the undesirable effects of irradiation could be reduced by using low temperatures during treatment, by the addition of antioxidants to the product, and by combining irradiation with other techniques and treatments such as mid-heating and use of antimicrobials.

There are numerous studies on the effect of ionizing radiation on the safety and quality of fruits and vegetables, too many to mention in this chapter. However, as pointed out by Niemira and Fan (2006) in their publication on low-dose irradiation of fresh and fresh-cut produce, irradiation does show promise for the improvement of the safety, sensory properties, and shelf life of a wide variety of fruits and vegetables, whether it is used singly or in combination with other treatments. Combining irradiation with other technologies such as calcium treatment, warm water dips, and MAP can further enhance shelf life and mitigate adverse effects on quality (Thomas 2001a, Prakash and Foley 2004). As noted by Prakash and Foley (2004), while the advantages of irradiation for fresh-cut applications have been clearly demonstrated, commercial applications for produce remain to be exploited.

13.2.4 TUBER AND BULB CROPS

Bulb and tuber crops are important fruit vegetables cultivated and consumed in most areas of the world. Onions, shallots, and garlic constitute the major bulb crops whereas potatoes and yams for the most economically important tuber crops (IAEA 1997). The main reason for treating tuber and bulb crops with ionizing radiation at low-dose levels is for inhibition of sprouting. Commercial irradiation of potatoes has been carried out in Japan since 1973 when the first industrial scale irradiation facility for treatment of potatoes was set up by the Shihoro Agricultural Cooperative Association. Back in 2001 it was reported that the plant was processing 15,000–20,000 tons annually (Thomas 2001b). The tubers are sold for industrial processing and for household use.

Factors that determine the efficacy of radiation treatment include the time interval between harvest and irradiation, radiation dose, cultivar, and initial product quality (IAEA 1997, Thomas 2001b). Good tuber and bulb crop handling and storage management practices are essential prerequisites for their successful irradiation on a commercial scale. Upon harvest, the crops must be dried well, cleaned of adhering soil, particularly in the case of tubers, and sorted to remove damaged and infected material (Thomas 2001b).

Generally irradiation has no significant effect on the carbohydrates and sugar content of onions whereas in garlic a lower content of water-soluble carbohydrates has been reported. For potatoes, the effect of irradiation on the sugar content of potatoes is variable, and this could be attributed to variations in cultivar, time of treatment, and storage temperature. It has been observed that there is a temporary rise in both reducing and nonreducing sugars immediately after irradiation, which, often, returns to normal levels during storage, followed by an increase on prolonged storage or senescent sweetening. Postirradiation storage temperature may influence the rise in sucrose and reducing sugars as demonstrated in early work by Metlitsky et al. (1957). They showed that tubers stored at 1.5°C for 7 months had higher levels than their nonirradiated counterparts, while at a higher temperature the irradiated tubers had fewer sugars than the controls. In a similar manner, increased levels of reducing sugars were observed for tubers treated at 100 and 250 Gy and stored at 0°C–4°C for 1 month compared to those stored at the higher temperature of 25°C. Work by Adesuyi and Mackenzie (1973) showed that the starch levels in yam tubers were almost identical in irradiated (150 Gy) and nonirradiated samples following storage for 5 months under ambient conditions. The starch content of tubers treated at 100 and 125 Gy was found to decrease upon treatment while those given doses of 25, 50, and 75 Gy had higher starch levels.

Some changes have been observed in the concentration of free amino acids in potato tubers upon irradiation without any alterations in the amino acid constituents of the protein (WHO 1977). Treatment of potatoes at 0.07–0.1 kGy 2 weeks postharvest or later has not been found to appreciably affect the nitrogenous substances in tubers except during the initial storage period when some of the nonprotein nitrogen increased at the expense of decomposition of protein nitrogen. However, with prolonged storage, levels of protein and nonprotein nitrogen were found to be equal in irradiated and control tubers (Metlitsky et al. 1968). According to Thomas (2001b), very little information is available for other tuber and bulb crops. Free amino nitrogen levels and crude protein have been found to decrease in onions irradiated at doses of 60–80 Gy while amino acid composition and protein content have been found to remain unchanged.

One of the most important components of potatoes is vitamin C (ascorbic acid); therefore, the study of its stability upon irradiation at doses used for sprout inhibition has been an important area of research. Generally, vitamin C is stable during and after irradiation and while a reduction has been observed during the early storage period after irradiation, its content in stored tubers is reported to be comparable to or even greater than that of nonirradiated tubers when stored under identical conditions (Thomas 1984, Stevenson and Graham 1995). When reporting vitamin C levels in irradiated food, some workers have not taken into consideration the fact that upon irradiation it is partially converted to dehydroascorbic acid. Since both compounds are biologically active it is important that each is measured in order to ensure that exaggerated losses of this vitamin are not recorded. Graham and Stevenson (1997) demonstrated that irradiated potatoes stored for 5 months had similar or marginally higher levels of vitamin C to their nonirradiated counterparts. In addition, their research showed that cooking did not markedly reduce the total ascorbic acid content of irradiated compared to nonirradiated potatoes. Thus, it appears that irradiation has no greater effect on the vitamin C content than other conventional processing techniques (Shirsat and Thomas 1998).

Onions and garlic are not normally consumed for their vitamin C content and generally results from various research studies have indicated that ionizing radiation does not adversely affect the levels of this vitamin. Any changes that do occur soon after irradiation are restored to the same levels that occur in nonirradiated samples during subsequent storage (Thomas 2001b). This was demonstrated for onions by Guo et al. (1981) who demonstrated that although there was a significant reduction in vitamin C content immediately after treatment with 100–500 Gy, the decrease in content was much lower in irradiated than in control onions during the subsequent 8 month storage period. Curzio et al. (1986) observed a significant increase in ascorbic acid during storage of irradiated garlic compared with nonirradiated samples.

The chlorophylls and glycoalkaloids present in potatoes are also affected by irradiation doses used to inhibit sprouting. Irradiation has been found to delay greening in potatoes due to the inhibition of chlorophyll formation. A study by Schwimmer and Weston (1958) showed that chlorophyll formation in potatoes given doses of 50 and 150 Gy was inhibited by 61% and 67%, respectively, when illuminated for 3 days after irradiation. Since glycoalkaloid formation accompanies greening, it is reasonable to assume that irradiation may also delay the formation of the toxic alkaloid solanine, the bitter compound in green potatoes. Dale et al. (1997) studied the effect of gamma irradiation on glycoalkaloid and chlorophyll synthesis in seven cultivars of potato. Following irradiation the potatoes were exposed to light after a period of storage, and it was found that there were significant genotype differences between cultivars in their response to irradiation, with some cultivars exhibiting dramatically reduced levels of glycoalkaloid synthesis compared with others. It has been shown that glycoalkaloid synthesis is generally more severely affected than chlorophyll synthesis (Thomas 2001b).

As onions and garlic are used mainly for their flavoring properties, many studies have been undertaken to determine the effect of irradiation at doses used for sprout inhibition on the development of flavor, odor, and lachrymatory compounds and related enzymes. The reduction of these compounds in irradiated onions may be due to partial inactivation of allinase. However, a recovery in the activity of this enzyme takes place during storage in onions treated at sprout inhibiting doses, while flavor characteristics are restored to levels comparable to those of nonirradiated onions stored under similar conditions (Thomas 2001b). Kwon et al. (1989) showed that irradiation at a dose of 100 Gy had no influence on the total sulfur and thiosulfonate contents of garlic bulbs during storage at 2°C–4°C for 10 months, although these compounds deceased after 6–8 month storage compared to initial levels.

More detailed information on the effect of ionizing radiation on the nutritional components and technological properties of tubers and bulbs can be found in the publication by Thomas (2001b) who earlier prepared a compilation of technical data on the subject for the IAEA (1997).

13.2.5 SPICES, HERBS, AND CONDIMENTS

Spices and herbs are important items of international trade for centuries, the majority of which are produced in equatorial and tropical countries by small landholders and local farmers. Generally, microbial contamination is inevitable under the prevailing production, harvesting, and postharvest handling conditions; therefore, most dried food ingredients of vegetable origin may contain large numbers of microorganisms that may cause spoilage or defects in composite food products into which they are incorporated, or more rarely, may cause food poisoning (Farkas 2001).

Until the early 1980s, the most commonly used method employed to destroy microorganisms in dried food ingredients was fumigation using ethylene oxide, or to a much lesser extent, with propylene oxide. However, because of toxicological reasons, the use of ethylene oxide has been discouraged or even banned for food uses, as is the case in the European Union (EU). Consequently, the use of ionizing radiation was considered as an alternative method for decontamination of dried food ingredients and it is now one of the main commercial applications of the technology (Farkas 1998, 2001).

Irradiation has a strong antimicrobial effect and depending on the initial number and type of microorganisms present, and on the chemical composition of the product, doses of 3–10 kGy can be used to reduce the total aerobic viable cell counts even in highly contaminated spices and other dried food ingredients, without having an adverse effect on their flavor, texture, or other properties (Farkas 1988). In a study of three types of paprika, Llorente-Franco et al. (1986) found that irradiation at 6.5 kGy for granulated or powdered paprika was found to be more effective than ethylene oxide treatment over 48 h at 25°C–30°C in a sterilizing chamber.

Work by Vajdi and Pereira (1973) and Farkas and Andrássy (1988) actually indicated that some spices rate better with regard to sensory quality than those treated with ethylene oxide while other research has shown that there are no major differences in sensory properties between irradiated and fumigated spices (Toffanian and Stegeman 1988, Diehl 1995, Farkas 2001). According to Sjövall et al. (1990) complex dried products such as spices are less affected chemically by ionizing radiation than their pure aroma compounds or high-moisture foods in general. Although sterilizing doses of 15–20 kGy used commercially may slightly or noticeably change the flavor characteristics of some spices, sensory analyses and practical food applications indicate that doses of 4–10 kGy used for pasteurization purposes do not result in significant differences between irradiated and nonirradiated samples for the majority of spices and herbs (Farkas 2001). Hansen (1966) and Farkas et al. (1973) have also shown that various meat products prepared with spices treated with doses of up to 20 kGy could not be distinguished by flavor from products prepared with corresponding nonirradiated spices.

Farag et al. (1995) carried out work on spices (marjoram, rhizomes of ginger, and powdered hot pepper) from Egyptian local markets irradiated with doses of

0, 5, 10, 20, and 30 kGy, studying the isolation and identification of microorganisms as well as gas chromatographic analysis for the presence and structure of volatile oils, pungent and pigment materials. Irradiation at 10, 20, and 30 kGy caused complete elimination of contaminating microorganisms, whereas a 5 kGy dose was less effective. Using gas chromatography (GC), 18 and 50 compounds were detected in the extracts of marjoram and ginger, respectively. Ginger was more sensitive to irradiation, especially at high doses, with moderate changes being detected at 5 and 10 kGy. A slight but significant effect on the capsaicin (pungent compound) in hot pepper was observed following irradiation, whereas no changes in total pigments occurred at any dose. The results showed that a dose of 10 kGy is a sufficiently high dose to eliminate the microorganisms in spices, causing only slight changes in the flavoring materials. Previous to this work other analytical studies showed that the volatile oils that determine flavoring are not significantly affected by the dose of ionizing radiation required for the treatment of spices (Lescano et al. 1991, Piggott and Othmann 1993). Ito and Islam (1994) demonstrated that the components of essential oils of rosemary and black pepper were not changed even at 50 kGy.

In an early work by Farkas et al. (1973), it was found that total carotenoids in ground paprika were unaffected by treatment with a dose of 5 kGy during 250 days of storage. A later study by Zachariev et al. (1991) determined 11 carotenoids in ground red paprika that had been left untreated, fumigated with ethylene oxide, or irradiated with 5 kGy, and stored for up to 5 months. It was found that irradiation had little effect on total carotenoid content or on carotenoid composition.

It has also been observed that the antioxidant properties of spices are unaltered by irradiation (Kuruppu et al. 1983). Kitazuru et al. (2004) studied the effect of irradiation on the natural antioxidants of cinnamon. Samples were treated with doses ranging from 5 to 25 kGy at room temperature and antioxidant activity determined by β-carotene/linoleic acid cooxidation. It was found that in the dose range applied, irradiation did not have any effect on the antioxidant potential of the cinnamon compounds.

The use of ionizing radiation to treat spices rather than ethylene oxide has increased steadily since the phasing out of ethylene oxide. While thermal treatments and extrusion processes can also be used to ensure the hygienic quality of these products, irradiation offers a broader spectrum for application, often at a more competitive price (Kovács et al. 1994, Farkas 1998). For further reference, a succinct outline of the effect of radiation contamination of dried food ingredients has been written by Farkas (2001).

13.2.6 NUTS

The major problems affecting the shelf life, quality, and safety of nuts are pest infestation and contamination with mold, which results in aflatoxin formation. To date, the nut industry has relied on chemical fumigants such as methyl bromide, ethylene oxide, and propylene oxide for control of postharvest insect infestation and microbial growth (Kwakwa and Prakash 2006). The use of gamma processing is being considered as an alternative to fumigation as it offers various advantages when the treatment of almonds in particular is being investigated. Treatment with ionizing radiation (1) can penetrate through the shell and provide homogeneous treatment of the surface of the nut kernel, (2) does not alter the general characteristics of raw almonds and does not raise the temperature of the kernel, (3) can be performed on bulk almonds as well as within the final packaging, and (4) can be used at dose levels which will control both microbial pathogens along with insect pests, mold, and other spoilage organisms (Kwakwa and Prakash 2006).

Narvaiz et al. (1992) looked at the effect of irradiation on almonds (1–2 kGy) and cashew nuts (1–4 kGy) followed by storage for 6 months at 5°C. Microbiological, chemical, and sensory tests were performed to verify the effectiveness of irradiation and its effect on product quality. It was found that the initial mould and yeast load were reduced to acceptable values being maintained throughout storage. The free fatty acid content and refractive index of the extracted oils were not altered by irradiation. Lipid peroxidation increased upon irradiation but this was not noticed organoleptically. No significant differences were found in the sensory quality of control and irradiated nuts at the beginning of the storage period with only a slight decrease in odor intensity being observed in almonds treated at 1.5 and 2 kGy after 6 months. The authors stated that although effects on insects were not studied, the applied radiation doses were high enough for disinfestation purposes.

Uthman et al. (1998) compared the effects of propylene oxide and irradiation treatments (6 and 10.5 kGy) on the lipid content of raw and roasted almonds stored for up to 16 weeks. Samples were analyzed for iodine number, peroxide value, and 2-thiobarbituric acid number. Overall, propylene oxide seemed to be a better decontamination treatment for almonds than irradiation to prevent rancidity which developed over the entire storage period. Result showed that dry-roasted almonds tolerated less irradiation as well as fumigation than raw almonds; however, this may have been related to the roasting process more than irradiation treatment. The increase in oxidation and formation of hydroperoxides and other rancid components during the storage period was significantly higher in roasted almonds than in raw almonds, and development of rancidity was significantly higher for propylene oxide treated and 6 kGy irradiated roasted almonds than for the same treated raw almonds. Propylene oxide seemed to be a better decontamination treatment for almonds than gamma irradiation to prevent rancidity which develops during storage.

Work by Sanchez-Bel et al. (2005) studied the changes in the lipid fraction and the deterioration of its quality in almonds (*Prunus amygdalus*) after treatment with accelerated electrons at doses of 3, 7, and 10 kGy during a storage period of 5 months. In almond oil, the most significant difference from the nutritional point of view was seen in the linolenic acid content. The initial content of this fatty acid was maintained during the whole storage period at 3 kGy, whereas at 7 and 10 kGy it disappeared upon irradiation. Peroxide values did not show changes at 3 and 7 kGy in nonirradiated samples, but increased significantly when the maximum dose of 10 kGy was used. These changes were reflected in the sensory analysis where the panelists did not find any sensory differences between the controls and samples treated at 3 or 7 kGy, whereas almonds irradiated at 10 kGy exhibited a rancid flavor and a significant decrease in general quality.

Further work by Sanchez-Bel et al. (2008) reported on the effects of irradiation (0, 3, 7, and 10 kGy) on the chemical composition (water content, proteins, neutral detergent fiber, sugars, lipid content, organic acids, and color) and sensorial properties (rancidity, sweetness, off-flavors and odors, texture, and whiteness) of the shelled almonds, aerobically packaged and stored for 5 months at 20°C. Irradiation was found to decrease the glucose content at all doses with an increase in citric acid being observed at doses above 3 kGy after which there was a decrease to values similar to those of the control. With respect to the sensorial analysis, there was no treatment effect on the sweetness, texture, or color but a marked rancidity in the samples treated with 10 kGy was noted which decreased the overall appreciation of the samples. The authors concluded that irradiation at doses up to 7 kGy seemed to be a suitable postharvest sanitation treatment since they did not cause significant changes in the sensorial quality or in the contents of protein, fiber, water, or lipid with respect to the control samples both following the treatments and after 5 months of storage.

Almond skins are a source of phenolic compounds and studies by Harrison and Were (2007) investigated the effect of gamma irradiation on total phenolic content and antioxidant capacity. Results showed that the yield of phenolics was significantly higher in almond skin extracts irradiated at doses greater than 4 or 12.7 kGy compared to controls. An increased antioxidant activity was observed in the irradiated extracts compared to the nonirradiated samples. Gamma irradiation of almond skins thus appears to increase the yield of total phenolic content as well as enhancing antioxidant activity.

Although this section has focused briefly on almonds, the effects of irradiation on other types of nut have been investigated. The recent review by Kwakwa and Prakash (2006) on the irradiation treatment of nuts and an earlier review on radiation preservation of foods of plant-origin including nuts by Thomas (1988) provide some detail on these investigations including the occurrence of lipid peroxidation and non-oxidative radiolytic reactions in irradiated nuts. Nuts with a higher moisture content, soft texture, and no shell, such as peanuts, are more vulnerable to lipid oxidation and degradation. Almonds and groundnuts, which have a harder texture, shells, and a lower moisture content, do not oxidize so easily (Satter et al. 1989, Kwakwa and Prakash 2006).

13.3 SAFETY OF IRRADIATED FOODS

Extensive studies undertaken on the safety of irradiated food have indicated that the compounds formed in irradiated foodstuffs are generally the same as those produced during other food processing technologies such as cooking, canning, and pasteurization and that any differences are not at risk to the consumer (GAO 2000).

During the Project in the Field of Food Irradiation (IFIP), which ran from 1970 to 1982, a large number of animal feeding studies were carried out with over 70 reports being generated. Overall, 100 compounds from irradiated beef, pork, ham, and chicken were examined. The generated data were reviewed at a series of international meetings organized by the World Health Organization (WHO), which were often jointly held with the Food and Agriculture Organization (FAO) of the United Nations and the International Atomic Energy Agency (IAEA). Subsequently, in 1980, a Joint FAO/IAEA/WHO Expert Committee on the Wholesomeness of Irradiated Food (JECFI) met in Geneva and their landmark report was published in 1981

(WHO 1981). The Committee concluded that the "irradiation of any food commodity up to an overall average dose of 10 kGy presents no toxicological hazard; hence, toxicological testing of foods so treated is no longer required." It also concluded that irradiation up to 10 kGy "introduces no special nutritional or microbiological problems."

The Raltech studies were amongst the most extensive toxicological studies carried out on irradiated food, the findings of which are summarized in a report written by Thayer et al. (1986). The studies took 7 years to complete at a cost of $8 million, and undoubtedly led to the most comprehensive safety evaluation ever undertaken on irradiated food (Thayer et al. 1987, Diehl 1995). During these studies, a total of 230,000 broiler chickens were processed producing the 134 metric tons of chicken meat required for this work and used to produce four diets: (i) frozen control (FC), canned in vacuo and frozen, (ii) thermally processed (TP) control, canned in vacuo and TP at 115.6°C, (iii) GAM containing enzyme-inactivated chicken meat, canned in vacuo, and sterilized by exposure to gamma radiation at −20°C ± 15°C from a cobalt-60 source giving a minimum absorbed dose of 46 kGy and a maximum dose of 68 kGy, and (iv) ELE containing the enzyme-inactivated chicken meat, vacuum packed in 26 mm thick slices in laminated foil packages and sterilized by exposure to 10 MeV electrons at −25°C ± 15°C giving an average dose of 58 kGy. Another diet, known as CLD, was used as the negative or husbandry control diet serving as a carrier for the chicken meat in the other four diets (Thayer et al. 1987, Diehl 1995). These diets underwent (1) nutritional studies, (2) teratology studies, (3) chronic toxicity, oncogenicity, and multigeneration reproductive studies, and (4) genetic toxicity studies.

With regard to the nutritional studies, results showed that all the diets containing the chicken meat had higher protein efficiency ratio values than the casein standard and were not significantly affected by any of the ways the chicken had been processed. A number of genetic toxicology studies of the diets were undertaken. The Ames test (*Salmonella*-microsomal mutagenicity test system) found that the way in which the chicken was processed, either irradiated or nonirradiated, did not affect the response of the test system to known mutagens and that no positive results were observed for any of the chicken diets in the absence of the known mutagens. Canton-S *Drosophila melanogaster* was used to test for sex-linked recessive lethal mutations and it was demonstrated that none of the diets produced evidence of sex-linked recessive lethal mutations. However, it was observed that there was a significant reduction in the egg hatchability of cultures of *D. melanogaster* reared on the GAM diet. Additional testing was carried out to confirm these results and it was concluded that although the irradiated chicken meat was not mutagenic in the test system used, the number of offspring from *D. melanogaster* which had been fed diets containing chicken was consistently reduced, particularly for those containing irradiated chicken.

In the series of teratology studies undertaken by Raltech, mice, hamsters, rats, and rabbits with pregnant females were exposed to the test diets as well as positive control substances (all-trans retinoic acid for mice, hamsters and rats, and thalidomide for rabbits). In vivo, it was found that either the irradiated or nonirradiated chicken diets induced a teratogenic response when consumed by the pregnant animals. When the positive controls were administered to the animals, significant incidences of resorbed embryos and congenital malformations were observed in both soft and skeletal body tissues. Such incidences were not found when any of the four processed chicken meat diets were consumed by the animals.

Chronic feeding studies were conducted in mice and beagle dogs with the test diets being provided to the animals ad lib. It was found that all five diets supported the growth of the beagles to maturity and there were no obvious signs of diet-related toxicity in any of the animals. Dogs fed GAM diets had lower body weights through-out adulthood than males fed the FC diet. It was noted, however, that many of the FC fed dogs became obese which indicated that the difference in body weight between the FC and GAM fed animals showed no evidence of toxicity. There was no evidence of any oncogenic effect from any of the diets as the ability to breed was greater in the female beagles which ate the GAM diet than in dogs fed the other diets. In addition, there was no evidence of reproductive toxicity.

The only impaired reproduction noted for mice was for comparatively decreased fertility in animals fed the TP diet. There were no significant differences in frequency of stillbirths, numbers of viable offspring born, and survival to weaning between groups of mice fed the irradiated chicken meat and those consuming the FC diet. Many mice became obese when fed the FC, TP, GAM, and ELE diets while animals consuming the CLD diet were found to have lower mean body weights throughout life. The study showed that male mice fed GAM had lower body weights than those fed the other meat diets while the mean body weights of the female mice consuming the chicken diets did not differ significantly. This was attributed to the decreased survival among heavier weight animals in the GAM group, although overall survival for the male mice was not significantly different among the four meat-containing diets. Overall, the highest incidence of tumor development was in both males and females fed the FC diet. Female mice fed the ELE diet exhibited the lowest incidence of tumors being significantly lower than the group fed the FC diet while the lowest occurrence of tumors for male mice was for those fed the GAM diet, although it should be noted that there were no significant differences among the groups consum-ing the chicken diets. Overall, the Raltech studies consistently produced negative results in all of the tests undertaken.

In the human feeding trial reported by Bhaskaram and Sadasivan (1975), 15 malnourished Indian children were fed freshly irradiated wheat (0.75 kGy) for 4–6 weeks. The authors reported increased polyploidy (cells containing twice or more the normal number of chromosomes) in lymphocytes of the children, although this result was not seen when wheat stored for 12 weeks after irradiation was used. In addition, the authors found no chromosomal aberrations like breaks, gaps, and deletions in the group of children consuming freshly irradiated wheat, which would have been typi-cal indicators of mutagenicity had the diet exerted a mutagenic effect (Diehl 1995). The Indian study was heavily criticized because of the small number of children used and the methods used in conducting the study (Diehl 1995, Louria 2001). As pointed out by Fielding (2007) although statistical significance was quoted there was no mention of statistical analysis as the sample size was too small to allow any robust statistics to be performed. Another larger human study was published in China in 1987 where healthy adults (mostly between 18 and 23 years) were fed irradiated foods for 3 months. Initially, this trial did not indicate any increase in chromosomal aberrations but upon reanalysis (Louria 1990, 2001) it appeared that subjects who consumed irradiated foods did have increased chromosome breaks at borderline statistical significance ($P = 0.07$). The results from both studies were inconclusive but resulted in doubts being cast on the safety of irradiated food.

An extensive evaluation of the safety of irradiated food was undertaken in 1992 by the WHO which had an expert committee review literature and data available since 1980 taking into account over 500 studies evaluating the safety of irradiated food (WHO 1994). The findings of this review indicated that "food irradiation is a thoroughly tested food technology" and that "safety studies have so far shown no deleterious effects" (WHO 1994, Diehl 1995). Furthermore in 1997, a Joint WHO/FAO/IEA Study Group reviewed data relating to irradiated foods irradiated to doses above 10 kGy and to consider whether a maximum irradiation dose needed to be specified. The studies reviewed were subdivided into subchronic, carcinogenicity and chronic toxicity, reproduction and teratology, mutagenicity, and human clinical studies. Subsequently, in their report published in 1999 in assessing risk (WHO 1999), the Study Group concluded that "irradiation to high doses is essentially analogous to conventional thermal processing, such as the canning of low-acid foods, in that it eliminates biological hazards (i.e., pathogenic and spoilage microorganisms) from food materials intended for human consumption, but does not result in the formation of physical or chemical entities that could constitute a risk."

In a literature review on the safety of irradiated foods Fielding (2007) drew attention to the study by DeRouchey et al. (2003) on the effect of irradiation on individual feed ingredients and the complete diet on nursery pig performance. When the pigs were fed a diet solely of irradiated food, their growth was lower than that of pigs fed a diet where selected ingredients were irradiated. When weaned pigs were fed on a diet of animal plasma (irradiated or nonirradiated) or no animal plasma, it was observed that the pigs fed on irradiated animal plasma were significantly heavier than either other group. Overall, the researchers concluded that while irradiation will reduce the bacterial load of diets fed to nursery pigs, the mechanism relating to the differences in growth rates depending on whether the whole diet or only ingredients of the diet are irradiated is unclear.

Recently, research has focused on the toxicological potential of 2-alkylcyclobutanones following the publication of a study by Delincée and Pool-Zoebel (1998). It has been proposed that the formation of the 2-alkylcyclobutanones in irradiated foods results from cleavage at the acyl–oxygen bond in triglycerides with the pathway involving a six-membered ring intermediate. The cyclobutanones so formed contain the same number of carbon atoms as the parent fatty acid and an alkyl group are located in ring position 2 (LeTellier and Nawar 1972, Stewart 2001b). To date, the cyclobutanones are the only cyclic compounds reported in the radiolytic products of saturated triglycerides. If the fatty acid composition of a lipid is known, then the products formed upon irradiation can be predicted to a certain degree and thus, for example, if the fatty acids palmitic, stearic, oleic, and linoleic acid are exposed to ionizing radiation then the respective 2-dodecyl-, 2-tetradecyl-, 2-tetradecenyl-, and 2-tetradecadienyl-cyclobutanones will be formed (Elliott et al. 1995).

Using the Comet Assay, Delincée and Pool-Zoebel (1998) investigated the effect of 2-dodecylcyclobutanone (2-DCB) on rat colon cells (in vitro and in vivo) and on colon cells from human biopsy samples. Results indicated that 2-DCB had a slight genotoxic potential in rat and in human colon cells. However, the actual identity and purity of the 2-DCB was not verified prior to the experimental work and later characterization of the compound was not possible. Given the equivocal findings of their work and the limitations of the test systems used, the authors cautioned against misinterpretation of the results, and concluded that a possible risk from 2-DCB must be at a very low level and that more experimental work would be required to assess and quantify the risk from the intake of 2-DCB with irradiated food. Subsequent to these initial investigations, results from a number of toxicological studies on 2-alkylcyclobutanones undertaken by Delincée et al. (2002), Burnouf et al. (2002), Raul et al. (2002), Horvatovich et al. (2002), Knoll et al. (2006), and Hartwig et al. (2007) indicated that these radiation-induced compounds have cytotoxic properties although the authors did note in some cases that the concentrations tested were much higher compared with assumed human intake. For example, Raul et al. (2002), using rats, studied whether the cyclobutanones could modulate carcinogenesis in an experimental animal model of colon carcinogenesis. The animals were given a daily solution of 2-tetradecylcyclobutanone or 2-tetradecenylcyclobutanone at a concentration of 0.005% in 1% ethanol as drinking fluid with the average consumption of the cyclobutanones being approximately 1.6 mg/rat/day. Control animals received 1% ethanol only. Following 2 weeks, all rats were injected with the chemical carcinogen azoxymethane (AOM) once a week for 2 weeks. Results from this study suggested that cyclobutanones may not initiate colon carcinogenesis per se but in the long term may be promoters of intestinal tumor formation. Six months after injection with AOM, it was found that the total number of tumors in the colon was threefold higher in the cyclobutanone treated rats than in the rats treated with AOM only. It should, however, be noted that the daily amount of pure cyclobutanones given to the rats corresponded to a pharmalogical dose of 3.2 mg/kg body weight, which was not comparable to the amount ingested by humans consuming irradiated foods.

Similarly, Knoll et al. (2006) explored the relative sensitivities of human colon cells, representing different stages of tumor development and healthy colon tissues, to 2-DCB. This group found that 2-DCB was genotoxic in healthy human colon epithelial cells and in cells representing preneoplastic colon adenoma. They also reported that 2-DCB induces chromosomal aberrations in a human colon adenoma which are associated with human cancer. However, they could not speculate if the cell-specific effects of the compound were as a result of differences in cellular uptake, metabolism, DNA repair, or other pathways. As for Raul et al. (2002), the authors noted the doses which caused the genetic alterations most likely far exceeded the normal exposure situation. As an example they cited that the amount of 2-DCB found in a 100 g beef burger equates to ~3.3 μg or ~14 nmol (Gadgil et al. 2002) is some 50,000-fold lower than effective concentrations in vitro, and possibly too low to significantly impact on human health.

Marchioni et al. (2004) presented the results of a collaborative study on the toxicology of 2-alkylcyclobutanones and concluded that although the results of the study pointed toward the toxic, genotoxic, and even tumor promoting activity of certain highly purified cyclobutanones, the experimental data were inadequate to characterize a possible risk associated with the consumption of irradiated fat-containing foods. The authors acknowledged that other food components may have an influence on the reactions of the cyclobutanones which was not evident from their experiments on the purified compounds and that further investigations are required on the kinetics and metabolism of these compounds in living organisms.

Other workers have reported findings which indicate the cyclobutanones are not detrimental to health. Research by Sommers (2003) showed that 2-DCB did not induce mutations in the *E. coli* Trp assay. This was substantiated by Sommers and Schiestl (2004) who demonstrated the absence of genotoxicity of purified 2-DCB using the *Salmonella* mutagenicity test and the yeast DEL assay to evaluate the genotoxic potential. Sommers and Mackay (2005) tested the ability of 2-DCB to increase the expression of DNA damage-inducible genes in *E. coli* that contained stress-inducible promoters fused to β-galactosidase reporter genes, and to induce the formation of 5-fluorouracil (5-FU) resistant mutants in *E. coli*. Data from this work showed that 2-DCB did not increase expression of DNA damage-inducible genes in *E. coli* or the formation of 5-FU-resistant mutants.

Further work by Gadgil and Smith (2004) evaluated the mutagenic potential of 2-DCB using the Ames assay and compared the acute toxicity of 2-DCB to the food additive cylohexanone and to *t*-2-nonenal (both carbonyl compounds like 2-DCB) using the Microtox assay. Results indicated that 2-DCB is not mutagenic to the *Salmonella* strains tested. It was found that 2-DCB is similar in toxicity to cyclohexanone, which has generally recognized as safe (GRAS) status in the United States, and is 10 times less toxic than *t*-2 nonenal, a normal food constituent of cooked ground beef and an approved food additive (GRAS status flavorant). Gadgil and Smith (2006) examined the fate of 2-DCB in rats after consumption by looking at its recovery from the feces and adipose tissue of rats and if any urinary metabolites could be identified. It was found that between 3% and 11% of the total amount of 2-DCB given to the rats was recovered from the feces with 0.33% being recovered from adipose tissue. Therefore, it would appear that most of the 2-DCB is metabolized and excreted or stored in tissues other than adipose.

Extensive studies on the cyto- and genotoxocity potential of 2-alkylcyclobutanones of varying chain length in different cell lines were undertaken by Hartwig et al. (2007). Results from this indicated that the cyclobutanones have cytotoxic properties both in bacteria and human cells. However, it was observed that the cyotoxic effects varied with the nature of the actual compound and the cells investigated. It was found that the shorter the chain of the cyclobutanone the higher the toxic effect, this observation being more pronounced in bacteria. Using the Ames test, no mutagenic potential was detected for any cyclobutanone which is in agreement with results previously published for 2-DCB (Sommers 2003, Smith and Pillai 2004). Data derived from alkaline unwinding experiments demonstrated a genotoxic potential of all compounds investigated with the intensity of DNA damage being dependent on the length of the fatty acid side chain, the degree of unsaturation, and the cell line applied. Subsequently, Hartwig et al. (2007) concluded that the data derived indicate a genotoxic potential of purified cyclobutanones in mammalian cells. However, the authors do state that the effects of the cyclobutanones need to be elucidated in more detail with clarification of mechanisms of action and an adequate risk assessment for human exposure being required.

The literature review on the safety of irradiated foods by Fielding (2007) also draws attention to the effect of irradiation on other compounds such as mycotoxins, furan, and acrylamide. Regarding mycotoxins, trials have indicated that irradiation can reduce the levels of mycotoxins present in foods such as tomato paste (Aziz et al. 1991), bread (Aziz et al. 1997), maize (Aziz et al. 2002), and spice paste (Refai et al. 2003). Research by Fan (2005) investigated production of furan in irradiated and heat-treated fruit juice (apple and orange). Juices were irradiated with doses up to 5 kGy in order to determine the amount of furan produced by irradiation or spiked with deuterated furan (d_4-furan) and irradiated up to 4 kGy to determine degradation of this compound. It was shown that the level of furan increased linearly with increasing irradiation dose, while the level of d_4-furan decreased on irradiation. Thermal processing of both types of juice gave similar results with orange juice submerged in boiling water for 5 min giving comparable levels of furan to an irradiation dose of 3.5 kGy. Fan and Mastovska (2006) investigated the effect of irradiation on both the furan and acrylamide levels in water, sausages, frankfurters, canned infant sweet potatoes, canola oil, and potato chips. It was found that the food matrix impacted on the amounts of furan present. The compound was highly sensitive to irradiation in water and meat products with levels increasing in carbohydrate and ascorbic rich foods. Acrylamide was also sensitive to irradiation in water but no significant reductions were found in the amounts in potato chips, even at a dose of 10 kGy. The authors concluded that in foods containing high levels of furan or acrylamide, an irradiation dose of 10 kGy will only partially reduce levels of these compounds.

Benzene and its derivatives are another group of compounds which have generated some concern with regard to irradiated food. Chinn (1979) reported on the results of studies carried out by the Federation of American Societies for Experimental Biology and reached the conclusion that the small amounts of benzene generated in irradiated beef (15 ppb treated at 56 kGy) compared to nonirradiated beef (3 ppb) do not constitute a significant risk. A benzene level of 3 ppb was reported in irradiated beef (1.5–4.5 kGy) by Health Canada (2002) and it was noted that this level of benzene was significantly lower than the naturally occurring levels of 200 ppb found in haddock and 62 ppb in eggs as reported by McNeal et al. (1993). Thus, according to Pillai and Smith (2004), the risk of benzene exposure from irradiated foods is considered as negligible.

Extensive reviews have been written on the assessment of the wholesomeness of irradiated foods. In 1985, a paper by Brynjolfsson (1985) reviewed the major findings in the wholesomeness studies on irradiated foods and concluded that the process of food irradiation was ready for industrial applications and "could be effectively regulated for the benefit of the consumer." Thayer (1994) reviewed many of the studies undertaken on chicken as well as looking at the nutritional adequacy of irradiated foodstuffs, while Diehl and Josephson (1994) covered the radiological safety, microbiological safety, nutritional adequacy, and toxicological safety of irradiated foods. As noted earlier in this chapter, the macronutrients in food (carbohydrate, protein, and fat) are largely unaffected by ionizing radiation as are minerals and most vitamins. Research has shown that vitamins E (alpha-tocopherol), carotene, B1 (thiamine), and C (ascorbic acid) are the most radiation sensitive vitamins. Irradiation of vitamins in solution or in model systems results in considerable destruction of these compounds, but in irradiated foods the effects observed are generally not as marked (Diehl et al. 1991, Thayer et al. 1991). Vitamin losses are affected by the dose applied and can be minimized by irradiation at freezing temperatures or by packaging the product in an inert atmosphere. It is worthy of note, however, that other food preservation methods, such as those involving heat, also destroy vitamins, so the effect of ionizing radiation on these minor components is not unique to the irradiation process (Stevenson 1994).

13.4 DETECTION OF IRRADIATED FOODS

Successful detection of irradiated products on sale in the marketplace and incorrectly labeled has been achieved using standardized analytical methods adopted by the EU (EC 2008) and subsequently by Codex as General Methods being referred to in the Codex General Standard for Irradiated Foods in Section 6.4 on "Postirradiation verification" (2003a). The 10 standardized methods (Table 13.1) are based on physical, chemical, biological, or microbiological changes that, although minimal, are induced in food during irradiation (Stewart 2001b).

Prior to the mid-1980s there was little emphasis on the development of methods to identify irradiated foods and this was partly because food products would be irradiated in licensed facilities and that appropriate documentation would accompany the irradiated food throughout the food chain. In addition, as the changes known to occur in foods upon irradiation are minimal and often similar to those produced by other food processes, the development of reliable detection methods

---

TABLE 13.1

European Standards for the Detection of Irradiated Foodstuffs

EN1784:2003	Foodstuffs—detection of irradiated food containing fat—gas chromatographic analysis of hydrocarbons
EN1785:2003	Foodstuffs—detection of irradiated food containing fat—gas chromatographic/mass spectrometric analysis of 2-alkylcyclobutanones
EN1786:1996	Foodstuffs—detection of irradiated food containing bone—method by ESR spectroscopy
EN1787:2000	Foodstuffs—detection of irradiated food containing cellulose, method by ESR spectroscopy
EN1788:2001	Foodstuffs—detection of irradiated food from which silicate minerals can be isolated, method by thermoluminescence
EN13708:2001	Foodstuffs—detection of irradiated food containing crystalline sugar by ESR spectroscopy
EN13751:2002	Detection of irradiated food using photostimulated luminescence (PSL)
EN13783:2001	Detection of irradiated food using direct epifluorescent filter technique/ aerobic plate count (DEFT/APC)—screening method
EN13784:2001	DNA comet assay for the detection of irradiated foodstuffs—screening method
EN14569:2004	Microbiological screening for irradiated foodstuffs—screening method (*Limulus* amoebocyte lysate/gram-negative bacteria [LAL/GNB])

Source: (http://ec.europa.eu/food/food/biosafety/irradiation/anal_methods_en.htm)

proved difficult. However, following a significant research effort over a 10 year period between 1985 and 1995, a wide range of methods are now available for the identification of irradiated foods (McMurray et al. 1996, Stewart 2001b, Delincée 2002, Marchioni 2006).

The technique of electron spin resonance (ESR) or electron paramagnetic resonance (EPR) spectroscopy is used to detect free radicals produced in food when irradiated. In foodstuffs with high moisture content, such as meat and vegetables, radiation-induced radicals disappear rapidly. However, if the food is of high dry matter or contains a component with a high dry matter, such as bone, seeds, or shells, the radicals may be trapped and remain sufficiently stable to be detected by ESR spectroscopy. Currently, there are three standardized methods for the identification of irradiated foods based on the use of ESR spectroscopy (Table 13.1). These are used for the detection of irradiated food containing bone (EN1786) (Anonymous 2000a), cellulose (EN1787) (Anonymous 2000b), and crystalline sugar (EN13708) (Anonymous 2001a).

Two specific methods are available for the identification of irradiated food containing fat. European Standard EN 1784 is based on the GC detection of radiation-induced hydrocarbons. As for all the standard methods, EN1784 was validated by a series of interlaboratory trials as a reliable test for the detection of irradiated products such as chicken meat, pork and beef, camembert, papaya, and mango (Anonymous 2003a). EN1785 is based on the detection of 2-alklycyclobutanones in fat-containing foods (Anonymous 2003b). The cyclobutanones have been identified in irradiated foods treated with irradiation doses as low as 0.1 kGy and to date have not been detected in nonirradiated foods or microbiologically spoiled produce. As for EN1784 the method is potentially applicable to a wide range of products such as chicken meat, ground beef, pork, liquid whole egg, Camembert cheese, mango, and papaya. Irradiated ingredients such as irradiated liquid whole egg in cakes can also be detected (Stewart et al. 2001, Obana et al. 2006).

Thermoluminescence (TL) (EN1788) (Anonymous 2001b) and PSL (EN13751) (Anonymous 2002) are the most sensitive methods for identifying irradiated herbs, spices, and seasonings. The luminescence signals detected actually originate from mineral grains adhering to the product even though they usually account for <1% of sample weight. In the case of TL the energy stored within the silicate minerals is released by controlled heating of isolated silicate minerals with light being emitted, the intensity of the emitted light being measured as a function of temperature resulting in the so-called glow curve. On the other hand PSL uses excitation spectroscopy for optical stimulation of minerals to release stored energy and there is no need to isolate the silicate minerals as whole samples can be measured. The methods can be applied to any foodstuff with silicate minerals attached including herbs, spices, their mixtures, fresh fruits, and vegetables (e.g., strawberries, mushrooms, papayas, and potatoes), dehydrated fruits and vegetables (sliced apples, carrots, leeks, and onions) as well as shellfish including shrimps and prawns. In the case of shrimps and prawns, for TL the mineral grains present in the intestinal gut are isolated while for PSL the signals from intestinally trapped silicates can be stimulated through the membranes of dissected guts and, in some cases, through the whole body of the creature (Sanderson 2003a,b).

The DNA Comet Assay EN 13784 (Anonymous 2001d) is a rapid and inexpensive screening test to identify irradiated food. The technique analyses the leakage of DNA from single cells or nuclei extracted from food material and embedded in agarose gel on microscopic slides. In irradiated samples, the fragmented DNA leaks from the nuclei during electrophoresis forming a tail in the direction of the anode giving the appearance of a "comet" when the gel is stained with a fluorescent dye and viewed with a microscope. Cells from nonirradiated samples will appear as nuclei with no or only slight tails. The DNA Comet Assay is not radiation-specific therefore positive results must be confirmed using specific standardized methods such as EN1784 or EN1785. The method has been validated by interlaboratory trials for identification of irradiated chicken bone marrow, chicken and pork muscle tissue given irradiation doses of 1, 3, or 5 kGy and plant foods (almonds, figs, lentils, linseed, rosé pepper, sesame seeds, soybeans, and sunflower seeds) given 0.2, 1, or 5 kGy (Anonymous 2001c).

Two screening methods were successfully developed, validated, and standardized for the identification of irradiated foods based on modification of the microbiological flora of samples. One microbiological method that has been developed, validated, and standardized as a screening method for irradiated foods is the DEFT/APC test (EN13783) (Anonymous 2001). Using the combined DEFT and APC the method can be used for the detection of irradiation treatment of herbs and spices and has been successfully validated for herbs and spices (including whole allspice, whole and powdered black pepper, whole white pepper, paprika powder, cut basil, cut marjoram, and crushed cardamom) by interlaboratory trials. The LAL/GNB test, European Standard EN14569 (Anonymous 2004), is another microbiological screening method comprising two procedures carried out in parallel to detect an abnormal microbiological profile of foods typically contaminated with a predominantly Gram negative. It is based on the principle that relatively low doses of irradiation can render large numbers of bacteria nonviable. The two procedures to be carried out are (1) enumeration of total resuscitated GNB in the test samples and (2) determination of lipopolysaccharide (bacterial endotoxin) concentration in the test sample using the LAL test. This screening method was validated by interlaboratory trials using boneless chicken breasts with skin and boneless chicken breast fillets. The method is generally applicable to whole parts of poultry, such as breast, legs, wings of fresh, chilled or frozen carcasses with or without skin. In addition, it can also provide useful information about the microbiological quality of a product prior to irradiation.

Within the EU, 10 Member States currently have facilities approved in accordance with Article 7(2) of Directive 1999/2/EC for the irradiation of food and in 2005 as only 8 Member States forwarded to the Commission the results of checks carried out in irradiation facilities, the precise amount of foodstuffs irradiated in the Union cannot be determined (European Union 2007). During 2005 the main products treated by ionizing radiation within the EU were dried herbs and spices, frog legs, poultry, and dried vegetables. In order to ensure that current labeling regulations for irradiated foodstuffs are being complied with, analytical checks are carried out on foods placed on the market. In 2005, 16 Member

States reported checks on foods placed on the market with a total of 7011 food samples being tested. About 4% of products tested from the marketplace were found to be illegally irradiated and/or not labeled (European Union 2007). The infringements were unevenly distributed over product categories with products from Asia, especially Asian-type noodles, and food supplements representing a significant proportion of the samples that were irradiated and not labeled as such. Only 6 of the 287 samples found to be irradiated complied with the regulations. In 2005, there were no irradiation facilities in Asia approved by the European Commission. Incorrectly labeled Asian products were found in Germany, the Republic of Ireland, and the United Kingdom with food supplements also incorrectly labeled also being detected in the same countries as well as in Finland and the Netherlands. In Germany, 47 samples out of 96 soups and sauces analyzed were found to be irradiated, with irradiation being unauthorized for these products and/or the samples not being correctly labeled. Other products found to be irradiated within the EU and not labeled correctly included dried herbs, spices and vegetable seasonings, fish and fisheries products, frog's legs, dried mushrooms, and tea and tea-like products. The detection methods TL (EN 1788) and PSL (EN 13751) were most commonly used for identification purposes with PSL being used for screening purposes and confirmation of positive results being undertaken using TL. The results of these tests within the EU provide good evidence of the reliability of the detection methods in place and can help reassure consumers that unlabeled irradiated foods on sale in the market place can be identified and regulations enforced.

13.5 CONCLUSIONS

Food irradiation, like other commonly using food preservation technologies, will make food safer to eat, last longer, and enhance international food trade (Stewart 2004). Taking into account the significant amount of research undertaken on the application and safety of the technology, and the fact that it is considered safe by national and international bodies such as the WHO, the FAO, the American Medical Association amongst others, it would be hard to conceive consuming irradiated foods as part of a healthy balanced diet as a risk (Sommers et al. 2006). However, if acceptability and commercialization of food irradiation is to increase, an awareness for understanding and appreciation of the relevant benefits of the technology needs to be raised and a strategy for taking this forward needs to be implemented (Roberts 2002). A study carried out by Nayga et al. (2005) in the United States showed that successful marketing can be achieved by educating consumers as to the benefits of the process. The study conducted, in which 484 consumers were interviewed, provided empirical evidence of the importance of scientific information on food irradiation on consumers' decisions to buy irradiated ground beef. Availability of such information led to positive changes in consumers' perceptions and buying decisions. The findings of Nayga et al. (2005) suggest that it is highly beneficial to have readily accessible information available for consumers about the nature and benefits of food irradiation through education campaigns or just before the consumer considers purchasing an irradiated product.

As for any other food processing technology the risks and benefits must be weighed up but, overall, scientific evidence does suggest that adoption of food irradiation would benefit society as a means of enhancement of human health and welfare (Borsa 2004) in the same manner as was achieved by the introduction of the comparable food processing technology of pasteurization (Stewart 2006).

REFERENCES

Adesuyi, S.A. and Mackenzie, J.A. 1973. The inhibition of sprouting in stored yams, *Dioscorea rotundata* Poir by gamma irradiation and chemicals. In *Radiation Preservation of Food*, ed. International Atomic Energy Agency, pp. 127–136. Vienna, Austria: IAEA.

Ahn, D.U. and Lee, E.J. 2004. Mechanisms and prevention of off-odor production and color changes in irradiated meat. In *ACS Symposium Series 875: Irradiation of Food and Packaging—Recent Developments*, eds. V. Komolprasert and K.M. Morehouse, pp. 43–76. Washington, DC: American Chemical Society.

Ahn, D.U. and Lee, E.J. 2006. Mechanisms and prevention of quality changes in meat by irradiation. In *Food Irradiation Research and Technology*, eds. C.H. Sommers and X. Fan, pp. 127–142. Ames, IA: Blackwell Publishing.

Ahn, D.U., Wolfe, F.H., Sim, J.S., and Kim, D.H. 1992. Packaging cooked turkey meat patties while hot reduces lipid oxidation. *Journal of Food Science* 57: 1075–1077, 1115.

Ahn, D.U., Wolfe, F.H., and Sim, J.S. 1993. Prevention of lipid oxidation in pre-cooked turkey meat patties with hot packaging and antioxidant combinations. *Journal of Food Science* 58: 283–287.

Andrews, L.S. and Grodner, R.M. 2004. Ionizing radiation of seafood. In *ACS Symposium Series 875: Irradiation of Food and Packaging—Recent Developments*, eds. V. Komolprasert and K.M. Morehouse, pp. 151–164. Washington, DC: American Chemical Society.

Anonymous. 2000a. Foodstuffs—detection of irradiated food containing bone—method by ESR spectroscopy, European Standard EN 1787:2000, European Committee for Standardization, Brussels.

Anonymous. 2000b. Foodstuffs—detection of irradiated food containing cellulose—method by ESR spectroscopy, European Standard EN 1786:1996, European Committee for Standardization, Brussels.

Anonymous. 2001a. Foodstuffs—detection of irradiated food containing crystalline sugar by ESR spectroscopy, European Standard EN 13708:2001, European Committee for Standardization, Brussels.

Anonymous. 2001b. Thermoluminescence detection of irradiated food from which silicate minerals can be isolated, European Standard EN 1788:2001, European Committee for Standardization, Brussels.

Anonymous. 2001c. DNA comet assay for the detection of irradiated foodstuffs—screening method, European Standard EN 1788:2001, European Committee for Standardization, Brussels.

Anonymous. 2001d. Detection of irradiated food using Direct Epifluorescent Filter Technique/ Aerobic Plate Count (DEFT/APC), European Standard, EN 13783:2001, European Committee for Standardization, Brussels.

Anonymous. 2002. Detection of irradiated food using photostimulated luminescence, European Standard EN 13751:2002, European Committee for Standardization, Brussels.

Anonymous. 2003a. Foodstuffs—detection of irradiated food containing fat—gas chromatographic analysis of hydrocarbons, European Standard EN1784:2003, European Committee for Standardization, Brussels.

Anonymous. 2003b. Foodstuffs—detection of irradiated food containing fat—gas chromato-graphic/mass spectrometric analysis of 2-alkylcyclobutanones, EN1785:2003, European Committee for Standardization, Brussels.

Anonymous. 2004. Microbiological screening for irradiated food using LAL/GNB procedures. EN 13784:2004, European Committee for Standardization, Brussels.

Aziz, N.H., Farag, S., and Hassain, M.A. 1991. Effect of gamma irradiation and water activity on mycotoxin production of Alternaria in tomato paste and juice. *Die Nahrung* 41: 34–37.

Aziz, N.H., Attia, E.S., and Farag, S.A. 1997. Effect of gamma-irradiation on the natural occur-rence of Fusarium mycotoxins in wheat, flour and bread. *Die Nahrung* 35: 359–362.

Aziz, N.H., El-Zeany, S.A., and Moussa, L.A.A. 2002. Influence of gamma-irradiation and maize lipids on the production of aflatoxin B-1 by *Aspergillus flavus*. *Nahrung—Food* 46: 327–331.

Batzer, O.F. and Doty, D.M. 1955. Nature of undesirable odours formed by gamma irradiation of beef. *Journal of Agricultural and Food Chemistry* 3: 64–69.

Bhaskaram, C. and Sadasivan, G. 1975. Effects of feeding irradiated wheat to malnourished children. *American Journal of Clinical Nutrition* 28: 130–135.

Borsa, J. 2004. Outlook for food irradiation in the 21st century. In *ACS Symposium Series 875: Irradiation of Food and Packaging—Recent Developments*, eds. V. Komolprasert and K.M. Morehouse, pp. 326–339. Washington, DC: American Chemical Society.

Brynjolfsson, A. 1985. Wholesomeness of irradiated foods: a review. *Journal of Food Safety* 7: 107–126.

Burnouf, D., Delincée, H., Hartwig, A., Marchioni, E., Miesch, M., and Werner, D. 2002. *Toxicological Study to Assess the Risk Associated with the Consumption of Irradiated Fat-Containing Food*. Bundesforschungsanstalt für Ernährung (BFE), Karlsruhe, Germany (in German and French with English Summary and Conclusion).

CAC. 2003a. Recommended International Code of Practice for Radiation Processing of Food (CAC/RCP 19-1979, Rev. 2-2003). Codex Alimentarius Commission (CAC), Rome, Italy.

CAC. 2003b. General Standard for Irradiated Foods (Codex Stan 106-1983, Rev. 1-2003). Codex Alimentarius Commission (CAC), Rome, Italy.

Chen, X., Jo, C., Lee, J.I., and Ahn, D.U. 1999. Lipid oxidation, volatiles and color changes in irradiated pork patties as affected by antioxidants. *Journal of Food Science* 64: 16–19.

Chen, Y.P., Andrews, L.S., and Grodner, R.M. 1996. Sensory and microbial quality of irradi-ated crab meat products. *Journal of Food Science* 61: 1239–1242.

Chinn, H.I. 1979. Further toxicological considerations of volatile products. In *Evaluation of the Health Aspects of Certain Compounds Found in Irradiated Beef*, ed. Life Sciences Research Office, pp. 1–29. Bethseda, MD: Federation of American Societies for Experimental Biology.

Cleland, M.R. 2006. Advances in gamma ray, electron beam and x-ray technologies for food irradiation. In *Food Irradiation Research and Technology*, eds. C.H. Sommers and X. Fan, pp. 11–35. Ames, IA: Blackwell Publishing.

Curzio, O.A., Croci, C.A., and Ceci, L.N. 1986. The effects of radiation and extended storage on the chemical quality of garlic bulbs. *Food Chemistry* 21: 153–159.

Dale, M.F.B., Griffiths, D.W., Bain, H., and Goodman, B.A. 1997. The effect of gamma irradiation on glycoalkaloid and chlorophyll synthesis in seven potato cultivars. *Journal of the Science of Food and Agriculture* 75: 141–147.

Delincée, H. 2002. Analytical methods to identify irradiated food—a review. *Radiation Physics and Chemistry* 63: 455–458.

Delincée, H. and Pool-Zobel, B.-L. 1998. Genotoxic properties of 2-dodecylcyclobutanone, a compound formed on irradiation of food containing fat. *Radiation Physics and Chemistry* 52: 39–42.

Delincée, H., Soika, C., Horvatovich, P., Rechkemmer, G., and Marchioni, E. 2002. Genotoxicity of 2-alkylcyclobutanones, markers for an irradiation treatment of fat-containing food— Part I: cyto- and genotoxic potential of 2-tetradecylcyclobutanone. *Radiation Physics and Chemistry* 63: 431–435.

DeRouchey, J.M., Tokach, M.D., Nelssen, J.L. et al. 2003. Effect of irradiation of individual feed ingredients and the complete diet on nursery pig performance. *Journal of Animal Science* 81: 1799–1805.

Diehl, J.F. 1995. *Safety of Irradiated Foods*. New York: Marcel Dekker, Inc.

Diehl, J.F. and Josephson, E.S. 1994. Assessment of wholesomeness of irradiated foods (a review). *Acta Alimentaria* 23: 195–214.

Ehlermann, D. (2001). Process control and dosimetry in food irradiation. In *Food Irradiation Principles and Applications*, ed. R. Molins, pp. 387–414. New York: John Wiley & Sons.

Elliott, C.T., Hamilton, L., Stevenson, M.H., McCaughey, W.J., and Boyd, D.R. 1995. Detection of irradiated chicken meat by analysis of lipid extracts for 2-substituted cyclobutanones using an enzyme linked immunosorbent assay. *Analyst* 120: 2337–2341.

European Commission. 2009. Food Irradiation—Analytical Methods. http://ec.europa.eu/food/food/biosafety/irradiation/anal_methods_en.htm

European Union. 2007. Report from the Commission on food irradiation for the year 2005. *Official Journal of the European Union*, 2007/C122/03, 2 June 2007.

Fan, X. 2005. Impact of ionizing radiation and thermal treatments on furan levels in fruit juice. *Journal of Food Science* 70: E409–E414.

Fan, X.T. and Mastovska, K. 2006. Effectiveness of ionizing radiation in reducing furan and acrylamide levels in foods. *Journal of Agricultural and Food Chemistry* 54: 8266–8270.

Fan, X.T. and Sokorai, K.B. 2007. Effects of ionizing radiation on sensorial, chemical, and microbiological quality of frozen corn and peas. *Journal of Food Protection* 70: 1901–1908.

Fan, X., Niemira, B.A., and Thayer, D.W. 2004. Low-dose ionizing radiation of fruit juices: benefits and concerns. In *ACS Symposium Series 875: Irradiation of Food and Packaging—Recent Developments*, eds. V. Komolprasert and K.M. Morehouse, pp. 138–150. Washington, DC: American Chemical Society.

Farag, S.E., Aziz, N.H., and Attia, E.S. 1995. Effect of irradiation on the microbiological status and flavouring materials of selected spices. *Zeitschrift fuer Lebensmittel-Untersuchung and Forschung* 201: 283–288.

Farkas, J. 1973. Radurization and radicidation of spices. In *Aspects of the Introduction of Food Irradiation in Developing Countries*, ed. International Atomic Energy Agency, pp. 43–59. Vienna: IAEA.

Farkas, J. 1988. *Irradiation of Dry Food Ingredients*. Boca Raton, FL: CRC Press.

Farkas, J. 1998. Irradiation as a method for decontaminating food: a review. *International Journal of Food Microbiology* 44: 189–204.

Farkas, J. 2001. Radiation decontamination of spices, herbs, condiments, and other dried food ingredients. In *Food Irradiation—Principles and Applications*, ed. R. Molins, pp. 291–312. New York: John Wiley & Sons.

Farkas, J. 2006. Irradiation for better foods. *Trends in Food Science and Technology* 17: 148–152.

Farkas, J. and Andrássy, É. 1988. Comparative analysis of spices decontaminated by ethylene oxide or gamma irradiation. *Acta Alimentaria* 17: 77–94.

Farkas, J., Saray, T., Mohacsi-Farkas, C., Horti, K., and Andrássy, E. 1997. Effects of low dose gamma radiation on shelf-life and microbiological safety of pre-cut/prepared vegetables. *Advances in Food Science* 19: 111–119.

Fielding, L. 2007. The safety of irradiated foods: a literature review. Food Standards Agency Project A05009, Technical Report, January 2007.

Foley, D.M. 2006. Irradiation of seafood with a particular emphasis on *Listeria monocytogenes* in ready-to-eat products. In *Food Irradiation Research and Technology*, eds. C.H. Sommers and X. Fan, pp. 185–197. Ames, IA: Blackwell Publishing.

Gadgil, P. and Smith, J.S. 2004. Mutagenicity and acute toxicity evaluation of 2-dodecylcyclobutanone. *Food Chemistry and Toxicology* 69: C713–C716.

Gadgil, P. and Smith, J.S. 2006. Metabolism of 2-dodecylcyclobutanone, a radiolytic compound present in irradiated beef. *Journal of Agricultural and Food Chemistry* 54: 4896–4900.

Gadgil, P., Hachmeister, K.A., Smith, J.S., and Kropf, D.H. 2002. 2-Alkylcyclobutanones as irradiation indicators in irradiated ground beef patties. *Journal of Agricultural and Food Chemistry* 50: 5746–5750.

GAO. 2000. Food irradiation. available research indicates that benefits outweigh risks. United States General Accounting Office, Report to Congressional Requesters, GAO/RCED-00-217, August 2000.

Goularte, L., Martins, C.G., Morales-Aizpurúa, I.C. et al. 2004. Combination of minimal processing and irradiation to improve the microbiological safety of lettuce (*Lactuca sativa*, L.). *Radiation Physics and Chemistry* 71: 155–159.

Graham, W.D. and Stevenson, M.H. 1997. Effect of irradiation on vitamin C content of strawberries and potatoes in combination with storage and with further cooking in potatoes. *Journal of the Science of Food and Agriculture* 75: 371–377.

Grant, I.R. and Patterson, M.F. 1991. Effect of irradiation and modified atmosphere packaging on the microbiological safety of minced pork stored under temperature abuse conditions. *International Journal of Food Science and Technology* 26: 507–519.

Grodner, R.M. and Andrews, L.S. 1991. Irradiation. In *Microbiology of Marine Food Products*, eds. D.R. Ward and C. Hackney, 429–440. New York: Van Nostrand Reinhold.

Guo, A.-X., Wang, G.Z., and Wang, Y. 1981. Biochemical effect of irradiation on potato, onion and garlic in storage. 1. Changes of major nutrients during storage. *Yuang Tzu Neng Nung Yeh Ying Yung* 1: 16–25.

Hansen, P.-I.E. 1966. Radiation treatment of meat products and animal by-products. In *Food Irradiation*, ed. International Atomic Energy Agency, pp. 411–425. Vienna: IAEA.

Harrison, K. and Were, L.M. 2007. Effects of ionizing radiation on sensorial, chemical, and microbiological quality of frozen corn and peas. *Journal of Food Protection* 70: 1901–1908.

Hartwig, A., Pelzer, A., Burnouf, D. et al. 2007. Toxicological potential in 2-alkylcyclobutanones—specific radiolytic products in irradiated fat-containing food—bacteria and human cell lines. *Food and Chemical Toxicology* 45: 2481–2501.

Health Canada. 2002. Irradiation of ground beef: summary of submission process. Food Directorate, Food Products and Health Branch, October 29, 2002, Ottawa. www.hc-sc. gc.ca/food-aliment/fpi-ipa/e_gbeef_submission.pdf

Horvatovich, P., Raul, F., Miesch, M. et al. 2002. Detection of 2-alkylcyclobutanones, markers for irradiated foods, in adipose tissues of animals fed with these substances. *Journal of Food Protection* 65: 1610–1613.

Hussain, P.R., Meena, R.S., Dar, M.A., and Wani, A.M. 2007. Studies on enhancing the keeping quality of peach (Prunus persica Bausch) Cv. Elberta by gamma-irradiation. *Radiation Physics and Chemistry* 77: 473–481.

IAEA. 1993. Irradiation of poultry meat and its products: a compilation of technical data for its authorization and control. IAEA-Tecdoc-688, February 1993, International Atomic Energy Agency, Vienna, Austria.

IAEA. 1996. Irradiation of red meat: a compilation of technical data for its authorization and control. IAEA-Tecdoc-688, August 1996, International Atomic Energy Agency, Vienna, Austria.

IAEA. 1997. Irradiation of bulb and tuber crops: a compilation of technical data for its authorization and control. IAEA Tecdoc-937, April 2007, International Atomic Energy Agency, Vienna, Austria.

Ito, H. and Islam, M. 1994. Effect of dose rate on inactivation of microorganisms in spices by electron-beams and gamma-rays irradiation. *Radiation Physics and Chemistry* 43: 545–550.

Kilgen, M.B. 2001. Irradiation processing of fish and shellfish products. In *Food Irradiation— Principles and Applications*, ed. R. Molins, pp. 193–212. New York: John Wiley & Sons.

Kiss, I. 1984. Pre-commercial scale irradiation experiments with spices, vegetables and fruits. *Food Irradiation Newsletter* 8: 10–11.

Kiss, I. and Farkas, J. 1972. Radurization of whole eviscerated chicken carcass. *Acta Alimentaria* 1: 73–86.

Kiss, I.F., Mészáros, L., and Kovács-Domján, H. 2001. Reducing microbial contamination including some pathogens in minced beef by irradiation. In *Irradiation for Food Safety and Quality*, eds. P. Loaharanu and P. Thomas, pp. 81–92. Lancaster, PA: Technomic Publishing Co. Inc.

Kitazuru, E.R., Moreira, A.V.B., Mancini-Filho, J., Delincée, H., and Villavicencio, A.L.C.H. 2004. Effects of irradiation on natural antioxidants of cinnamon (*Cinnamomum zeylanicum* N.). *Radiation Physics and Chemistry* 71: 39–41.

Knoll, N., Weise, A., Claussen, U. et al. 2006. 2-Dodecylcyclobutanone, a radiolytic product of palmitic acid, is genotoxic in primary human colon cells and in cells from preneoplastic lesions. *Mutation Research* 594: 10–19.

Komolprasert, V. and Morehouse, K. 2004. *ACS Symposium Series 875: Irradiation of Food and Packaging—Recent Developments*. Washington, DC: American Chemical Society.

Kovács, A., Hargittai, P., Kaszanyiczki, L., and Földiák, G. 1994. Evaluation of multipurpose electron irradiation of packaged and bulk spices. *Applied Radiation and Isotopes* 45: 783–788.

Kuruppu, D.F., Langerak, D.I., and van Duren, M.D.A. 1983. Effect of gamma irradiation, fumigation and storage time on volatile oil content of some spices. IFFIT Report 41, International Facility for Food Irradiation Technology, Wageningen, the Netherlands.

Kwakwa, A. and Prakash, A. 2006. Irradiation of nuts. In *Food Irradiation Research and Technology*, eds. C.H. Sommers and X. Fan, pp. 221–235. Ames, IA: Blackwell Publishing.

Kwon, J.H., Choi, J.U., and Yoon, H.S. 1989. Sulfur-containing components of gamma-irradiated garlic bulbs. *Radiation Physics and Chemistry* 34: 969–972.

Lebepe, S., Molins, R.A., Charoen, S.P., Farrar, IV, H., and Skowronski, R.P. 1990. Changes in microflora and other characteristics of vacuum-packaged pork loins irradiated at 3.0 kGy. *Journal of Food Science* 55: 918–924.

Lee, E.J. and Ahn, D.U. 2003. Production of off-odour volatiles from fatty acids and oils by irradiation. *Journal of Food Science* 69: C485–C490.

Lee, N.Y., Jo, C., Shin, D.H., Kim, W.G., and Byun, M.W. 2006. Effect of gamma-irradiation on pathogens inoculated into ready-to-use vegetables. *Food Microbiology* 23: 649–656.

Lescano, G., Narvaiz, P., and Kairiyama, E. 1991. Sterilization of spices and vegetable seasonings by gamma radiation. *Acta Alimentaria* 20: 233–242.

LeTellier, P.R. and Nawar, W.W. 1972. 2-Alkylcyclobutanones from radiolysis of triglycerides. *Lipids* 7: 75–76.

Llorente-Franco, S., Giménez, J.L., Martinez Sánchez, F., and Romojaro, D. 1986. Effectiveness of ethylene oxide and gamma irradiation on the microbiological population of three types of paprika. *Journal of Food Science* 51: 1571–1572.

Louria, D.B. 1990. Zapping the food supply. *Bulletin of the Atomic Scientists* 46: 34–36.

Louria, D.B. 2001. Food irradiation: unresolved issues. *Clinical Infectious Diseases* 33: 378–380.

Luchsinger, S.E., Kropf, D.H., Garcia Zepeda, C.M. et al. 1996. Sensory analysis and consumer acceptance of irradiated boneless pork chops. *Journal of Food Science* 61: 1261–1266.

Marchioni, E. 2006. Detection of irradiated food. In *Food Irradiation Research and Technology*, eds. C.H. Sommers and X. Fan, pp. 85–104. Ames, IA: Blackwell Publishing.

Marchioni, E., Raul, F., Burnouf, D. et al. 2004. Toxicological study on 2-alkylcyclobutanones—results of a collaborative study. *Radiation Physics and Chemistry* 71: 145–148.

McMurray, C.H., Stewart, E.M., Gray, R., and Pearce, J. 1996. Detection methods for irradiated foods—current status. Royal Society of Chemistry, Special Publication No. 171. Cambridge, U.K.: RSC.

Merritt, C. Jr., Bresnick, S.R., Bazinet, M.L., Walsh, J.T., and Angelini, P. 1959. Determination of volatile components of foodstuffs. Techniques and their application to studies in irradiated beef. *Journal of Agricultural and Food Chemistry* 7: 784–787.

Merritt, C. Jr., Angelini, P., Wierbicki, E., and Shuts, G.W. 1975. Chemical changes associated with flavour in irradiated meat. *Journal of Agricultural and Food Chemistry* 23: 1037–1043.

Metlitsky, L.V., Rubin, B.A., and Krushchev, V.G. 1957. Use of γ-radiation in lengthening storage time of potatoes. *Proceedings of the All Union Conference on the Application of Radioactive and Stable Isotopes and Radiation in the National Economy and Science*, USAEC Report—AEC-tr 2925.

Miller, S.J., Moss, B.W., Macdougall, D.B., and Stevenson, M.H. 1995. The effect of ionising radiation on the CIE-LAB colour co-ordinates of chicken breast meat as measured by different instruments. *International Journal of Food Science and Technology* 30: 663–674.

Molins, R.A. 2001a. *Food Irradiation—Principles and Applications*. New York: John Wiley & Sons.

Molins, R.A. 2001b. Irradiation of meats and poultry. In *Food Irradiation—Principles and Applications*, ed. R. Molins, pp. 131–191. New York: John Wiley & Sons.

Moreno, M.A., Castell-Perez, M.E., Gomes, C., Da Silva, P.F., and Moreira, R.G. 2007. Quality of electron beam irradiation of blueberries (*Vaccinium corymbosum* L.) at medium dose levels (1.0–3.2 kGy). *Lebensmittel-Wissenschaft und-Technologie/Food Science and Technology* 40: 1123–1132.

Nam, K.C. and Ahn, D.U. 2002a. Carbon monoxide–heme pigment is responsible for the pink colour in irradiated raw turkey breast meat. *Meat Science* 60: 25–33.

Nam, K.C. and Ahn, D.U. 2002b. Combination of aerobic and vacuum packaging to control lipid oxidation and off-odor volatiles of irradiated raw turkey breast. *Meat Science* 63: 389–395.

Nam, K.C. and Ahn, D.U. 2003. Effects of ascorbic acid and antioxidants on the color of irradiated beef patties. *Journal of Food Science* 68: 1686–1690.

Nam, K.C., Kim, Y.H., and Ahn, D.U. 2002. Off-odour and pink colour development in pre-cooked, ionizing-radiated turkey breast during frozen storage. *Poultry Science* 81: 269–275.

Nanke, K.E., Sebranek, J.G., and Olson, D.G. 1998. Color characteristics of irradiated vacuum-packaged pork, beef, and turkey. *Journal of Food Science* 63: 1001–1006.

Narvaiz, P., Lescano, G., and Kairiyma, E. 1992. Irradiation of almonds and cashew nuts. *Lebensmittel-Wissenschaft und-Technologie/Food Science and Technology* 25: 232–235.

Nayga, R.M. Jr., Aiew, W., and Nicols, P. 2005. Informative effects on consumers' willingness to purchase irradiated food products. *Review of Agricultural Economics* 27: 37–48.

Niemira, B.A. 2007. Relative efficacy of sodium hypochlorite wash versus irradiation to inactivate Escherichia coli O157:H7 internalized in leaves of romaine lettuce and baby spinach. *Journal of Food Protection* 70: 2526–2532.

Niemira, B.A. 2008. Irradiation compared with chlorination for elimination of *Escherichia coli* O157:H7 internalized in lettuce leaves: influence of lettuce variety. *Journal of Food Science* 73: M208–M213.

Niemira, B.A. and Fan, X. 2006. Low-dose irradiation of fresh and fresh-cut produce: safety, sensory, and shelf-life. In *Food Irradiation Research and Technology*, eds. C.H. Sommers and X. Fan, pp. 169–184. Ames, IA: Blackwell Publishing.

O'Bryan, C., Crandall, P.G., Ricke, S.C., and Olson, D.G. 2008. Impact of irradiation on the safety and quality of poultry and meat products: a review. *Critical Reviews in Food Science and Nutrition* 48: 442–457.

Obana, H., Furuta, M., and Tanaka, Y. 2006. Detection of 2-alkylcyclobutanones in irradiated meat, poultry and egg after cooking. *Journal of Health Science* 52: 375–382.

Patterson, M.F. 2001. Combination treatments involving food irradiation. In *Food Irradiation—Principles and Applications*, ed. R. Molins, pp. 313–327. New York: John Wiley & Sons.

Patterson, M.F. 2001. Role of irradiation in a multiple-hurdle approach to food safety. In *Irradiation for Food Safety and Quality*, eds. P. Loaharanu and P. Thomas, pp. 71–80. Lancaster, PA: Technomic Publishing Co. Inc.

Patterson, R.L. and Stevenson, M.H. 1995. Irradiation-induced off-odour in chicken and its possible control. *British Poultry Science* 36: 425–441.

Piggot, J.R. and Othman, Z. 1993. Effect of irradiation on volatile oils in black pepper. *Food Chemistry* 46: 115–119.

Poole, S.E., Mitchell, G.E., and Mayze, J.L. 1994. Low-dose irradiation affects microbiological and sensory quality of sub-tropical seafood. *Journal of Food Science* 59: 85–87.

Prakash, A. and Foley, D. 2004. Improving safety and extending shelf life of fresh-cut fruits and vegetables using irradiation. In *ACS Symposium Series 875: Irradiation of Food and Packaging—Recent Developments*, eds. V. Komolprasert and K.M. Morehouse, pp. 90–106. Washington, DC: American Chemical Society.

Prakash, A., Guner, A.R., Caporaso, E., and Foley, D.M. 2000a. Effects of low-dose gamma irradiation on the shelf life and quality characteristics of cut romaine lettuce packaged under modified atmosphere. *Journal of Food Science* 65: 549–553.

Prakash, A., Inthajak, P., Huibregtse, H., Caporaso, E., and Foley, D.M. 2000b. Effects of low-dose gamma irradiation and conventional treatments on shelf life and quality characteristics of diced celery. *Journal of Food Science* 65: 1070–1075.

Raul, F., Gossé, F., Delincée, H. et al. 2002. Food-borne radiolytic compounds (2-alkylcyclobutanones) may promote experimental colon carcinogenesis. *Nutrition and Cancer* 44: 188–191.

Refai, M.K., Niazi, Z.M., Aziz, H.H., and Khafaga, N.E. 2003. Incidence of aflatoxin B1 in the Egyptian cured meat basterma and control by gamma-irradiation. *Die Nahrung* 47: 377–382.

Reyes, L.F. and Cisneros-Zevallos, L. 2007. Electron-beam ionizing radiation stress effects on mango fruit (*Mangifera indica* L.) antioxidant constituents before and during postharvest storage. *Journal of Agricultural and Food Chemistry* 55: 6132–6139.

Roberts, P.B. 2002. *Food Irradiation at a Crossroads: Background for Strategic Options*. IAEA, Vienna, 6–10 May 2002.

Sanchez-Bel, P., Martinez-Madrid, M.C., Egea, I., and Romojaro, F. 2005. Oil quality and sensory evaluation of almond (*Prunus amygdalus*) stored after electron beam processing. *Journal of Agricultural and Food Chemistry* 53: 2567–2573.

Sanchez-Bel, P., Egea, I., Romojaro, F., and Martinez-Madrid, M.C. 2008. Sensorial and chemical quality of electron beam irradiated almonds (Prunus amygdalus). *Lebensmittel-Wissenschaft und-Technologie/Food Science and Technology* 41: 442–449.

Sanderson, D.C.W., Carmichael, L.A., and Fisk, S. 2003a. Photostimulated luminescence detection of irradiated herbs, spices and seasonings: International interlaboratory trial. *Journal of AOAC International* 86: 990–998.

Sanderson, D.C.W., Carmichael, L.A., and Fisk, S. 2003b. Photostimulated luminescence detection of irradiated shellfish: International interlaboratory trial. *Journal of AOAC International* 86: 983–989.

Satter, A., Jan, M., Ahmad, A., Wahid, M., and Khan, I. 1989. Irradiation disinfestation and biochemical quality of dry nuts. *Acta Alimentaria* 18: 45–52.

Schindler, M., Solar, S., and Sontag, G. 2005. Phenolic compounds in tomatoes. Natural variations and effect of gamma-irradiation. *European Food Research and Technology* 221: 439–445.

Schwinner, S. and Weston, W.J. 1958. Chlorophyll formation in potato tubers as influenced by gamma irradiation and by chemicals. *American Potato Journal* 35: 534–539.

Shahidi, F. and Pegg, R.B. 1994. Hexanal as an indicator of the flavor deterioration of meat and meat products. In *ACS Symposium Series 558: Lipids in Food Flavors*, eds. C.-T. Ho and T.G. Hartman, pp. 165–175. Washington, DC: American Chemical Society.

Shirsat, S.G. and Thomas, P. 1998. Effect of irradiation and cooking methods on ascorbic acid levels of four potato cultivars. *Journal of Food Science and Technology* 35: 509–514.

Sjövall, O., Honkanen, E., Kallio, H., Latva-Kala, K., and Sjöberg, A.-M. 1990. The effects of gamma-irradiation on some pure aroma compounds of spices. *Zeitschrift fuer Lebensmittel-Untersuchung and Forschung* 191: 181–183.

Smith, J.S. and Pillai, S. 2004. Irradiation and food safety. *Food Technology* 58: 48–55.

Sommers, C.H. 2003. 2-Dodecylcyclobutanone does not induced mutations in the *Escherichia coli* tryptophan reverse mutation assay. *Journal of Agricultural and Food Chemistry* 51: 6367–6370.

Sommers, C.H. and Mackay, W.J. 2005. DNA damage-inducible gene expression and formation of 5-fluorouracil-resistant mutants in *Escherichia coli* exposed to 2-dodecylcyclobutanone. *Food Chemistry and Toxicology* 70: C254–C257.

Sommers, C.H. and Schiestl, R.H. 2004. 2-Dodecylcyclobutanone does not induce mutations in the Salmonella mutagenicity test or intrachromosomal recombination in Saccharomcyes cerevisiae. *Journal of Food Protection* 67: 1293–1298.

Sommers, C.H., Delincée, H., Scott Smith, J., and Marchioni, E. 2006. Toxicological safety of irradiated foods. In *Food Irradiation Research and Technology*, eds. C.H. Sommers and X. Fan, pp. 43–62. Ames, IA: Blackwell Publishing.

Song, H.P., Byun, M.W., Jo, C., Lee, C.H., Kim, K.S., and Kim, D.H. 2007. Effects of gamma irradiation on the microbiological, nutritional, and sensory properties of fresh vegetable juice. *Food Control* 18: 5–10.

Stevenson, M.H. 1990. The practicalities of food irradiation. *Food Technology International Europe 1990* 2: 73–77.

Stevenson, M.H. 1992. Irradiation of meat and poultry. In *The Chemistry of Muscle-Based Foods*, eds. D.A. Ledward, D.E. Johnston and M.K. Knight, pp. 308–324. Royal Society of Chemistry, Special Publication No. 106. London: RSC.

Stevenson, M.H. 1994. Nutritional and other implications of irradiating meat. *Proceedings of the Nutrition Society* 53: 317–325.

Stewart, E.M. 2001a. Food irradiation chemistry. In *Food Irradiation—Principles and Applications*, ed. R. Molins, pp. 37–76. New York: John Wiley & Sons.

Stewart, E.M. 2001b. Detection methods for irradiated foods. In *Food Irradiation—Principles and Applications*, ed. R. Molins, pp. 347–386. New York: John Wiley & Sons.

Stewart, E.M. 2004. Food irradiation: more pros than cons? *Biologist* 51: 91–96.

Stewart, E.M. 2006. Food Irradiation in the 21st Century. FoodInfo Online Features 20 February 2006. http://www.foodsciencecentral.com/fsc/ixid14262

Stewart, E.M., McRoberts, W.C., Hamilton, J.T.G., and Graham, W.D. 2001. Isolation of lipid and 2-alkylcyclobutanones from irradiated food by supercritical fluid extraction. *Journal of AOAC International* 84: 976–986.

Thayer, D.W. 1994. Wholesomeness of irradiated foods. *Food Technology* 48: 132–135.

Thayer, D.W. 1995. Use of irradiation to kill enteric pathogens on meat and poultry. *Journal of Food Safety* 15: 181–192.

Thayer, D.W., Christopher, J.P., Campbell, L.A. et al. 1987. Toxicology studies of irradiation-sterilized chicken. *Journal of Food Protection* 50: 278–288.

Thayer, D.W., Fox, J.B. Jr., and Lakritz, L. 1991. Effects of ionizing radiation on vitamins. In *Food Irradiation*, ed. S. Thorpe, pp. 285–325. London: Applied Science Publishers Ltd.

Thomas, P. 1988. Radiation preservation of foods of plant origin. Part VI. Mushrooms, tomatoes, minor fruits and vegetables, dried fruits, and nuts. *CRC Critical Reviews in Food Science and Nutrition* 26: 313–358.

Thomas, P. 2001a. Irradiation of fruits and vegetables. In *Food Irradiation—Principles and Applications*, ed. R. Molins, pp. 213–240. New York: John Wiley & Sons.

Thomas, P. 2001b. Irradiation of tuber and bulb crops. In *Food Irradiation—Principles and Applications*, ed. R. Molins, pp. 241–272. New York: John Wiley & Sons.

Toffanian, F. and Stegeman, H. 1988. Comparative effect of ethylene oxide and gamma irradiation on the chemical, sensory and microbiological quality of ginger, cinnamon, fennel and fenugreek. *Acta Alimentaria* 17: 271–281.

Uthman, R.S., Toma, R.B., Garcia, R., Medora, N., and Cunningham, S. 1998. Lipid analyses of fumigated vs irradiated raw and roasted almonds. *Journal of the Science of Food and Agriculture* 78: 261–266.

Vajdi, M. and Pereira, N.N. 1973. Comparative effects of ethylene oxide, gamma irradiation and microwave treatment on selected spices. *Journal of Food Science* 38: 893–895.

Venugopal, V., Doke, S.N., and Thomas, P. 1999. Radiation processing to improve the quality of fishery products. *Critical Reviews in Food Science and Nutrition* 39: 391–440.

WHO. 1977. *Wholesomeness of Irradiated Food.* Technical Report Series 606, World Health Organization, Geneva.

WHO. 1981. *Wholesomeness of Irradiated Food: A Report of a Joint FAO/IAEA/WHO Expert Committee on Food Irradiation.* WHO Technical Report Series, p. 659, World Health Organization, Geneva.

WHO. 1994. *Safety and Nutritional Adequacy of Irradiated Food.* World Health Organization, Geneva.

WHO. 1999. *High-Dose Irradiation: Wholesomeness of Food Irradiated with Doses above 10 kGy.* Report of a Joint FAO/IAEA/WHO Study Group, WHO Technical Report Series 890. World Health Organization, Geneva.

Wick, E.L., Murray, E., Mizutani, J., and Koshika, M. 1967. Irradiation flavour and the volatile components of beef. In *Radiation Preservation of Foods (Advances in Chemistry Series)*, ed. American Chemical Society, pp. 12–25. Washington, DC: American Chemical Society.

Zachariev, G., Kiss, I., Szabolcs, J., Tóth, G., Molnár, P., and Matus, Z. 1991. HPLC analysis of carotenoids in irradiated and ethylene oxide treated red pepper. *Acta Alimentaria* 20: 115–122.

Zhang, L., Lu, Z., Lu, F., and Bie, X. 2006. Effect of γ irradiation on quality-maintaining of fresh-cut lettuce. *Food Control* 17: 225–228.

14 Improving Food Safety and Quality by High-Pressure Processing

J. Antonio Torres, Pedro D. Sanz, Laura Otero,
María Concepción Pérez Lamela, and
Marleny D. Aranda Saldaña

CONTENTS

This chapter presents the process engineering considerations for high-pressure processing (HPP) applications including examples of commercial processing units and products. HPP technology will be reviewed including some process and product design considerations for commercial viability and market success. Also highlighted will be the promising future of pressure-assisted thermal processing (PATP) technology to be used at moderate temperature for the reduction of pressure holding time needed for food pasteurization, and at high temperature for the production of shelf-stable low-acid foods. This will be followed by a review of engineering principles including the mechanisms and kinetics for the reduction of microbial load and the inactivation of enzymes by pressure and temperature. A fundamental understanding of the inactivation of microorganisms and enzymes by pressure and temperature is required for the reliable commercial implementation of HPP and PATP technology. To date, much research has been conducted in model systems, particularly in buffer solutions. Therefore, future efforts must consider the effect of the food matrix including pH, water activity, interactions, and the use of food additives.

14.1   INTRODUCTION

Today, new processing technologies are needed to deliver at the point of food consumption near absolute safety and perfect quality. Quality can be defined as the degree of excellence in the totality of features and characteristics that have a bearing on a product's ability to satisfy a need. Moreover, high quality means that this superiority is delivered consistently. A most important quality factor for the food industry is the reputation of their brands. In an era of rapid local and global communications, a product recall destroys years of market penetration efforts. Companies facing this situation must provide to government agencies and reporters detailed information on the cause for a safety failure and the actions taken to ensure that the event is unlikely to occur again. In a 2008 survey conducted by the National Cattlemen's Beef Association,[*] 67% of beef consumers had seen, heard, or read about *Escherichia coli* bacteria in food during the previous month. Since 83% acknowledged not using a thermometer to ensure safe cooking procedures, it can be concluded that improving existing and developing new technologies is needed by food processors to reduce the incidence of pathogens in our foods, particularly in raw, fresh, and minimally processed foods. Such an effort is necessary to raise consumer confidence in foods which has dropped dramatically in recent years, falling from 82% in 2006 to 66% in 2007 for grocery food purchases.[†]

A new and difficult challenge for industry is the means of information delivery used today by consumers to widely disseminate personal experiences and opinions that impact negatively on the quality and safety image of the food supply. Food processors need means to ensure the robustness of the product image by implementing improved or new processing steps reducing safety risks while enhancing quality. An example is the need to make sliced processed meats safer by postpackaging treatments to inactivate pathogens and spoilage organisms without altering sensory properties or adding chemical additives. Finally, another key driver for developing new technologies is the need for "invisible processing." Consumers demand foods minimally affected by processing so as to preserve nutrients and desirable life-quality

---

[*] http://www.meatami.com/Education/Presentations/2008/Ecoli/RickMcCarty.pdf, consulted 2/2/2008
[†] http://www.meatami.com/Education/Presentations/2008/Ecoli/JillHollingsworth.pdf, consulted 2/2/2008

enhancing factors, and most importantly, to protect the sensory properties that provide eating pleasure.

HPP, a relatively new technology to the food industry (Velazquez et al. 2002, Torres and Velazquez 2005, 2008, Torres and Rios 2006), inactivates microorganisms without causing significant flavor and nutritional changes to foods (Cheftel and Culioli 1997, Beslin et al. 1999). On the other hand, the effectiveness of thermal processing technologies explains why they remain a prevailing method to achieve microbial safety and the inactivation of enzymes and microorganisms responsible for food spoilage. New packaging (e.g., Recart, a retortable square carton package) and retort technologies (e.g., Shaka, a retort with a reciprocating actuator to shake foods in pouches and other containers for faster heat transfer) are improving the quality and efficiency of thermal processing. However, thermal processing causes significant chemical changes in foods. Particularly important are thermal degradation reactions leading to off-flavors, destruction of nutrients, and other product quality losses. For example, high-temperature short-time (HTST) pasteurization treatments ($72°C$ for $15 s$) impart a slight cooked, sulfurous note that has become acceptable to milk consumers but its refrigerated shelf life is only ~20 days. Ultra pasteurization, a process similar to HTST pasteurization using more severe treatments (e.g., $1 s$ at $89°C$, $0.1 s$ at $96°C$, or $0.01 s$ at $100°C$), lowers flavor quality and causes more nutrient damage but yields milk with a refrigerated shelf life of approximately 30 days (Anonymous 2006a,b).

In the particular case of milk processing, HPP treatments of $400 MPa$ for $15 min$ or $500 MPa$ for $3 min$ at room temperature achieve microbiological reductions similar to thermal pasteurization (Rademacher and Kessler 1996; Rademacher et al. 2002) but it is not used commercially because long pressure processing times are not financially viable. A short pressure-holding time, typically $0–3 min$, is an important process design consideration in HPP applications. One effective means to reduce this time is to treat foods at temperatures higher than ambient. For example, pressure treatments ($586 MPa$ for 3 and $5 min$) at moderate temperature ($55°C$) extend the refrigerated shelf life of milk to over 45 days (Tovar-Hernandez et al. 2005) while retaining milk volatile profiles similar to those observed after conventional HTST treatments (Pérez Lamela and Torres 2008a,b). Finally, ultrahigh temperature (UHT) processing ($135°C–150°C$ for $3–5 s$) yields milk that is stable at room temperature for 6 months; however, this process induces strong "cooked" off-flavor notes (Shipe 1980, Colahan-Sederstrom and Peterson 2005) thus limiting its consumer acceptance in important markets (Steely 1994). No commercial pressure processing has been developed for milk sterilization as an alternative to UHT. However, future advances are expected from the synergistic effects of using high-pressure and high-temperature combinations in the rapidly evolving PATP technology. This is not yet a commercially used application and it will require more complex safety validation procedures than HPP, particularly for the case of low-acid foods (pH > 4.5). A PATP-sterilization process for mashed potatoes is under review and could becomes the world's first PATP-sterilized product (Stewart 2008; Ramirez et al. 2009).

PATP conditions are sufficiently severe to achieve the inactivation of bacterial spores and it appears that pressure can lower the degradation rate of product quality caused by high-temperature treatments (Vazquez Landaverde et al. 2007, Pérez

Lamela and Torres 2008a,b, Torres and Velazquez 2008). The use of pressure to reduce the impact of thermal degradation reactions could reduce losses of quality factors and of constituents with important health benefits to consumers. It may also inhibit formation reactions for potential toxicants. However, it is also possible that pressure may induce the formation of toxicants not observed in conventional thermal processes. Therefore, the extent to which the severity of PATP conditions is increased to enhance microbial inactivation and shelf life must be carefully approached. Unfortunately very few reports have been published on PATP effects on chemical changes in foods.

14.2 HIGH-PRESSURE PROCESSING TECHNOLOGY

The key HPP unit components are the high hydrostatic pressure vessel, the pressure-generating pump or pressure intensifier, and the yoke (Figure 14.1). Oil at ~20 MPa is fed on the high-pressure oil side of the main pump piston, which has an area ratio of ~30:1 with respect to the high-pressure fluid piston displacing into the high-pressure vessel a food-grade contact fluid at ~600 MPa, typically purified water. When the main piston reaches the end of its displacement, the system is reversed and high-pressure oil is then fed to the other side of the main pump piston and the high-pressure fluid exits on the other side of the pump. Casting constraints limit the size of single block pressure vessels to ~25 L for operating pressures greater than 400 MPa.

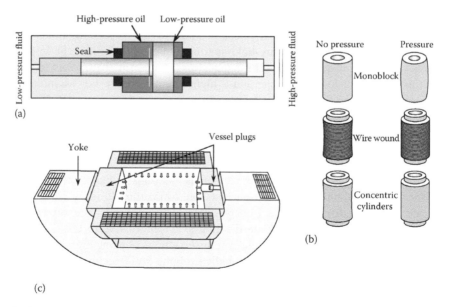

FIGURE 14.1 High-pressure processing technology components. (a) Pressure intensifier; (b) Pressure vessel technologies; (c) Horizontal wire wound vessel with vessel plugs supported by wire-wound yoke. (Adapted from Torres, J.A. and Velazquez, G. *Food Processing Operations Modeling: Design and Analysis*, eds., Jun, S. and Irudayaraj, J., CRC Press/Taylor & Francis, Boca Raton, FL, 2008.)

Prestressing by wire-winding and other technologies allows the construction of ~100 to 500 L commercial-size vessels operating at higher pressures. Typically, the same technology is used for the yoke holding the top and bottom seals. Vessel reinforcement increases equipment costs leading to the present definition of low-cost operations such as oyster shucking requiring 200–400 MPa separated by a technology barrier at ~400 MPa from higher cost operations such as guacamole salsa production at ~600 MPa. A second technology barrier exists at ~650 MPa, that is, the typical maximum pressure for vessels available for commercial applications (Torres and Velazquez 2008). However, the next generation of equipment is expected to reach ~700 MPa and operate at temperatures higher than 100°C to inactivate bacterial spores (Ting 2003).

Although the original HPP equipment designs were for vertical units, the current trend is to supply horizontal units (Figure 14.2). A horizontal orientation avoids sublevel construction requirements, eliminates height and floor load restrictions, makes system installation and relocation more feasible, facilitates product flow in the plant, and reduces the risk of confusing treated and unprocessed products. Another design trend is the use of multiple intensifiers, up to nearly eight pressure intensifiers to reduce pressure come-up time and working independently from each

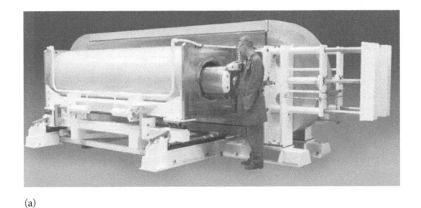

(a)

(b)

FIGURE 14.2 Examples of large commercial pressure processing vessels. (a) Model Quintus Type QFP 687L-310. (Courtesy of Avure Technologies, Kent, WA.) (b) Model Wave 6000-55. (Courtesy of NC Hyperbaric, Burgos, Spain.)

other to facilitate equipment maintenance and repairs. It is also possible to share intensifiers between two or more pressure vessels to reduce capital investment. Finally, unlike thermal processing and other preservation technologies, HPP effects are uniform and nearly instantaneous throughout the food and thus independent of food geometry and equipment size. These observations have facilitated the scale-up of laboratory findings to full-scale production and changes to product packaging required during the commercial life of a product. This ease of scale-up will not hold true for the implementation of PATP technology.

HPP technology has been adopted at an exponential rate as reflected by the world-wide number of units installed all over the world. The cost of the capital needed to implement HPP has been overcome in part by designing larger units as shown by a larger average volume (Figure 14.3a). Another positive development has been the 60% reduction in the equipment cost to process 1 L/h at 600 MPa reflecting commercial competition and the economy of scale gained from larger units (Figure 14.3b). At present, the number of units installed in the United States is greater than the units installed in Europe and in Asia (Figure 14.3c). The HPP impact on product cost for a specific application depends on multiple factors: (1) plant operation schedule (two shifts, 300 days/year is recommended); (2) pressure come-up time (investing in multiple intensifiers reduces it); (3) holding time (3 min desirable maximum for commercial viability); (4) vessel filling ratio (50% minimum recommended, improved by packaging design modifications); (5) product handling time (automatic loading/unloading recommended when feasible); and (6) equipment downtime (minimized by personnel training and ample supply of spare parts).

The rapid and worldwide commercial implementation of HPP technology responds to a strong demand for fresh or minimally processed products from consumers seeking high-quality foods and superior eating pleasure experiences. Examples of commercial products pasteurized by HPP include juices and other beverages, avocado puree, oysters, shrimps, lobsters, and processed meats of superior quality than those possible by conventional technologies (Figure 14.4). Indeed, the successful introduction of a new technology requires competitive advantages over existing ones. Consumers are purchasing HPP products because they retain life-enhancing components and provide more eating pleasure. However, commercial success depends also on the ability of the HPP entrepreneur to find desirable applications (e.g., formulating an apple–tropical fruit mix to deliver a unique flavor experience vs apple-only juice which competes with juices commercialized in large and low-cost volumes), manage the financial risk of adopting a new technology, and invest with reasonable expectations of the capital recovery period (5 years is highly desirable).

14.3 PRESSURE EFFECTS ON FOOD TEMPERATURE AND pH

Early HPP researchers failed to point out the need to consider the pressure-induced effects on pH and temperature when interpreting pressure-processing effects on foods. Adiabatic heating of the sample and of the pressure-transmitting fluid during compression (Figure 14.5a) is a well-described phenomenon (Denys 2000, Denys et al. 2000a,b, Ting et al. 2002, Ardia et al. 2004) and vessels equipped with multiple thermocouples are now used to obtain sample and pressurizing fluid temperature data.

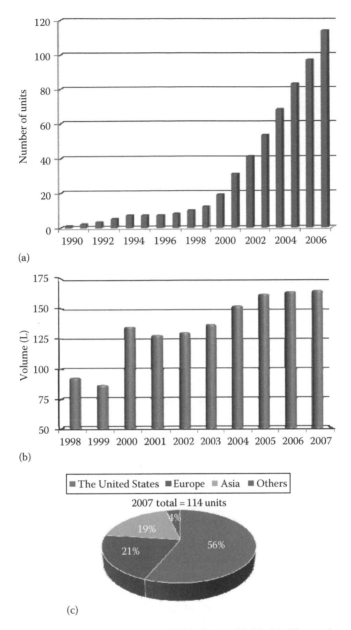

FIGURE 14.3 Worldwide growth of the HPP industry. (a) Worldwide number of installed pressure vessels (b) Average size of worldwide installed units (c) 2007 distribution of worldwide installed units. (Courtesy of NC Hyperbaric, Burgos, Spain.)

The temporary pH change under pressure reflects the effect on the dissociation constant of the weak acid and base under pressure. As pressure increases, the columbic field of the ions produces an alignment of molecules resulting in a more compact arrangement of water molecules around the charged species. This rearrangement

Avocado Paste Simply Avo Fresherized Foods, USA	Fruit puree Sonatural Frubaca, Cooperativa de Hortifruticultores, C.R.L., Portugal	Cooked Spanish sausages "Tapas al minuto" product line Esteban Espuña, Spain

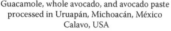

Prepared meals
Spinach in a cream sauce
Carretilla, Grupo Alimentario IAN, Spain

Guacamole, whole avocado, and avocado paste
processed in Uruapán, Michoacán, México
Calavo, USA

Sliced meats
"Vuelta y Vuelta" product line
CAMPOFRIO ALIMENTACION S.A., Spain

Chicken fajitas
"Menu Fresh" product line
Fresherized Foods, USA

FIGURE 14.4 Worldwide examples of pressure-processed foods. (Adapted from Ulloa-Fuentes, P.A. et al., *Industria Alimentaria (México)*, 30, 20, 2008a; Ulloa-Fuentes, P.A. et al., *Industria Alimentaria (México)*, 30, 19, 2008b.)

changes the acid–base dissociation constant and leads to the dissociation of uncharged acid (HA) in aqueous solution forming additional H^+ and A^- ions, a phenomenon known as electrostriction. The pressure-induced pH shift cannot be measured experimentally as there are no pH probes currently unavailable for operations at high pressure. However, the extent of the pH-shift can be predicted if values for the empirical constants involved in Equation 14.1 are available (Neuman et al. 1973, El'Yanov and Hamann 1975):

$$\left(pK_a\right)_p = \left(pK_a\right)_0 + \frac{p\left(\Delta V^0\right)}{\log\left(RT\left(1 + bp\right)\right)} \tag{14.1}$$

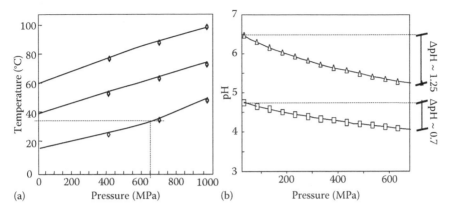

FIGURE 14.5 Examples of pressure effects on sample temperature and pH. (a) Pressure effects on water temperature. (Adapted from Ting, E.Y. et al., *Food Technol.*, 56, 31, 2002.) (b) Pressure effects on citric acid buffer. (Adapted from Paredes-Sabja, D. et al., *J. Food Sci.*, 72, M202, 2007.)

where

$(pK_a)_p$ is the pressure-shifted dissociation constant

$(pK_a)_0$ is the dissociation constant at the reference pressure (0.1 MPa)

p is the pressure (MPa)

ΔV^0 is the partial molar volume change of the dissociating acid at 0.1 MPa (m³/mol)

R is the universal gas constant, that is, 8.31×10^{-6} (MPa m³)/(K mol)

T is the absolute temperature (K)

b is the 9.2×10^{-4} MPa (assumed constant for all acids)

Values for the effect of pH on citric acid buffer predicted by Equation 14.1 are shown in Figure 14.5b (Paredes-Sabja et al. 2007). In this example, pH is reduced from an initial pH value of 4.75 and 6.5 to a final pH value of 4.05 and 5.25, respectively, when the pressure is increased to 650 MPa. At 550 MPa, the corresponding pH shift values are 0.6 and 1.1, respectively. Reestablishment of the initial pH value occurs upon pressure release. Unfortunately, there are no data to estimate the effect of temperature on this pressure-induced pH shift. Furthermore, in many studies, it is desirable to use experimental buffer systems that are pressure independent. Also, a combination of buffers that exhibit positive and negative pressure-induced pH shifts can be used. But, this is not possible when investigating food systems which make data interpretation of pressure effects more challenging.

14.4   PRESSURE INACTIVATION OF MICROORGANISMS

HPP inactivates microorganisms by interrupting cellular functions responsible for reproduction and survival (Torres and Velazquez 2008). HPP can damage microbial membranes interfering with nutrient uptake and disposal of cell waste. Intracellular fluid compounds found in the cell suspending fluid after pressure treatment indicate that leaks occur while cells are held under pressure (e.g., Shimada et al. 1993). Increasing the knowledge of cell damage and repair mechanisms could lead to new HPP applications (e.g., Hauben et al. 1996, Chilton et al. 2001). For example, lysis of starter bacteria induced by HPP treatments could promote the release of intracellular proteases important in cheese ripening. Viability, morphology, lysis, and cell wall hydrolase activity measurements suggest that HPP can cause inactivation, physical damage, and lysis in *Lactobacillus lactis* (Malone et al. 2002). Knowing the behavior of bacterial membrane proteins when subjected to pressure treatments under different conditions (e.g., temperature, pH, or a_w) can lead to effective hurdle preservation technologies. For another example, electrophoretic profiles of the outer membranes of untreated *Salmonella typhimurium* reveal 3 major and 12 minor protein bands but only 2 major bands after pressure treatments (Ritz et al. 2000). One band is more pressure resistant in acidic pH media suggesting a different protein conformation at this condition. Pressure (200 or 400 MPa) resistance of exponential-phase *E. coli* NCTC 8164 cells is highest for cells grown at 10°C and decreases with growth temperature up to 45°C (Casadei et al. 2002). In contrast, pressure resistance of stationary-phase cells is the lowest in cells grown at 10°C and increases with growth temperature reaching a maximum at 30°C–37°C before decreasing at 45°C. This pressure effect can be correlated to the proportion of unsaturated fatty acids in the membrane lipids, which decreases with growth temperature in both exponential- and stationary-phase cells. In exponential-phase cells, pressure resistance increased with greater membrane fluidity, whereas in stationary-phase cells, no simple relationship between membrane fluidity and pressure resistance was observed (Casadei et al. 2002).

14.4.1 Pressure Resistance of Bacterial Spores

Although the application of 400–800 MPa inactivates the vegetative forms of pathogenic and spoilage bacteria (San Martin et al. 2002, Smelt et al. 2002, Velazquez et al. 2002, Moermann 2005, Velazquez et al. 2005a,b, Torres and Velazquez 2008), the inactivation of bacterial spores has been a major challenge as they are extremely resistant to pressure. Therefore, current HPP products in the market rely on refrigeration, reduced water activity, or low pH to prevent bacterial spore outgrowth. For example, spores of six *Bacillus* species pressurized at 980 MPa for 40 min at room temperature showed no significant inactivation (Nakayama et al. 1996); however, pressure treatments at temperatures higher than 50°C can be very effective for some bacterial spores (Paredes-Sabja et al. 2007). For example, treating *Bacillus subtilis* spores at 404 MPa and 70°C for 15 min can achieve 5 decimal reductions (DRs) at neutral pH (Stewart et al. 2000). However, subjecting spores of *Clostridium sporogenes*, considered a nontoxigenic equivalent to proteolytic *C. botulinum* and an important food spoilage bacteria, to 400 MPa at 60°C for 30 min at neutral pH yields only 1 DR (Mills et al. 1998). To date, three PATP strategies are being explored to attempt the inactivation of bacterial spores (Table 14.1). In all strategies, processing conditions are always restricted by technology restrictions (~700 MPa maximum) and cost considerations (e.g., ~3 and ~15 min maximum for commercial pasteurization and sterilization applications, respectively).

TABLE 14.1
PATP Strategies for the Inactivation of Bacterial Spores

(a) Combined Processes at Moderate Temperature (<100°C)	(b) Combined Processes at Moderate Temperature (<100°C) plus germination control	(c) Combined Processes at High Temperature (>100°C)

Processing factors
  Compression heating and
    decompression cooling
  Pressure pulsing
  Pressure-shifted pH
	Processing factors in
	  addition to those in (a)
	  Control of bacterial spore
	    germination
		Processing factors in addition to
		  those in (a)
		  Temperature sufficiently high
		    to achieve bacterial spore
		    inactivation

Conclusion
  Insufficient to achieve
    inactivation
	Conclusion
	  Germinant identification
	  Kinetics of germination
		Conclusion
		  Kinetics for sensory and
		    nutritional quality retention
		    needs to be studied in foods

Example
  Paredes-Sabja et al. (2007)
	Example
	  Akhtar et al. (2008, 2009)
		Example
		  Pérez Lamela and Torres
		    (2008b) and Ramirez et al.
		    (2009)

The first strategy ((a) in Table 14.1) explored by several researchers included moderate temperature preheating, compression heating and decompression cooling, food formulation for the lowest pH under pressure, and pressure pulsing to promote spore germination. For example, 15 min treatments at 550 and 650 MPa, preheating to 55°C and 75°C before compression heating, citric acid buffer at 4.75 and 6.5 pH before pressure-induced pH shift, were tested on spores of five isolates of *C. perfringens* type A carrying the gene that encodes the *C. perfringens* enterotoxin (*cpe*) on the chromosome (C-cpe), four isolates carrying the *cpe* gene on a plasmid (P-cpe), and two strains of *C. sporogenes* (Paredes-Sabja et al. 2007). Treatments at 650 MPa, 75°C, and pH 6.5 were moderately effective against spores of P-cpe (approximately 3.7 DR) and *C. sporogenes* (approximately 2.1 DR) but not for C-cpe spores (approximately 1.0 DR). Treatments at pH 4.75 were moderately effective against spores of P-cpe (approximately 3.2 DR) and *C. sporogenes* (approximately 2.5 DR) but not of C-cpe (approximately 1.2 DR) when combined with 550 MPa at 75°C. However, under the same conditions but at 650 MPa, high inactivation of P-cpe (approximately 5.1 DR) and *C. sporogenes* (approximately 5.8 DR) spores and moderate inactivation of C-cpe (approximately 2.8 DR) spores were observed. In most cases, these inactivation levels would be insufficient to meet the needs of commercial food processors (Paredes-Sabja et al. 2007).

Many studies confirm that it is not possible to assume that the most heat-resistant spore is also the most baroresistant. In the case of *C. perfringens* spores, C-cpe strains are ~60 times more heat resistant than P-cpe strains (Sarker et al. 2000, Raju et al. 2006) but this information cannot be used to select strains for safety evaluations of PATP-treated foods. For example, when *C. perfringens* spores were treated at 650 MPa and 75°C in pH 6.5 buffer, C-cpe strain SM101 with a $D_{100°C}$ of 90 min showed significantly higher pressure inactivation ($DR_{15 min, 650 MPa, 75°C} = 3.1$) than C-cpe strain E13 ($DR_{15 min, 650 MPa, 75°C} = 0.1$) and P-cpe F5603 ($DR_{15 min, 650 MPa, 75°C} = 2.4$) with $D_{100°C}$ of 30 and 0.6 min, respectively (Sarker et al. 2000, Raju and Sarker 2005, Paredes-Sabja et al. 2007). These and other studies show the risks of following the strategy of using the extensive knowledge of thermal resistance to select barore-sistant spores for PATP tests of food safety.

14.4.2 Bacterial Spore Germination

Advances in high-pressure treatment strategies are needed to inactivate bacterial spores more efficiently since tests at moderately high temperatures are insufficient to reach a desired inactivation level. An alternative approach under investigation is to induce spore germination to reduce the processing time needed for their inactivation (see (b) in Table 14.1). This application requires (1) identifying germinants effective on the spores of interest; (2) determining the kinetics of germination as affected by pH, temperature, germinant concentration, food matrix, etc.; and (3) investigating the molecular mechanisms involved to reach the full benefits of this strategy. Particularly important are models for germination kinetics to determine what process and product formulation conditions should be used in combination with PATP treatments (Paredes-Sabja and Torres 2008).

The optical density (OD) decrease, $S(t)$, reflecting the overall loss in spore refractivity during germination has been used to describe the germination process using several mathematical functions including the Weibull model (Rode and Foster 1962, Vary and Halvorson 1965, Collado et al. 2006, Paredes-Sabja and Torres 2008):

$$S(t) = 1 - \frac{OD_t - OD_f}{OD_i - OD_f} = 1 - e^{-\left(\frac{t}{\alpha}\right)^{\beta}} \qquad (14.2)$$

where
OD$_i$, OD$_f$, and OD$_t$ are the initial, final, and time t OD values
α and β are the Weibull model parameters

This model was used to describe the germination of *C. perfringens* type A food poisoning isolates in buffer solutions as affected by pH (5.8–8), germinant concentration (10–100 mM KCl), and spore germination temperature (T_{SG}, 30°C–50°C). Estimations of model accuracy (A_f) and bias (B_f) showed an excellent fit of the Weibull model to the experimental data (Paredes-Sabja and Torres 2008).

A predictive model of the effect of T_{SG} on germination extent can be developed by assuming a constant behavior index β (Fernandez Garcia et al. 2002, Mafart et al. 2002) but improved results are obtained when using the following expression:

$$1/\beta = a_0 + a_1 T_{SG} + a_2 T_{SG}^2 + a_3 T_{SG}^3 + \cdots + a_n T_{SG}^n \qquad (14.3)$$

where a_i ($i = 1$ to n) are model parameters obtained experimentally. The effect of T_{SG} on the rate index α can be assumed to follow an Arrhenius behavior:

$$\ln(\alpha) = \ln(\alpha_{ref}) - \frac{E_a}{R} \times \frac{1}{273.15 + T_{SG}} \tag{14.4}$$

where

α_{ref} is the scale parameter (min)
E_a is the activation energy (kJ/mol)
R is the ideal gas constant (8.31 J mol^{-1} K^{-1})

Combining Equations 14.2 through 14.4 yields the following expression to predict the germination of *C. perfringens* type A food poisoning isolates as a function of T_{SG} (Paredes-Sabja and Torres 2008):

$$S = \exp\left[-\left[\frac{t}{\alpha_{ref} \cdot \exp\left[-\frac{E_a}{R} \frac{1}{273.15 + T_{SG}} \right]} \right]^{\left[\frac{1}{a_0 + a_1 T_{SG} + a_2 T_{SG}^2 + a_3 T_{SG}^3 + \cdots + a_n T_{SG}^n} \right]} \right] \tag{14.5}$$

This model was reported to fail only once by overpredicting germination for one isolate (Paredes-Sabja and Torres 2008). This may reflect the need to obtain the experimental data needed to construct the germination prediction expression using as samples the product of interest instead of buffer solutions and other simple model systems. This is important as the bioavailability of the germinants is likely to affect the germination response. A simple method to estimate the bioavailability of nutrients, for example, the effect of KCl in meat products on the germination of *C. perfringens* spores, might incubate them in the product for 60 min, and measure their survival after a thermal treatment (70°C–80°C, 20 min). A bioavailability near the optimum, for example, ~50 to 100 mM KCl, should yield more than 1 DR in spore counts, indicating that more than 90% of the spores have germinated.

The approach of promoting bacterial spore germination by the addition of L-asparagine and KCl to develop a more effective PATP spore inactivation process for use by poultry and meat processors was investigated by Akhtar et al. (2008). The FDA authorizes the use of L-asparagine as a nutrient or dietary supplement for which a regulation of use must be issued, while KCl is an additive in the generally recognized as safe category. As stated before, the success of this novel strategy requires the bioavailability of L-asparagine and KCl in an amount necessary for rapid spore germination, that is, greater than 50 mM. The following processing strategy was shown to achieve at least 4 DR of *C. perfringens* spores: (1) primary heat treatment to pasteurize and denature meat proteins for palatability and to activate *C. perfringens* spores for germination (e.g., 80°C for 10 min); (2) product cooling in about 20 min to 55°C and incubation at this temperature for spore germination (e.g., 15 min); and (3) inactivation of germinated spores by PATP (e.g., 586 MPa at 73°C for 10 min). Finally, PATP at ~100°C or higher temperature ((c) in Table 14.1) inactivates bacterial spores in reasonably short time and is currently being explored by many researchers (e.g., Balogh et al. 2004, Vazquez Landaverde et al. 2007, Pérez Lamela and Torres 2008b). However, its application requires an examination of the potential chemical changes to the product under these extreme processing conditions and will be covered in Chapter 15, under the section on chemical reaction kinetics at high pressure and high temperature.

14.4.3 Microbial Inactivation Kinetics

Commercial processors and government agencies require models of microbial survival to predict the microbiological consequences of food processing and storage and distribution conditions with respect to spoilage and safety. These models can serve three goals, that is, understanding, prediction, and control (Haefner 2005). As for understanding, models are used when applying the scientific method, and in that sense it can contribute to our understanding of the phenomena occurring in a given food subjected to a certain processing technology (van Boekel 2008). Thermal processing is frequently cited as an example of where mathematical modeling has been used traditionally for many years to predict the outcome of different processing conditions on the survival of microorganisms and the quality of the foods obtained (Doona and Feeherry 2007). The model must respond to changes in specific food properties and to the processing conditions selected (Wedzicha et al. 1993). These properties are essential when the model is used to optimize a process or to control a process when it is necessary to generate process corrections when deviations are detected. An often ignored consideration is that in all models the statistical variability of their parameters needs to be considered as they will impact our confidence on our assessment of process or storage consequences (Halder et al. 2007, Almonacid Merino and Torres 2009). Finally, alternative models are being increasingly considered as more accurate representations of microbial growth and lethality; however, it appears that a more important consideration than developing new models is the need to include variability estimations. This is facilitated when using conventional models for which ample data have been accumulated in the literature. In the food science and engineering literature, both the activation energy (E_a) and the thermal resistance constant (z-value) are utilized to describe the influence of temperature on the microbial inactivation kinetics and the loss of food quality attributes. The calculations and mathematical relations between these parameters are extensively documented (Adams and Moss 1995, Mossel et al. 2002, Morales Blancas and Torres 2003a–c).

Microbial inactivation kinetics has been extensively studied when using high temperature to preserve food safety and eliminate spoilage organisms (McClure et al. 1994, Adams and Moss 1995, Mossel et al. 2002). Although much research has been published on microbial inactivation by pressurization treatments (e.g., Guerrero Beltran et al. 2005), much information is still needed because the capabilities of commercial pressure-processing units changed with respect to maximum pressure (~400 to 500 MPa in the 1980s to ~600 to 700 MPa to date) and maximum operating temperature (next generation expected to exceed 100°C). Early research focusing on long moderate-pressure processes, typically 5–30 min at ~400 MPa, reflected the limitation of laboratory units available at the time and not current equipment technology (e.g., Aleman et al. 1994, 1996). Current vessels operate at higher pressure allowing processing times in the 1–3 min range reducing processing costs (Torres and Velazquez 2008).

14.4.3.1 Conventional Microbial Inactivation Models

Since pressure-treated foods first appeared in the market, significant advances in food microbial modeling have been accomplished (e.g., McMeekin et al. 1993, Baranyi and Roberts 1995, Peleg and Cole 1998, Xiong et al. 1999a, Ahn et al. 2007, Corradini and Peleg 2007, Klotz et al. 2007, Koseki and Yamamoto 2007b, Black and Davidson 2008, Pérez Lamela and Torres 2008b). However, most information is still focused on thermal processing. The inactivation of microorganisms by heat became a fundamental food preservation operation during the twentieth century and will remain so in the foreseeable future (Morales Blancas and Torres 2003a–c). In the 1920s, the principles of thermobacteriology were established assuming that microbial inactivation follows a first-order reaction kinetics, that is, a logarithmic-linear behavior is observed when survivors (N_t) of an initial population N_0 are plotted as a function of time at a constant lethal temperature T (Adams and Moss 1995, Peleg and Cole 1998, Mossel et al. 2002, Morales Blancas and Torres 2003a–c). This easy-to-use model developed in the early 1900s assumes that all microorganisms of the same strain in the same medium have the same sensitivity to heat and thus the same probability of inactivation. Assuming k_T as the rate of death at the constant temperature T leads to the following mathematical expressions (Bigelow 1921, McKellar and Xuewen 2003):

$$\ln \frac{N_t}{N_0} = -k_T t \qquad (14.6)$$

or

$$\log \frac{N_t}{N_0} = -\frac{t}{D} \qquad (14.7)$$

When using Equation 14.7, the important model parameters are

1. DR time (D_T) defined as the time at a constant lethal temperature T to achieve one logarithmic reduction (90% inactivation) in the microbial population of interest (Figure 14.6a)
2. Thermal resistance constant (z) defined as the temperature increase needed to achieve one logarithmic reduction in D_T (90% reduction), that is, $D_{T+z} = 0.1\ D_T$ (Figure 14.6b)

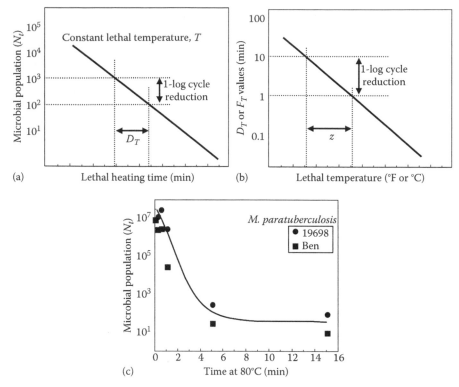

FIGURE 14.6 First-order microbial inactivation kinetics: (a) DR time D_T, (b) thermal resistance z, and (c) model deviation example. (Adapted from Ulloa-Fuentes, P.A. et al., *Industria Alimentaria (México)*, 30, 19, 2008.)

3. Number of DRs (S) in the microbial population of interest defined as $S = \log(N_0/N_f)$ where N_0 and N_f are the initial and final microbial counts in the container with the processed food
4. Sterilizing value at a reference temperature (typically, $T_0 = 250°F = 121.1°C$) defined as $F_{T = T_0} = SD_{T = T_0}$, an expression that is often simplified to read $F_0 = SD_0$

The number of DRs (S) is established by a risk assessment taking into account initial contamination levels, storage and distribution conditions, and the consumer population at risk (Smelt et al. 2002). If the microorganism is a pathogen (e.g., *C. botulinum*), the decision is made by regulatory agencies responsible for food safety. If it is a spoilage organism, the decision is made by producers comparing the market impact of spoiled foods reaching a consumer, processing costs, and the processing impact on food quality.

Many authors have reported that in thermal processes the survival curves for microorganisms deviate from the expected linear behavior (Kamau et al. 1990, Bhaduri et al. 1991, Linton et al. 1995, Anderson et al. 1996, Linton et al. 1996, Buzrul 2007). An example of significant tailing deviation is shown for the inactivation of *Mycobacterium paratuberculosis* in milk (Figure 14.6c), which may be the etiological agent for Crohn's disease in humans (Stabel et al. 1997). Very significant deviations from first-order kinetics have been observed also after non-thermal treatments such as irradiation (Juneja and Thayer 2000, Quintero-Ramos et al. 2004) and pressure treatments (Torres and Velazquez 2008). Finally, microbial inactivation models developed in buffers and other laboratory systems should be used for foods with caution (Dilek Avsaroglu et al. 2006). Researchers must continually question the use of a model and the quality of data to support any conclusions (Black and Davidson 2008).

Microbial inactivation can be described as a cumulative form of a temporal distribution of lethal events and this approach has been demonstrated for the lethal effect on pathogens and spores for preservation food technologies such as irradiation (Juneja and Thayer 2000), pulsed electric fields (Rodrigo et al. 2000), heat alone, and in combination with other factors including pH (Murphy et al. 1996, Fernandez et al. 1997, Janssen et al. 2007) and high pressure (Peleg 2006). Models proposed to describe this nonlinear behavior include the Buchanan (Xiong et al. 1999b), log-logistic (Peleg et al. 2002), Gompertz (Bhaduri et al. 1991, Patterson and Kilpatrick 1998), Weibull (Corradini and Peleg 2007) models. Experimental data obtained for microorganisms subjected to pressure treatments have been analyzed using also linear (Chen 2007), Weibull (Dilek Avsaroglu et al. 2006, Koseki and Yamamoto 2007a,b), biphasic (Panagou et al. 2007), fuzzy (Ganzle et al. 2007) and other mathematical models. Some authors have described modifications of the linear model considering multicomponent kinetics and extended microbial inactivation models to enzyme inactivation (Fujikawa and Itoh 1996).

14.4.3.2 Weibull Model

This model considers that the microbial death probability depends on the biological variation or heterogeneity within the population of a microorganism. This assumption leads to the following expression (Buzrul and Alpas 2004):

$$\log \frac{N_t}{N_0} = \log S = -bt^n \tag{14.8}$$

where b and n are the scale and shape factors, respectively, which are temperature and pressure dependent. The coefficient b can be considered as a "rate parameter" while the exponent n is a measure of the shape of the isothermal and of the isobaric semilogarithmic survival curve (Doona et al. 2007). This model enables to simulate a variety of survived-injury patterns in thermal and nonthermal food preservation processes (Corradini and Peleg 2007). While remaining simple, this alternative model is sufficiently robust. It can describe downward concave survival curves ($n > 1$) when damage accumulating in the survivors sensitizes them to further injury or inactivation. Upward concave curves ($n < 1$) suggest that the sensitive members of the population die quickly, leaving progressively more resistant survivors and previously described as "tailing." Finally, the model includes the survival curve for the conventional first-order model ($n = 1$) (Mafart et al. 2002, Buzrul et al. 2005). Several authors have reported that the Weibull model performs much better than the classical linear inactivation model (Chen and Hoover 2003, Chen 2007, Pina Perez et al. 2007, van Boekel 2008).

14.4.3.3 Log-Logistic Model

The log-logistic model is described by the following mathematical expression (Cole et al. 1993, Pina Perez et al. 2007):

$$\log N(t) = \alpha + \frac{\omega - \alpha}{1 + e^{\left[\frac{4\sigma[z - \log(t)]}{(\omega - \alpha)}\right]}} \tag{14.9}$$

where
α is the upper asymptote (log CFU/mL)
ω is the lower asymptote (log CFU/mL)
σ is the maximum rate of inactivation (log CFU/mL)/log min)
z is the log time to the maximum rate of inactivation (log min)

This model is adequate to describe thermal inactivation (Little et al. 1994) but few applications to pressure processing have been reported (Guan et al. 2005).

14.4.3.4 Modified Gompertz Equation

This equation form was originally proposed for microbial growth curves (Gibson et al. 1988) but has been modified for inactivation modeling as follows (Pina Perez et al. 2007):

$$\log N(t) = \log N_0 + C \left(e^{-e^{BM}} - e^{-e^{B(t-M)}} \right) \tag{14.10}$$

where
C is the difference in value of the upper and lower asymptotes
B is the relative death rate at its maximum M

14.4.3.5 Baranyi Equation

This equation form was also originally proposed for microbial growth curves (Baranyi et al. 1993) but can be used in its modified form to describe microbial inactivation using the following expressions (Pina Perez et al. 2007):

$$\log N(t) = \log N_0 + \log \left\{ \frac{N_{min}}{N_0} + \left(1 - \frac{N_{min}}{N_0} \right) e^{-k_{max}(t - B_t)} \right\} \tag{14.11}$$

$$B_t = \frac{r}{3} \left\{ 0.5 \ln \frac{(r+t)^2}{r^2 - rt - t^2} \right\} + \sqrt{3} \left\{ \arctan \frac{2t - r}{r\sqrt{3}} + \arctan \frac{1}{\sqrt{3}} \right\} \tag{14.12}$$

where
N_{min} is the microbial population remaining in the tail phase so that if $N_{min} = 0$, there is no tailing
k_{max} is the maximum relative death rate
r is the model parameter

In conclusion, several models are available to model microbial inactivation but the general limitation is the lack of experimental values for all the model parameters. Also missing are comprehensive comparisons of the different models.

14.5   PRESSURE INACTIVATION OF ENZYMES

Pressure processing can inactivate food enzymes to a varying extent, reducing in many cases the degradation of quality properties and nutritional value. However, effects of HPP on enzymes vary extensively with the processing pressure, temperature, time, pH, moisture content, and food matrix. Relatively low pressure (~100 to 200 MPa) may activate some enzymes while high pressure (400–1000 MPa) may induce their inactivation. The largest contribution of pressure to enzyme inactivation comes from structural rearrangements of proteins under high pressure (Hendrickx et al. 1998) such as hydration changes that accompany other intramolecular noncovalent interactions (Mozhaev et al. 1996a,b). Pressure-treated proteins retain their primary structure because covalent bonds are unaffected by pressure.

14.5.1   Inactivation of Specific Food Enzymes

Enzymes found in fruits, vegetables, milk, fish, and meat products include polyphenol oxidase (PPO), lipoxygenase (LOX), pectinmethylesterase (PME), peroxidase (POD), lipases, and proteases. A recent review on the pressure-processing effects on enzymes present in fruits and vegetables can be found in Ludikhuyze et al. (2002). The thermal resistance of enzymes cannot be used to predict their pressure resistance (Hendrickx et al. 1998). Enzymes such as PPO and LOX are inactivated at 300 MPa, while others such as PME and POD are very difficult to inactivate within the pressure range of the commercial units available today.

14.5.1.1   Polyphenol Oxidase

PPO is considered a moderately heat-stable enzyme (Yemenicioglu et al. 1997). Its inactivation is highly desirable as it causes enzymatic browning of fruits and vegetables. In fruits such as apples, apricots, pears, plums, strawberry, bananas, peppers, and grapes and in vegetables such as mushrooms, onions, avocados, and potatoes, pressures necessary to induce inactivation of PPO vary from 200 to 1000 MPa, depending on food composition, pH, and the use of additives (Weemaes et al. 1997a,b, 1998, 1999). PPO is pressure sensitive in apples, strawberries, apricots, and grapes and pressure resistant in other fruits such as pears and plums. A low pressure may protect PPO from thermal inactivation and even enhance its activity in apple (Jolibert et al. 1994, Anese et al. 1995), pear (Asaka et al. 1994), and onion (Butz et al. 1994). However, apricot and strawberry PPO are inactivated by pressures exceeding 100 and 400 MPa, respectively (Jolibert et al. 1994, Amati et al. 1996). Depending on pH, pressures from 100 to 700 MPa are required for the inactivation of apple PPO (Anese et al. 1995). At a lower pH, pressure induced a faster inactivation of PPO (Jolibert et al. 1994, Weemaes et al. 1997a,b). Besides pH, pressure inactivation is also influenced by the addition of salts, sugars, and compounds such as benzoic acid, glutathione, or $CaCl_2$. For example, pressure inactivation of apple PPO and mushroom PPO is enhanced by the addition of $CaCl_2$ (Jolibert et al. 1994) and the addition of benzoic acid or glutathione, respectively (Weemaes et al. 1997a,b). PPO in white grape must is only partly inactivated at 300–600 MPa for 2–10 min (Castellari et al. 1997) with a clearly lower enzymatic activity only at 900 MPa. Since this pressure level exceeds the capacity of commercial units, complete inactivation can be achieved by pressure treatments combined with a mild thermal treatment (~40°C to 50°C). Studies on the residual PPO activity in fruit purees suggest that inhibition is possible by pressure treatments combined with pH reduction, blanching, or a low refrigeration storage temperature (Ludikhuyze et al. 2000, Palou et al. 2000). For example, the PPO activity in banana puree adjusted to pH 3.4 and water activity of 0.97 was reduced after steam blanching and then further reduced by a pressure treatment (Palou et al. 1999). However, extrapolating from findings in one matrix to another is not possible. For example, studies on the combination of thermal blanching and 100–200 MPa for 10–20 min showed that the PPO in red pepper fruits is more stable to pressure and temperature than the PPO in green peppers (Castro et al. 2008). Mushroom, potato, and avocado PPO are pressure resistant since treatments at 800–900 MPa are necessary to reduce enzyme activity at room temperature (Gomes and Ledward 1996, Weemaes et al. 1997a,b). However, the effect of pressure on the PPO activity in mushrooms and potatoes is different (Gomes and Ledward 1996). In mushrooms, considerable browning is observed immediately after pressurization to 200 MPa, while in potatoes limited browning is observed at 600 MPa and even at 800 MPa. In another study, treatments at 600 MPa and 35°C (including adiabatic compression heat) for 10 min decreased PPO activity in mushroom by 7% while extending the treatment to 20 min resulted in no further PPO activity reduction (Sun et al. 2002). However, PPO activity decreased to 28% and 43% after 800 MPa treatments for 10 and 20 min, respectively. Furthermore, PPO isolated from apple, grape, avocado, and pear is inactivated slowly at 600 MPa but is completely inactivated at 900 MPa (Weemaes et al. 1998).

14.5.1.2 Lipoxigenase

At atmospheric pressure, the thermal stability of this enzyme catalyzing the oxygenation of fatty acids and inducing changes in flavor, color, and nutritional value varies significantly with the enzyme source and medium. Inhibition of LOX activity by 400–600 MPa pressure treatments has been investigated in tomato, soybean, green bean, and pea (Heinisch et al. 1995, Ludikhuyze et al. 1998, Indrawati et al. 1999, Tangwonchai et al. 1999). Inactivation kinetics was obtained also in soybean, green bean, and pea in the 0.1 to 650 MPa and −10°C to 80°C range. For green beans, inactivation was faster in the intact vegetable than in a crude extract (Indrawati et al. 2000). However, for soybean, higher LOX pressure stability was observed in milk than in a buffer solution (Seyderhelm et al. 1996). In the case of LOX activity in soybean and green bean, an antagonistic effect between temperature below 30°C and pressure above 500 MPa was observed (Ludikhuyze et al. 1998, Indrawati et al. 1999). In addition, pressure resistance of the enzyme increased with increasing enzyme concentration and decreased with pH (Hendrickx et al. 1998). LOX is protected by increasing the concentration of soluble solids (Ogawa et al. 1990).

14.5.1.3 Pectinmethylesterase

PME is responsible for cloud destabilization in juices and the loss of consistency in many foods. This enzyme, one of the most abundant pectinases in plants, has high thermoresistance. Temperatures between 80°C and 95°C are required to induce significant inactivation and even then PME may remain active. This resistance reflects the presence of heat-labile and heat-stable PME isoforms (Wicker and Temelli 1988, van den Broeck et al. 2000). The pressure required for the inactivation at room temperature of PME from different sources has been reported to vary significantly from ~150 to 1200 MPa. Most studies report only partial inactivation of PME and this reflects the presence of pressure-resistant isoforms. Pressure inactivation studies of PME in orange, grapefruit, guava, and tomato in the 0.1–800 MPa and 15°C–65°C range revealed a slight antagonistic effect of low pressure and high temperature (van den Broeck et al. 2000). However, pressure-treated orange juice resulted in a stable and higher quality product as ~600 MPa reduces PME activity irreversibly by 90% (Ogawa et al. 1990, Irwe and Olson 1994). Lower pressure treatments at room temperature (200 to 400 MPa) induced PME activation in freshly squeezed orange juice (Cano et al. 1997). Compared to orange PME, tomato PME is more pressure resistant. PME inactivation in tomato occurs at smaller rates than in orange at high pressure (Weemaes et al. 1999, van den Broeck et al. 2000). Tomato PME is less pressure stable in the presence of Ca^{2+} ions or in citric acid buffer (pH 3.5–4.5) than in water. The antagonistic effect, that is, lower inactivation at elevated pressure than at atmospheric pressure, was less pronounced in citric acid buffer (pH 3.8–4.5) and in the presence of $CaCl_2$ than in water. In the absence or presence of Ca^{2+} ions, the optimal pressure for enzyme activation was 100 and 400 MPa, respectively (van den Broeck et al. 1999a,b, 2000). A study of PME of carrot extract (purified form), carrot juice, and carrot pieces showed that PME in carrot pieces is much more resistant to thermal and pressure treatments than in the carrot juice or the extract. This could be attributed to the presence of a stabilizing factor in the carrot matrix including the fact that PME in carrot pieces is bound to the cell wall (Balogh et al. 2004). PME in carrots is inactivated by temperatures above 50°C but resist pressures up to 600 MPa (Balogh et al. 2004, Sila et al. 2007). For example, to reduce PME activity by 90% it is necessary to treat carrot juice at 800 MPa and 10°C for 36 min.

14.5.1.4 Peroxidase

POD induces unfavorable flavors during storage and is considered to be the most heat stable vegetable enzyme. In some cases, it is also extremely pressure resistant. For example, in green beans at room temperature, a pressure of 900 MPa induced only a slight inactivation, while a significant inactivation was achieved at 600 MPa at elevated temperature (Quaglia et al. 1996). However, POD in strawberry puree and orange juice at 20°C can be inactivated in 15 min at 300 and 400 MPa, respectively (Cano et al. 1997). When orange POD was treated at room temperature for 15 min and up to 400 MPa, pressure decreased activity continuously with the highest inactivation rate (50%) observed at 32°C. However, in orange juice at 32°C–60°C, pressure increased POD activity (Cano et al. 1997).

14.5.1.5 Lipase and Protease

This enzyme is responsible for the hydrolysis of animal and vegetable fats and oils. In pressure-treated foods, large differences in lipase activity have been reported (Macheboeuf and Basset 1934, Weemaes et al. 1998). In some cases, lipase has been reported to be stable at 1100 MPa (Macheboeuf and Basset 1934), while in others inactivation has been achieved at 600 MPa (Seyderhelm et al. 1996). The effect of HPP on the proteolytic enzymatic activity in milk, meat, and fish products has been reported (Lakshmanan et al. 2005, Quiros et al. 2007). Inactivation of proteolytic enzymes at pressures over 800 MPa suggests that HPP may have a beneficial effect on the sensory properties of cheese (Reps et al. 1998). Also, the enzyme activity in crude enzyme extracts prepared from cold-smoked salmon muscles was reduced by high pressure. Proteolytic enzyme activity in cold-smoked salmon muscles was decreased by 20 min treatments at 9°C and pressures equal or higher than 300 MPa. Also, proteolytic activity was substantially reduced by pressure treatments of protease extract solutions (Lakshmanan et al. 2005).

14.5.2 ENZYME INACTIVATION KINETICS

Information on enzyme inactivation kinetics is essential to model, design, and optimize preservation processes. Ludikhuyze et al. (2002) determined the pressure and temperature required for 90% inactivation of key enzymes, six DRs in microbial counts, and 90% loss in total chlorophyll content for 15 min treatments. The region defined by the reduction in spoilage organisms and the loss of quality expressed by chlorophyll loss corresponds to pressure and temperature combinations that must achieve enzyme inactivation if the process is to be feasible (Figure 14.7a). In the case of several enzymes, pressure and temperature tend to have synergistic effects at high but not at low pressure (Figure 14.7b). Furthermore, enzyme inactivation depends on the enzyme source (green beans, peas, and pears) and test matrix (in situ or in juice) as shown for the case of LOX (Figure 14.7c). This means that substantial more research is required to determine for each enzyme source and specific physicochemical environment effects on its inactivation rate.

Recently, kinetic studies reported have been used to develop inactivation models for some food spoilage enzymes (van Loey et al. 1998, van den Broeck et al. 1999a,b, Indrawati et al. 2000, Reyns et al. 2000, Balogh et al. 2004). Some of the most used kinetic models for pressure–temperature inactivation of enzymes are described in detail as follows. For example, pressure inactivation of soybean LOX and tomato PME was accurately described by a first-order kinetic model (Indrawati et al. 2000, Crelier et al. 2001):

$$A = A_0 \exp(-kt) \tag{14.13}$$

where
A is the enzyme activity at time t
A_0 is the initial enzyme activity
k is the inactivation rate constant

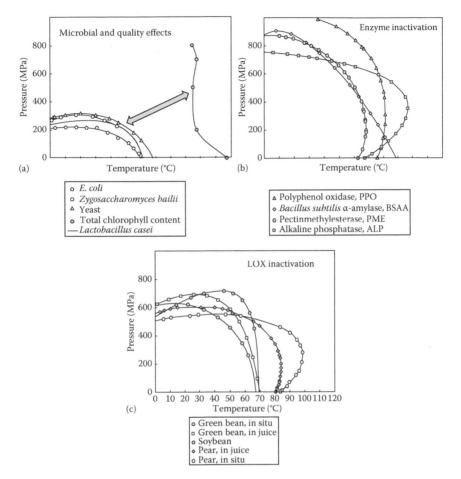

FIGURE 14.7 Pressure and temperature conditions required for 15 min treatments achieving desired levels of microbial and enzyme inactivation: (a) conditions for 90% total chlorophyll loss and six DRs in microbial load, (b) conditions for 90% inactivation of multiple enzymes, and (c) conditions for 90% inactivation of LOX, in situ and in juice. (Adapted from Ludikhuyze, L. et al., *Fruits and Vegetable Processing: Improving Quality*, Ed., Jongen, W., CRC Press, Boca Raton, FL, 2002.)

A special case of first-order kinetic inactivation model is the fractional conversion model, which is applied when a nonzero residual activity after prolonged treatment time is observed (van den Broeck et al. 1999a,b):

$$A = A_{\infty} + (A_0 - A_{\infty})\exp(-kt) \tag{14.14}$$

where A_{∞} is the residual activity after prolonged treatment time. The two parameters, k and A_{∞}, can be estimated by nonlinear regression. The dependence on pressure p of the inactivation rate constant k at an absolute temperature T can be calculated as follows (Eyring et al. 1946):

$$k = k_{ref} \exp\left[\frac{-V_a}{RT}(p - p_{ref})\right] \qquad (14.15)$$

where

k_{ref} is the inactivation rate constant at a reference pressure (p_{ref})
V_a is the activation volume
R (= 8.31 J/mol K) is the universal gas constant

The parameter V_a is estimated by linear regression. The enzyme inactivation rate k at a high pressure p (above the value p_{ref}) decreases or increases above the value k_{ref} if V_a is positive or negative, respectively. Finally, the temperature dependence at constant pressure of the inactivation rate constant can be estimated using the Arrhenius model (Arrhenius 1889):

$$k = k_{ref} \exp\left[\frac{-E_a}{R}\left(\frac{1}{T} - \frac{1}{T_{ref}}\right)\right] \qquad (14.16)$$

where

E_a is the activation energy
k_{ref} is the inactivation rate constant at the reference temperature T_{ref}
E_a is the estimated by linear regression
R is the universal gas constant

Inactivation of enzymes can be described by a first-order kinetic model as in the case of purified carrot PME solutions, PME in carrot juice, and PME in carrot pieces (Balogh et al. 2004). However, in food processing calculations, first-order reactions are also commonly expressed in terms of DR time (D), thermal (z_T), and pressure resistance constant (z_p) values:

$$\log \frac{A_t}{A_0} = -\frac{t}{D} \qquad (14.17)$$

with the following expressions used to estimate pressure and temperature effects on D values:

$$z_T = -\frac{T - T_{ref}}{\log D - \log D_{ref}} \qquad (14.18)$$

$$z_p = -\frac{p - p_{ref}}{\log D - \log D_{ref}} \qquad (14.19)$$

Table 14.2 shows the enzyme activity of purified carrot, apple, and strawberry PME during thermal and high-pressure treatments, showing an increase of PME activity

TABLE 14.2

Effect of Pressure and Temperature on PME Activity

P (MPa)	Apple		Strawberry		Carrot		E_a^2 (kJ/mol)
	40°C	50°C	40°C	50°C	40°C	50°C	
200	3.5 ± 0.1	3.4 ± 0.2	3.7 ± 0.1	4.2 ± 0.1	1.6 ± 0.1	2.6 ± 0.1	31.6 ± 0.9
400	2.4 ± 0.1	3.8 ± 0.1	2.6 ± 0.1	3.2 ± 0.2	3.5 ± 0.3	5.9 ± 0.7	47.2 ± 6.0
500	2.2 ± 0.1	2.8 ± 0.1	2.1 ± 0.1	2.6 ± 0.1	4.7 ± 0.6	6.0 ± 0.3	81.4 ± 6.5
ΔV (cm³/mol)					−7.8 ± 0.7	−6.9 ± 1.4	

Sources: Adapted from Fraeye, I. et al., *Innovative Food Science and Emerging Technologies*, 8, 93, 2007; Sila, D.N. et al., *J. Food Eng.*, 755, 2007.

TABLE 14.3

Pressure Inactivation of Carrot PME Activity

P (MPa)	$D_{10°C}$ (min)		$D_{40°C}$ (min)
	Extract (pH 6)	Juice	Pieces
700	188.0 ± 9.6	171.6 ± 4.0	391.5 ± 28.1
750	55.7 ± 1.9	77.9 ± 3.6	161.4 ± 15.1
800	30.7 ± 0.5	36.2 ± 1.9	75.8 ± 3.1
z_p (MPa)	127.1 ± 25.2	147.9 ± 1.2	140.3 ± 6.4

Source: Adapted from Balogh, T. et al., *Innov. Food Sci. Emerg. Technol.*, 5, 429, 2004.

with temperature. However, incrementing pressure caused a PME activity decrease in apple and strawberry but the opposite behavior in carrot. Table 14.3 shows the kinetic parameters for the inactivation of carrot PME in a purified pH 6 extract, and in carrot juice and carrot pieces subjected to temperature and high-pressure treatments (Balogh et al. 2004). At 10°C, PME inactivation is slightly more pressure dependent in the extract (z_p = 127 MPa) than in the juice (z_p = 148 MPa). Also, PME in purified carrot and in carrot juice at 700 MPa and 10°C inactivates at a similar rate as in carrot pieces at 750 MPa and 40°C. Balogh et al. (2004) reported that at 10°C and 25°C, D values increased with pH and decreased with pressure.

Other inactivation kinetics models for PATP have been proposed (Ludikhuyze et al. 1998, Denys et al. 2000a,b, van den Broeck et al. 2000, Sampedro et al. 2008). For example, soybean LOX also follows first-order kinetics but five additional constants are required to improve the original equation (Denys et al. 2000a,b):

$$\ln k = \left(a_2 T^2 + b_2 T + c_2\right) - \left[\left(\frac{a_1 T e^{-b_1 T}}{R(T + 273.15)}\right)\left(p - p_{\text{ref}}\right)\right] \quad (14.20)$$

where

k (min^{-1}) is the inactivation rate constant at pressure p (MPa) and temperature T (°C)

p_{ref} are the reference pressure (500 MPa)

R is the universal gas constant (8.31 J mol^{-1} K^{-1})

a_i, b_i, and c_i are constants with the following values (Ludikhuyze et al. 1998).

$a_1 = -15.6 \pm 1.4\,\text{cm}^3\,\text{mol}^{-1}$

$b_1 = 7.1 \pm 0.28 \times 10^{-2}\text{°C}^{-1}$

$a_2 = 2.66 \pm 0.27 \times 10^{-3}\text{°C}^{-2}$

$b_2 = -1.39 \pm 0.18 \times 10^{-1}\text{°C}^{-1}$

$c_2 = -3.12 \pm 0.28$

Also proposed in the literature is a biphasic model that considers a stable and a labile fraction following each enzyme first-order kinetics inactivation (van den Broeck et al. 2000, Sampedro et al. 2008) as in the following example for the inactivation kinetics of PME in an orange juice–milk beverage:

$$A = A_L \exp(-k_L t) + A_S \exp(-k_S t) \tag{14.21}$$

where

A_L and A_S are the activities of the labile and stable fraction, respectively

k_L and k_S are the inactivation rate constants of the labile and stable fraction, respectively

These parameters are estimated by nonlinear regression. The effect of temperature and pressure on these rate constants is analyzed using the Arrhenius (1889) and Eyring et al. (1946) equations, respectively. Table 14.4 shows the biphasic model kinetic parameters for the PME inactivation in the orange juice–milk beverage when subjected to PATP treatments. This model shows that at 65°C–73°C, only the labile fraction is inactivated. Estimations of ΔV values at 25°C–55°C for the labile and stable fractions show that the labile fraction is more pressure dependent than the stable

TABLE 14.4

Biphasic Kinetic Parameters for PME Inactivation in Orange Juice–Milk Beverage

P (MPa)	25°C		65°C	
	k_L (min^{-1})	k_S (min^{-1})	k_L (min^{-1})	k_S (min^{-1})
500	—	—	0.35 ± 0.07	—
600	0.29 ± 0.03	0.014 ± 0.001	0.78 ± 0.05	0.031 ± 0.002
700	0.71 ± 0.08	0.023 ± 0.001	—	—
	ΔV_L (cm^3/mol)	ΔV_S (cm^3/mol)	ΔV_L (cm^3/mol)	ΔV_S (cm^3/mol)
	-66.6 ± 4.8	-50.0 ± 4.5	-44.2 ± 3.9	-52.8 ± 4.2

Source: Sampedro, F. et al., *J. Food Eng.*, 86, 133, 2008.

fraction while the opposite behavior is observed at 65°C (Table 14.3). In addition, the inactivation parameters using the biphasic and fractional models were similar (Sampedro et al. 2008).

14.6 FINAL REMARKS

Kinetic information on microbial and enzyme inactivation is essential to ensure the continuing commercial success of preservation processes by HPP. This information is also necessary to optimize the new PATP applications under development in many research laboratories. The effect of processing conditions when implementing these new technologies on the chemical composition of foods is also necessary to ensure the retention of nutritional and sensory properties demanded by consumers. This topic and the control of process temperature in large-scale pressure vessels are covered in Chapter 15.

Conventional mathematical models do not describe the experimental data well for enzyme and microbial inactivation, and in many cases, complete information on the kinetic constant describing the effects of new processing alternatives on foods has not been generated yet. Although there are significant data on several food enzymes, the literature lacks data on key enzymes such as LOX, POD, lipases, and proteases. Particularly needed are more consistent kinetic data to determine activation volume and energy values. As in microbial inactivation, knowledge of the heat resistance for enzymes cannot be extrapolated to HPP and PATP. The enzyme inactivation rate is also strongly dependent on the enzyme source and the food matrix and there are both synergistic and antagonistic effects of pressure. Finally, the design of processes ensuring the inactivation of microbial pathogens, particularly bacterial spore formers, requires advances in mathematical modeling to reduce the need for overprocessing as a means to compensate for the lack of confidence in predicted inactivation values. The past reliance on a single model in thermal processing calculations, first-order microbial inactivation kinetics, was useful as it generated a large database of D and z values. This chapter has shown several new mathematical approaches for the microbial and enzyme inactivation effect of HPP and PATP, but information on the constants required for their use with real food systems is extremely scarce. However, the quality advantages of these new processing technologies are driving the efforts to fulfill this knowledge gap.

REFERENCES

Adams, M. R. and Moss, M. O. 1995. The microbiology of food preservation. In *Food Microbiology*, 55–75 London, U.K.: The Royal Society of Chemistry.

Ahn, J., Balasubramaniam, V. M., and Yousef, A. E. 2007. Inactivation kinetics of selected heat resistant aerobic and anaerobic bacterial surrogate spores by pressure-assisted thermal processing. *International Journal of Food Microbiology* 113: 321–329.

Akhtar, S., Paredes-Sabja, D., and Sarker, M. R. 2008. Inhibitory effects of polyphosphates on *Clostridium perfringens* growth, sporulation and spore outgrowth. *Food Microbiology* 25:802–808.

Akhtar, S., Paredes-Sabja, D., Torres, J. A., and Sarker, M. R. 2009. Strategy to inactivate *Clostridium perfringens* spores in meat products. *Food Microbiology* 26: 272–277.

Aleman, G., Farkas, D. F., Torres, J. A., Wilhelmsen, E., and McIntyre, S. 1994. Ultra-high pressure pasteurization of fresh cut pineapple. *Journal of Food Protection* 57: 931–934.

Aleman, G. D., Ting, E. Y., Mordre, S. C., Hawes, A. C. O., and Torres, J. A. Fresh cut pineapple processing by ultra-high pressure; 1996.

Almonacid Merino, S. F. and Torres, J. A. 2009. Uncertainty of microbial shelf-life estimations for refrigerated foods due to the experimental variability of the model parameters. *Journal of Food Process Engineering* 32 (In Press).

Amati, A., Castellari, M., Matricardi, L., Arfelli, G., and Carpi, G. 1996. Modificazione indotte in mosti d'uva da trattamenti con alte pressioni idrostatische (Effects of high pressure on grape musts composition). *Industrie delle Bevande* 25: 324–328.

Anderson, W. A., McClure, P. J., Baird Parker, A. C., and Cole, M. B. 1996. The application of a log-logistic model to describe the thermal inactivation of *Clostridium botulinum* 213B at temperatures below 121.1°C. *Journal of Applied Bacteriology* 80: 283–290.

Anese, M., Nicoli, M. C., Dall'Aglio, G., and Lerici, C. R. 1995. Effect of high pressure treatments on peroxidase and polyphenoloxidase activities. *Journal of Food Biochemistry* 18: 285–293.

Anonymous. 2006a. Milk pasteurization. In *Wikipedia, the Free Encyclopedia*. www.en.wikipedia. org/wiki/Raw_milk

Anonymous. 2006b. *Pasteurization: Definition and Methods*. International Dairy Foods Association (IDFA). www.idfa.org/facts/milk/pasteur.cfm

Anonymous. 2009a. FDA accepts novel food sterilization process. The Weekly, from the publishers of Food Technology. members.ift.org/IFT/Pubs/Newsletters/weekly/nl_030409.htm

Anonymous. 2009b. The PATS process paves the way for advanced processing of next-generation shelf-stable foods, says national research consortium. www.nafwa.org/blog/

Ardia, A., Knorr, D., and Heinz, V. 2004. Adiabatic heat modeling for pressure build up during high-pressure treatment in liquid-food processing. *Food and Bioproducts Processing* 82(C1): 89–95.

Arrhenius, S. 1889. Uber die reaktionsgeschwindigkeit bei der inversion von rohrzucker durch sauren. *Zeitschrift fur Physikalische Chemie* 4: 226–248.

Asaka, M., Aoyama, Y., Ritsuko, N., and Hayashi, R. 1994. Purification of a latent form of polyphenoloxidase from La France pear fruit and its pressure-activation. *Bioscience, Biotechnology, and Biochemistry* 58: 1486–1489.

Balogh, T., Smout, C., Ly Nguyen, B., van Loey, A., and Hendrickx, M. E. 2004. Thermal and high pressure inactivation kinetics of carrot pectinmethylesterase (PME): From model systems to real foods. *Innovative Food Science and Emerging Technologies* 5: 429–436.

Baranyi, J. and Roberts, T. A. 1995. Mathematics of predictive food microbiology. *International Journal of Food Microbiology* 26: 199–218.

Baranyi, J., McClure, P. J., Sutherland, J. P., and Roberts, T. A. 1993. Modelling bacterial growth responses. *Journal of Industrial Microbiology* 12: 190–194.

Berlin, D. L., Herson, D. S., Hicks, D. T., and Hoover, D. G. 1999. Response of pathogenic Vibrio species to high hydrostatic pressure. *Applied Environmental Microbiology* 65: 2776–2780.

Bhaduri, S., Smith, P. W., Palumbo, S. A. et al. 1991. Thermal destruction of *Listeria monocytogenes* in liver sausage slurry. *Food Microbiology* 8: 75–78.

Bigelow, W. D. 1921. The logarithmic nature of thermal death time curves. *Journal of Infectious Diseases* 29: 528–536.

Black, G. and Davidson, M. 2008. Use of modelling to enhance the microbiological safety of the food system. *Comprehensive Reviews of Food Science and Food Safety* 7: 159–167.

Butz, P., Koller, D., and Tauscher, B. 1994. Ultra-high pressure processing of onions: Chemical and sensory changes. *Lebensmittel Wissenschaft und Technologie* 27: 463–467.

Buzrul, S. 2007. A suitable model of microbial survival curves for beer pasteurization. *Lebensmittel Wissenschaft und Technologie* 40: 1330–1336.

Buzrul, S. and Alpas, S. 2004. Modeling the synergistic effect of high pressure and heat on inactivation kinetics of *Listeria innocua*: A preliminary study. *FEMS Microbiology Letters* 238: 29–36.

Buzrul, S., Alpas, H., and Bozoglu, F. 2005. Use of Weibull frequency distribution model to describe the inactivation of *Alicyclobacillus acidoterrestris* by high pressure at different temperatures. *Food Research International* 38: 151–157.

Cano, M. P., Hernandez, A., and de Ancos, B. 1997. High pressure and temperature effects on enzyme inactivation in strawberry and orange products. *Journal of Food Science* 62: 85–88.

Casadei, M. A., Manas, P., Niven, G., Needs, E., and Mackey, B. M. 2002. Role of membrane fluidity in pressure resistance of *Escherichia coli* NCTC 8164. *Applied and Environmental Microbiology* 68: 5965–5972.

Castellari, M., Matricardi, L., Arfelli, G., Rovereb, P., and Amati, A. 1997. Effects of high pressure processing on polyphenoloxidase enzyme activity of grape musts. *Food Chemistry* 60: 647–649.

Castro, S. M., Saraiva, J. A., Lopes-Da-Silva, J. A., Delgadillo, I., van Loey, A., Smout, C., Hendrickx, M. E. 2008. Effect of thermal blanching and of high pressure treatments on sweet green and red bell pepper fruits (*Capsicum annuum* L.). *Food Chemistry* 107: 1436–1449.

Cheftel, J. C. and Culioli, J. 1997. Effect of high pressure on meat: A review. *Meat Science* 46: 211–236.

Chen, H. 2007. Use of linear, Weibull, and log-logistic functions to model pressure inactivation of seven foodborne pathogens in milk. *Food Microbiology* 24: 197–204.

Chen, H. and Hoover, D. G. 2003. Pressure inactivation kinetics of *Yersinia enterocolitica* ATCC 35669. *International Journal of Food Microbiology* 87: 161–171.

Chilton, P., Isaacs, N. S., Manas, P., and Mackey, B. M. 2001. Biosynthetic requirements for the repair of membrane damage in pressure-treated *Escherichia coli*. *International Journal of Food Microbiology* 71: 101–104.

Colahan-Sederstrom, P. M. and Peterson, D. G. 2005. Inhibition of key aroma compound generated during ultra-high temperature processing of bovine milk via epicatechin addition. *Journal of Agricultural and Food Chemistry* 53: 398–402.

Cole, M. B., Davies, K. W., Munro, G., Holyoak, C. D., and Kilsby, D. C. 1993. A vitalistic model to describe thermal inactivation of *L. monocytogenes*. *Journal of Industrial Microbiology* 12: 232–235.

Collado, J., Fernandez, A., Rodrigo, M., and Martinez, A. 2006. Modelling the effect of a heat shock and germinant concentration on spore germination of a wild strain of *Bacillus cereus*. *International Journal of Food Microbiology* 106: 85–89.

Corradini, M. G. and Peleg, M. 2007. A Weibullian model for microbial injury and mortality. *International Journal of Food Microbiology* 119: 319–328.

Crelier, S., Robert, M., Claude, J., and Juillerat, M. 2001. Tomato (*Lycopersicon esculentum*) pectin methylesterase and polygalacturonase behaviours regarding heat and pressure induced inactivation. *Journal of Agricultural and Food Chemistry* 49: 5566–5575.

Denys, S. 2000. Process calculations for design of batch high hydrostatic pressure processes without and with phase transitions [Ph. D.]. Leuven, Belgium: Katholieke Universiteit.

Denys, S., Ludikhuyze, L. R., van Loey, A. M., and Hendrickx, M. E. 2000a. Modeling conductive heat transfer and process uniformity during batch high-pressure processing of foods. *Biotechnology Progress* 16: 92–101.

Denys, S., van Loey, A. M., and Hendrickx, M. E. 2000b. A modeling approach for evaluating process uniformity during batch high hydrostatic pressure processing: combination of a numerical heat transfer model and enzyme inactivation kinetics. *Innovative Food Science and Emerging Technologies* 1: 5–19.

Dilek Avsaroglu, M., Buzrul, S., Alpas, H., Akcelik, M., and Bozoglu, F. 2006. Use of the Weibull model for lactococcal bacteriophage inactivation by high hydrostatic pressure. *International Journal of Food Microbiology* 108: 78–83.

Doona, C. J. and Feeherry, C. P. 2007. *High Pressure Processing of Foods*. Malden, MA: Blackwell Publishing.

Doona, C. J., Feeherry, F. E., Ross, E. W., Corradini, M., and Peleg, M. 2007. The quasi-chemical and Weibull distribution models of nonlinear inactivation kinetics of *Escherichia coli* ATCC 11229 by high pressure processing. In *High Pressure Processing of Foods*, Eds. C. J. Doona and F. E. Feeherry, 115–144. Ames, Iowa, USA: Blackwell Publishing.

El'Yanov, B. S. and Hamann, S. D. 1975. Some quantitative relationships for ionization reactions at high pressure. *Australian Journal of Chemistry* 28: 945–954.

Eyring, H., Johnson, F. H., and Gensler, R. L. 1946. Pressure and reactivity of proteins, with particular interest to invertase. *Journal of Physical Chemistry* 50: 453–464.

Fernandez, P. S., George, S. M., Sills, C. C., and Peck, M. W. 1997. Predictive model of the effect of CO_2, pH, temperature and NaCl on the growth of *Listeria monocytogenes*. *International Journal of Food Microbiology* 37: 37–45.

Fernandez Garcia, A., Collado, J., Cunha, L. M., Ocio, M. J., and Martinez, A. 2002. Empirical model building based on Weibull distribution to describe the joint effect of pH and temperature on the thermal resistance of *Bacillus cereus* in vegetable substrate. *International Journal of Food Microbiology* 77: 147–153.

Fraeye, I., Duvetter, T., Verlent, I., Ndaka Sila, D., Hendrickx, M. E., and van Loey, A. 2007. Comparison of enzymatic de-esterification of strawberry and apple pectin at elevated pressure by fungal pectinmethylesterase. *Innovative Food Science and Emerging Technologies* 8:93–101.

Fujikawa, H. and Itoh, T. 1996. Characteristics of a multicomponent first-order model for thermal inactivation of microorganisms and enzymes. *International Journal of Food Microbiology* 31: 263–271.

Ganzle, M. G., Kilimann, K. V., Hartmann, C., Vogel, R., and Delgado, A. 2007. Data mining and fuzzy modelling of high pressure inactivation pathways of *Lactococcus lactis*. *Innovative Food Science and Emerging Technologies* 8: 461–468.

Gibson, A. M., Bratchell, N., and Roberts, T. A. 1988. Predicting microbial growth: Growth responses of Salmonellae. *International Journal of Food Microbiology* 6: 155–178.

Gomes, M. R. A. and Ledward, D. A. 1996. Effect of high-pressure treatment on the activity of some polyphenoloxidase. *Food Chemistry* 56: 1–5.

Guan, D., Chen, H., and Hoover, D. 2005. Inactivation of *Salmonella typhimurium* DT 104 in UHT whole milk by high hydrostatic pressure. *International Journal of Food Microbiology* 104: 145–153.

Guerrero Beltran, J. A., Barbosa-Canovas, G. V., and Swanson, B. 2005. High hydrostatic pressure processing of fruit and vegetables products. *Food Reviews International* 21: 411–425.

Haefner, J. W. 2005. Modeling Biological Systems: Principles and Applications. 2nd ed. New York: Springer.

Halder, A., Datta, A. K., and Geedipalli, S. S. R. 2007. Uncertainty in thermal process calculations due to variability in first-order and Weibull kinetic parameters. *Journal of Food Science* 72: E155–E167.

Hauben, K. J. A., Wuytack, E. Y., Soontjens, C. C. F., and Michiels, C. W. 1996. High-pressure transient sensitization of *Escherichia coli* to lysozyme and nisin by disruption of outer-membrane permeability. *Journal of Food Protection* 59: 350–355.

Heinisch, O., Kowalski, E., Goossens, K., Frank, J., Heremans, K., Ludwig, H., Tauscher, B. 1995. Pressure effects on the stability of lipoxygenase: Fourier Transform-Infrared Spectroscopy (FT-IR) and enzyme activity studies. *Zeitschrift für Lebensmittel Untersuchung und Forschung* 201: 562–565.

Hendrickx, M. E., Ludikhuyze, L. R., van den Broeck, I., and Weemaes, C. A. 1998. Effects of high-pressure on enzymes related to food quality. *Trends in Food Science and Technology* 9: 197–203.

Indrawati, A. M., van Loey, A. M., Ludikhuyze, L. R., and Hendrickx, M. E. 1999. Soybean lipoxygenase inactivation by pressure at subzero and elevated temperatures. *Journal of Agricultural and Food Chemistry* 47: 2468–2474.

Indrawati, A. M., Ludikhuyze, L. R., van Loey, A. M., and Hendrickx, M. E. 2000. Lipoxygenase inactivation in green beans (*Phaseolus vulgaris* L.) due to high pressure treatment at subzero and elevated temperatures. *Journal of Agricultural and Food Chemistry* 48: 1850–1859.

Irwe, S. and Olson, I. 1994. Reduction of pectinarase activity in orange juice by high pressure treatment. In *Minimal Processing of Foods and Process Optimization*, Eds. R. P. Singh and F. A. R. Oliveira, 35–42. Boca Raton, FL: CRC Press.

Janssen, M., Geeraerd, A. H., Cappuyns, A., Garcia Gonzalez, L., Schockaert, G., van Houteghem, N., Vereecken, K. M., Debevere, J., Devlieghere, F., van Impe, J. F. 2007. Individual and combined effects of pH and lactic acid concentration on *Listeria innocua* inactivation: Development of a predictive model and assessment of experimental variability. *Applied and Environmental Microbiology* 73: 1601–1611.

Jolibert, F., Tonello, C., Sagegh, P., and Raymond, J. 1994. Les effets des hautes pressions sur la polyphenol oxydase des fruits. *Bios Boissons* 251: 27–35.

Juneja, V. K. and Thayer, D. W. 2000. Irradiation and other physically-based control strategies for foodborne pathogens. In *Microbial Food Contamination*, Eds. C. Wilson and S. Droby, 171–186. Boca Raton, FL: CRC Press.

Kamau, D. N., Doores, S., and Pruitt, K. M. 1990. Antibacterial activity of the lactoperoxidase system against *Listeria monocytogenes* and *Staphylococcus aureus* in milk. *Journal of Food Protection* 53: 1010–1014.

Klotz, B., Pyle, D. L., and Mackey, B. M. 2007. New mathematical modeling approach for predicting microbial inactivation by high hydrostatic pressure. *Applied and Environmental Microbiology* 73: 2468–2478.

Koseki, S. and Yamamoto, K. 2007a. Modelling the bacterial survival/death interface induced by high pressure processing. *International Journal of Food Microbiology* 116: 136–143.

Koseki, S. and Yamamoto, K. 2007b. A novel approach to predicting microbial inactivation kinetics during high pressure processing. *International Journal of Food Microbiology* 116: 275–282.

Lakshmanan, R., Patterson, M. F., and Piggott, J. R. 2005. Effects of high-pressure processing on proteolytic enzymes and proteins in cold-smoked salmon during refrigerated storage. *Food Chemistry* 90: 541–548.

Linton, R. H., Carter, W. H., Pierson, M. D., and Hackney, C. R. 1995. Use of a modified Gompertz equation to model nonlinear survival curves for *Listeria monocytogenes* Scott A. *Journal of Food Protection* 58: 946–954.

Linton, R. H., Carter, W. H., Pierson, M. D., Hackney, C. R., and Eifert, J. D. 1996. Use of a modified Gompertz equation to predict the effects of temperature, pH, and NaCl on the inactivation of *Listeria monocytogenes* Scott A heated in infant formula. *Journal of Food Protection* 59: 16–23.

Little, C. L., Adams, M. R., Anderson, W. A., and Cole, M. B. 1994. Application of a log-logistic model to describe the survival of *Yersinia enterocolitica* at sub-optimal pH and temperature. *International Journal of Food Microbiology* 22: 63–71.

Ludikhuyze, L., Claeys, W., and Hendrickx, M. E. 2000. Combined pressure-temperature inactivation of alkaline phosphatase in bovine milk: A kinetic study. *Journal of Food Science* 65: 155–160.

Ludikhuyze, L., Indrawati, A. M., van den Broeck, I., Weemaes, C., and Hendrickx, M. E. 1998. Effect of combined pressure and temperature on soybean lipoxygenase. II. Modeling inactivation kinetics under static and dynamic conditions. *Journal of Agricultural and Food Chemistry* 46: 4081–4086.

Ludikhuyze, L., van den Broeck, I., Indrawati, A. M., and Hendrickx, M. E. 2002. High pressure processing of fruits and vegetables. In *Fruits and Vegetable Processing: Improving Quality*, Ed. W. Jongen, 346–362. New York, NY: CRC Press Inc.

Macheboeuf, M. A. and Basset, J. 1934. Die Wirkung sehr hoher Drucke auf enzyme. *Ergebnisse der Enzymforschung* 3: 303–308.

Mafart, P., Couvert, O., Gaillard, S., and Leguerinel, I. 2002. On calculating sterility in thermal preservation methods: Application of the Weibull frequency distribution model. *International Journal of Food Microbiology* 72: 107–113.

Malone, A. S., Shellhammer, T. H., and Courtney, P. D. 2002. Effects of high pressure on the viability, morphology, lysis and cell wall hydrolase activity of *Lactococcus lactis* subsp. cremoris. *Applied and Environmental Microbiology* 68: 4357–4363.

McClure, P. J., Blackburn, C. W., Cole, M. B. et al. 1994. Modeling the growth, survival and death of microorganisms in foods: The U.K. food micromodel approach. *International Journal of Food Microbiology* 23: 265–275.

McKellar, R. C. and Xuewen, L. 2003. *Modeling Microbial Response in Food*. Boca Raton, FL, USA: CRC Press.

McMeekin, T. A., Olley, J. N., Ross, T., and Ratkowsky, D. A. 1993. *Predictive Microbiology: Theory and Application*. Somerset, England: Research Studies Press, Ltd.

Mills, G., Earnshaw, R., and Patterson, M. F. 1998. Effect of high hydrostatic pressure on *Clostridium sporogenes* spores. *Letters in Applied Microbiology* 26: 227–230.

Moermann, F. 2005. High hydrostatic pressure inactivation of vegetative microorganisms, aerobic and anaerobic spores in pork Marengo, a low acidic particulate food product. *Meat Science* 69: 225–232.

Morales Blancas, E. F. and Torres, J. A. 2003a. Activation energy (Ea). In *Encyclopedia of Agricultural, Food and Biological Engineering*, Ed. D. R. Heldman, 1–4. New York, NY: Marcel Dekker, Inc.

Morales Blancas, E. F. and Torres, J. A. 2003b. Measurement of kinetic parameters. In *Encyclopedia of Agricultural Food and Biological Engineering*, Ed. D. R. Heldman, 1038–1043. New York, NY: Marcel Dekker, Inc.

Morales Blancas, E. F. and Torres, J. A. 2003c. Thermal resistance constant (z). In *Encyclopedia of Agricultural Food and Biological Engineering*, Ed. D. R. Heldman, 1030–1037. New York, NY: Marcel Dekker, Inc.

Mossel, D. A. A., Moreno, B., and Sruijk, C. B. 2002. Factores que influencian el destino y las actividades metabólicas de los microorganismos en los alimentos. In *Microbiología de los Alimentos*, 77–83. Zaragoza, Spain: Acribia.

Mozhaev, V. V., Heremans, K., Frank, J., Masson, P., and Balny, C. 1996a. High pressure effects on protein structure and function. *Proteins, Structure, Function and Genetics* 24: 81–91.

Mozhaev, V. V., Lange, R., Kudryashova, E. V., and Balny, C. 1996b. Application of high hydrostatic pressure for increasing activity and stability of enzymes. *Biotechnology and Bioengineering* 52: 320–331.

Murphy, P. M., Rea, M. C., and Harrington, D. 1996. Development of a predictive model for growth of *Listeria monocytogenes* in a skim milk medium and validation studies in a range of dairy products. *Journal of Applied Bacteriology* 80: 557–564.

Nakayama, A., Yano, Y., Kobayashi, S., Ishikawa, M., and Sakai, K. 1996. Comparison of pressure resistances of spores of six Bacillus strains with their heat resistance. *Applied and Environmental Microbiology* 62: 3897–3900.

Neuman, R. C., Kauzmann, W., and Zipp, A. 1973. Pressure dependence of weak acid ionization in aqueous buffers. *The Journal of Physical Chemistry* 77: 2687–2691.

Ogawa, H., Fukuhisa, K., Kubo, Y., and Fukumoto, H. 1990. Pressure inactivation of yeast, mould and pectinesterase in satsuma mandarin juice: Effects of juice concentration, pH, and organic acids and comparison with heat sanitation. *Agricultural and Biological Chemistry* 5: 1219–1225.

Palou, E., Hernandez-Salgado, C., Lopez-Malo, A., Barbosa-Cánovas, G. V., Swanson, B. G., and Welti, J. 2000. High pressure-processed guacamole. *Innovative Food Science and Emerging Technologies* 1: 69–75.

Panagou, E. Z., Tassou, C. C., Manitsa, C., and Mallidis, C. 2007. Modelling the effect of high pressure on the inactivation kinetics of a pressure-resistant strain of *Pediococcus damnosus* in phosphate buffer and gilt-head seabream (*Sparus aurata*). *Journal of Applied Microbiology* 102: 1499–1507.

Paredes-Sabja, D. and Torres, J. A. 2008. Modeling of the germination of spores from *Clostridium perfringens* food poisoning isolates. *Journal of Food Process Engineering* (In press).

Paredes-Sabja, D., Gonzalez, M., Sarker, M. R., and Torres, J. A. 2007. Combined effects of hydrostatic pressure, temperature, and pH on the inactivation of spores of *Clostridium perfringens* Type A and *Clostridium sporogenes* in buffer solutions. *Journal of Food Science* 72: M202–M206.

Patterson, M. F. and Kilpatrick, D. J. 1998. The combined effect of high hydrostatic pressure and mild heat on inactivation of pathogens in milk and poultry. *Journal of Food Protection* 61: 432–436.

Peleg, M. 2006. *Advanced Quantitative Microbiology: Models for Predicting Growth and Inactivation*. Boca Raton, FL: CRC Press/Taylor & Francis.

Peleg, M. and Cole, M. B. 1998. Reinterpretation of microbial survival curves. *Critical Reviews in Food Science and Nutrition* 38: 353–380.

Peleg, M., Engel, R., Gonzalez Martinez, C., and Corradini, M. G. 2002. Non-Arrhenius and non-WLF kinetics in food systems. *Journal of the Science of Food and Agriculture* 82: 1346–1355.

Pérez Lamela, C. and Torres, J. A. 2008a. Pressure-assisted thermal processing: A promising future for high flavour quality and health-enhancing foods. *AgroFOOD Industry Hi-Tech* 19: 60–62.

Pérez Lamela, C. and Torres, J. A. 2008b. Pressure processing of foods: Microbial inactivation and chemical changes in pressure-assisted thermal processing (PATP). *AgroFOOD Industry Hi-Tech* 19: 34–36.

Pina Perez, M. C., Rodrigo Aliaga, D., Saucedo Reyes, D., and Martinez Lopez, A. 2007. Pressure inactivation kinetics of *Enterobacter sakazakii* in infant formula milk. *Journal of Food Protection* 70: 2281–2289.

Quaglia, G. B., Gravina, R., Paperi, R., and Paoletti, F. 1996. Effect of high pressure treatments on peroxidase activity, ascorbic acid content and texture in green peas. *Lebensmittel Wissenschaft und Technologie* 29: 552–555.

Quintero-Ramos, A., Churey, J. J., Hartman, P., Barnard, J., and Worobo, R. W. 2004. Modeling of *Escherichia coli* inactivation by UV irradiation at different pH values in apple cider. *Journal of Food Protection* 67: 1153–1156.

Quiros, A., Chichon, A., Recio, I., and Lopez-Fandino, R. 2007. Analytical, nutritional and clinical methods: The use of high hydrostatic pressure to promote the proteolysis and release of bioactive peptides from ovalbumin. *Food Chemistry* 104: 1734–1739.

Rademacher, B. and Kessler, H. G. 1996. High pressure inactivation of microorganisms and enzymes in milk and milk products. *European High Pressure Research Conference*, Leuven, Belgium.

Rademacher, B., Werner, F., and Pehl, M. 2002. Effect of the pressurizing ramp on the inactivation of Listeria innocua considering thermofluidynamical processes. *Innovative Food Science and Emerging Technologies* 3: 13–24.

Raju, D. and Sarker, M. R. 2005. Comparison of the heat sensitivities of vegetative cells and spores of wild-type, cpe knock-out and cpe plasmid-cured *Clostridium perfringens*. *Applied and Environmental Microbiology* 71: 7618–7620.

Raju, D., Waters, M., Setlow, P., and Sarker, M. R. 2006. Investigating the role of small, acid-soluble spore proteins (SASPs) in the resistance of *Clostridium perfringens* spores to heat. *BMC Microbiology* 6: 50–53.

Ramirez, R., Saraiva, J. A., Perez Lamela, C., and Torres, J. A. 2009. Reaction kinetics analysis of chemical changes in pressure-assisted thermal processing, PATP. *Food Engineering Reviews* (In press).

Reps, A., Kolakowski, P., and Dajnowiec, F. 1998. The effect of high pressure on microorganisms and enzymes of ripening cheeses. In *High Pressure Food Science, Bioscience and Chemistry*, Ed. N. S. Isaacs, 265–270. Cambridge, U.K.: The Royal Society of Chemistry.

Reyns, K. M. F. A., Soontjens, C. C. F., Cornelis, K., Weemaes, C. A., Hendrickx, M. E., and Michiels, C. W. 2000. Kinetic analysis and modelling of combined high-pressure-temperature inactivation of the yeast *Zygosaccharomyces bailii*. *International Journal of Food Microbiology* 56: 199–210.

Ritz, M., Freulet, M., Orange, N., and Federighi, M. 2000. Effects of high hydrostatic pressure on membrane proteins of *Salmonella typhimurium*. *International Journal of Food Microbiology* 55: 115–119.

Rode, L. J. and Foster, J. W. 1962. Ionic germination of spores of *Bacillus megaterium* QM B 1551. *Archiv fur Mikrobiologie* 43: 183–200.

Rodrigo, M., Martínez, A., and Rodrigo, D. 2000. Inactivation kinetics of microorganisms by pulsed electric fields. In *Novel Food Processing Technologies*, Eds. G. V. Barbosa Cánovas, M. S. Tapia, and M. P. Cano, 69–86. Boca Raton, FL: CRC Press Inc.

Sampedro, F., Rodrigo, D., and Hendrickx, M. E. 2008. Inactivation kinetics of pectin methyl esterase under combined thermal-high pressure treatment in an orange juice-milk beverage. *Journal of Food Engineering* 86: 133–139.

San Martin, M. F., Barbosa-Canovas, G. V., and Swanson, B. G. 2002. Food processing by high hydrostatic pressure. *Critical Reviews in Food Science and Nutrition* 42: 627–645.

Sarker, M. R., Shivers, R. P., Sparks, S. G., Juneja, V. K., and McClane, B. A. 2000. Comparative experiments to examine the effects of heating on vegetative cells and spores of *Clostridium perfringens* isolates carrying plasmid genes versus chromosomal enterotoxin genes. *Applied and Environmental Microbiology* 66: 3234–3240.

Seyderhelm, I., Boguslawski, S., Michaelis, G., and Knorr, D. 1996. Pressure induced inactivation of selected food enzymes. *Journal of Food Science* 61: 308–310.

Shimada, S., Andou, M., Naito, N., Yamada, N., Osumi, M., and Hayashi, R. 1993. Effects of hydrostatic pressure on the ultrastructure and leakage of internal substances in the yeast *Saccharomyces cerevisiae*. *Applied Microbiology and Biotechnology* 40: 123–131.

Shipe, W. F. 1980. Analysis and control of milk flavor. In *The Analysis and Control of Less Desirable Flavors in Foods and Beverages*, Ed. G. Charalambous, 201–239. New York: Academic Press Inc.

Sila, D. N., Smout, C., Satara, Y., Truong, V., van Loey, A., and Hendrickx, M. E. 2007. Combined thermal and high pressure effect on carrot pectinmethylesterase stability and catalytic activity. *Journal of Food Engineering* 78: 755–764.

Smelt, J. P. P. M., Hellemons, J. C., Wouters, P. C., and van Gerwen, S. J. C. 2002. Physiological and mathematical aspects in setting criteria for decontamination of foods by physical means. *International Journal of Food Microbiology* 78: 57–77.

Stabel, J. R., Steadham, E. M., and Bolin, C. A. 1997. Heat inactivation of *Mycobacterium paratuberculosis* in raw milk: Are current pasteurization conditions effective? *Applied and Environmental Microbiology* 63: 4975–4977.

Steely, J. S. 1994. Chemiluminiscence detection of sulfur compounds in cooked milk. In: Mussinan CJ, Keelan ME, editors. Sulfur compounds in foods. Chicago, Illinois: American Chemical Society. p. 22–35.

Stewart, C. 2008. Spore inactivation by high pressure processing and pressure assisted thermal sterilization. Nonthermal Processing Workshop. Portland, OR.

Stewart, C. M., Dunne, C. P., Sikes, A., and Hoover, D. G. 2000. Sensitivity of spores of *Bacillus subtilis* and *Clostridium sporogenes* PA 3679 to combinations of high hydrostatic pressure and other processing parameters. *Innovative Food Science and Emerging Technologies* 1: 49–56.

Sun, N., Seunghwan, L., and Kyung, B. S. 2002. Effect of high-pressure treatment on the molecular properties of mushroom polyphenoloxidase. *Lebensmittel Wissenschaft und Technologie* 35: 315–318.

Tangwonchai, R., Ledward, D. A., and Ames, J. M. 1999. Effect of high pressure on lipoxygenase activity in cherry tomatoes. In *Advances in High Pressure Bioscience and Biotechnology*, Ed. H. Ludwig, 435–438. Berlin and Heidelberg, Germany: Springer-Verlag.

Ting, E. Y. 2003. Personal communication.

Ting, E. Y., Balasubramaniam, V. M., and Raghubeer, E. 2002. Determining thermal effects in high-pressure processing. *Food Technology* 56(2): 31–35.

Torres, J. A. and Rios, R. A. 2006. Alta presión hidrostática: Una tecnología que irrumpirá en Chile. *Agro Económico* Febrero 2006: 40–43.

Torres, J. A. and Velazquez, G. 2005. Commercial opportunities and research challenges in the high pressure processing of foods. *Journal of Food Engineering* 67: 95–112.

Torres, J. A. and Velazquez, G. 2008. Hydrostatic pressure processing of foods. In: Jun S, Irudayaraj J, editors. Food Processing Operations Modeling: Design and Analysis. Second ed. Boca Ratón, FL: CRC Press Inc. p. 173–212.

Tovar-Hernandez, G., Peña, H. R., Velazquez, G., Ramirez, J. A., and Torres, J. A. 2005. Effect of combined thermal and high pressure processing on the microbial stability of milk during refrigerated storage. *The Institute of Food Technologists (IFT) Annual Meeting*. New Orleans, LA.

Ulloa-Fuentes, P. A., Galotto, M. J., and Torres, J. A. 2008a. Procesos térmicos asistidos por presión (PTAP), el futuro de una nueva tecnología ya instalada en México – Parte I. *Industria Alimentaria* (México) 30: 20, 22, 24, 26, 28, 29.

Ulloa-Fuentes, P. A., Galotto, M. J., and Torres, J. A. 2008b. Procesos térmicos asistidos por presión (PTAP), el futuro de una nueva tecnología ya instalada en México – Parte II. *Industria Alimentaria* (México) 30: 19–23.

van Boekel, M. A. J. S. 2008. Kinetic modeling of food quality: A critical review. *Comprehensive Reviews in Food Science and Food Safety* 7: 144–158.

van den Broeck, I., Ludikhuyze, L. R., van Loey, A. M., Weemaes, C. A., and Hendrickx, M. E. 1999a. Thermal and combined pressure–temperature inactivation of orange pectinesterase: Influence of pH and additives. *Journal of Agricultural and Food Chemistry* 47: 2950–2958.

van den Broeck, I., Ludikhuyze, L. R., Weemaes, C. A., van Loey, A. M., and Hendrickx, M. E. 1999b. Thermal inactivation kinetics of pectinarase extracted from oranges. *Journal of Food Processing and Preservation* 23: 391–406.

van den Broeck, I., Ludikhuyze, L. R., van Loey, A. M., and Hendrickx, M. E. 2000. Inactivation of orange pectinesterase by combined high-pressure and temperature treatments: A kinetic study. *Journal of Agricultural and Food Chemistry* 48: 1960–1970.

van Loey, A., Ooms, V., Weemaes, C., van den Broeck, I., Ludikhuyze, L., Indrawati, Denys, S., and Hendrickx, M. E. 1998. Thermal and pressure-temperature degradation of chlorophyll in broccoli (*Brassica oleracea L. italica*) juice: A kinetic study. *Journal of Agricultural and Food Chemistry* 46: 5289–5294.

Vary, J. C. and Halvorson, H. O. 1965. Kinetics of germination of Bacillus spores. *Journal of Bacteriology* 89: 1340–1347.

Velazquez, G., Gandhi, K., and Torres, J. A. 2002. High hydrostatic pressure: A review. *Biotam* 12: 71–78.

Velazquez, G., Vazquez, P., Vazquez, M., and Torres, J. A. 2005a. Aplicaciones del procesado de alimentos por alta presión. *Ciencia y Tecnología Alimentaria* 4: 343–352.

Velazquez, G., Vazquez, P., Vazquez, M., and Torres, J. A. 2005b. Avances en el procesado de alimentos por alta presión. *Ciencia y Tecnología Alimentaria* 4: 353–367.

Vazquez Landaverde, P. A., Qian, M. C., and Torres, J. A. 2007. Kinetic analysis of volatile formation in milk subjected to pressure-assisted thermal treatments. *Journal of Food Science* 72: E389–E398.

Wedzicha, B. L., Goddard, S. J., and Zeb, A. 1993. Approach to the design of model systems for food additive–food component interactions. *Food Chemistry* 47: 129–132.

Weemaes, C. A., de Cordt, S. V., Ludikhuyze, L. R., van den Broeck, I., Hendrickx, M. E., and Tobback, P. P. 1997a. Influence of pH, benzoic acid, EDTA, and glutathione on the pressure and/or temperature inactivation kinetics of mushroom polyphenoloxidase. *Biotechnology Progress* 13: 25–32.

Weemaes, C. A., Rubens, P., de Cordt, S., Ludikhuyze, L., van den Broeck, I., Hendrickx, M. E., Heremans, K., Tobback, P. 1997b. Temperature sensitivity and pressure resistance of mushroom polyphenoloxidase. *Journal of Food Science* 62: 261–266.

Weemaes, C. A., Ludikhuyze, L., van den Broeck, I., and Hendrickx, M. E. 1998. High pressure inactivation of polyphenoloxidase. *Journal of Food Science* 63: 873–877.

Weemaes, C. A., Ludikhuyze, L., van den Broeck, I., and Hendrickx, M. E. 1999. Kinetic study of antibrowning agents and pressure inactivation of avocado polyphenoloxidase. *Journal of Food Science* 64: 823–827.

Wicker, L. and Temelli, F. 1988. Heat inactivation of pectinarase in orange juice pulp. *Journal of Food Science* 53: 162–164.

Xiong, R., Xie, G., Edmondson, A. E., and Sheard, M. A. 1999a. A mathematical model for bacterial inactivation. *International Journal of Food Microbiology* 46: 45–55.

Xiong, R., Xie, G., Edmondson, A. S., Linton, R. H., and Sheard, M. A. 1999b. Comparison of the Baranyi model with the modified Gompertz equation for modelling thermal inactivation of *Listeria monocytogenes* Scott A. *Food Microbiology* 16: 269–279.

Yemenicioglu, A., Ozkan, M., and Cemeroglu, B. 1997. Heat inactivation kinetics of apple polyphenoloxidase and activation of its latent form. *Journal of Food Science* 62: 508–510.

15 Temperature Distribution and Chemical Reactions in Foods Treated by Pressure-Assisted Thermal Processing

J. Antonio Torres, Pedro D. Sanz, Laura Otero,
María Concepción Pérez Lamela, and
Marleny D. Aranda Saldaña

CONTENTS

15.1 INTRODUCTION

The exponential growth of the pressure-processing industry and the many products found all over the world illustrate the commercial success of the high pressure processing (HPP) technology. The previous chapter described the essential elements of pressure processing including the mechanisms and inactivation kinetics of microorganisms and enzymes. A new equipment generation is expected in the near future to allow the commercial application of pressure-assisted thermal processing (PATP) research. Moderate-temperature PATP will be used when HPP is not commercially feasible due long processing times when using pressure alone, for example, pasteurized milk. PATP at higher temperatures will deliver shelf-stable, low-acid foods with higher quality than the conventional thermal treatments used today for the commercial sterilization of foods. In this chapter, a reaction kinetics approach will be followed to explain PATP effects on food composition as compared to conventional thermal treatments. The design of PATP processes will require knowledge of the temperature distribution in the high-pressure vessel and the food being processed. Also important will be to ensure homogenous temperature profile to ensure that all foods are subjected to treatments that are not dependent upon location within the vessel. Modeling approaches proposed to explore solutions to these two technological challenges are presented in this chapter.

15.2 MECHANISMS AND KINETICS FOR CHEMICAL REACTIONS UNDER HIGH PRESSURE

A reaction kinetics approach to interpret chemical reactions at high pressure and high temperature will be followed to examine chemical changes in PATP foods. The information supporting this analysis is based on the formation of "cooked-flavor" compounds in milk (Vazquez Landaverde et al. 2007; Pérez Lamela and Torres 2008a,b). Many chemical reactions are involved in the formation of the 27 volatile compounds associated with this product defect covered in this study. Although, Pérez Lamela & Torres (2008a,b) concluded that PATP technology is a promising alternative to satisfy the current consumer demand for increased food safety and eating pleasure in milk, PATP research needs to be expanded to other foods and to the many health-enhancing compounds considered important by consumers in many markets of the world. At present, it appears that the beneficial effects of PATP are highly food matrix dependent. The complex effects of food matrix, pH, dissolved oxygen and presence of antioxidants show that optimization of vitamin, pigment and flavor retention while ensuring PATP microbial and enzyme inactivation will require substantially more chemical reaction kinetics research. A summary of chemical changes observed in PATP research is presented in Table 15.1 (Ramirez et al. 2009). This comprehensive review of published reports shows that most work has focused on the destruction of vitamins and pigment content of fruits and vegetables with a few additional studies in milk and other foods of animal origin. Most of this research has been conducted in Europe reflecting the impact of novel food laws. Novel foods were originally defined as all foods and food ingredients not used for human consumption to a significant degree within the EU prior to May 1997. In Europe and other countries following similar regulations (e.g., Canada), information on the loss of nutrients and the formation of new, possibly toxic compounds, is now required (Hepburn et al. 2008).

Although the pressure range covered in the studies summarized in Table 15.1 is generally consistent with pressure levels expected to be used for sterilization in commercial PATP units (~700 MPa), the same is not true for temperature. None of the reaction kinetics studies cover the range above 100°C which will be required for most PATP sterilization processes. In addition, relatively few studies cover the 70–100°C range. Another limitation of the experimental data available is that only one study reported oxygen concentrations and a few do not report sample pH. This

TABLE 15.1
Chemical Reaction Studies on PATP-Treated Foods

Compound	Matrix	P MPa	T °C	t min	pH	O2 present?	Source
[6S] & [6R,S] 5-Methyltetrahydrofolic acid	Phosphate buffer	200–700	30–45	0–80	7	Yes	(Indrawati et al. 2005)
Folic acid	Phosphate buffer	0.1–800	20–90	0–130	7	Yes	(Nguyen et al. 2003)
5-Methyltetrahydrofolic acid							
5-Methyltetrahydrofolic acid	Acetate buffer	0.1–800	10–65	–	3–9	Yes	(Indrawati et al. 2004)
	Phosphate buffer						
	Citric phosphate buffer						
	Sodium borate solution						
	Orange juice						
	Carrot juice						
	Kiwi pure						
	Asparagus pieces						
Tetrahydrofolate	Phosphate buffer	600	25, 80	0–24	7	nr[a,b]	(Butz et al. 2004)
5-Methyltetrahydrofolate [6R,S]	Juice model						
5-Formyltetrahydrofolate							
[6R,S] 5-Formyltetrahydrofolic acid	Acetate buffer	100–800	30–70	0–300	3.4–9.2	nr[c]	(Nguyen et al. 2006)
	Phosphate buffer						
	Borax buffer						
5-Methyltetrahydrofolic acid	Phosphate buffer	0.1–600	40–50	–	7	2.1 & 8.1 ppm	(Oey et al. 2006)
Folylpolyglutamate	Broccoli	0.1–600	25–45	30	nr	Vacuum	(Verlinde et al. 2008)
Chlorophyll	Broccoli juice	0.1–850	70–90	0–180	nr	nr	(Weemaes et al. 1999)
Chlorophyll	Broccoli juice	200–800	30–80	0–140	nr	nr	(van Loey et al. 1998)

(continued)

TABLE 15.1 (continued)
Chemical Reaction Studies on PATP-Treated Foods

Compound	Matrix	P MPa	T °C	t min	pH	O2 present?	Source
L-Ascorbic acid	Phosphate buffer, Orange juice, Tomato juice	850	65–80	0–400	4–8	Yes	(van den Broeck et al. 1998)
L-Ascorbic acid	Na acetate buffer, Pineapple juice, Grapefruit juice	300–600	40–75	0–40	3.5–4	Yes	(Taoukis et al. 1998)
Cyanidin-3-O-glucoside	Acetate buffer with 6.56g/L pyruvate	600	70	30	4.4	nr	(Corrales et al. 2008)
α and β Carotene	Carrot juice	300–500	50–70	10	nr	nr	(Kim et al. 2001)
Methyl chavicol	Basil leaves	700, 850	75, 85	Pulse	–	–	(Krebbers et al. 2002)
Linalool							
Vitamin A acetate	Ethanol solutions	650	70	4–960	nr	nrd	(Kübel et al. 1997)
Retinol	Orange juice	50–300	30–60	2.5–15	nr	Yes	(de Ancos et al. 2002)
Retinol	Orange juice	400	40	1	3.6	Vacuum	(Sanchez Moreno et al. 2005)
Retinol	Cream	0.1–600	75	40	nr	nr	(Butz and Tauscher 2000)
Cooked flavor volatiles except aldehydes	Milk	482–620	45–75	1–10	nr	nr	(Vazquez Landaverde et al. 2007)
Aldehyde volatiles							
Volatile sulfur compounds	Buffer	0.1–600	95	20–120	nr	nr	(Butz and Tauscher 2000)
Thiamin	Buffer, Pork	0.1, 600	25–100	15–1080	5.5	Yes	(Butz et al. 2007)
Riboflavin	Pork	0.1, 600	25–100	15–1080	5.5	Yes	(Butz et al. 2007)

a nr = value not reported.
b Oxygen excluded whenever possible.
c Minimum direct contact with air.
d Air-tight bags.

is unfortunate as the emerging literature is beginning to show. For example, Oey et al. (2006) showed that the dissolved oxygen concentration affects the ascorbic acid stability in PATP-treated foods since it affects the amount degraded under the faster aerobic pathway. Another example is the work reported by Oey et al. (2006) on the effect of 2.11 and 8.11 ppm dissolved oxygen concentration on the stability of folates. Unfortunately, most studies do not report the dissolved oxygen concentration and only a few reports indicate whether efforts were made to remove oxygen during sample preparation for the studies summarized in Table 15.1.

15.3 CHEMICAL REACTION KINETICS

Pressure preserves foods by disrupting hydrogen and other weak bonds without affecting covalent bonds. Therefore, HPP treatments in the 0°C to ~50°C range induce minor chemical compositions changes in foods (Hoover et al. 1989; Torres and Velazquez 2005; 2008). The formation of chemical compounds can be investigated by expressing the change in concentration (c) with respect to time (t) as follows:

$$\frac{dc}{dt} = kc^n \tag{15.1}$$

where
k is the reaction rate constant at a given pressure and temperature
n is the reaction order

Integrating Equation 15.1 yields the following linearized kinetic expressions:
Zero order

$$c - c_0 = kt \tag{15.2}$$

First order

$$\log(c) - \log(c_0) = kt \tag{15.3}$$

Second order

$$\frac{1}{c} - \frac{1}{c_0} = kt \tag{15.4}$$

The regression curve at constant pressure and temperature with the best correlation coefficient (R^2) as obtained by simple linear regression for each of the three kinetic models is then used to calculate the rate constant k. In the case of PATP treatments, the experimental steps required to bring the sample from room temperature and atmospheric pressure to the pressure and temperature condition of interest imply that the c_0 value in these equations is a "pseudo-initial concentration." The difference between this concentration and the one found in untreated samples represents the effect of time when the food sample is not at the vessel temperature and pressure including sample loading/unloading, heating/cooling, and pressure come-up/come-down time. Multiple linear regression analysis can be used to test the difference between the linearized intercepts at the same temperature for each pressure level tested to confirm that all regression lines start at the same pseudo-initial concentration c_0, which is then reported as an average value for that initial temperature level (Figure 15.1).

As an application example of this reaction kinetic analysis, concentration values for 27 volatiles determined for raw milk samples subjected to combinations of pressure (482, 586, 620, and 655 MPa), initial temperature (45°C, 55°C, 60°C, and 75°C before compression), and pressure-holding time (1, 3, 5, and 10 min) will be used (Vazquez Landaverde et al. 2007, Pérez Lamela and Torres 2008a,b). All samples were processed with vessel loading (1 min) and unloading times (1.5 min) kept constant for all runs. After treatment, samples were placed immediately in a saturated salt slurry and ice bath before storage at −38°C until analysis. The 27 volatiles monitored in PATP milk were affected by the initial preheating and the temperature increase due to adiabatic heating during sample compression. In the 400–1000 MPa range, milk temperature increases approximately 3°C for every 100 MPa (Harvey et al. 1996, Ting et al. 2002). Also, even though milk is a well-buffered food system (e.g., Datta and Deeth 1999), reaction rates may have been affected by the temporary pH shift induced by pressure, a value that cannot be measured as there are no probes available that operate at high pressure. In the particular case of the reactions leading to the 27 volatiles monitored in the milk study, their formation followed either zero- or first-order kinetics (Vazquez Landaverde et al. 2007). Pseudo-initial concentration c_0 values, except those for 2-methylpropanal and 2,3-butanedione, increased significantly with test temperature above the amount present in raw milk. The observed differences between the pseudo-initial concentration (c_0) and the one found in untreated milk reflected the cumulative effect of the time that the milk was at high temperature but not at the test pressure. This included time to equilibrate the milk sample to each experimental temperature, time to open and close the pressure vessel, time for pressurization and decompression, and finally time to remove the sample from the vessel and its subsequent cooling down for frozen storage before analysis. Values of c_0 at each initial temperature did not change with pressure confirming that regardless of the pressure treatment applied, sample heating before and after each HPP run was responsible for most c_0 values being above the amount present in raw milk.

Chemical reaction theory postulates that before a reactant participates in a reaction it needs to reach an activated state requiring an energy increase

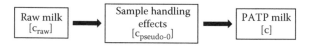

FIGURE 15.1 Definition of pseudo-0 concentration. Raw milk is subjected to thermal treatments before and after each PATP test at given pressure and temperature as required for sample preheating, loading, unloading and cooling.

independent of temperature and called the Arrhenius activation energy (E_a) (Figure 15.2a). E_a values for volatile formation in milk at constant pressure can be calculated using the Arrhenius expression (Equation 15.5). The slope of this curve ($-E_a/R$ with R as the universal gas constant) and the intercept (ln k_0, with k_0 = pre-exponential rate constant) are calculated from the linearized form of the Arrhenius equation (Equation 15.6).

$$k = k_0\, e^{\frac{-E_a}{RT}} \tag{15.5}$$

$$\ln(k) = \ln(k_0) - \frac{E_a}{RT} \tag{15.6}$$

A quantity derived from the pressure dependence of the rate constant k Equation 15.7 is the partial activation volume (V_a) (Figure 15.2b). V_a is defined as the difference between the partial molar volumes of the transition state and the sums of the partial volumes of the reactants at the same temperature and pressure (McNaught and Wilkinson 1997). When foods are treated under pressure, those reactions with $V_a < 0$ will increase in reaction rate while $V_a > 0$ has the opposite effect. The greater the magnitude of V_a (positive or negative) the higher the sensitivity of

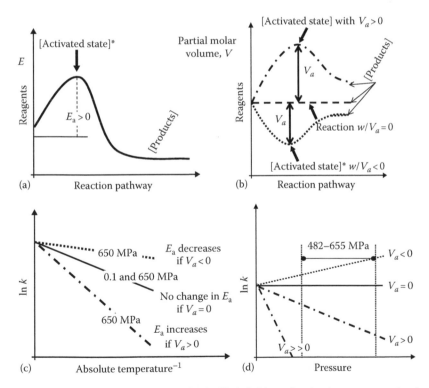

FIGURE 15.2 Reaction kinetic analysis: (a, b) definition of activation energy and volume (c, d) pressure effects on activation energy and activation volume. (Adapted from Pérez Lamela, C. and Torres, J. A., *AgroFOOD Industry Hi-Tech*, 19, 34, 2008b.)

a chemical reaction to pressure while reactions with $V_a = 0$ are pressure independent (Mussa and Ramaswamy 1997). The corresponding effects on E_a values are a
decrease, no change and an increase if $V_a < 0$, $= 0$, or > 0, respectively (Figure 15.2c).
Equation 15.7 can be integrated to obtain Equation 15.8 where $\ln A$ is the integration
constant. Values for V_a at constant temperature are obtained by linear regression of
$\ln k$ versus pressure p (Figure 15.2d).

$$V_a = -RT \left(\frac{\partial \ln k}{\partial p} \right)_t \tag{15.7}$$

$$\ln k = \ln A - \left(\frac{(V_a)p}{RT} \right) \tag{15.8}$$

In the case of the volatiles analyzed in PATP-treated milk, the first order kinetic constants observed for straight chain aldehydes fitted well ($R^2 > 0.9$) the Arrhenius model
with activation energy (E_a) values decreasing significantly with pressure. Hexanal
formation had the lowest E_a value decreasing nearly 40 times from 35.2 kJ mol^{-1} at
482 MPa to 0.9 kJ mol^{-1} at 655 MPa. Reactions following zero-order kinetics showed
also a good fit to the Arrhenius model ($R^2 > 0.9$); however, E_a values were affected
differently by pressure, increasing for 2-methylpropanal and 2,3-butanedione and
remaining practically unchanged for hydrogen sulfide regardless of the pressure level.
This reflected the experimental observation that pressure increased the formation of
straight-chain aldehydes, decreased that of 2-methylpropanal and 2,3-butanedione,
and did not affect that for hydrogen sulfide.

A remarkable observation was that the concentration of the remaining 18 volatiles
analyzed in this study on PATP milk did not increase during pressurization time for
all pressure and temperature levels tested (i.e., $V_a \gg 0$). As expected, the pseudo-
initial concentration c_0 values for some of these compounds increased with treatment temperature due to the time that these samples were at atmospheric pressure
and high temperature. However, the concentration of these volatiles remained stable
during pressurization time. This indicated that only the heating time at atmospheric
pressure before and after each pressure treatment was responsible for the increase in
these volatiles above the level found in raw milk.

The lack of contribution to the concentration of most milk volatiles by heating
time under pressure, 18 of a total of 27 compounds analyzed, was the most interesting finding. This showed that high hydrostatic pressure inhibited the formation of
these volatiles even though initial vessel temperature varied from 45°C to 75°C, and
compression to 482 to 655 MPa for 1–10 min further increased milk temperature due
to adiabatic heating. This was surprising because the concentration increase with
processing temperature using conventional technologies has been well demonstrated
for these volatiles (Scanlan et al. 1968, Shibamoto et al. 1980, Calvo and de la Hoz
1992, Christensen and Reineccius 1992, Contarini et al. 1997, Vazquez Landaverde
et al. 2005, Vazquez Landaverde et al. 2006b). Additional evidence of this severe
inhibition by pressure could be obtained by PATP experiments in the range 0.1–
482 MPa, a range outside the pressure range selected in the PATP-milk case study,

482–655 MPa. This range was chosen to approach conditions achieving microbial inactivation including bacterial spores (Figure 15.2d).

As a confirmation of the pressure effects on the formation of undesirable volatiles, milk subjected to 620 MPa for 5 min at initial temperature 75°C was compared to those observed in samples of the same milk subjected to heat treatments at atmospheric pressure replicating the temperature and time during the PATP treatment, that is, 3.5 min at 75°C to account for sample handling, plus 5 min at 93.6°C, a temperature chosen to account for milk adiabatic heating during pressurization of milk initially at 75°C. As expected, straight-chain aldehydes were present at higher concentrations in PATP-treated samples while methyl ketones and some sulfur compounds were present at lower concentrations than in the heat-only samples (Table 15.2). Volatiles with final higher concentrations for PATP-treated milk than in milk subjected to the equivalent thermal process at atmospheric pressure are highlighted by the symbol ↑ and the cells are shown in grey. Volatiles with lower concentration for PATP-treated milk have the symbol ↓ and the cells are shown in black. Also shown are the pseudo-initial concentration values, c_0, for the same compounds confirming that volatile formation reactions with $V_a > 0$ are inhibited by high pressure. This analysis shows that most volatiles, perceived by consumers as cooked-milk off-flavors, are formed in much larger amounts in conventionally treated milk as compared to experimental PATP-milk samples (Pérez Lamela and Torres 2008b).

These findings provided an explanation to the observation that PATP effects on the volatile profile of milk is different to that of conventional commercial pasteurization as suggested by principal component analysis (PCA). This statistical analysis was used to group milk samples with a similar profile in the concentration of the 27 volatiles considered responsible for the cooked flavor in milk (Vazquez Landaverde et al. 2006a) (Figure 15.3). PCA showed that PATP milk treated at moderate temperature and moderate pressure (Group A, moderate T PATP-milk samples) yielding milk with a refrigerated shelf life longer than 7 weeks had a volatile profile similar to thermally pasteurized milk (Group A, commercial samples PA and PB) with a shelf life of only 2–3 weeks. This observation suggests that it should be possible to use moderate temperature PATP to produce pasteurized milk with sensory properties similar to the ones obtained by conventional pasteurization but with much longer shelf life. This finding opens many important commercial opportunities. PATP pasteurization would have advantages for low-cost milk producers wishing to distribute fresh milk in markets at long distances from their production and processing facilities location. PATP technology could help also organic milk producers needing a longer refrigerated shelf life to cover a broad distribution area with the desired number of consumers. Even more interesting is the observation that severe PATP processes, that is, applying pressure and temperature conditions approaching those needed for commercial milk sterilization (Group B, high T PATP-milk samples), the PCA showed a smaller and different-direction volatile profile shift to the one observed for UHT milk (Group C, commercial UHT samples UA and UB) (Figure 15.3). This result is promising since it indicates that PATP sterilization will produce milk with a different flavor profile to that of UHT milk, currently rejected by many consumers in important markets.

TABLE 15.2
Comparative Effects in the Formation of Volatile Compounds Associated with "Cooked" Flavor in Milk Treated at High Temperature under High (PATP Technology) or Low Pressure (Conventional Thermal Technology)[a, b]

Treatment	Hexanal	Heptanal	Octanal	2-Heptanone	2-Octanone	2-Nonanone	MeSH	DMS	DMDS
620 MPa, 75°C, 5 min	44.7↑	7.3↑	7.0↑	4.0↓	0.97↓	0.64↓	7.7↓	3.9↓	0.03↓
Thermal treatment[c]	16.9	2.6	5.1	5.2	4.8	8.6	24.8	8.4	0.06
Pseudo-initial conc., c_0	12.4	2.1	1.2	4.0	0.97	0.64	7.8	3.9	0.03

Source: Adapted from Pérez Lamela, C. and Torres, J.A., *AgroFOOD Industry Hi-Tech*, 19, 34, 2008.

[a] Concentration in µg/kg.

[b] MeSH, methanethiol; DMS, dimethyl sulfide; DMDS, dimethyl disulfide.

[c] Thermal treatment at atmospheric pressure, equivalent to time and temperature values for PTAP-treated (620 MPa, 75°C, 5 min) milk samples.

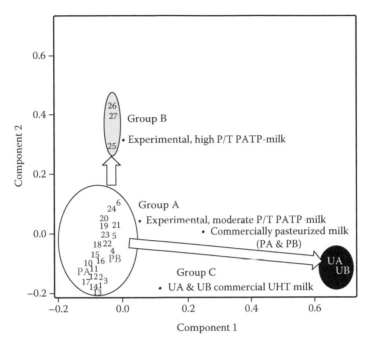

FIGURE 15.3 Principal component analysis grouping milk with a similar profile in the concentration of 27 flavor volatiles. Numbered samples are PATP-treated samples while UA, UB and PA, PB correspond to commercial UHT and pasteurized milk, respectively. (Adapted from Pérez Lamela, C. and Torres, J. A., *AgroFOOD Industry Hi-Tech*, 19, 60, 2008a.)

However, consumer sensory studies on PATP-treated milk are needed because the flavor profiles obtained are so different from those produced during conventional thermal treatments. The inhibitory effect of pressure on the formation of several sulfur compounds offers a promising improvement in milk processing technology. However, the increase in the concentration of saturated aldehydes with low sensory thresholds and thought to cause the stale off-flavor in milk (Rychlik et al. 1998) could affect the consumer acceptance of PATP milk. On the other hand, it has been reported that volatile sulfur compounds are mainly responsible for the development of the cooked flavor defect in heated milk (Christensen and Reineccius 1992, Steely 1994, Simon et al. 2001, Datta et al. 2002). Methanethiol is probably the most powerful sulfur-containing aroma compound in heated milk with a low sensory threshold and an unpleasant rotten cabbage aroma (Vazquez Landaverde et al. 2006b). Dimethyl sulfide is also an important compound commonly present in milk at concentrations above its sensory threshold and has a sulfury aroma (Rychlik et al. 1998, Vazquez Landaverde et al. 2005). Hydrogen sulfide also has an unpleasant eggy, sulfury aroma, but a recent analysis using improved quantification techniques indicated that hydrogen sulfide may not be as important to the aroma of heated milk (Vazquez Landaverde et al. 2006a,b). Inhibition of methyl ketones formation by pressure is also of importance since their concentration increase has been associated to the development of stale-heated flavor in UHT milk (Contarini and Povolo 2002). Although their high sensory thresholds suggest that they could be less important than previously thought (Rychlik et al. 1998), some researchers have indicated that methyl ketones could act in a synergistic manner to impart a perceptible flavor (Langler and Day 1964). The favorable impact of pressure on the formation of all these compounds is a promising aspect of PATP-treated milk.

There is another important conclusion from the analysis of volatile formation in milk. The increase, decrease, or lack of change caused by pressure and temperature on the formation of volatiles in PATP milk could be explained with no need to assume alternative reaction pathways. When comparing conventional heated and PATP-treated milk, it was not necessary to assume new unique reaction pathways to interpret the large differences in chemical composition observed. This is important as it will facilitate the implementation of PATP technologies in countries requiring safety studies when a new processing technology is adopted (Hepburn et al. 2008). However, reactions *never* detected in conventionally processed foods, because of very low reaction rates at low pressures, may occur at significantly higher rates in PATP foods if these reactions have large negative V_a values (Figure 15.4). This risk needs to be considered since a pressure acceleration of thermal degradation reactions have been observed in some food matrices (Ramirez et al. 2009).

15.4 TEMPERATURE PREDICTION MODELS

Many future commercial applications of high-pressure food processing will be pressure and temperature dependent. Therefore, temperature distribution inside the pressure vessel needs to be known and preferably as homogeneous as possible to extract valid conclusions about the effects that a given treatment produces in a specific food.

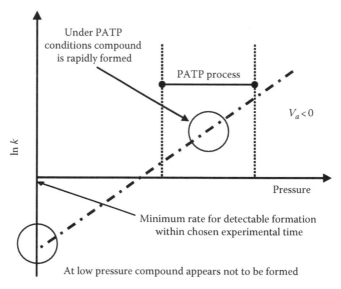

FIGURE 15.4   Potential new chemical pathways at high pressure and temperature. (Adapted from Pérez Lamela, C. and Torres, J. A., *AgroFOOD Industry Hi-Tech*, 19, 34, 2008b.)

A uniform temperature distribution is particularly important is some industrial processes that must take place in a maximum–minimum temperature threshold to avoid altering some properties of the product (e.g., protein stability, gelling of some components, fat migration, and others) or to assure a uniform distribution of the pursued pressure effect (Farkas and Hoover 2000, Ting et al. 2002). Several authors (Denys et al. 2000a, de Heij et al. 2002, Hartmann and Delgado 2003) have shown that small temperature differences have significant effects on the pressure inactivation rate of microorganisms and enzymes.

Securing temperature uniformity during the HPP of foods is a difficult challenge because of the complex heat and mass transfer phenomena involved. There are a considerable number of variables related to the pressure equipment, pressurizing fluid used, food product properties, and process parameters employed that will have strong effects on heat transfer phenomena in the pressure vessel. Analyzing the relative contribution of each variable to design high-pressure processes of high thermal uniformity is a complex challenge but mathematical modeling can be a useful tool to solve the problem. However, modeling a high-pressure process is a difficult task for several reasons including (1) lack of complete information on the thermal properties of the materials under pressure; (2) need for mathematical expressions on the pressure and temperature dependency of these thermal properties; (3) calculation methods for the temperature increase and decrease of the pressurizing fluid and the food as a result of the pressure build up and release, respectively; (4) estimation of the convective heat transfer that takes place between the inner wall of the vessel and the pressurizing fluid; and (5) heat transfer from the fluid to the product involving convection or conduction heating if the product can be considered liquid or solid, respectively (Otero and Sanz 2003). All these issues will be analyzed in this chapter.

15.4.1 Thermal Properties under Pressure

One of the main difficulties when developing temperature prediction models for high-pressure processes is the lack of thermophysical properties of the pressure-transmitting media, food products, and their packaging materials as affected by pressure. In spite of enormous efforts made in the last years for determining thermophysical properties under pressure of food and compressing fluids, more research in this area is urgently needed and knowledge of these properties is essential for the optimization of high-pressure technology. Thermophysical properties of interest include density, heat capacity, thermal conductivity, thermal expansion coefficient, and viscosity, which are generally unavailable in the literature. Technological problems hinder their experimental determination and it has been necessary to adapt existing techniques and devices and develop new instruments capable of working under high-pressure conditions (Barbosa-Cánovas and Rodríguez 2005). At present, only a few thermal properties have been measured under high pressures for a limited number of food products including density and volume expansion coefficient in tomato paste, apple sauce, and agar gel (Denys 2000, Denys et al. 2000a,b), diffusivity and volumetric heat capacity in fresh potato and Cheddar cheese (Zhu et al. 2007) and thermal conductivity in tomato paste and apple pulp (Denys and Hendrickx 1999), aqueous sugar solutions (Werner et al. 2007), and in apple juice, canola oil, clarified butter, honey, and high fructose corn syrup (Ramaswamy et al. 2007). Shariaty-Niassar et al. (2000) determined the thermal conductivity of gelatinized potato starches in a rather wide range of temperatures but only at relatively low pressures (0.2–10 MPa) and thus not very useful for high-pressure food processing. The methods employed for the experimental determinations are described in depth in the publications cited for each product.

Different approaches have been used in the literature for their prediction. Denys et al. (1997) shifted the atmospheric pressure conductivity data of tylose according to the freezing point depression associated with the pressure applied. Similarly, Schlüter et al. (1999) fitted cumulative and density Weibull distributions to thermal conductivity and apparent specific heat data for potato at atmospheric conditions. Then, they modified these distributions according to the freezing point depression induced by pressure, maintaining the scale and shape parameters. Hartmann and Delgado (2003) and Ghani and Farid (2007) multiplied the physical property of interest for food at atmospheric pressure, $X_F(P_{atm}, T)$, by a constant representing the ratio of the physical property for water at high pressure, $X_W(P, T)$, to that of water at atmospheric pressure, $X_W(P_{atm}, T)$:

$$X_F(P, T) = X_F(P_{atm}, T) \frac{X_W(P, T)}{X_W(P_{atm}, T)} \qquad (15.9)$$

where
  P is pressure
  P_{atm} is atmospheric pressure
  T is temperature

The thermophysical properties for water from the melting-pressure curve to 1000°C and pressures up to 1000 MPa can be obtained "for general and scientific use" from the International Association for the Properties of Water and Steam (1996). They are also available from calculation routines reviewed by Otero et al. (2002), which include water in the supercooled state.* In addition, Otero et al. (2006) developed an additive predictive model for the thermophysical properties of frozen and unfrozen foods under pressure based on their water content. The model uses the properties of ice and water calculated at the desired pressure/temperature conditions and considers the intrinsic properties of the dry matter to be independent of pressure and temperature.

The methods described above need as a starting point thermophysical data at atmospheric conditions to predict the corresponding values under high pressure for the food involved and accurate predictions are not always guaranteed (Otero et al. 2006). Some determinations have been using reverse methods, that is, experiments are performed in which the sample temperature is recorded during the pressure treatment and then numerical simulations are carried out varying the values of the thermophysical parameter involved to obtain predicted temperatures in optimal agreement with the experimental data. Kowalczyk et al. (2005) obtained density,

---

* These calculation routines are available through http:/www.if.csic.es/programas/ifiform.htm

specific heat, and thermal conductivity values of potatoes, pork, and cod at different pressure in unfrozen and frozen states assuming nontemperature dependent properties. A similar approach was used by Chen et al. (2007) to obtain the thermal conductivity of pork meat up to 200 MPa.

15.4.2 Temperature Prediction Models for High-Pressure Processes

Several mathematical models have been used to predict the temperature evolution during HPP. An artificial neural network approach was used to predict two descriptive process parameters, that is, the maximum temperature reached in the sample after pressurization and the time needed to thermal re-equilibration in the high-pressure system (Torrecilla et al. 2004, 2005). Both variables together give a rough picture of the temperature evolution during the high-pressure process. The key advantage of artificial neural networks is their capability to learn from examples without requiring a previous knowledge of the relationships between process parameters. Therefore, equations describing the physical relations between them and appropriate thermal properties under pressure are not needed. Nevertheless, they require an important training phase and the information that can be extracted from the model is rather limited. On the contrary, physical-based models require equations suitable for describing all the heat transfer phenomena occurring during the compression, pressure-holding, and decompression phases. This type of models is much more complicated to build but they can provide a complete and accurate vision of all the parameters involved and also of their relationships. These models can become powerful tools for process optimization and will be the focus of this section. This will begin with a comprehensive vision of the heat transfer phenomena which take place in the different phases of a typical high-pressure treatment presented along with a review of the physical-based models describing them.

Compression/decompression phases are mainly characterized by a pressure change that induces a temperature increase/decrease due to the work of compression/expansion in the food and pressurizing fluid. If the pressure-dependent thermal properties of the food and pressurizing fluid are known, the calculation of their temperature increase/decrease after an adiabatic pressure change can be modeled by solving the following equation:

$$\frac{dT}{dP} = \frac{T \cdot V \cdot \alpha}{c_p} \tag{15.10}$$

where
  T is the temperature (K)
  P is the pressure (Pa)
  V is the specific volume (m³/kg)
  α is the thermal expansion coefficient (K⁻¹)
  c_p is the specific heat capacity (J/kg·K)

Under adiabatic conditions, dT/dP in Equation 15.2 depends only on the initial temperature of the product and its composition defined by the corresponding specific volume, thermal expansion coefficient and specific heat capacity. Since these parameters depend on T and P, the calculation of dT/dP is complex. Theoretically calculated dT/dP values for liquid water after quasiadiabatic expansions were compared to experimental data and found to be in good agreement (Otero et al. 2000). In a study of the compression heating of a wide range of real foods, including vegetable oils, fish, meat, egg, and milk products, fats and oils were found to high compression heating factors with temperature increasing up to 8.7°C/100 MPa as compared to 2°C–3°C/100 MPa for water (Shimizu 1992, Rasanayagam et al. 2003, Patazca et al. 2007).

Adiabatic conditions are not observed in most processes, particularly when pressure is increased or decreased slowly. The heat transfer between the sample, its packaging, and the pressure medium can be very important when oil-based pressurizing fluids are employed due to their high heat compression factors in comparison to those of most food products. Moreover, after pressurization, the temperature of the food to be treated and that of the pressurizing fluid will be higher than the temperature of the high-pressure vessel. As a result, heat will flow to the vessel walls and the product fraction close to it will cool down and will differ from the temperature reached in the center of the vessel (de Heij et al. 2002, Ting et al. 2002). The impact of the different components can be analyzed using a model based on ordinary differential equations that take into account all the thermal exchanges, that is, sample, pressurizing fluid, high-pressure vessel, and thermoregulating system. This approach revealed the enormous significance of the steel mass of the high-pressure vessel on the thermal control of the process (Otero et al. 2002). The model allowed the assessment of several conditions to secure stable temperature during the pressure treatment including the initial sample temperature, thermoregulation system temperatures, and heating capacity or number of baths. However, the ordinary differential equations model did not reproduce the thermal gradients established inside the pressure vessel. To define these thermal gradients established during compression, pressure-holding, and decompression phases, it is necessary to employ partial differential equations and develop a numerical heat transfer model.

Numerical approaches are of special interest when sample and pressure medium undergo very different temperature changes during compression and decompression. Also, they are a useful tool to model thermal histories in foods with nonhomogeneous composition. For example, temperature will increase more in the adipose depots of pressurized meat than in other meat locations with lower fat content. Moreover, a numerical approach allows the handling of nonlinearity problems caused by the pressure and temperature dependency of thermal and physical properties. A numerical approach for modeling conductive heat transfer during batch HPP of food products based on a numerical solution of the Fourier equation for a conductive finite cylinder has been developed (Denys et al. 2000a,b):

$$\rho \cdot C_p \frac{\partial T}{\partial t} - \nabla \cdot (k \nabla T) = Q_p \quad \text{in } \Omega \tag{15.11}$$

where
   Ω is the cylindrical domain defined by the pressure chamber (sample and pressure medium included)
   ρ is the density (kg/m^3)

k is the thermal conductivity (W/m·K)
t is the time (s)

The calculation of the temperature rise/fall during the pressure buildup/release was solved by incorporating Equation 15.2 in the Fourier equation as an internal heat generation/reduction term (Q_p). This numerical model could accurately describe the conductive heat transfer during different high-pressure experiments but it ignores an important heat exchange mechanism in pressure processing. Experimental work revealed the important contribution of convection currents in the high-pressure medium layer located between the solid sample and the inner wall of the high-pressure vessel (Denys et al. 2000a,b). To account for convection currents rather than assuming by pure conduction, an "apparent" thermal conductivity value was assumed (Denys et al. 1997). An alternative was to work with samples with the maximum radial dimensions allowed by the pressure chamber and thus minimize the impact of convective heat transfer in the pressure medium (Denys et al. 2000a,b).

During the compression phase in most systems, pressurizing fluid, frequently at ambient temperature, is introduced into the treatment chamber increasing the pressure to the target value and creating a fluid flow governed by forced convection. Moreover, the compression heat generated during the pressurization produces heterogeneous temperature distributions inside the pressure vessel. The temperature fields established in liquid water after pressurization can be visualized and photographed using thermochromic liquid crystals (Pehl et al. 2000). These temperature fields yield a heterogeneous density distribution in the liquid sample inducing a buoyancy fluid motion resulting in free convection during the pressure-holding phase. Periods of forced convection during the compression and expansion phases with the pressure-holding phase characterized by free convection have been observed (Rademacher et al. 2002). These observations have led to the need of using computational fluid dynamics (CFD) models to predict the flow behavior of fluid food and pressure mediums taking into account the governing equations, that is, the Navier–Stokes' continuity and energy equations. These models take into account the redistribution of momentum and energy and cannot be described only by the thermodynamic parameters density, pressure, and temperature. When free convection is considered, the global mass and momentum balances for the pressure medium (and also for the food sample if it is in liquid state) are given by

$$\rho \frac{\partial u}{\partial t} - \nabla \cdot \eta (\nabla u + (\nabla u)^T) + \rho (u \cdot \nabla) u + \nabla P = g\rho \text{ in } \Omega_p \text{ and } \Omega_{LS} \quad (15.12)$$

$$\nabla \cdot (\rho u) = 0 \text{ in } \Omega_p \quad (15.13)$$

where
Ω_p is the part of Ω filled by the pressurizing fluid
Ω_{LS} is the part of Ω corresponding to the liquid sample
u is the velocity (m/s)
η is the dynamic viscosity (Pa·s)
g is the gravity vector (m/s^2)

The momentum and energy balances are coupled in the pressure medium (and within the sample if it is liquid) and transport occurs by convection and conduction, according to

$$\rho C_p \frac{\partial T}{\partial t} + \nabla \cdot (-k\nabla T) + \rho C_p u \nabla \cdot T = Q_p \quad \text{in } \Omega \qquad (15.14)$$

Using this kind of numerical simulations, several authors have shown how the fluid motion generated by free and forced convection, especially during the early stages of compression, strongly influences the temporal and spatial distribution of the temperature (Hartmann 2002, Hartmann and Delgado 2002, 2003, Hartmann et al. 2003, 2004, Ghani and Farid 2007, Otero et al. 2007). The simulation of convective and diffusive transport effects allows to accurately evaluate the relative importance of process parameters such as the compression rate (Hartmann 2002), the size of the high-pressure vessel (Hartmann and Delgado 2002, 2003), or the effects of additional compression cycles during the pressure-holding phase (Hartmann et al. 2004). Otero et al. (2007) showed how solid foods treated under pressure are also influenced by the free convection induced in the pressure medium after pressurization. Convection currents in the pressure medium play an important role in the thermal evolution of the processed solid samples particularly when the filling ratio in the pressure vessel is low (Figure 15.5).

Several authors have utilized CFD models to show that high-pressure induced effects such as microbial and enzyme inactivation are highly influenced by the thermal–hydraulic phenomena that occur during HPP of foods. Differences in the pressure–temperature–time profiles applied at different locations in the high-pressure vessel may result in a pronounced nonuniform distribution of residual enzyme and/or microbial activity and in nutrient and sensorial quality degradation within the pressure-processed product (Denys et al. 2000a). Depending on the pressure–temperature degradation kinetics of the component under examination, this nonuniformity will be more or less pronounced. De Heij et al. (2002) developed an integrated mathematical model combining thermodynamics and inactivation kinetics to illustrate the effect of the temperature distribution in the high-pressure vessel on the inactivation of *Bacillus stearothermophilus*. Hartmann and Delgado (2002) used dimensional analyses to determine the timescales of convection, conduction, and bacterial inactivation to determine which relative values contribute to the efficiency and uniformity of pressure processing. Hartmann et al. (2003) studied the inactivation of *E. coli* suspended in packed UHT milk showing the importance of the geometrical scale of the high-pressure vessel and how the package material of the food sample strongly influences the microbial inactivation achieved. Low thermal conductivity materials allowed the preservation of the high temperature reached after milk compression leading to increased microbial reductions. The combination of CFD models with kinetic models of pressure–temperature inactivation of microorganisms and enzymes can be used as powerful tools to quickly optimize PATP processes representing a promising alternative to experimental testing, which can be restricted to validation of model predictions.

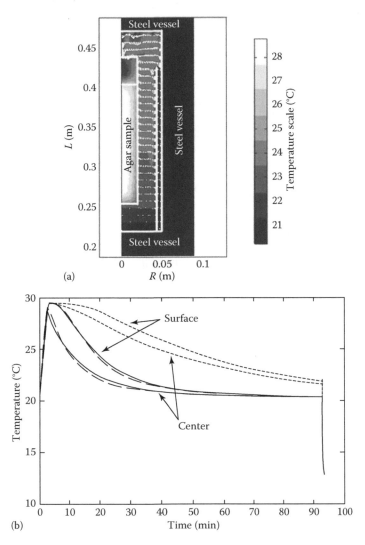

FIGURE 15.5 Temperature profile in high pressure vessels. Processing to 350 MPa of an agar gel using watr as pressurizing fluid (initial sample and pressure medium temperature = 20°C). (a) Temperature and velocity fields inside the high-pressure vessel 5 minutes after reaching 350 MPa; (b) Agar sample surface and center temperature. (—): Experimental data, (– – –): Predicted values assuming heat-conduction is the only heat transfer mechanisms in the pressure medium, (- - -): Predicted values when free convection in the pressure medium is also taken into account. (Adapted from Otero, L. et al., *J. Food Eng.*, 78, 1463, 2007.)

CFD models have also been employed to seek for strategies to reach a high degree of thermal uniformity since they can accurately describe all the phenomena of heat and mass transfer taking place during the high-pressure treatment of food. The thermal insulation of the high-pressure chamber inner wall and the use of pressurizing fluids of low viscosity are examples of successful strategies (Hartmann et al. 2004).

Using this kind of models, Otero et al. (2007) designed high-pressure processes of high thermal uniformity by determining the initial temperature for the sample and pressurizing fluid, considering and taking advantage of the temperature increase resulting from the compression phase.

Finally, most studies reported were performed using laboratory-scale pressure vessels. The temperature histories may become quite different when the same treatments are carried out in large-scale equipments (Ting et al. 2002). An accurate modeling of the temperature evolution would help to determine the correct scale-up to industrial size of processes developed in HPP units of laboratory size.

15.5 RECOMMENDATIONS FOR PROPER EXPERIMENTAL PROCEDURES

The previous section revealed the many difficulties in developing high-pressure processes of high thermal uniformity. A considerable number of variables specific to the pressure equipment and pressurizing fluid used, the food product treated, and the process parameters employed are involved and each one has an enormous influence on the results obtained. These difficulties along with an inadequate reporting of the experimental methodologies used combined with a poor understanding of the phenomena involved may hinder comparison of microbial and enzymatic inactivation experiments between laboratories. To overcome these limitations, several authors have made recommendations on practical means to control the thermal effects during pressure processing and of the parameters that must be reported to characterize a specific pressure treatment as completely as possible (Meyer et al. 2000, Ting et al. 2002, Balasubramaniam et al. 2004). Following these guidelines will facilitate comparisons of experimental findings and allow their confirmation even when using different pressure-processing equipment.

The high-pressure equipment used in the experimentation must be adequately described from a process engineering point of view. As previously indicated, the vessel size, chamber dimensions, material of construction, wall thickness, pressure-transmitting fluid, and the heating and cooling system are factors that will strongly influence the heat and mass transfer phenomena taking place during the pressure processing of food. The composition of the sample, its size, shape, and location are also essential variables. Moreover, it is desirable to provide a sample graph illustrating the time–temperature evolution of the processed product. The come-up, processing and decompression times and the process pressure also need to be reported. Initial temperatures of the product, pressure-transmitting fluid, and the vessel must be documented and special care must be taken to ensure uniform thermal initial conditions. Moreover, since after compression the sample temperature may change considerably, it is desirable to have thermocouple sensors at the surface and at the center of the product. In noninsulated pressure vessels, the coldest region of the sample is located near the wall and not in the axial center of the pressure chamber where the temperature sensors are usually located. When several temperature sensors are not possible, it is essential to specify the location of the thermocouple in the sample tested. The importance of the composition of the pressure medium (determining the compression heat and its viscosity and diffusivity) along with the mass ratio between the fluid and the sample has already been commented. It is important also to monitor and report the temperature of the pressurizing fluid pumped into the vessel during the compression phase as it may enter the system at a much lower temperature than that of the thermoregulated chamber. Depending on the total amount introduced, which is a function of the global compressibility coefficient of the different materials inside the vessel, it may cause important thermal effects. Therefore, means should be considered to control its temperature before entering the pressure vessel. Finally, the importance of the samples packing must not be underestimated. Authors should report the geometry, composition, and thickness of the packaging materials employed.

15.6 FINAL REMARKS

Findings on the formation kinetics in PATP foods represent a significant improvement in the understanding of the effect of temperature and pressure on reaction rates in foods and support the need for further PATP research to develop products meeting current consumer demand for foods with minimal effects of processing. The likelihood that PATP produces new compounds of unknown safety needs to be investigated too. An equally important is to continue the development of models to predict and manipulate the temperature distribution inside the vessel and in the food to generate the estimation of food quality and safety required by commercial processors and regulatory agencies.

REFERENCES

Balasubramaniam, V. M., Ting, E. Y., Stewart, C. M., and Robbins, J. A. 2004. Recommended laboratory practices for conducting high-pressure microbial inactivation experiments. *Innovative Food Science and Emerging Technologies* 5: 299–306.
Barbosa-Cánovas, G. V. and Rodríguez, J. J. 2005. Thermodynamic aspects of high hydrostatic pressure food processing. In Barbosa-Cánovas, G. V., Tapia, M. S., and Cano, M. P., editors. Novel Food Processing Technologies. New York, NY: Marcel Dekker, Inc. pp. 183–205.
Butz, P., Bognar, A., Dieterich, S., and Tauscher, B. 2007. Effect of high-pressure processing at elevated temperatures on thiamin and riboflavin in pork and model systems. *Journal of Agricultural and Food Chemistry* 55: 1289–1294.
Butz, P., Serfert, Y., and Fernandez Garcia, A. et al. 2004. Influence of high-pressure treatment at 25°C and 80°C on folates in orange juice and model media. *Journal of Food Science* 79: SNQ 117–121.
Butz, P. and Tauscher, B. 2000. Recent studies on pressure-induced chemical changes in food constituents. *High Pressure Research* 19: 11–18.
Calvo, M. M. and de la Hoz, L. 1992. Flavour of heated milks: A review. *International Dairy Journal* 2: 69–81.
Chen, C. R., Zhu, S. M., Ramaswamy, H. S., Marcotte, M., and Le Bail, A. 2007. Computer simulation of high pressure cooling of pork. *Journal of Food Engineering* 79: 401–409.
Christensen, K. R. and Reineccius, G. A. 1992. Gas chromatographic analysis of volatile sulfur compounds from heated milk using static headspace sampling. *Journal of Dairy Science* 75: 2098–2104.
Contarini, G. and Povolo, M. 2002. Volatile Fraction of Milk: Comparison between purge and trap and solid phase microextraction techniques. *Journal of Agricultural and Food Chemistry* 50: 7350–7355.

Contarini, G., Povolo, M., Leardi, R., and Toppino, P. M. 1997. Influence of heat treatment on the volatile compounds of milk. *Journal of Agricultural and Food Chemistry* 45: 3171–3177.

Corrales, M., Butz, P., and Tauscher, B. 2008. Anthocyanin condensation reactions under high hydrostatic pressure. *Food Chemistry* 110: 627–635.

Datta, N. and Deeth, H. C. 1999. High pressure processing of milk and dairy products. *Australian Journal of Dairy Technology* 54: 41–48.

Datta, N., Elliot, A. J., Perkins, M. L., and Deeth, H. C. 2002. Ultra-high-temperature (UHT) treatment of milk: comparison of direct and indirect modes of heating. *Australian Journal of Dairy Technology* 57: 211–227.

de Ancos, B., Sgroppo, S., Plaza, L., and Cano, M. P. 2002. Possible nutritional and health-related value promotion in orange juice preserved by high-pressure treatment. *Journal of the Science of Food and Agriculture* 82: 790–796.

de Heij, W., van Schepdael, L., van den Berg, R., and Bartels, P. V. 2002. Increasing preservation efficiency and product quality through control of temperature profiles in high pressure applications. *High Pressure Research* 22: 653–657.

Denys S. 2000. Process calculations for design of batch high hydrostatic pressure processes without and with phase transitions [Ph. D.]. Leuven, Belgium: Katholieke Universiteit.

Denys, S. and Hendrickx, M. E. 1999. Measurement of the thermal conductivity of foods at high pressure. *Journal of Food Science* 64: 709–713.

Denys, S., van Loey, A. M., Hendrickx, M. E., and Tobbak, P. P. 1997. Modeling heat transfer during high-pressure freezing and thawing. *Biotechnology Progress* 13: 416–423.

Denys, S., Ludikhuyze, L. R., van Loey, A. M., and Hendrickx, M. E. 2000a. Modeling conductive heat transfer and process uniformity during batch high-pressure processing of foods. *Biotechnology Progress* 16: 92–101.

Denys, S., van Loey, A. M., and Hendrickx, M. E. 2000b. A modeling approach for evaluating process uniformity during batch high hydrostatic pressure processing: Combination of a numerical heat transfer model and enzyme inactivation kinetics. *Innovative Food Science and Emerging Technologies* 1: 5–19.

Farkas, D. F. and Hoover, D. G. 2000. High pressure processing. *Journal of Food Science* 65 (Supplement: Kinetics of Microbial Inactivation for Alternative Food Processing Technologies): 47–64.

Ghani, A. G. and Farid, M. M. 2007. Numerical simulation of solid-liquid food mixture in a high pressure processing unit using computational fluid dynamics. *Journal of Food Engineering* 80: 1031–1042.

Hartmann, C. 2002. Numerical simulation of thermodynamic and fluid-dynamic processes during the high-pressure treatment of fluid food systems. *Innovative Food Science and Emerging Technologies* 3: 11–18.

Hartmann, C. and Delgado, A. 2002. Numerical simulation of convective and diffusive transport effects on a high-pressure-induced inactivation process. *Biotechnology and Bioengineering* 79: 94–104.

Hartmann, C. and Delgado, A. 2003. The influence of transport phenomena during high-pressure processing of packed food on the uniformity of enzyme inactivation. *Biotechnology and Bioengineering* 82: 725–735.

Hartmann, C., Delgado, A., and Szymczyk, J. 2003. Convective and diffusive transport effects in a high pressure induced inactivation process of packed food. *Journal of Food Engineering* 59: 33–44.

Hartmann, C., Schuhholz, J. P., Kitsubun, P., Chapleau, N., le Bail, A., and Delgado, A. 2004. Experimental and numerical analysis of the thermofluid dynamics in a high-pressure autoclave. *Innovative Food Science and Emerging Technologies* 7: 1–12.

Harvey, A. E., Peskin, A. P., and Klein, S. A. 1996. NIST/ASME steam program. Physical and Chemical Properties Div., Natl. Inst. of Standards and Technology, US. Dept. of Commerce, Boulder, CO.

Hepburn, P., Howlett, J., Boeing, H., Cockburn, A., Constable, A. D., de Jong, N., Moseley, B., Oberdörfer, R., Robertson, C., Walk, J. M., Samuels, F. 2008. The application of post-market monitoring to novel foods. *Food Chemistry and Toxicology* 46: 9–33.

Hoover, D. G., Metrick, C., Papineau, A. M., Farkas, D. F., and Knorr, D. 1989. Biological effects of high hydrostatic pressure on food microorganisms. *Food Technology* 43: 99–107.

Indrawati, Arroqui, C., Messagie, I., Nguyen, M. T., van Loey, A., and Hendrickx, M. E. 2004. Comparative study on pressure and temperature stability of 5-methyltetrahydrofolic acid in model systems and in food products. *Journal of Agricultural and Food Chemistry* 52: 485–492.

Indrawati, van Loey, A., and Hendrickx, M. E. 2005. Pressure and temperature stability of 5-methyltetrahydrofolic acid: a kinetic study. *Journal of Agricultural and Food Chemistry* 53: 3081–3087.

International Association for the Properties of Water and Steam (IAPWS). 1996. Release on the IAPWS Formulation 1995 for the thermodynamic properties of ordinary water substance for general and scientific use, Fredericia, Denmark, 8–14 September, 1996.

Kim, Y. S., Park, S. J., Cho, Y. H., and Park, J. 2001. Effects of combined treatment of high hydrostatic pressure and mild heat on the quality of carrot juice. *Journal of Food Science* 66: 1355–1360.

Kowalczyk, W., Hartmann, C., Luscher, C., Pohl, M., Delgado, A., and Knorr, D. 2005. Determination of thermophysical properties of foods under high hydrostatic pressure in combined experimental and theoretical approach. *Innovative Food Science and Emerging Technologies* 6: 318–326.

Krebbers, B., Matser, A., Koets, M., Bartels, P. V., and van den Berg, R. 2002. High pressure-temperature processing as an alternative for preserving basil. *High Pressure Research* 22: 711–714.

Kübel, J., Ludwig, H., and Tauscher, B. 1997. Influence of UHP on vitamin A. In: Heremans, K., editor. High pressure research in the biosciences and biotechnology. Leuven: University Press. pp. 331–334.

Langler, J. E. and Day, E. A. 1964. Development and flavor properties of methyl ketones in milk fat. *Journal of Dairy Science* 47: 1291–1296.

McNaught, A. D. and Wilkinson, A. 1997. *Compendium of Chemical Terminology: IUPAC Recommendations*. Malden, MA: Blackwell Science. p. 450.

Meyer, R. S., Cooper, K. L., Knorr, D., and Lelieveld, H. L. M. 2000. High-pressure sterilization of foods. *Food Technology* 54: 67, 68, 70, 72.

Mussa, D. M. and Ramaswamy, H. S. 1997. Ultra high pressure pasteurization of milk: Kinetics of microbial destruction and changes in physico-chemical characteristics. *Lebensmittel Wissenschaft und Technologie* 30: 551–557.

Nguyen, M. T., Indrawati, and Hendrickx, M. E. 2003. Model studies on the stability of folic acid and 5-methyltetrahydrofolic acid degradation during thermal treatment in combination with high hydrostatic pressure. *Journal of Agricultural and Food Chemistry* 51: 3352–3357.

Nguyen, M. T., Oey, I., Hendrickx, M. E., and van Loey A. 2006. Kinetics of (6R,S) 5-formyltetrahydrofolic acid isobaric–isothermal degradation in a model system. *European Food Research and Technology* 223: 325–331.

Oey, I., Verlinde, P., Hendrickx, M. E., and van Loey, A. 2006. Temperature and pressure stability of L-ascorbic acid and/or [6s] 5-methyltetrahydrofolic acid: A kinetic study. *European Food Research and Technology* 223: 71–77.

Otero, L., Molina-García, A. D., and Sanz, P. D. 2000. Thermal effect in foods during quasi-adiabatic pressure treatment. *Innovative Food Science and Emerging Technologies* 1: 119–126.

Otero, L., Molina-García, A. D., and Sanz, P. D. 2002. Some interrelated thermophysical properties of liquid water and ice I: A user-friendly modelling review for high-pressure processing (http://www.if.csic.es/programs/ifiform.htm). *Critical Reviews in Food Science and Nutrition* 42: 339–352.

Otero, L., Ousegui, A., Guignon, B., le Bail, A., and Sanz, P. D. 2006. Evaluation of the thermophysical properties of tylose gel under pressure in the phase change domain. *Food Hydrocolloids* 20: 449–460.

Otero, L., Ramos, A. M., de Elvira, C., and Sanz, P. D. 2007. A model to design high-pressure processes towards a uniform temperature distribution. *Journal of Food Engineering* 78: 1463–1470.

Otero, L. and Sanz, P. D. 2003. Modelling heat transfer in high pressure food processing: A review. *Innovative Food Science and Emerging Technologies* 4: 121–134.

Patazca, E., Koutchma, T., and Balasubramaniam, V. M. 2007. Quasi-adiabatic temperature increase during high pressure processing of selected foods. *Journal of Food Engineering* 80: 199–205.

Pehl, M., Werner, F., and Delgado, A. 2000. First visualization of temperature fields in liquids at high pressure. *Experiments in Fluids* 29: 302–304.

Pérez Lamela, C. and Torres, J. A. 2008a. Pressure-assisted thermal processing: A promising future for high flavour quality and health-enhancing foods – Part 1. *AgroFOOD Industry Hi-Tech* 19: 60–62.

Pérez Lamela, C. and Torres, J. A. 2008b. Pressure-assisted thermal processing: 2. Microbial inactivation kinetics and pressure and temperature effects on chemical changes. *AgroFOOD Industry Hi-Tech* 19: 34–36.

Rademacher, B., Werner, F., and Pehl, M. 2002. Effect of the pressurizing ramp on the inactivation of *Listeria innocua* considering thermofluid dynamical processes. *Innovative Food Science and Emerging Technologies* 3: 13–24.

Ramaswamy, R., Balasubramaniam, V. M., and Sastry, S. K. 2007. Thermal conductivity of selected liquid foods at elevated pressures up to 700 MPa. *Journal of Food Engineering* 83: 444–451.

Ramirez, R., Saraiva, J. A., Perez Lamela, C., and Torres, J. A. 2009. Reaction kinetics analysis of chemical changes in pressure-assisted thermal processing, PATP. *Food Engineering Reviews* (In press).

Rasanayagam, V., Balasubramaniam, V. M., Ting, E. Y., Sizer, C. E., Bush, C., and Anderson, C. 2003. Compression heating of selected fatty food materials during high-pressure processing. *Journal of Food Science* 68: 254–259.

Rychlik, M., Schieberle, P., and Grosch, W. 1998. *Compilation of Odor Thresholds, Odor Qualities and Retention Indices of Key Food Odorants.* Garching, Germany: Deutsche Forschungsanstalt für Lebensmittelchemie and Institut für Lebensmittelchemie der Technischen Universität München.

Sanchez Moreno, C., Plaza, L., Elez Martinez, P., de Ancos, B., Martin Belloso, O., and Cano, M. P. 2005. Impact of high pressure and pulsed electric fields on bioactive compounds and antioxidant activity of orange juice in comparison with traditional thermal processing. *Journal of Agricultural and Food Chemistry* 53: 4403–4409.

Scanlan, R. A., Lindsay, R., Libbey, L. M., and Day, E. A. 1968. Heat-induced volatile compounds in milk. *Journal of Dairy Science* 51: 1001–1007.

Schlüter, O., George, S., Heinz, V., and Knorr, D. 1999. Pressure assisted thawing of potato cylinders. In *Advances in High Pressure Bioscience and Biotechnology*, Ed. H. Ludwing, 475–480. Berlin, Germany: Springer-Verlag.

Shariaty-Niassar, M., Hozawa, M., and Tsukada, T. 2000. Development of probe for thermal conductivity measurements of food materials under heated and pressurized conditions. *Journal of Food Engineering* 43: 133–139.

Shibamoto, T., Mihara, S., Nishimura, O., Kamiya, Y., Aitoku, A., and Hayashi, J. 1980. Flavor volatiles formed by heated milk. In *The Analysis and Control of Less Desirable Flavors in Foods and Beverages*, Ed. G. Charalambous, 260–263. New York: Academic Press Inc.

Shimizu, T. 1992. High pressure experimental apparatus with windows for optical measurements up to 700 MPa. *High Pressure Bioscience and Biotechnology* 224: 525–527.

Simon, M., Hansen, A. P., and Young, C. T. 2001. Effect of various dairy packaging materials on the shelf life and flavor of ultrapasteurized milk. *Journal of Dairy Science* 84: 784–791.

Steely, J. S. 1994. Chemiluminiscence detection of sulfur compounds in cooked milk. In *Sulfur Compounds in Foods*, Eds. C. J. Mussinan and M. E. Keelan, 22–35. Chicago, IL: American Chemical Society.

Taoukis, P. S., Panagiotidis, P., Stoforos, N. G., Butz, P., Fister, H., and Tauscher, B. 1998. Kinetics of vitamin C degradation under high pressure-moderate temperature processing in model systems and fruit juices. Special Publication – Royal Society of Chemistry 222: 310–316.

Ting, E. Y., Balasubramaniam, V. M., and Raghubeer, E. 2002. Determining thermal effects in high-pressure processing. *Food Technology* 56: 31–35.

Torrecilla, J. S., Otero, L., and Sanz, P. D. 2004. A neural network approach for thermal/pressure food processing. *Journal of Food Engineering* 62: 89–95.

Torrecilla, J. S., Otero, L., and Sanz, P. D. 2005. Artificial neural networks: A promising tool to design and optimize high-pressure food processes. *Journal of Food Engineering* 69: 299–306.

Torres, J. A. and Velazquez, G. 2005. Commercial opportunities and research challenges in the high pressure processing of foods. *Journal of Food Engineering* 67: 95–112.

Torres, J. A. and Velazquez, G. 2008. Hydrostatic pressure processing of foods. In *Food Processing Operations Modeling: Design and Analysis*, Eds. S. Jun and J. Irudayaraj, 173–212. Boca Raton, FL: CRC Press/Taylor & Francis.

van den Broeck, I., Ludikhuyze, L., Weemaes, C., van Loey, A., and Hendrickx, M. E. 1998. Kinetics for isobaric–isothermal degradation of L-ascorbic acid. *Journal of Agricultural and Food Chemistry* 46: 2001–2006.

van Loey, A., Ooms, V., Weemaes, C., van den Broeck, I., Ludikhuyze, L., Indrawati, Denys, S., Hendrickx, M. E. 1998. Thermal and pressure–temperature degradation of chlorophyll in broccoli (*Brassica oleracea L. italica*) juice: A kinetic study. *Journal of Agricultural and Food Chemistry* 46: 5289–5294.

Vazquez Landaverde, P. A., Velazquez, G., Torres, J. A., and Qian, M. C. 2005. Quantitative determination of thermally derived off-flavor compounds in milk using solid-phase microextraction and gas chromatography. *Journal of Dairy Science* 88: 3764–3772.

Vazquez Landaverde, P. A., Torres, J. A., and Qian, M. C. 2006a. Effect of high-pressure-moderate-temperature processing on the volatile profile of milk. *Journal of Agricultural and Food Chemistry* 54: 9184–9192.

Vazquez Landaverde, P. A., Torres, J. A., and Qian, M. C. 2006b. Quantification of trace volatile sulfur compounds in milk by solid-phase microextraction and gas chromatography-pulsed flame photometric detection. *Journal of Dairy Science* 89: 2919–2927.

Vazquez Landaverde, P. A., Qian, M. C., and Torres, J. A. 2007. Kinetic analysis of volatile formation in milk subjected to pressure-assisted thermal treatments. *Journal of Food Science* 72: E389–E398.

Verlinde, P., Indrawati Oey, Hendrickx, M. E., and van Loey, A. 2008. High-pressure treatments induce folate polyglutamate profile changes in intact broccoli (*Brassica oleraceae L. cv. Italica*) tissue. *Food Chemistry* 111: 220–229.

Weemaes, C., Ooms, V., Indrawati et al. 1999. Pressure-temperature degradation of green color in broccoli juice. *Journal of Food Science* 64: 504–508.

Werner, M., Baars, A., Werner, F., Eder, C., and Delgado, A. 2007. Thermal conductivity of aqueous sugar solutions under high pressure. *International Journal of Thermophysics* 28: 1161–1180.

Zhu, S., Ramaswamy, H. S., Marcotte, M., Chen, C., Shao, Y., and le Bail, A. 2007. Evaluation of thermal properties of food materials at high pressure using a dual-needle line-heat method. *Journal of Food Science* 72: E49–E56.

16 Food Quality and Safety Issues during Pulsed Electric Field Processing

Michael O. Ngadi, LiJuan Yu, Malek Amiali,
and Enrique Ortega-Rivas

CONTENTS

16.1　INTRODUCTION

The increasing consumer demand for fresh-like and safe products is a limiting constraint on selection of food processing technologies. The well-established thermal-based methods are available and provide high degree of microbial safety. However, they also degrade product quality. Therefore, there is currently intense search for nonthermal technologies as alternatives or complementary to thermal processes. Although nonthermal technologies such as high electric field pulses have been demonstrated to have some advantages over the conventional thermal technologies, they have only recently started gaining recognition in the food industry.

Application of these nonthermal technologies offers interesting opportunities for mildly processed safe products with preserved sensory and nutritional qualities. However, the technologies are still mostly in the development stages. This chapter presents some recent advances in application of high-voltage, pulsed-electric fields (PEFs) in food processing with respect to quality and safety issues. Some of the critical parameters of PEF processing are identified and described. Potential applications of PEF in processing for food safety and quality are discussed.

Thermal processing has been used in the food industry for several decades with various degrees of success to inactivate pathogenic microbial load in food products. Heat effectively destroys microorganisms. However, it is the balancing act of preserving food quality and maintaining safety that makes thermal processing not very attractive for some products. The current trend in consumers' preference is for fresh-like, minimally processed, and high-quality products. Emerging technologies for nonthermal (or minimal thermal) processing include high-pressure processing, UV light irradiation, magnetic fields, electron irradiation, ozonation, and PEFs. Interest in these technologies has arisen from the desire to overcome the problems of traditional thermal processing. The technologies enable unique modes of energy transfer to foods and target biological cells to achieve inactivation or modification without significantly heating the products. These methods have their strengths and weaknesses depending on process objectives and the type of food to be processed. Among the various emerging technologies, ultrahigh hydrostatic pressure (UHP), PEF, and UV light methods seem to be the most promising. Studies are ongoing to improve understanding of these technologies. This chapter provides an up-to-date description of the PEF technology, review of developments, trends, and applications of the emerging technology with emphasis on food quality and process kinetics.

PEF processing involves the application of externally generated electric field across a food product with the intent of inactivating pathogenic microorganisms, modifying enzymes, intensifying some processes, or achieving some specific transformation in the product. The technology has long been used for cell hybridization and electrofusion in genetic engineering and biotechnology. Its application is based on the transformation or rupture of cells under a sufficiently high external electric field, resulting in increased permeability and electrical conductivity of the cellular material. This effect named dielectric breakdown (Zimmermann et al. 1976) or electroplasmolysis (McLellan et al. 1991) can be explained by two main factors: (1) electroporation, that is electroinduced formation and growth of pores in biomembranes as a result of their polarization; (2) denaturation of cell membranes as a result of their ohmic heating caused by the electric resistance of membranes which is typically much higher than that of cell sap content. Beside these, the physiological impact and electroosmotic effect may also influence electroplasmolysis efficiency (Weaver and Chizmadzhev 1996).

PEF technologies have been demonstrated to be a viable alternative to high temperature inactivation of microbial load in liquid foods such as fruit juices and milk (Knorr et al. 1994, Barbosa-Cánovas et al. 1999). The majority of research effort on PEF has been on liquid food pasteurization. However, these technologies have also been shown to be applicable for microstructural modification of vegetable, fish, and meat tissues (Wu and Pitts 1999, Angersbach et al. 2000, Gudmundsson and Mafsteinsson 2001), for intensification of juice yield and for increasing product quality in juice production (Bazhal 2001), for processing of vegetable raw materials (Papchenko et al. 1988a), and for winemaking and sugar production (Gulyi et al. 1994). PEF treatment significantly enhances certain food processing unit operations such as pressing (Bazhal and Vorobiev 2000), diffusion (Jemai 1997), osmotic dehydration (Rastogi et al. 1999), and drying (Ade-Omowaye et al. 2000). These processes generally impact quality of foods. Besides, PEF is known to impact some unique quality attributes to food products that may not be possible with other technologies.

PEF processing involves a short burst of high-voltage application to a food placed between two electrodes (Qin et al. 1995). When high electric voltage is applied, a large flux of electric current flows through food materials, which may act as electrical conductors due to the presence of electrical charge carriers such as large concentration of ions (Barbosa-Cánovas et al. 1999). In general, a PEF system consists of a high-voltage power source, an energy storage capacitor bank, a charging current limiting resistor, a switch to discharge energy from the capacitor across the food, and a treatment chamber. The bank of capacitors is charged by a direct current power source obtained from amplified and rectified regular alternative current main source. An electrical switch is used to discharge energy (instantaneously in millionth of a second) stored in the capacitor storage bank across the food held in the treatment chamber. Apart from those major components, some adjunct parts are also necessary. In case of continuous systems a pump is used to convey the food through the treatment chamber. A chamber cooling system may be used to diminish the ohmic heating effect and control food temperature during treatment. High-voltage and high-current probes are used to measure the voltage and current delivered to the chamber (Barbosa-Cánovas et al. 1999, Amiali et al. 2004, 2006b, Floury et al. 2005). Figure 16.1 shows a basic PEF treatment unit, while Figure 16.2 presents different chamber designs.

The type of electrical field waveform applied is one of the important descriptive characteristics of a PEF treatment system. The exponentially decaying or square waves are among the most common waveforms used. To generate an exponentially decaying voltage wave, a DC power supply charges the bank of capacitors that are connected in series with a charging resistor. When a trigger signal is applied, the charge stored in the capacitor flows through the food in the treatment chamber. Exponential waveforms are easier to generate from the generator point of view. Generation of square waveform generally requires a pulse-forming network (PFN)

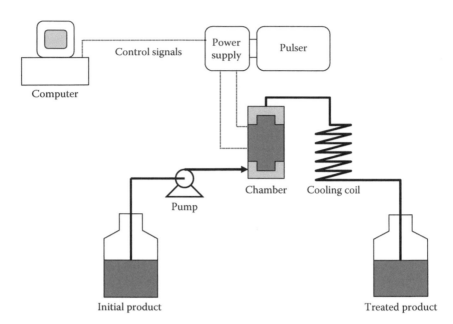

FIGURE 16.1 Schematic diagram of a PEF operation.

consisting of an array of capacitors and inductors. It is more challenging to design a square waveform system compared to an exponential waveform system. However, square waveforms may be more lethal and energy efficient than exponentially decaying pulses (Zhang et al. 1995a, Evrendilek et al. 2005, Amiali et al. 2006b). In order to produce effective square waveform using a PFN, the resistance of the food must be matched with the impedance of the PFN. Therefore, it is important to determine the resistance of the food in order to treat it properly.

The discharging switch also plays a critical role in the efficiency of the PEF system. The type of switch used will determine how fast it can perform and how much current and voltage it can withstand. In increasing order of service life, suitable switches for PEF systems include ignitrons, spark gaps, trigatrons, thyratrons, and semiconductors. Solid-state semiconductor switches are considered as the future of high-power switching. They present better performance and are easier to handle, require fewer components, allow faster switching times, and are more economically sound (Gongóra-Nieto et al. 2002).

16.2 APPLICATION OF PEF IN THE FOOD INDUSTRY

There is a growing interest in the application of PEF in food processing (Barbosa-Cánovas et al. 1999, Dutreux et al. 2000, Fleischman et al. 2004, Floury et al. 2005, Huang et al. 2006, Sobrino-Lopez et al. 2006). Generally, applications of PEF in food processing have been directed at two main categories namely microbial inactivation and preservation of liquid foods, as well as enhancement of mass transfer and texture in solids and liquids.

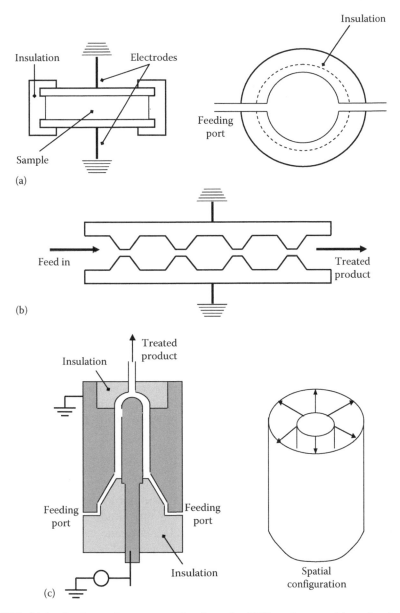

FIGURE 16.2 Designs of treatment chambers for PEF equipment: (a) static chamber, (b) side view of a basic continuous design, and (c) coaxial chamber.

Large portion of work on PEF has been focused on reducing microbial load in liquid or semisolid foods in order to extend their shelf life and assure their safety. The products that have been mostly studied include milk (Dunn and Pearlman 1987, Grahl and Märkl 1996, Sensoy et al. 1997, Reina et al. 1998, Dutreux et al. 2000, Fleischman et al. 2004, Evrendilek et al. 2005), apple juice (Vega-Mercado et al.

1997, Ortega-Rivas et al. 1998, Zárate-Rodríguez et al. 2000, Charles-Rodríguez et al. 2007), orange juice (Zhang et al. 1997, Yeom et al. 2000, Jia et al. 1999), and liquid egg (Jeantet et al. 1999, 2004, Amiali et al. 2004, 2006a,b, Hermawan et al. 2004). These studies and others have reported successful PEF-inactivation of pathogenic and food spoilage microorganisms as well as selected enzymes, resulting in better retention of flavors and nutrients and fresher taste compared to heat pasteurized products (Barbosa-Cánovas et al. 1999, Ho and Mittal 2000, Barsotti et al. 2002, Bendicho et al. 2002, Van Loey et al. 2002, Espachs-Barroso et al. 2003, Sepúlveda-Ahumada et al. 2005a,b, Sobrino-Lopez et al. 2006).

Another area that is showing a great potential is applying PEF on plant tissues as a pretreatment to enhance subsequent processes such as juice extraction (Bazhal and Vorobiev 2000, Eshtiaghi and Knorr 2002) and dehydration (Angersbach and Knorr 1997, Rastogi et al. 1999, Ade-Omowaye et al. 2000, Taiwo et al. 2002, Lebovka et al. 2007).

16.3 MECHANISM OF MICROBIAL INACTIVATION BY PEF

PEF treatment causes electroporation (generation of pores) of the cell membrane, leading consequently to microbial destruction and inactivation (Tsong 1991, Knorr et al. 1994, Ho and Mittal 1996, Pothakamury et al. 1996, García et al. 2007). Although it is still unclear whether the pore formation occurs in the lipid or the protein matrices, it is believed that electric fields induce structural changes in the membranes of microbial cells based on the transmembrane potential, electromechanical compression, and the osmotic imbalance theories (Zimmermann 1986, Barbosa-Cánovas et al. 1999, Gongóra-Nieto et al. 2002, Ohshima and Sato 2004).

16.3.1 TRANSMEMBRANE POTENTIAL

The membrane in a biological cell acts as an insulator to the cytoplasm, whose electrical conductivity is six to eight orders of magnitude greater than that of the membrane (Chen and Lee 1994). The cell membrane can be regarded as a capacitor filled with a low dielectric constant material ($\varepsilon \approx 2$). When a certain electric field is applied to the cell suspension, the ions inside the cell move along the field until the free charges are accumulated at both membrane surfaces. This accumulation of charge increases the electromechanical stress or transmembrane potential (V_t), to a value that is much greater than the applied electric field (Zimmermann 1986). The V_t gives rise to a pressure that causes the membrane thickness to decrease. A further increase in the electric field intensity reaching a critical transmembrane potential (V_c) leads to a reversible membrane breakdown (pore formation). When the size and number of the pore became larger compared to the membrane surface, irreversible breakdown occurs (Zimmermann 1986, Chen and Lee 1994, Sepúlveda-Ahumada et al. 2005a,b). For a given PEF treatment conditions, the induced potential across the cell membrane is proportional to the size of the microorganism.

16.3.2 Electromechanical Compression

Naturally, the charges on the capacitor plates of the biological cell membrane attract each other. This causes a thinning of the membrane provided that the membrane is compressible (Ho and Mittal 1996). The membrane thickness attained at a given membrane potential is determined by the equilibrium between the electric compression forces and the resulting electric restoring forces. With increasing membrane potential, a critical membrane thickness is reached at which an electric compressive force changed more rapidly than the generated electric restoring forces. The membrane becomes unstable and may be broken through. The emerging pores fill up the internal and external solution, both of which are highly conducting. The resulting increase in the electrical permeability of the membrane leads to a very rapid discharge of the membrane capacitor. An increase in the intensity of the external field will lead first to membrane breakdown at the poles of the cells. The required field strength for this transmembrane breakdown is in the range of 1–20 kV/cm depending on cell radius. The breakdown voltage itself is of the order of 1 V depending on temperature, field duration, and so on. At higher field strength, the breakdown voltage is reached for other membrane sites (Coster and Zimmerman 1975, Ohshima and Sato 2004).

16.3.3 Osmotic Imbalance

It is believed that the cause of membrane rupture is due to the osmotic imbalance generated by the leakage of ions and small molecules induced by the PEF treatment (Kinosita and Tsong 1977). Due to the osmotic pressure of the cytoplasmic content, the cell begins to swell and the pores gradually shrink. When the volume of the cell reaches 155% of its normal volume, rupture of the cell membrane and lysis of the cell occur (Tsong 1990).

Vega-Mercado (1996) further confirmed the osmotic imbalance theory. The author investigated pH, ionic strength effect, and PEF combined effect on *Escherichia coli* inactivation and found that the inactivation of microorganisms is caused mainly by an increase in their membrane permeability due to compression and poration. More than 2 log reductions in plate counts are observed when both pH and electric field are modified: pH from 6.8 to 5.7 and electric field from 20 to 55 kV/cm. Similar results are obtained when the ionic strength is reduced from 168 to 28 mM. The authors concluded that the electric field and ionic strength are more likely related to the poration rate and physical damage of the cell membranes, while pH is more likely related to changes in the cytoplasmic conditions due to the osmotic imbalance caused by the poration.

16.4 FACTORS AFFECTING PEF EFFECTIVENESS

Major factors that affect PEF effectiveness during food processing can be grouped as process factors (electric field intensity, pulse type, treatment time, and treatment temperature), product factors (pH, ionic strength, electric conductivity, and constituents of foods), and microbial factors (type, concentration, and growth stage of microorganisms).

16.4.1 PROCESS FACTORS

16.4.1.1 Electric Field Intensity

Electric field intensity is one of the main factors that influence the microbial inactivation (Dunn 1996). It is defined as electric potential difference V between two given points in space divided by the distance d between them:

$$E = \frac{V}{d} \tag{16.1}$$

To achieve microbial inactivation the applied electric field needs to be greater than the critical electric field for a particular microorganism (Castro et al. 1993). The electric field should be evenly distributed in the treatment chamber in order to achieve an efficient treatment. An electric field of 16 kV/cm or greater is usually sufficient to reduce the viability of Gram-negative bacteria by 4–5 log cycles and Gram-positive bacteria by 3–4 log cycles. In general, the electric field required to inactivate microorganisms in foods is in the range of 12–45 kV/cm. However, some studies have reported that electric fields of up to 90 kV/cm could be applied to food under a continuous treatment conditions (Zhang et al. 1994a, Dunn 1996, Liang et al. 2002). The fact that microbial inactivation increases with increasing applied electric field (EF) strength is consistent with the electroporation theory, in which the induced potential difference across the cell membrane is proportional to the applied electric field.

The most accepted model showing the relationship between the survival ratio $(S = N/N_0)$ of microorganisms and the field strength E was proposed by Hülsheger et al. (1981) and is represented by the relationship

$$\ln(S) = -b_E(E - E_c) \tag{16.2}$$

where
  S is the microbial survival rate in fraction given as the ratio of microbial count after treatment
  N is microbial count before treatment N_0
  b_E is the regression coefficient (cm/kV)
  E is the applied electric field
  E_c is the critical electric field obtained by the extrapolated value of E for 100% survival

Grahl and Märkl (1996) found that the critical value of electric field E_c was a function of cell size; the bigger the size of a cell, the lower the E_c. They attributed this phenomenon to the transmembrane potential experienced by the cell, which is proportional to the cell size. Also, Hülsheger et al. (1983) found that E_c for Gram-negative bacteria was lower than that of Gram-positive bacteria, which may be explained by the smaller resistance of the former.

16.4.1.2 Treatment Time and Frequency

Apart from electric field intensity, treatment time and pulse frequency are important factors. PEF treatment time is calculated by multiplying the pulse number by the pulse duration. An increase in any of these variables increases microbial inactivation (Sale and Hamilton 1967). Sepúlveda-Ahumada (2003) proposed that electric field pulses between 1 and 5 μs produced the best results for microbial inactivation. Martín-Belloso et al. (1997) found that pulse width influenced microbial reduction by affecting E_c. Longer widths decreased E_c, resulting, therefore, in higher inactivation. An increase in pulse duration may, however, result also in an undesirable food temperature increase and promotion of electrolytic reactions and electrode position at the electrode surfaces (Zhang et al. 1995a).

Normally, the inactivation of microorganisms increases with an increase in the pulse number, up to a certain number (Hülsheger et al. 1983). Grahl and Märkl (1996) reported that the log reduction of *E. coli* in UHT milk increased from 1 to 4 with the pulse number increasing from 5 to 20 at less than 45°C and 22.4 kV/cm of electric field intensity. Zhang et al. (1994b) also reported that the log reduction of *E. coli* in skim milk increased from 1 to 4 with the pulse number increasing from 16 to 64 at 15°C and 40 kV/cm. Liu et al. (1997) found that microbial inactivation was usually achieved during the first several pulses, additional pulses display a lesser lethality. Zhang et al. (1994a) also noticed that the inactivation of *Saccharomyces cerevisiae* by PEF in apple juice reached saturation up to 10 pulses at an electric field of 25 kV/cm. Elez-Martinez and Martin-Belloso (2007) evaluated the effects of PEF processing conditions on vitamin C and antioxidant capacity of orange juice. The treatments were performed at 25 kV/cm and 400 μs with square bipolar pulses of 4 μs and pulse frequency from 50 to 450 Hz. The retention of vitamin C in orange juice and gazpacho increased with a decrease of pulse frequency.

16.4.1.3   Pulse Shape and Polarity

Exponential decaying and square wave pulses are the two commonly used pulse shapes. Other waveforms such as bipolar, instant charge reversal, or oscillatory pulses have been used depending on the circuit design. An exponential decay voltage wave is a unidirectional voltage that rises rapidly to a maximum value and decays slowly to zero. Therefore, food is subjected to the peak voltage for short period of time. Hence, exponential decay pulses have a long tail with a low electric field, during which excess heat is generated in the food without an antimicrobial effect (Zhang et al. 1995a). Oscillatory decay pulses are the least efficient as they prevent the cell from being continuously exposed to high intensity electric field for an extended period of time, thus preventing the cell membrane from irreversible breakdown over a large area (Jeyamkondan et al. 1999).

The square waveform may be generated by using a PFN consisting of an array of capacitors and inductors or by using long coaxial cable and solid-state switch devices. The disadvantage of using high-voltage square waves lies in trying to match the load resistance of the food with the characteristic impedance of the transmission line. By matching the impedances, a higher energy transfer to the treatment chamber can be obtained. Zhang et al. (1994a) reported 60% more inactivation of *S. cerevisiae* when using square pulses than exponentially decaying pulses.

Although results in literature are not conclusive, bipolar pulses are more lethal than monopolar pulses (square or exponential decay) because bipolar pulses cause the alternating changes in the movement of charged molecules which lead to extra stress in the cell membrane and enhance its electric breakdown (Qin et al. 1994,

Barbosa-Cánovas et al. 1999, Evrendilek and Zhang 2005). Bipolar pulses also offer the advantages of minimum energy utilization, reduced deposition of solids on the electrode surface, and decreased food electrolysis. These advantages were tested by Qin et al. (1994) on *Bacillus subtilis* and Evrendilek and Zhang (2005) on *E. coli* O157:H7 in skim milk.

Ho et al. (1995) proposed instant reversal pulses where the charge is partially positive at first and partially negative immediately thereafter. The inactivation effect of an instant reversal pulse is believed to be due to a significant alternating stress on microbial cell which causes structural fatigue. Amiali et al. (2006b) used instant reversal square wave pulses and found this kind of waveforms to be more efficient than others in terms of egg product pasteurization, since it combines instant reversal charge and square waveform pulses.

16.4.1.4   Treatment Temperature

Treatment temperature influences the effectiveness of PEF on cellular materials. Since PEF treatment normally increases the product temperature (largely due to ohmic heating components), a cooling device is sometimes used to maintain the temperature at levels that maintain the nutritional, sensory, or functional properties of products. On the other hand, application of PEF at mild temperatures tends to enhance the microbial inactivation. Dunn and Pearlman (1987) found that a combination of PEF and heat was more efficient than conventional heat treatment alone. A higher level of inactivation was obtained using a combination of 55°C temperature and PEF to treat milk. Dunn (1996) obtained a 6 log reduction of *Listeria innocua* inoculated in milk after few seconds at 55°C accompanied with PEF. Increasing treatment temperature from 7°C to 20°C significantly increased PEF inactivation of *E. coli* in simulated milk ultrafiltrate (SMUF). However, additional increase in temperature from 20°C to 33°C did not result in any further increase in PEF inactivation (Zhang et al. 1995b). Reina et al. (1998) reported a higher inactivation rate of *L. monocytogenes* in milk with a temperature increase from 25°C to 50°C. At 30°C and 30 kV/cm, a 3.5 log reduction of *L. monocytogenes* was obtained after 600 µs of treatment, whereas at 50°C more than 4 log reductions were obtained. Sepúlveda-Ahumada et al. (2005a) observed a marked increase of PEF inactivation at 55°C on *L. innocua* suspended in a buffer. The electric field intensity and number of pulses were applied in the range of 31–40 kV/cm and 5–35 pulses. These authors found synergy between thermal and PEF treatment. It was suggested that the marked increase of PEF inactivation effectiveness at 55°C may be due to the occurrence of phase transition on the cell membrane of *L. innocua* at this temperature, since it is possible that a thinning of the bacterial membrane would render bacterial cells more susceptible to disruption by electric fields (Jayaram et al. 1992). Ravishankar et al. (2002) also investigated *E. coli* O157:H7 at electric field strength (15–30 kV/cm), pulse number (1–20), and temperature (5°C–65°C) using a static chamber and gellan gum gel as a suspension medium. The authors found that thermal energy began taking effect at 55°C. At this temperature, a 1 log reduction was attributable to thermal energy. Above this temperature, all reductions were attributed entirely to thermal energy. The authors suggested that there was no synergy between the concurring thermal and PEF energies.

Bazhal et al. (2006) investigated the combined effect of heat treatment with PEF on the inactivation of *E. coli* O157:H7 in liquid whole egg. The electric field strength was varied from 9 to 15 kV/cm and the treatment temperatures were 50°C, 55°C, or 60°C. At 60°C, a 2 log reduction of *E. coli* O157:H7 was obtained using thermal treatment alone, while a combination of heat and PEF resulted in a 4 log reduction. These results indicated a synergy between temperature and electric field.

Increase in the rate of inactivation with temperature is attributed to reduced transmembrane breakdown potentials at higher temperatures (Zimmermann 1986). Stanley (1991) proposed that phospholipid molecules in cell membrane undergo temperature-related transitions, changing from a firm gel-like structure to a less-ordered liquid crystal phase at higher temperature, thus reducing the mechanical resistance of the cell membrane. Another deduction proposed by Schwan (1957) was that the higher lethal effect of PEF combined with heat might be due to the increase in electrical conductivity of the medium, making it similar to electrolytic conduction (Barbosa-Cánovas et al. 1999, Sepúlveda-Ahumada et al. 2005a).

16.4.2 Operation Mode

In general, continuous processes for liquids reached higher inactivation rates than those in batch mode. Martín et al. (1997) reported that in order to achieve a 2 log reduction of *E. coli* inoculated in milk by a batch mode, 64 pulses and 35 kV/cm were needed, while for a continuous mode, 25 pulses and 25 kV/cm were enough. Operation of PEF in continuous mode for solids is much more challenging.

16.4.3 Product Factors

16.4.3.1 pH and Ionic Strength

Vega-Mercado (1996) studied the effect of pH and ionic strength of SMUF medium during PEF treatment. The authors reported that the lower the pH and ionic strength, the higher the inactivation rate. When the ionic strength decreased from 168 to 28 mM, the inactivation ratio increased from no detectable to 2.5 log cycles. Also, when the pH reduced from 6.8 to 5.7, the inactivation ratio increased from 1.5 to 2.2 log cycles. The PEF treatment and ionic strength were responsible for electroporation and compression of the cell membrane, whereas the pH of the medium affected the cytoplasm when the electroporation was complete. Alvarez et al. (2000) also studied the influence of pH of treatment medium on the inactivation of *Salmonella senftenberg* by PEF treatment. The authors found that at the same electric conductivity, inactivation of *S. senftenberg* was greater at neutral (7.0) than acidic pH (3.8).

16.4.3.2 Electrical Conductivity

The electrical conductivity of a medium (σ, s/m), which is defined as the ability to conduct electric current, is an important variable in PEF treatment:

$$\sigma = \frac{d}{RA} = \frac{1}{\rho} \tag{16.3}$$

where

σ is the electrical conductivity of medium (s/m)

R is the resistance of the medium (Ω)

A is the electrode surface area (m^2)

d (m) and ρ are the gap between electrodes and the resistivity, respectively

The electrical conductivity of a medium depends on treatment temperature as defined by

$$\sigma = \alpha T + \beta \tag{16.4}$$

where α and β are constants depending on the composition and concentration of the medium.

At constant temperature conditions, foods with high electrical conductivities (low resistivity) exhibit smaller electric fields across the treatment chamber and therefore are difficult to be treated with PEF process. An increase in electrical conductivity increases the ionic strength of the food, resulting in a decrease in the inactivation rate. Furthermore, an increase in the difference between the electrical conductivity of a medium and microbial cytoplasm weakens the membrane structure due to an increased flow ionic substance across the membrane (Jayaram et al. 1992).

Alvarez et al. (2000) studied the influence of conductivity of treatment medium on the inactivation of *S. senftenberg* by PEF treatment. The authors found that at constant input voltage, electric field strength obtained in the treatment chamber depended on medium conductivity. At the same electric field strength, conductivity did not influence *S. senftenberg* inactivation.

16.4.3.3 Composition

Food components such as fat and protein may influence the effect of PEF on the food product. These effects may be related to the capacity of some substances to shield microorganisms from applied field, or the ability of some chemical species to stabilize or prevent ion migration.

Martín et al. (1997) found that inactivation of *E. coli* in milk was more limited than in a buffer solution, because of the presence of milk proteins. Grahl and Märkl (1996) subjected different media (milk with 1.5% and 3.5% fat, solutions of sodium-alginate) inoculated with *E. coli* and other microorganisms to PEF. The treatment conditions are 5–15 kV/cm, 1–22 Hz, and the temperatures did not exceed 45°C–50°C. The authors noticed that the fat particles of milk seemed to protect the bacteria against electric pulses. Picart et al. (2002) also claimed that whole milk with a higher fat content (3.6%) appeared to reduce *L. innocua* inactivation compared to skim milk at temperatures between 25°C and 45°C, a pulse repeat frequency of 1.1 Hz and electric intensity of 29 kV/cm.

There is currently no agreement on the possible influence of fat content on PEF inactivation. Reina et al. (1998) compared the effect of PEF treatment under 25°C at 30 kV/cm and frequency of 1700 Hz in milk with different fat content. The authors inoculated *L. monocytogenes* into skim milk, 2% fat milk, and whole milk, and evaluated the effects of the fat content on the inactivation rates; no differences were observed among the results. Manas et al. (2001) used 33% emulsified fat cream to test fat effect on the inactivation of *E. coli* by PEF treatment. The treatment was conducted under 34 kV/cm with a pulse frequency of 1.1 Hz and temperatures less than 30°C. The result was that the emulsified lipids do not appear to protect against microbial inactivation by electric pulses. Sobrino-Lopez et al. (2006) also claimed that fat content of the milk did not modify the resistance of *Staphylococcus aureus* to a PEF treatment. Three types of milk (whole, 1.5% and skim) were treated under 25°C, 30–35 kV/cm, and frequency of 100 Hz.

16.5   MICROBIAL INACTIVATION KINETICS

When microorganisms are treated with heat the logarithm of cell population decreases, linearly, with the treatment time for constant treatment intensity. Alternative food processing technologies are also believed to inactivate microorganisms logarithmically. Dose–response models are derived from kinetic data to predict efficiency of variables of alternative processes. For the case of PEF, such models are based on sigmoid inactivation plots, as the one shown in Figure 16.3. As previously stated, the lethal effect of PEF has been described as a function of field intensity, treatment time (pulse duration and number of pulses), and a model constant determined by the microorganism and its physiological status. Microbial inactivation in liquid medium has been reported to follow the first-order kinetics (Hülsheger et al. 1981, Grahl and Märkl 1996) as follows:

$$\log S = B_E (E - E_c) \tag{16.5}$$

where
   S is the microbial survival rate
   B_E is an electric field constant obtained as the coefficient of regression of the
      straight survival curves
   E is applied electric field
   E_c is critical value of electric field below which there will be no inactivation
      (that is 100% survival)

Treatment time can be calculated as product of number of pulses and pulse width. Inactivation kinetics in terms of treatment time has been given as (Hülsheger et al. 1981, Grahl and Märkl 1996)

$$\log S = B_t \frac{t}{t_c} \tag{16.6}$$

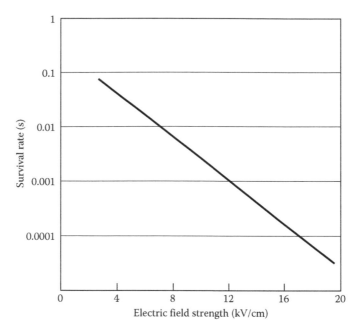

FIGURE 16.3   Typical curve of inactivation rate of microorganisms by PEF treatment.

where
  B_t is electric field constant or coefficient of the regression of the straight
    survival curves
  t is treatment time
  t_c is the critical time below which there will be no inactivation

Combining Equations 16.5 and 16.6 yields

$$S = \left(\frac{t}{t_c}\right)^{\frac{E-E_c}{k}}$$

(16.7)

where k is a constant factor as can be expressed as follows:

$$k = \frac{E-E_c}{B_{t(E)}} = \frac{\log\left(\dfrac{t}{t_c}\right)}{B_{E(t)}}$$

(16.8)

If logarithms to base 10 on both sides of Equation 16.7 are taken, the left-hand side of such equation would represent the inactivation ratio or log reduction, which refers to 90% reduction in the initial microorganism's population (Figure 16.3). The transformed equation would also indicate that inactivation ratio depends linearly on the applied field strength and logarithmically on the treatment time. Although electric field strength would represent a more pronounced effect than treatment time, both are important elements. Apart from electric field strength and treatment time, some other variables such as pulse characteristics can also influence the microbial inactivation ratio and reaction kinetics in PEF processing. In a particular process, a high-voltage field will be discharged to a food material placed between two electrodes, and it will be released in pulsed form. Operating field strengths range from 10 to 70 kV/cm, while duration of the pulses can vary from μs to ms range. The frequency of the pulses can be as low as a single one, or as high as 2000 pulses per second.

Considering the kinetic equations, smaller values of E_c or t_c and higher values of B_E and B_t indicate greater susceptibility of the particular microorganism to PEF processing. The above first-order kinetics may be simplistic for most PEF applications. Tailing phenomena in microbial inactivation kinetics have been observed (Sensoy et al. 1997). A two-phase kinetic model may be required to adequately describe PEF inactivation kinetics. Some other more complicated inactivation kinetic models have been reported (Anderson et al. 1996, Raso et al. 2000). The search for the most robust kinetic model for PEF inactivation of microorganisms is currently ongoing. Bacteria cells are generally more resistant to PEF inactivation than yeast cells. However, spores are most resistant to PEF. Vegetative cells at the logarithmic growth stage are more susceptible to PEF inactivation. Increasing the conductivity of medium generally decreases effectiveness of PEF for microbial inactivation since the higher the conductivity of a medium, the lower the achievable peak electric field at a constant input energy.

16.6   QUALITY ASPECTS OF PEF PROCESSING

PEF treatment may influence physical and chemical properties of products. The nature and extent of PEF influence on quality changes are still being actively discussed. Barsotti et al. (2002) indicated that PEF treatment of model emulsions and liquid dairy cream may result in dispersal of oil droplets and dissociation of fat globule aggregates. Qin et al. (1995) reported no apparent change in the physical and chemical attributes of PEF processed milk. PEF treatment of various liquid foods including apple juice, orange juice, and milk has not shown any significant physicochemical changes (Jia et al. 1999, Charles-Rodríguez et al. 2007, Shin et al. 2007). PEF processed cranberry juice and chocolate milk retained its physical and chemical characteristics (Evrendilek et al. 2001). There was a slight decrease in vitamin C content in PEF-treated orange juice compared to heat-treated orange juice (Zhang et al. 1997).

Gallardo-Reyes et al. (2008) conducted a comparative study of orange juice pasteurized by ultrahigh temperature (UHT) (processing at 110°C, 120°C, and 130°C for 2 and 4 s) and PEF (20 and 25 kV/cm for 2 ms). The authors concluded that although there was no difference in pH and soluble solids obtained with both treatments and freshly squeezed control samples, the color of PEF-treated sample was closer to the control. There have been attempts to improve juice extraction from plants and fruit products using PEF (Bazhal and Vorobiev 2000, Schilling et al. 2007). PEF may enhance the extraction of higher amounts of valuable compounds into the extraction juice resulting in high quality. Torregrosa et al. (2006) reported higher vitamin A contents of orange–carrot juices, compared to untreated control samples, when the product was treated at 25 kV/cm. Apple juice contains several phenolic compounds including chlorogenic acid (5-CQA), catechins, procyanidins, quercetin glycosides, and phloridzin. Schilling et al. (2007) monitored the phenolic compositions of juices obtained from PEF-treated mash and untreated samples. The PEF treatments were at the field intensities of 1, 3, and 5 kV/cm. There was no significant difference between treated and untreated samples. However, enzymatic maceration of the apple mash resulted in a marked increase in quercetin glycosides. The authors also reported that there is no difference in the antioxidative capacity of the apples juices. It was then postulated that the fact that the antioxidative potential was not affected indicates that radical formation has not taken place during PEF treatment.

Table 16.1 shows the influence of PEF treatment on the quality of juice expressed from apple and sugar beet. Data on the Table 16.1 were obtained from Lazarenko et al. (1977), Bazhal and Vorobiev (2000), and Bazhal (2001). Bazhal and Vorobiev (2000) reported that juice from the samples treated at E = 800 V/cm was lighter (absorbance = 0.02) than juice expressed from apple tissue pulsed at E = 150 V/cm (absorbance = 0.07). The significant change in juice color may be attributed to the inhibition of polyphenoloxidase by electric fields (Giner et al. 2001). Lazarenko et al. (1977) suggested that electric field can break the chains of pectin molecules resulting in deceased pectin concentrations and thus reducing the kinematic viscosity of the extracted juices. Lower viscosity improves juice filtration. The reduction in transmittance of juice after electroplasmolysis indicates a reduction in suspended particle contents because of improved tissue filtration properties resulting from the additional pores formed in the cell walls after PEF treatment. PEF treatment of sugar beet resulted in increased purity of the extracted juice compared to the traditional processing by thermal plasmolysis. Despite lower temperature leaching for

TABLE 16.1

Qualitative Parameters of Apple Juice from Control and PEF-Treated Samples

Parameter	Apple Juice		Sugar Beet Juice	
	PEF	**Control**	**PEF**	**Control**
Density (kg/m³)	1059.4	1057.7	1.0392	1.038
Brix	13.8	13.1	9.8	9.5
pH	3.91	3.84	5.6	5.65
Pectin (mg/L)	290	517		
Kinematic viscosity (10⁻⁶ m²/s)	5.917	6.747	2.284	2.595
Absorbance (wavelength of 520 nm)				
Filtered	0.02	0.39		
Nonfiltered	0.03	1.18		
Transmittance	0.67	0.33		

PEF-treated samples, sucrose loss in pulp decreased from 0.62% (thermally treated beet) to 0.57% (Knorr et al. 2001). It is uncertain if all necessary chemical analysis has been performed to fully ascertain the effect of PEF on quality of processed foods. However, the clear consensus is that liquid food products generally retain their fresh-like quality after PEF treatment.

Solid food products undergo significant changes when treated with PEF. Changes in electrical conductivity of the treated vegetable samples indicated increasing cell permeability (Lebovka et al. 2000, 2001). Table 16.2 shows that diffusion coefficient of sugar from the beetroot increases from 0.68×10^{-9} up to 1.2×10^{-9} m²/s after PEF treatment (Gulyi et al. 1994, Jemai 1997). Elastic modulus of sugar beet

TABLE 16.2

Some Vegetable Tissues Properties Estimated for Untreated (Control) and PEF Pulsed Samples

Parameter	Value of Parameter		Material	Operation	Reference
	Control	PEF Treatment			
Electrical conductivity (S/m)	0.003–0.007	0.035–0.070	Apple	Conductivity measurement	Lebovka et al. (2001)
Electrical conductivity (S/m)	0.03	0.41	Carrot	Conductivity measurement	Rastogi et al. (1999)
Electrical conductivity (S/m)	0.06	0.53	Potato	Conductivity measurement	Knorr and Angersbach (1998)
Porosity (%)	67	75	Apple	Using of penetrometer	Bazhal et al. (2003b)
Water diffusion coefficient (m²/s)	0.98×10^{-9}	1.55×10^{-9}	Carrot	Osmotic dehydration	Rastogi et al. (1999)
Sugar diffusion coefficient (m²/s)	0.68×10^{-9}	1.2×10^{-9}	Sugar beet	Leaching	Gulyi et al. (1994)
Mass transfer coefficient (kg/m² s)	0.043	0.058	Paprika	Drying	Ade-Omowaye et al. (2002)
Constant drying rate (kg/m² s)	9.68×10^{-4}	13.02×10^{-4}	Paprika	Drying	Ade-Omowaye et al. (2002)
Heat transfer coefficient (W/m² s)	73.13	98.36	Paprika	Drying	Ade-Omowaye et al. (2002)
Elastic modulus (MPa)	12.5	6.5	Sugar beet	Compression test	Matvienko (1996)
Elastic modulus (MPa)	1.53	0.32	Apple	Compression test	Bazhal et al. (2003b)
Failure stress (MPa)	1.26	0.53	Apple	Compression test	Bazhal et al. (2003b)

decreased after PEF treatment (Bazhal 2001). The microstructure of salmon and chicken changed considerably due to PEF treatment as the muscle cells decreased in size and gaping occurred (Gudmundsson and Mafsteinsson 2001). Electric field treatment generally affects biological cell membranes whereas heating destructs the cell walls (Calderón-Miranda et al. 1999). There is a potential of inducing rheological changes in a product as a result of PEF treatment. This phenomenon depends on the type of product involved and requires detailed investigations.

16.6.1 PEF Effects on Milk and Cheese Quality

In recent years, with the demand of high-quality milk and milk products, more and more researchers have focused on studies of loss of sensory and physicochemical characteristics in milk and milk products following treatment with pulse electric field (Qin et al. 1995, Dunn 1996, Bendicho et al. 1999, Evrendilek et al. 2001, Li et al. 2003, Michalac et al. 2003, Sepúlveda-Ahumada 2003, Sampedro et al. 2005, Shin et al. 2007, Yeom et al. 2007, Shamsi et al. 2008). As for the PEF effect on cheese process and quality, limited research work was found (Sepúlveda-Ahumada et al. 2000). Dunn (1996) reported that milk treated with PEF ($E = 20–80\,kV/cm$) suffered less flavor degradation. The author proposed the possibility of manufacturing dairy products such as cheese, butter, and ice cream using PEF-treated milk although no detailed information was given in his report. Qin et al. (1995) carried out a study of shelf life, physicochemical properties, and sensory attributes of milk with 2% milk fat, treated with $40\,kV/cm$ of electric field and of 6–7 pulses. No physicochemical or sensory changes were observed after treatment, in comparison with a sample treated with thermal pasteurization. Bendicho et al. (1999) studied the destruction of riboflavin, thiamine (water soluble), and tocopherol (liposoluble) in milk by treatment with PEF ($E = 16–33\,kV/cm$; $N = 100$ pulses). The vitamin concentrations before and after treatment were determined by HPLC. The authors observed no destruction of vitamins by treatment with pulses. Michalac et al. (2003) studied the variation in color, pH, proteins, moisture, and particle size of UHT skim milk subjected to treatment with PEF ($E = 35\,kV/cm$; $W = 3\,\mu s$; and time = $90\,\mu s$). The authors saw no differences in the parameters studied (color, pH, proteins, moisture, and particle size) before and after treatment.

Sepúlveda-Ahumada et al. (2000) compared the textural properties and sensory attributes of Cheddar cheese made with heat-treated milk, PEF-treated milk ($E = 35\,kV/cm$; $N = 30$ pulses), and untreated milk. In the hardness and springness study, the cheeses made from milk pasteurized by any method were harder than those made from untreated milk. In the sensory evaluation, the panelists also found differences between the cheeses made from untreated milk and milk treated by PEF or heat. Regardless of the differences, the authors still considered using PEF-treated milk to obtain cheese as a feasible option in order to improve the product quality.

Yeom et al. (2007) studied a commercial, plain, low-fat yogurt mixed with fruit jelly and syrup. They observed the changes in physical attributes (pH, color, and Brix) and sensory attributes during storage at 4°C after treatment with PEF (electric field = $30\,kV/cm$; treatment time = $32\,\mu s$) and heat ($T = 65°C$; time = $30\,s$). The sensory evaluation indicated that there were no changes between the control samples and the treated samples. There was also no variation in the color, pH, and Brix.

Evrendilek et al. (2001) studied color, pH, Brix, and conductivity at 4°C, 22°C, and 37°C in milk with chocolate using treatment with PEF (E = 35 kV/cm; W = 1.4 μs; time = 45 μs), and PEF + heat (112°C and 105°C, 33 s). They compared the results with a control sample not treated by PEF or heat. Measurement of the a, b, and L parameters at 4°C revealed that the treatments of PEF at 105°C and PEF at 112°C did not cause changes in color. Sepúlveda-Ahumada (2003) treated HTST pasteurized milk with electric field of 35 kV/cm and 2.3 μs of pulse width, at a temperature of 65°C for less than 10 s. PEF treatments were applied either immediately after thermal pasteurization to produce an extended shelf life product, or 8 days after thermal pasteurization to simulate processing after bulk shipping. Application of PEF immediately after HTST pasteurization extended the shelf life of milk to 60 days, while PEF-processing after 8 days caused a shelf life extension of 78 days, both were proving to be successful strategies to extend the shelf life of milk.

Li et al. (2003) investigated the effects of PEFs and thermal processing on the stability of bovine immunoglobulin G (IgG) in enriched soymilk. PEF at 41 kV/cm for 54 μs caused a 5.3 log reduction of natural microbial flora, with no significant change in bovine IgG activity. Analysis using circular dichroism spectrometry revealed no detectable changes in the secondary structure or the thermal stability of secondary structure of IgG after the PEF treatment (Li et al. 2005). However, in an experiment investigating the effect of temperature on the stability of IgG during PEF treatment (30 kV/cm, 54 μs), up to 20% of IgG was inactivated when the temperature was increased to 41°C (Li et al. 2003).

Shin et al. (2007) applied PEFs with square wave pulse to whole milk inoculated with *E. coli*, *Pseudomonas fluorescens*, and *Bacillus stearothermophilus*. The samples were exposed to 30–60 kV/cm electric field intensity with 1 μs pulse width and 26–210 μs treatment time in a continuous PEF treatment system. Eight log reductions were obtained for *E. coli* and *P. fluorescens* and 3 logs reduced for *B. stearothermophilus* under 210 μs treatment time, 60 kV/cm pulse intensity at 50°C. There was no significant change in pH and titration acidity of milk samples after PEF treatment.

Shamsi et al. (2008) determined the effects of PEF treatments on the inactivation of alkaline phosphatase (ALP), total plate count (TPC), *Pseudomonas*, and *Enterobacteriaceae* counts in raw skim milk at field intensities of 25–37 kV/cm and final PEF treatment temperatures of 15°C and 60°C. At 15°C, PEF treatments of 28–37 kV/cm resulted in 24%–42% inactivation in ALP activity and <1 log reduction in TPC and *Pseudomonas* count, whereas the *Enterobacteriaceae* count was reduced by at least 2.1 log units to below the detection limit of 1 CFU/mL. PEF treatments of 25–35 kV/cm at 60°C resulted in 29%–67% inactivation in ALP activity and up to 2.4 log reduction in TPC, while the *Pseudomonas* and *Enterobacteriaceae* counts were reduced by at least 5.9 and 2.1 logs, respectively, to below the detection limit of 1 CFU/mL. Kinetic studies suggested that the effect of field intensity on ALP inactivation at the final PEF treatment temperature of 60°C was more than twice that at 15°C. A combined effect was observed between field intensity and temperature in the inactivation of both ALP enzyme and the natural microbial flora in raw

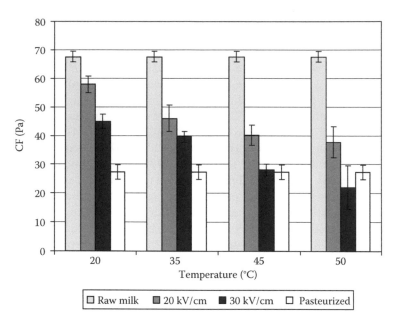

FIGURE 16.4   Effect of electric field intensity and temperature on CF. The applied pulse width is 2 µs, pulse frequency is 2 Hz, and pulse number is 120.

skim milk. The results of this study suggest that PEF as a nonthermal process can be employed for the treatment of raw milk in mild temperature to achieve adequate safety and shelf life while preserving the heat-sensitive enzymes, nutrients, and bioactive compounds.

Figure 16.4 shows that electric field intensity E and treatment temperature significantly affected rennet coagulation properties of milk in terms of curd firmness (CF) (Yu et al. 2008). Increasing both E and temperature decreased CF. Raw milk coagulum showed the highest CF (67.5 Pa), while pasteurized milk gave the lowest CF value (27.4 Pa) and PEF-treated milk presented values in between. For PEF-treated samples, all the CF values obtained at 20 kV/cm were significantly higher than the CF values for pasteurized milk samples. Also, most CF data obtained under 30 kV/cm were significantly higher than the CF values for pasteurized milk samples, except those treated at 50°C. The result implied that treating milk with PEF and mild temperature impacts less changes in terms of milk coagulation properties compared to heat pasteurized milk. This result is consistent with the finding of Dunn (1996) who studied the PEF-treated raw milk ($E = 20$–80 kV/cm; pulse width = 1–10 µs) and concluded that no significant physicochemical changes were observed within similar experimental range. Qin et al. (1995) also reported similar result for 2% fat milk (treated using 40 kV/cm and 6–7 pulses) in comparison with a sample treated with thermal pasteurization.

The Yu et al. (2008) data showed that applying electric field of 30 kV/cm combined temperature of 50°C may lead to similar coagulating effect as thermal pasteurization ($p > 0.05$). Fox et al. (2000) showed that heat treatment of milk at temperatures above 65°C adversely affects its rennet coagulability. Thus, for the purpose of cheese production, the treatment temperature under higher electrical intensity (30 kV/cm)

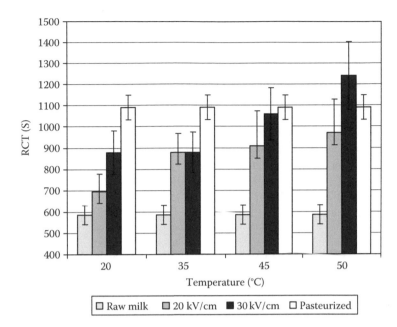

FIGURE 16.5   Effect of electric field intensity and temperature on CF (RCT). The applied pulse width is 2 μs, pulse frequency is 2 Hz, and pulse number is 120.

and longer pulses (120 pulse numbers) may not exceed 50°C in order to obtain desirable curd formation.

Figure 16.5 shows that treatment temperature and electric field intensity E also significantly affected rennet coagulation time (RCT) (Yu et al. 2008). Higher E and temperature led to longer RCT. RCT is a good index of the gelation potential of milk. A low RCT usually shows potentially good gel formation and high gel strength. Raw milk obviously obtained the lowest RCT (586.2 s). Pasteurized milk obtained highest RCT (1091.0 s) while PEF-treated samples got the medium RCT values.

The values of RCT obtained using an electric field intensity of 20 kV/cm and temperatures from 18°C to 50°C were significantly lower than that of heat pasteurized samples. However, at a 30 kV/cm electric field intensity, in order to keep the RCT value lower than that of heat pasteurized milk, the treatment temperature could not exceed 45°C. The RCT results again confirm that minimal changes were impacted to the milk product when appropriate PEF intensities and temperatures were used for processing. Evrendilek et al. (2001) studied a yogurt drink prepared using combined PEF (E = 30 kV/cm; treatment time = 32 μs) and heat (60°C, 32 s). The authors reported no significant differences between the control sample and the treated samples in terms of color, soluble solids, and pH. However, when milk was subjected to long-duration pulses (Perez and Pilosof 2004), or high-intensity electric fields (45–55 kV/cm) as described by Floury et al. (2005), the structure of milk protein was apparently modified.

Perez and Pilosof (2004) attributed the effects of PEF on milk proteins as due to polarization of the protein molecule, dissociation of noncovalently linked protein subunits involved in quaternary structure, and changes in the protein conformation so that buried hydrophobic amino acids or sulfydryl groups are exposed. If the duration of the electric pulse is high enough, hydrophobic interactions or covalent bonds may occur, forming aggregates. Similarly, Floury et al. (2005) explained the effect of PEF on milk protein as due to the modification of the apparent charge after exposure to intense electric fields and then modification of ionic interactions between the proteins.

The modification of milk protein structure may lead to changes in milk functional properties such as coagulation, foaming, and emulsifying. Different authors have reported varying levels of electric field strengths and temperatures beyond which changes in milk properties will occur. Coagulation properties of raw milk may be better preserved by using lower electric field strength (≤ 30 kV/cm) and temperature ($\leq 50°$C) combinations.

16.6.2 EFFECTS OF PEF ON COLOR AND TEXTURE OF FRUITS AND VEGETABLES

Very limited information is available in the literature on the effect of PEF on color and texture of vegetables and fruits. Knorr and Angersbach (1998) reported an increase of the enzyme polyphenoloxiase (PPO) from potato-cultured cells. The results were explained as due to cell membrane rupture and the subsequent decompartmentalization of the enzymes. The authors reported that the application of PEF beyond the optimum conditions (15–30 pulses at electric field strengths between 1.5 and 3.0 kV/cm, with a pulse width of 500 μs) resulted in the enzymatic browning of potato-cultured cells. Arevalo (2003) observed an increased color change reaction rate (up to 2.5 times) after subjecting apple slices to an electric field in the range from 0.75 to 1.5 kV/cm. Up to 60 pulses of 100 μs width square wave was applied. The compressive strength of apple tissue was reduced by 21%–47% as a result of the application of electric field pulses (Figure 16.6). In another study Arevalo (2003), using 0.75 and 1.5 kV/cm field intensity and 5, 30, or 120 pulses of 100, 200, and 300 μs pulse, the impact of PEF on color changes rate of potato slices also increased but the effect of pulse width did not produce any significant effect. The maximum compressive strength of potato was not affected by the PEF treatment. Thus there may be a saturation point beyond which the effect of PEF treatment is no longer relevant. Galindo et al. (2008) also studied the effect of PEF application on the potato tissues. The authors observed the effect on diffusion of fluorescent dye FM1-43 through the cell wall. The electric field strength was varied from 30 to 500 V/cm with 1 ms rectangular pulse. The result showed a slower diffusion of FM1-43 in the electropulsed tissue when compared with the untreated sample. They also suggested that electric field decreased the cell wall permeability. This response was mimicked by exogenous H_2O_2 and blocked by sodium azide, an inhibitor of production of H_2O_2 by peroxidase.

In a study on dehydration properties of carrots, Rastogi et al. (1999) reported changes in the textural characteristics of samples that were subjected to PEF pretreatments. The study showed the loss of turgor pressure and a softening of the tissue due to the damage induced by PEF pretreatments. It was also reported that further softening was very limited with increasing electric field above 1.09 kV/cm. Taiwo

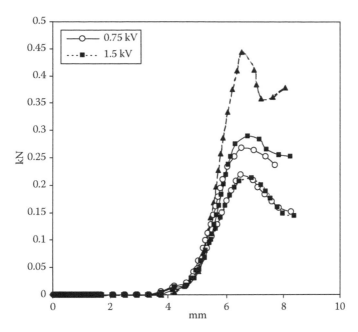

FIGURE 16.6 Maximum compressive strength force (kN) in apple slices. *n* represents the number of pulses.

et al. (2001) studied the effect of different pretreatments on texture and color changes of apple slices. The PEF applied as a pretreatment in the study consisted of 20 pulses of 800 μs with an electric field strength of 1.4 kV/cm. The authors reported decay in compressive strength of the apple tissue and a browning reaction (therefore a subsequent decrease in lightness values) due to application of PEF. The *L* color parameter was measured directly after application of PEF pretreatment before samples were osmotically dehydrated. In a subsequent study, Taiwo et al. (2002) compared quality characteristics of apple slices subjected to PEF and osmotically rehydrated. The authors studied color and textural changes of the apple slices after rehydration. Results showed that PEF-treated samples presented higher deformation force which was explained by the fact PEF-treated samples retained more solids. Results also showed that longer rehydration times and higher temperatures of the immersion fluid produced darker products. No conclusive results were obtained on the effect of PEF on the color of rehydrated samples. Bazhal et al. (2003a,b) studied the influence of PEF treatment on morphological changes in apple tissues. The later were subjected up to 60 pulses of 1 kV/cm electric field strength at 1 Hz pulse frequency and 300 μs pulse duration. The authors reported an increase of porosity from 63% to 69.4% after electroplasmolysis (Figure 16.7). The sizes of the induced pores were smaller compared to the pores of untreated samples and were comparable with the tissue cell wall thickness. The overall average mean size of the PEF-induced pores was 5.86 μm, which is lower than 7.81 μm obtained for untreated samples. In addition, by determining electrical conductivity, disintegration index, and failure stress of apple samples, a linear dependency was observed between failure stress and degree of

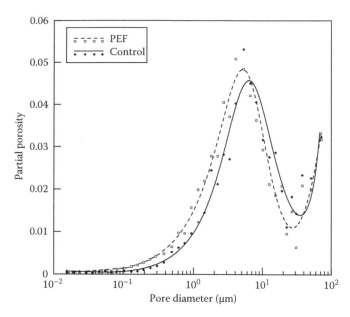

FIGURE 16.7 Distribution of pores in untreated and PEF-treated apple samples with conductive disintegration index in the range of 0.9–1 (n = 60 pulses, E = 1000 V/cm, t_i = 300 μs, and f = 1 Hz). Points represent averaged data from three determinations. Lines were obtained with least square polynomial fitting of experimental data.

electroplasmolysis. They concluded that electroplasmolysis affects not only plasmalemma membrane but also cell wall integrity of samples. Also, the failure stress decreased with intensification of electrical treatment.

Other authors have focused on inactivation of the enzymes that are responsible for the browning reaction observed in various fruits and vegetables (Ho et al. 1997, Giner et al. 2001, Giner et al. 2002, Aguiló-Aguyo et al. 2008). PEF treatments were applied on the enzyme extracts of apple, pear, tomato, and mushrooms. The strength of the electric field ranged from 2.4 to 35 kV/cm with 20 μs pulse width. Though the range of PEF parameters used in the studies differs from the PEF parameters used for the cell membrane permeabilization of plant tissues, valuable information can be drawn from the results.

16.6.3 EFFECTS OF PEF ON QUALITY OF FRUIT JUICES

Fruit juices are perceived as nutritive, healthy and, even, functional contributors to the human diet. The general consumer seems to imply that intake of some valuable nutrients, such as vitamin C, can be guaranteed by drinking any type of fruit juice on a regular basis. Consumers tend to choose any particular fruit juice due to its unique combination of sensory attributes, such as color, aroma, and flavor. Orange juice is said to be the most consumed juice worldwide while apple juice runs, apparently, in a stable second position. Tropical fruit juices are also in high demand, but tend to be more related to its use in mixtures of the cocktail type. Some other fruit juicy products, like peach or mango nectar, may also be considered healthy and nutritive, but their availability is more limited globally, due to the smaller massive scale of the fruits used as their raw materials.

In statistics terms, citrus juices are the most popular fruit juices with more than 50% of the international commercialization volume of juices (Varnam and Sutherland 1999). Within citrus juices, orange juice represents an approximate 60% of all Western Europe consumption of juices and juice-based drinks and a similar amount (60%) of all fruit juice sales in the United States (Parker 2006b). Apple juice, on the other hand, represents roughly a 60% of the global production of orange juice (Parker 2006a).

According to the figures presented above, orange and apple juice would be considered the most important type of fruit juices and investment should be likely to occur in order to promote research and technology transfer to expand their markets. There might be, therefore, a vested interest in research groups and fruit growers to team up and try to obtain orange and apple juices of the highest possible quality to compete and gain consumer acceptance globally. Nutritive and sensory quality should be considered paramount in all technology transfer efforts, since fruit juices are the type of products in which the consumer associates freshness with quality. Research and development is also important and needs to be promoted for other type of juices, but the priority will be inevitably higher in the more demanded products.

Pasteurization of fruit juices has been traditionally performed by thermal processing either as a low temperature–long time (LTLT) treatment or a high temperature–short time (HTST) technique. UHT has been also employed for treatment of fruit juices. HTST pasteurization is a continuous process that shows several advantages over batch pasteurization, which is the conventional way of applying the LTLT method. In HTST, fresh juice flows through a holding tube or flow arrangement in which it is heated at a given temperature for a specific time. UHT processing, also known as aseptic processing, involves the production of a sterile product by rapid heating to high temperatures followed by a short holding time and ending with a rapid cooling. The processed product is filled into a presterilized container within a sterile environment, to provide a prolonged shelf life.

HTST pasteurization is, possibly, the most widely employed technique for heat treatment of orange and apple juices. In terms of industrial scale, HTST is normally carried out in a plate heat exchanger, which represents the advantage of having the stages of preheating, heating, holding, and cooling in a single passage. Commercially, orange juice is pasteurized by HTST at 90°C–95°C for 15–30 s (Braddock 1999), while apple juice is processed at 77°C–88°C for a 25–30 s by the same method (Moyer and Aitken 1980). Aseptic processing has gained popularity as a thermal pasteurization technique for both juices. Temperature and holding time in this case could be as much as 138°C for at least 2 s for both types of juice (Lewis 1993). Aseptic processing produces shelf-stable juices and other products with shelf lives as high as 8 months without refrigeration (Ellis 1982). There is, however, a typical cooked flavor detected in aseptically processed juices.

Thermal pasteurization is quite efficient for preventing microbial spoilage of fruit juices but the applied heat may also cause undesirable biochemical and nutritious changes, which may affect overall quality of the final product. Orange juice can undergo quality degradation due to microbiological and enzymatic activities and chemical reactions (Chen et al. 1993). Spoilage microorganisms and native enzymes can be inactivated by thermal treatment, but thermal treatment causes the irreversible loss of fresh juice flavor (Braddock 1999) as well as a reduction of nutrients and the initiation of undesirable browning reactions in the juices (Chen et al. 1993). Thermal pasteurization of fruit juices can also cause undesirable changes, such as loss of vitamin content and color changes due to browning, mainly triggered by enzymatic reactions (Yeom et al. 2000).

With increasing demand to obtain processed foods with better attributes than have been available to date, food researchers have pursued the discovery and development of improved preservation processes with minimal impact on the fresh taste, texture, and nutritional value of food products. As previously discussed, fruit juices may present a series of undesirable effects when processed by conventional, thermal methods of pasteurization. Alternatives to traditional treatment, which do not involve direct heat, have been investigated in order to obtain fruit juices safe for consumption, but with sensory attributes resembling the fresh product. The first attempt to improve processed fruit juice quality could be considered, in fact, the variation of the temperature–time relationship in aseptic processing. As previously mentioned, such variations gave rise to a new generation of fruit juices with extended shelf life, which represents an obvious advantage of course, but some sensory impairment still remained due to the heat involved in the treatment. This is what could be considered as a "third generation" of processing alternatives, which seek to eliminate heat completely in pasteurization of fruit juices and other fluid foods such as milk and dairy products. PEF has already been successfully applied in treatment of fruit juices; the main developments are presented below.

16.6.3.1 Orange Juice

The flavor of orange is due to more than 200 chemical compounds (Maarse 1991), and is comprised of hydrocarbons, aldehydes, esters, ketones, and alcohols. Limonene is the most important flavor compound in quantity, although not in quality (Siezer et al. 1988). It has been reported (Ahmed et al. 1978) that acetaldehyde, citral, ethyl butyrate, limonene, linalool, octanal, and α-pinene are the major contributors to orange juice flavor. The development of off-flavors in orange juice has been attributed to the degradation of limonene to α-terpineol and other compounds (Tatum et al. 1975).

Thermal processing is conventionally used to pasteurize orange juice, but also reduces nutritional and flavor qualities and produces undesirable off-flavor compounds (Ekasari et al. 1986). The citrus industry has been exploring alternative methods with minimal heat treatment to increase markets by improving quality, and PEF may be one of such alternatives. The case for orange juice and its possible treatment with PEF, as a function of quality and stability, has been extensively studied at Ohio State University (Jia et al. 1999, Yeom et al. 2000). Effects of PEF (35 kV/cm for 59 μs) on the quality of orange juice were investigated. The PEF-treated juice was compared with juice pasteurized by heat at 94.6°C for 30 s. PEF pasteurization prevented the growth of microorganisms at 4°C, 22°C, and 37°C for 112 days and inactivated 88% of activity of the enzyme pectin methyl esterase. The juice treated by PEF retained greater amounts of vitamin C and some representative flavor compounds, than the juice pasteurized by heat during storage at 4°C. The PEF pasteurized orange juice also presented lower browning index than the thermally processed juice. Pertaining sweetness, expressed as Brix, and pH, no significant difference was observed for any of the pasteurization methods.

In terms of specific flavor compounds, it was found that 40% of decanal was lost by heat treatment at 90°C for 3 min while no loss was observed by PEF treatment at 30 kV/cm, either at 240 or 480 μs (30). Octanal showed a loss of 9.9% for the heat treatment and 0% for any of the two PEF treatments. Some compounds suffered losses for the PEF treatments, but always in less proportion than the heat pasteurized juice. For example, 5.1% and 9.7% of ethyl butyrate were lost for the 240 and 480 μs treatments, respectively, but 22.4% was lost in the thermal process. The loss of these volatile compounds in orange juice treated by PEF was attributed to the vacuum degassing system of the PEF unit (Jia et al. 1999).

The advantage of PEF compared with thermal processing was also observed in nutritive aspects. PEF-treated orange juice retained a significantly higher content of ascorbic acid than heat pasteurized juice during storage at 4°C (Yeom et al. 2000). Although more research needs to be completed before considering PEF as the sole treatment to retain completely all flavor and color components of orange juice, it can be stated that PEF pasteurized juice retains more flavor and shows less browning than conventionally pasteurized juice. Under certain conditions, PEF-treated orange juice retains ascorbic acid better than heat-treated juice. All these findings are important and may prove invaluable for the adaptation of PEF as a real alternative for orange juice pasteurization.

16.6.3.2 Apple Juice

As stated earlier, apple juice is a popular beverage worldwide, which is perceived as a wholesome and nutritious product. Overall quality of apple juice is an important factor to consider in processing, since some attributes such as aroma, color, and flavor are appreciated by the final consumer and are associated with freshness and authenticity. Flavor components in apple juice are numerous, so flavor identification may be considered more complex than the correspondent to orange juice due to the aromatic nature of apples. Eight odor-active volatiles have been, however, identified as the most important contributors for the aroma–flavor authenticity of apple juice (Cunningham et al. 1986).

Some research on the application of PEF to pasteurize apple juice has been carried out. The use of PEF has achieved satisfactory microbial inactivation in several applications and has proved to be energy efficient also. An investigation of PEF inactivation has demonstrated that, to achieve a 7 log reduction in survivability of *S. cerevisiae* in apple juice, PEF utilized less than 10% of the electric energy for heat treatment (Qin et al. 1994). It has been also reported (Mittal 1998) that a PEF low energy pulser with an instant-charge-reversal pulse waveform was successfully used in apple cider treatment. The consumed energy was as low as 5.76 J/mL at 20°C, compared with the 50 J/mL normally required in conventional thermal processing. It has been reported that 6 log reductions in survivability of the indigenous aerobic bacteria of apple juice were obtained using PEF at 50 kV/cm. PEF has been compared directly with HTST in pasteurization of apple juice (Charles-Rodríguez et al. 2007) finding that PEF was efficient in microbial inactivation, as well as in preserving

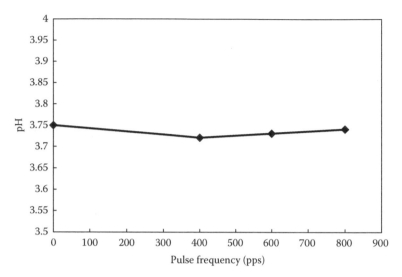

FIGURE 16.8   pH as a function of pulse frequency for treatment at 36 kV/cm in PEF-pasteurized apple juice. (Adapted from Charles-Rodríguez, A.V. et al., *Food Bioprod. Process.*, 85C, 93, 2007.)

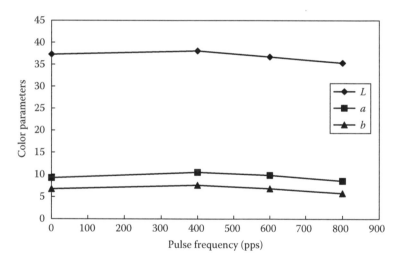

FIGURE 16.9   Color parameters as a function of pulse frequency for treatment at 36 kV/cm in PEF-pasteurized apple juice. (Adapted from Charles-Rodríguez, A.V. et al., *Food Bioprod. Process.*, 85C, 93, 2007.)

some quality attributes, such as pH and color (Figures 16.8 and 16.9). HTST, on the other hand, had apparent effects on pH and browning increase.

Comparative studies of PEF and other nonthermal techniques have been also reported. In a particular investigation (Zárate-Rodríguez et al. 2000), PEF and ultra-filtration (UF) were utilized for pasteurization of apple juice. No significant changes were observed in variables such as pH, Brix, and acidity, expressed as malic acid, for the PEF-treated juice and the ultrafiltered one. Color was the quality attribute that did show change for membrane treatments. The observed trend was for the juice to become darker as a function of applied transmembrane pressure. Similarly to UF treatments, relative color changes were observed but the registered effect was opposite, i.e., the treated juices became paler as a function of applied field strength. Color changes were independent of pulse number but dependent on field strength. The different color ratio perception in UF and PEF-treated juices could be due to haze formation, which may be caused by tannins, proteins, and carbohydrate polysaccharides. It has been reported that proteins, independently and in association with phenols, are responsible for fruit juice turbidity, as well as for postclarification haze and sediment formation (Flores et al. 1988). Haze is formed quite often by interaction of these biochemical compounds in many types of fruit juices. Since enzyme inactivation capability of PEF has been reported (Vega-Mercado et al. 1995) a tannin–protein or carbohydrate–protein bonding would not be possible due to the lack of the protein fraction, and little or no haze formation would be likely to occur in the PEF-treated juice. On the other hand, in the UF pasteurized juice the presence of proteins may have caused an incipient haze formation, because all the interactions mentioned above would be important. In such a case, haze or turbidity might have been registered as a deviation to the darker side of the specific absorbance value considered as the control or reference. The observed browning in the UF processes is considered a quality problem, but there is some evidence that might be controlled by membrane pore size effects (Zárate-Rodríguez et al. 2000).

Direct effects of PEF on volatiles of apple juice, and comparison with a conventional thermal treatment, have been also been investigated (Aguilar-Rosas et al. 2007). PEF and HTST were tested in order to determine decrease in concentration of eight odor responsible volatiles. In general terms, PEF retained better most of the volatile compounds responsible for color and flavor of the apple juice. For example, as shown in Table 16.3, hexanal and hexyl acetate were only lost in 7% and 8.4%, respectively, when using PEF, while they were virtually lost by HTST. Also,

TABLE 16.3

Percentage of Volatiles Losses, Compared with Untreated Sample, in Apple Juice Treated by Two Methods

Compound	Loss Percentage for PEF	Loss Percentage for HTST
Acetic acid	39.792 ± 20.84	100
Hexanal	7.042 ± 9.32	62.348 ± 5.35
Butyl hexanoate	18.108 ± 7.72	36.273 ± 24.86
Ethyl acetate	77.458 ± 29.23	67.126 ± 39.33
Ethyl butyrate	60.190 ± 17.80	88.398 ± 12.46
Methyl butyrate	30.081 ± 31.37	51.200 ± 19.56
Hexyl acetate	8.408 ± 16.12	22.910 ± 21.99
1-hexanal	14.101 ± 7.65	69.307 ± 5.62

Source: Adapted from Aguilar-Rosas, S.F. et al., *J. Food Eng.*, 83, 41, 2007.

Note. Differences by a Student t-test for independent samples ($p < 0.05$, $n = 3$).

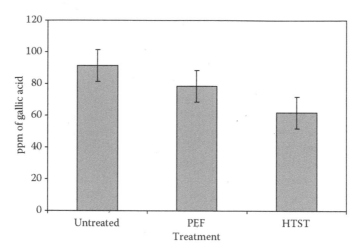

FIGURE 16.10 Effect of treatment method on total phenol compounds of pasteurized apple juice treated by HTST pasteurization and PEF technique. (Adapted from Aguilar-Rosas, S.F. et al., *J. Food Eng.*, 83, 41, 2007.)

important biochemical substances in apple juice, such as phenol compounds, were better retained by PEF than by HTST treatment (Figure 16.10).

16.6.3.3 Other Types of Fruit Juices

Reports on some other types of juices are not abundant in the literature. Peach nectar was pasteurized by PEF (Gutierrez-Becerra et al. 2002). The effects of field strength and pulse frequency were evaluated for reduction in microbial count, as well as for variation in pH and color of the treated juice. An appropriate microbial inactivation was shown, while no significant changes were observed in pH and color. The results obtained suggest that PEF may be used as alternative for peach juice pasteurization, to preserve quality and sensory attributes.

16.7 OPTIMIZATION OF PEF PARAMETERS

As obvious from the preceding discussion, pertinent parameters of PEF from the point of view of processing include pulse characteristics, electric field strength, and treatment time. The pulse characteristic factors include waveform, frequency, width, and number. Also microbial and product characteristics and design of treatment chamber may influence effectiveness of PEF treatment since it determines distribution of electric fields across product. Successful application of PEF for a product depends on rational and innovative optimization of these parameters. PEF applications allow for better control of electric power input and effective permeabilization of cellular membranes without significant temperature elevation (Weaver and Chizmadzhev 1996). In general, the transmembrane voltage u_m induced on the cell membrane due to an external electric field is given as (Zimmermann 1986)

$$u_m \sim \alpha d_c E \cos\theta \qquad (16.9)$$

where

d_c is cell diameter

E is electric field strength

θ is the angle between a point on membrane surface and direction of electric field strength

α is a parameter depending on the cell shape ($\alpha = 0.75$ for spherical cell and $\alpha = 1$ for rectangular cell)

The smaller the exposed cells are in the electric field, the higher the field strength required for creating a critical transmembrane potential needed for the cell membrane's plasmolysis. Mean diameters of microorganisms and biological tissue cells are in the ranges of $10\,nm^{-1}$ to $1\,\mu m$ and $10\,\mu m^{-1}$ to $1\,mm$, respectively (Aguilera and Stanley 1999). Therefore, understandably high electric field pulses with voltages in the range of 20–$50\,kV/cm$ are used to kill microorganisms for PEF pasteurization. However, for solid materials such as vegetables tissues, for which cells are usually larger than microbial cells, electroplasmolysis can be obtained at much lower electric field strengths. A number of publications report that electroplasmolysis of vegetable cells may be achieved at moderate electric field pulses with voltages in the range of 0.3–$3\,kV/cm$ (Ngadi et al. 2001). Lower electric field pulses in the range of 0.1–$0.3\,kV/cm$ have also been reported (Kupchik et al. 1998).

It has been established experimentally that the electrical treatment time needed for electroplasmolysis is inversely proportional to electric field strength; the higher the field strength, the less specific energy consumption needed for achieving the same degree of plasmolysis (Lebovka et al. 2000). In liquids, the extent to which electric field strength can be increased is limited by dielectric breakdown of the products and uncontrolled temperature increase in the products. Therefore, it is vital to balance the need for higher electric field with product response. Different authors have reported different effectiveness of PEF application for various products. There are currently inherent practical difficulties involved in optimizing PEF processing of foods. It is difficult to compare treatment results obtained by different authors using different PEF systems. PEF systems are very expensive and it has not been possible to build systems that allow wide variation of pulse parameters. Therefore, studies have been conducted using equipment with restricted range of parameters.

Two experimental methods have been proposed for determination of the optimal field strength E for PEF treatment of different solid food tissues. One of the methods is based on estimation of the characteristic damage times τ and the total energy consumption factor during treatment τE^2 as a function of electric field strength (Lebovka et al. 2002). The other approach is based on estimation of the maximal change of sample disintegration index caused by the energy input during each pulse (Bazhal et al. 2003a). The optimal E value for electroplasmolysis depends on the type of tissue and is higher for cells with developed secondary cell wall. The overall goal of PEF treatment objective (for instance microbial inactivation in liquid medium vs. textural enhancement in solid medium), the material to be treated and all pulse parameters must be taken into consideration in order to optimize PEF treatment for a product.

16.8 CONCLUSIONS

PEF applications hold several promises for modifying and improving quality and safety of foods. In particular, since it is essentially a nonthermal technology, PEF treatment can be used to minimally process products and preserve their delicate sensory and nutritional qualities. Although impressive progress has been made in understanding the mechanism and influence of PEF processing, much remains to be done. The process parameters need to be optimized for specific products to match equipment used. There is need to develop a common equipment-independent platform for assessing PEF processing.

The use of PEF technology to pasteurize a wide variety of foodstuffs and obtain high-quality, competitive products in world markets can be considered a plausible reality. Scientific validation of pasteurization efficiency has been carried out and the technique renders results comparable with the conventional, thermal way of pasteurizing many food products. Safety is definitely preserved, while quality appears virtually unaltered by use of PEF. The engineering challenge of making equipments commercially viable has also advanced. It can be, therefore, considered that the technology is ready for industrial scale application and that a number of PEF-pasteurized food products will be seen in supermarket shelves in a foreseeable future.

REFERENCES

Ade-Omowaye, B.I.O., Angersbach, A., Eshtiaghi, N.M., and Knorr, D. 2000. Impact of high intensity electric field pulses on cell permeabilisation and as pre-processing step in coconut processing. *Innovative Food Science and Emerging Technologies* 1: 203–209.

Ade-Omowaye, B.I.O., Rastogi, N.K., Angersbach, A., and Knorr, D. 2002. Osmotic dehydration of bell peppers: influence of high intensity electric field pulses and elevated temperature treatment. *Journal of Food Engineering* 54: 35–43.

Aguilar-Rosas, S.F., Ballinas-Casarrubias, M., Nevarez-Moorillon, G.V., Martin-Belloso, O., and Ortega-Rivas, E. 2007. Thermal and pulsed electric fields pasteurization of apple juice: effects on physicochemical properties and flavour compounds. *Journal of Food Engineering* 83: 41–46.

Aguilera, J.M. and Stanley, D.W. 1999. *Microstructural Principles of Food Processing and Engineering*. Gaithersburg, MD: Aspen Publishers Inc.

Aguiló-Aguyo, I., Odrizola-Serrano, I., Quintãa-Teixeira, L.J., and Martín-Belloso, O. 2008. Inactivation of tomato juice peroxidase by high-intensity pulsed electric fields as affected by process conditions. *Food Chemistry* 107: 949–955.

Ahmed, E.M., Dennison, R.A., and Shaw, P.E. 1978. Effects of selected oil and essence volatile components on flavor quality of pumpout orange juice. *Journal of Agricultural and Food Chemistry* 26: 368–372.

Alvarez, I., Raso, J., Palpo, A., and Sala, F.J. 2000. Influence of different factors on the inactivation of *Salmonella senftenberg* by pulsed electric fields. *International Journal of Food Microbiology* 55: 143–146.

Amiali, M., Ngadi, M.O., Smith, J.P., and Raghavan, G.S.V. 2004. Inactivation of *Escherichia coli* O157:H7 in liquid dialyzed egg using pulsed electric fields. *Food and Bioproducts Processing* 82C: 151–156.

Amiali, M., Ngadi, M.O., Raghavan, G.S.V., and Nguyen, D.H. 2006a. Electrical conductivities of liquid egg products and fruit juices exposed to high pulsed electric fields. *International Journal of Food Properties* 9: 533–540.

Amiali, M., Ngadi, M.O., Raghavan, G.S.V., and Smith, J.P. 2006b. Inactivation of *Escherichia coli* O157:H7 and *Salmonella enteritidis* in liquid egg white using pulsed electric field. *Journal of Food Science* 71: 88–94.

Anderson, W.A., McClure, P.J., Baird-Parker, A.C., and Cole, M.B. 1996. The application of a log-logistic model to describe the thermal inactivation of Clostridium botulinum 213B at temperatures below 121.1°C. *Journal of Applied Microbiology*, 80: 238–290.

Angersbach, A. and Knorr, D. 1997. High intensity electric field pulses as pre-treatment for affecting dehydration characteristics and rehydration properties of potato cubes. *NahrungFood* 41: 194–200.

Angersbach, A., Heinz, V., and Knorr, D. 2000. Effects of pulsed electric fields on cell membranes in real food systems. *Innovative Food Science and Emerging Technology* 1: 135–149.

Arevalo, P. 2003. effect of pulsed electric field on physical properties of apples and potatoes. Master Thesis. Montreal, Quebec, Canada: McGill University.

Barbosa-Cánovas, G.V., Góngora-Nieto, M.M., Pothakamury, U.R., and Swanson, B.G. 1999. *Preservation of Foods with Pulsed Electric Fields*. San Diego, CA: Academic Press.

Barsotti, L., Dumay, E., Mu, T.H., Diaz, M.D.F., and Cheftel, J.C. 2002. Effects of high voltage electric pulses on protein-based food constituents and structures. *Trends in Food Science and Technology* 12: 136–144.

Bazhal, M.I. 2001. Etude du mécanisme d'électroperméabilisation des tissus végétaux. Application à l'extraction du jus des pommes. Thèse de Doctorat. Compiègne, France: Université de Technologie de Compiègne.

Bazhal, M.I. and Vorobiev, E. 2000. Electrical treatment of apple cossettes for intensifying juice pressing. *Journal of the Science of Food and Agriculture* 80: 1668–1674.

Bazhal, M.I., Ngadi, M.O., Raghavan, G.S.V., and Nguyen, D.H. 2003a. Textural changes in apple tissue during pulsed electric field treatment. *Journal of Food Science* 68: 249–253.

Bazhal, M.I., Ngadi, M.O., and Raghavan, G.S.V. 2003b. Influence of pulsed electroplasmolysis on the porous structure of apple tissue. *Biosystems Engineering* 86: 51–57.

Bazhal, M.I., Ngadi, M.O., Smith, J.P., and Raghavan, G.S.V. 2006. Kinetics of *Escherichia coli* in liquid whole egg using combined PEF and thermal treatments. *Lebensmittel-Wissenschaft und-Technologie—Food Science and Technology* 39: 420–426.

Bendicho, S., Espachs, A., Stevens, D., Arantegui, J., and Martin, O. 1999. Effect of high intensity pulsed electric fields on vitamins of milk, p. 108. European Conference on Emerging Food Science and Technology.

Bendicho, S., Barbosa-Cánovas, G.V., and Martín, O. 2002. Milk processing by high intensity pulsed electric fields. *Trends in Food Science and Technology* 13: 195–204.

Braddock, R.J. 1999. Single-strength orange juice and concentrates. In *Handbook of Citrus By-Products and Processing Technology*, ed. R.J. Braddock, pp. 53–83. New York: John Wiley & Sons.

Calderón-Miranda, M.L., Barbosa-Cánovas, G.V., and Swanson, B.G. 1999. Transmission electron microscopy of *Listeria innocua* treated by pulsed electric fields and nisin skimmed milk. *International Journal of Food Microbiology* 51: 31–38.

Castro, A., Barbosa-Cánovas, G.V., and Swanson, B.G. 1993. Microbial inactivation of foods by pulsed electric fields. *Journal of Food Preservation* 17: 47–73.

Charles-Rodríguez, A.V., Nevárez-Moorillón, G.V., Zhang, Q.H., and Ortega-Rivas, E. 2007. Comparison of thermal processing and pulsed electric fields treatment in pasteurization of apple juice. *Food and Bioproducts Processing* 85C: 93–97.

Chen, W. and Lee, R. 1994. Altered ion channel conductance and ionic selectivity induced by large imposed membrane potential pulse. *Biophysics Journal* 67: 603–612.

Chen, C.S., Shaw, P.E., and Parish, M.E. 1993. Orange and tangerine juices. In *Fruit Juice Processing Technology*, eds. S. Nagy, C.S. Chen, and P.E. Shaw, pp. 110–165. Auburndale, FL: Agscience.

Coster, H.G.L. and Zimmerman, U. 1975. The mechanisms of electrical breakdown in the membrane of *Valonia utricularis*. *Journal of Membrane Biology* 22: 73–90.

Cunningham, D.G., Acree, T.E., Barnard, J., Butts, R.M., and Braell, P.A. 1986. Charm analysis of apple volatiles. *Food Chemistry* 19: 137–147.

Dunn, J. 1996. Pulsed light and pulsed electric field for foods and eggs. *Poultry Science* 75: 1133–1136.

Dunn, J.E. and Pearlman, J.S. 1987. Methods and apparatus for extending the shelf life of fluid food products. Maxwell Laboratories, Inc., U.S. Patent No. 4695472.

Dutreux, N., Notermans, W.T., Gongóra-Nieto, M.M., Barbosa-Cánovas, G.V., and Swanson, B.G. 2000. Pulsed electric fields inactivation of attached and free-living *Escherichia coli* and *Listeria innocua* under several conditions. *International Journal of Food Microbiology* 54: 91–98.

Ekasari, I., Jongen, M.F., and Pilnik, W. 1986. Use of a bacterial mutagenicity assay as a rapid method for the detection of early stage of Maillard reactions in orange juices. *Food Chemistry* 21: 125–131.

Elez-Martinez, P. and Martin-Belloso, O. 2007. Effects of high intensity pulsed electric field processing conditions on vitamin C and antioxidant capacity of orange juice and gazpacho, a cold vegetable soup. *Food Chemistry* 102: 201–209.

Ellis, R.F. 1982. Aseptically packed juice and milk taking hold on U.S. market. *Food Processing (Chicago)* 43: 96–100.

Eshtiaghi, M.N. and Knorr, D. 2002. High electric field pulse pretreatment: potential for sugar beet processing. *Journal of Food Engineering* 52: 265–272.

Espachs-Barroso, A., Barbosa-Cánovas, G.V., and Martín-Belloso, O. 2003. Microbial and enzymatic changes in fruit juice induced by high-intensity pulsed electric fields. *Food Reviews International* 19: 253–273.

Evrendilek, G.A. and Zhang, Q.H. 2005. Effects of pulse polarity and pulse delaying time on pulsed electric fields-induced pasteurization of *E. coli* O157:H7. *Journal of Food Engineering* 68: 271–276.

Evrendilek, G.A., Dantzer, W.R., Streaker, C.B., Ratanatriwong, P., and Zhang, Q.H. 2001. Shelf-life evaluations of liquid foods treated by pilot plant pulsed electric field system. *Journal of Food Processing and Preservation* 25: 283–297.

Fleischman, G.J., Ravishankar, S., and Balasubramaniam, V.M. 2004. The inactivation of *Listeria monocytogenes* by pulsed electric field (PEF) treatment in static chamber. *Food Microbiology* 21: 91–95.

Flores, J.H., Heatherbell, D.A., Hsu, J.C., and Watson, B.T. 1988. Ultrafiltration (UF) of white riesling juice-effect of oxidation and pre-UF juice treatment on flux, composition, and stability. *American Journal of Enology and Viticulture* 39: 180–187.

Floury, J., Grosset, N., Leconte, N., Pasco, M., Madec, M., and Jeantet, R. 2005. Continuous raw skim milk processing by pulsed electric field at non-lethal temperature: effect on microbial inactivation and functional properties. *Lait* 86: 43–57.

Fox, P.F., Guinee, T.P., Cogan, T.M., and McSweeney, P.L.H. 2000. *Fundamentals of Cheese Science*. Gaithersburg, MD: Aspen Publishers.

Galindo, F.G., Vernier, P.T., Dejmek, P., Vicente, A., and Gundersen, M.A. 2008. Pulsed electric field reduces the permeability of potato cell wall. *Bioelectromagnetics* 29: 296–301.

Gallardo-Reyes, E.D., Valdez-Fragoso, A., Nevarez-Moorillon, G.V., Ngadi, M.O., and Ortega-Rivas, E. 2008. Comparative quality of orange juice as treated by pulsed electric fields and ultra high temperature. *AgroFood Industry Hi-Tech* 19: 35–36.

García, D., Gómez, N., Mañas, P., Raso, J., and Pagán, R. 2007. Pulsed electric fields cause bacterial envelopes permeabilization depending on the treatment intensity, the treatment medium pH and the microorganism investigated. *International Journal of Food Microbiology* 113: 219–227.

Giner, J., Gimeno, V., Barbosa-Cánovas, G.V., and Martín, O. 2001. Effects of pulsed electric field processing on apple and pear polyphenoloxidases. *Food Science and Technology International* 7: 339–345.

Giner, J., Ortega, M., Mesegue, M., Gimeno, V., Barbosa-Cánovas, G.V., and Martín, O. 2002. Inactivation of peach polyphenoloxidase by exposure to pulsed electric fields. *Journal of Food Science* 67: 1467–1472.

Gongóra-Nieto, M.M., Sepúlveda-Ahumada, D.R., Pedrow, P., Barbosa-Cánovas, G.V., and Swanson, B.G. 2002. Food processing by pulsed electric fields: treatment delivery, inactivation level and regulatory aspects. *Lebensmittel-Wissenschaft und-Technologie—Food Science and Technology* 35: 375–388.

Grahl, T. and Märkl, H. 1996. Killing of microorganisms by pulsed electric fields. *Applied Microbiology and Biotechnology* 45: 148–157.

Gudmundsson, M. and Mafsteinsson, H. 2001. Effect of electric field pulses on microstructure of muscle foods and roes. *Trends in Food Science and Technology* 12: 122–128.

Gulyi, I.S., Lebovka, N.I., Mank, V.V., Kupchik, M.P., Bazhal, M.I., Matvienko, A.B., and Papchenko, A.Y. 1994. *Scientific and Practical Principles of Electrical Treatment of Food Products and Materials.* Kiev: UkrINTEI (in Russian).

Gutierrez-Becerra, L.E., Li, S., Ortega-Rivas, E., and Zhang, Q.H. 2002. Cold pasteurization of peach nectar using pulsed electric fields. IFT Annual Meeting, Anaheim, CA, June 2002.

Hermawan, G.A., Evrendilek, W.R., Zhang, Q.H., and Richter, E.R. 2004. Pulsed electric field treatment of liquid whole egg inoculated with *Salmonella enteritidis. Journal of Food Safety* 24: 1–85.

Ho, S.Y. and Mittal, G.S. 1996. Electroporation of cell membranes: a review. *Critical Reviews in Biotechnology* 16: 349–362.

Ho, S.Y. and Mittal, G.S. 2000. High voltage pulsed electrical field for liquid food pasteurization. *Food Reviews International* 16: 395–434.

Ho, S.Y., Mittal, G.S., Cross, J.D., and Griffiths, M.W. 1995. Inactivation of *Pseudomonas fluorescens* by high voltage electric pulses. *Journal of Food Science* 60: 1337–1340, 1343.

Ho, S.Y., Mittal, G.S., and Gross, J.D. 1997. Effect of high electric pulses on activity of selected enzymes. *Journal of Food Engineering* 31: 69–84.

Huang, E., Mittal, G.S., and Griffiths, M.W. 2006. Inactivation of *Salmonella enteritidis* in liquid whole egg using combination treatments of pulsed electric field, high pressure and ultrasound. *Biosystems Engineering* 94: 403–413.

Hülsheger, H., Potel, J., and Niemann, E.G. 1981. Killing of bacteria with electric pulses of high field strength. *Radiation and Environmental Biophysics* 20: 53–65.

Hülsheger, H., Potel, J., and Niemann, E.G. 1983. Electric field effects on bacteria and yeast cells. *Radiation and Environmental Biophysics* 22: 149–162.

Jayaram, S., Castle, G., and Margaritis, A. 1992. Kinetics of sterilization of *Lactobacillus brevis* by the application of high voltage pulses. *Biotechnology and Bioengineering* 40: 1412–1420.

Jeantet, R., Baron, F., Nau, F., Roignant, M., and Brulé, G. 1999. High intensity pulsed electric fields applied to egg white: effect on *Salmonella enteritidis* inactivation and protein denaturation. *Journal of Food Protection* 62: 1381–1386.

Jeantet, R., Mc Keag, J.R., Fernández, J.C., Grosset, N., Baron, F., and Korolczuk, J. 2004. Pulsed electric field continuous treatment of egg products. *Sciences Des Aliments* 24: 137–158.

Jemai, A.B. 1997. Contribution a l'étude de l'effet d'un Traitement électrique sur les Cossettes de Betterave a Sucre. Incidence sur le Procède d'extraction. Thèse de Doctorat. Compiègne, France: Université de Technologie de Compiègne.

Jeyamkondan, S., Jayas, D.S., and Holle, R.A. 1999. Pulsed electric field processing of foods. *Journal of Food Protection* 62: 1088–1096.

Jia, M., Zhang, Q.H., and Min, D.B. 1999. Pulsed electric field processing effects on flavor compounds and microorganisms of orange juice. *Food Chemistry* 65: 445–451.

Kinosita, K.J. and Tsong, T.Y. 1977. Hemolysis of erythrocytes by transient electric field. *Proceedings of the National Academy of Science of the U.S.A.* 74: 1923–1927.

Knorr, D. and Angersbach, A. 1998. Impact of high intensity electric field pulses on plant membrane permeabilization. *Trends in Food Science and Technology* 9: 185–191.

Knorr, D., Geulen, M., Grahl, T., and Sitzmann, W. 1994. Food application of high electric field pulses. *Trends in Food Science and Technology* 5: 71–75.

Knorr, D., Angersbach, A., Eshtiaghi, M.N., Heinz, V., and Lee, D.U. 2001. Processing concepts based on high intensity electric field pulses. *Trends in Food Science and Technology* 12: 129–135.

Kupchik, M.P., Bazhal, M.I., Guliy, L.S., Lebovka, N.I., and Mank, V.V. 1998. Pulsed electric treatment effect on food stuff. In *Physics of Agro and Food Products ICPAFP'98*. 1998, Lublin, Poland.

Lazarenko, B.R., Fursov, S.P., Scheglov, Y.A., Bordiyan, V.V., and Chebanu, V.G. 1977. *Electroplasmolysis*. Karta Moldovenaske: Kishinev.

Lebovka, N.I., Bazhal, M.I., and Vorobiev, E. 2000. Simulation and experimental investigation of food material breakage using pulsed electric field treatment. *Journal of Food Engineering* 44: 213–223.

Lebovka, N.I., Bazhal, M.I., and Vorobiev, E. 2001. Pulsed electric field breakage of cellular tissues: visualization of percolative properties. *Innovative Food Science and Emerging Technologies* 2: 113–125.

Lebovka, N.I., Bazhal, M.I., and Vorobiev, E. 2002. Estimation of characteristic damage time of food cellular materials in pulsed electric fields. *Journal of Food Engineering* 54: 337–346.

Lebovka, N.I., Shynkaryk, N.V., and Vorobiev, E. 2007. Pulsed electric field enhanced drying of potato tissue. *Journal of Food Engineering* 78(2): 606–613.

Lewis, M.J. 1993. UHT processing: safety and quality aspects. In *Food Technology International Europe*, ed. A. Turner, pp. 47–51. London: Sterling Publications.

Li, S.Q., Zhang, Q.H., Lee, Y.Z., and Pham, T.V. 2003. Effects of pulsed electric fields and thermal processing on the stability of bovine immunoglobulin G (IgG) in enriched soymilk. *Journal of Food Science* 68: 1201–1207.

Li, S.Q., Bomser, J.A., and Zhang, Q.H. 2005. Effects of pulsed electric fields and heat treatment on stability and secondary structure of bovine immunoglobulin G. *Journal of Agricultural and Food Chemistry* 53: 663–670.

Liang, Z., Mittal, G.S., and Griffiths, M.W. 2002. Inactivation of *Salmonella typhimurium* in orange juice containing antimicrobial agents by pulsed electric field. *Journal of Food Protection* 65: 1081–1087.

Liu, X., Yousef, A.E., and Chism, G.W. 1997. Inactivation of *Escherichia coli* O157:H7 by the combination of organic acids and pulsed electric field. *Journal of Food Safety* 16: 287–299.

Maarse, H. 1991. *Volatile Compounds in Foods and Beverages*. New York: Marcel Dekker.

McLellan, M.R., Kime, R.L., and Lind, L.R. 1991. Electroplasmolysis and other treatments to improve apple juice yield. *Journal of the Science of Food Agriculture* 57: 303–306.

Manas, P., Barsotti, L., and Cheftel, J.C. 2001. Microbial inactivation by pulsed electric fields in a batch treatment chamber: effects some electric parameters and food constituents. *Innovative Food Science and Emerging Technologies* 2: 239–249.

Martín, O., Qin, B.L., Chang, F.J., Barbosa-Cánovas, G.V., and Swanson, B.G. 1997. Inactivation of *Escherichia coli* in skim milk by high intensity pulsed electric fields. *Journal of Food Process Engineering* 20: 317–336.

Martín-Belloso, O., Vega-Mercado, H., Qin, B.L., Chang, F.J., Barbosa-Cánovas, G.V., and Swanson, B.G. 1997. Inactivation of *Escherichia coli* suspended in liquid egg using pulsed electric fields. *Journal of Food Processing and Preservation* 21: 193–208.

Matvienko, A.B. 1996. Intensification of the extraction of soluble substances by electrical treatment of plant materials and water. PhD Thesis. Kiev: Ukrainian State University of Food Technologies (in Ukrainian).

Michalac, S.L., Alvarez, V.B., and Zhang, Q.H. 2003. Inactivation of selected microorganisms and properties of pulsed electric field processed milk. *Journal of Food Processing and Preservation* 27: 137–151.

Mittal, G.S. 1998. A new approach to enhance efficiency of PEF treatments. IFT Annual Meeting, Atlanta, GA, June 1998.

Moyer, J.C. and Aitken, H.C. 1980. Apple juice. In *Fruit and Vegetable Juice Processing Technology*, eds. P.E. Nelson and D.K. Tressler, pp. 212–267. Westport, CT: AVI Publishing Co.

Ngadi, M.O., Arevalo, P., Raghavan, G.S.V., and Nguyen, D.H. 2001. Pulse electric fields in food processing. Paper Presented at the ASAE Annual International Meeting, July 29 to August 1, 2001, Sacramento, CA.

Ohshima, T. and Sato, M. 2004. Bacterial sterilization and intracellular protein release by a pulsed electric field. *Advanced Biochemical Engineering and Biotechnology* 90: 113–133.

Ortega-Rivas, E., Zarate-Rodríguez, E., and Barbosa-Canovas, G.V. 1998. Apple juice pasteurization using ultrafiltration and pulsed electric fields. *Food and Bioproducts Processing*, 76C: 193–198.

Papchenko, A.Y., Bologa, M.K., and Berzoi, S.E. 1998. Apparatus for processing vegetable raw material, U.S. Patent No. 4787303.

Parker, P.M. 2006a. *The World Market for Apple Juice: A 2007 Global Trade Perspective.* Sand Diego, CA: ICON Group International Inc.

Parker, P.M. 2006b. *The World Market for Orange Juice: A 2007 Global Trade Perspective.* Sand Diego, CA: ICON Group International Inc.

Perez, O.E. and Pilosof, A.M.R. 2004. Pulsed electric fields effects on the molecular structure and gelation of β-lactoglobulin concentrate and egg white. *Food Research International* 37: 102–110.

Picart, L., Dumay, E., and Cheftel, J.C. 2002. Inactivation of *Listeria innocua* in dairy fluids by pulsed electric fields: influence of electric parameters and food composition. *Innovative Food Science and Emerging Technologies* 3: 357–369.

Pothakamury, U.R., Vega, H., Zhang, Q.H., Barbosa-Cánovas, G.V., and Swanson, B.G. 1996. Effect of growth stage and processing temperature on the inactivation of *E. coli* by pulsed electric fields. *Journal of Food Protection* 59: 1167–1171.

Qin, B.L., Barbosa-Cánovas, G.V., Swanson, B.G., and Pedrow, P.D. 1994. Inactivation of microorganisms by pulsed electric field of different voltage waveforms. *IEEE Transactions on Dielectrics and Electrical Insulation* 1: 1047–1057.

Qin, B.L., Chang, F., Barbosa-Cánovas, G.V., and Swanson, B.G. 1995. Nonthermal inactivation of *S. cerevisiae* in apple juice using pulsed electric fields. *Lebensmittel-Wissenschaft und-Technologie—Food Science and Technology* 28: 564–568.

Raso, J., Alvarez, I., Condon, S., and Sala, F.J. 2000. Predicting inactivation of *Salmonella seftenberg* by pulsed electric fields. *Innovative Food Science and Emerging Technologies* 1: 21–30.

Rastogi, N.K., Eshtiaghi, M.N., and Knorr, D. 1999. Accelerated mass transfer during osmotic dehydration of high intensity electrical field pulse pretreated carrots. *Journal of Food Science* 64: 1020–1023.

Ravishankar, S., Fleischman, G.J., and Balasubramaniam, V.M. 2002. The inactivation of *Escherichia coli* O157:H7 during pulsed electric field (PEF) treatment in a static chamber. *Food Microbiology* 19: 351–361.

Reina, L.D., Jin, Z.T., Yousef, A.E., and Zhang, Q.H. 1998. Inactivation of *Listeria monocytogenes* in milk by pulsed electric field. *Journal of Food Protection* 61: 1203–1206.

Sale, A.J.H. and Hamilton, W.A. 1967. Effects of high electric fields on microorganisms: I. Killing bacteria and yeast. *Biochimica et Biophysica Acta* 148: 781–788.

Sampedro, F., Rodrigo, A., Martínez, A., and Rodrigo, D. 2005. Quality and safety aspects of PEF application in milk and milk products. *Critical Reviews in Food Science and Nutrition* 45: 25–47.

Schilling, S., Alber, T., Toepfl, S., Neidhart, S., Knorr, D., Andreas Schieber, A., and Reinhold, C. 2007. Effects of pulsed electric field treatment of apple mash on juice yield and quality attributes of apple juices. *Innovative Food Science and Emerging Technologies* 8: 127–134.

Schwan, H. 1957. Electrical properties of tissue and cell suspensions. *Advances in Biological and Medical Physics* 5: 147–209.

Sensoy, I., Zhang, Q.H., and Sastry, S. 1997. Inactivation kinetics of *Salmonella dublin* by pulsed electric field. *Journal of Food Process Engineering* 20: 367–381.

Sepúlveda-Ahumada, D.R. 2003. Preservation of fluid foods by pulsed electric fields in combination with mild thermal treatments. PhD Thesis. Pullman, WA: Washington State University.

Sepúlveda-Ahumada, D.R., Ortega-Rivas, E., and Barbosa-Cánovas, G.V. 2000. Quality aspects of cheddar cheese obtained with milk pasteurized by pulsed electric fields. *Food and Bioproducts Processing* 78C: 65–71.

Sepúlveda-Ahumada, D.R., Gongóra-Nieto, M.F., San-Martin, J.A., and Barbosa-Cánovas, G.V. 2005a. Influence of treatment temperature on the inactivation of *Listeria innocua* by pulsed electric fields. *Lebensmittel-Wissenschaft und-Technologie—Food Science and Technology* 38: 167–172.

Sepúlveda-Ahumada, D.R., Góngora-Nieto, M.M., Guerrero, J.A., and Barbosa-Cánovas, G.V. 2005b. Production of extended-shelf life milk by processing pasteurized milk with pulsed electric fields. *Journal of Food Engineering* 67: 81–86.

Shamsi, K., Versteeg, C., Sherkat, F., and Wan, J. 2008. Alkaline phosphatase and microbial inactivation by pulsed electric field in bovine milk. *Innovative Food Science and Emerging Technologies* 9: 217–223.

Shin, J.K., Jung, K.J., Pyun, Y.R., and Chung, M.S. 2007. Application of pulsed electric fields with square wave pulse to milk inoculated with *E. coli, P. fluorescens*, and *B. stearothermophilus*. *Food Science and Biotechnology* 16: 1082–1084.

Siezer, C.E., Waugh, P.L., Edstam, S., and Ackermann, P. 1988. Maintaining flavor and nutrient quality of aseptic orange juice. *Food Technology* 43: 152–159.

Sobrino-Lopez, A., Raybaudi-Massilia, R., and Martin-Belloso, O. 2006. High-intensity pulsed electric field variables affecting *Staphylococcus aureus* inoculated in milk. *Journal of Dairy Science* 89: 3739–3748.

Stanley, D.W. 1991. Biological membrane deterioration and associated quality losses in food tissues. *CRC Critical Reviews in Food Science and Nutrition* 31: 235–245.

Taiwo, K.A., Angersbach, A., Ade-Omowaye, B.I.O., and Knorr, D. 2001. Effects of pretreatments on the diffusion kinetics and some quality parameters of osmotically dehydrated apple slices. *Journal of Agricultural Food Chemistry* 49: 2804–2811.

Taiwo, K.A., Angersbach, A., and Knorr, D. 2002. Influence of high intensity electric field pulses and osmotic dehydration on the rehydration characteristics of apple slices at different temperatures. *Journal of Food Engineering* 52: 185–192.

Tatum, J.H., Steven, N., and Roberts, B. 1975. Degradation products formed in canned single-strength orange juice during storage. *Journal of Food Science* 40: 707–709.

Torregrosa, F., Esteve, M.J., Frigola, A., and Cortes, C. 2006. Ascorbic acid stability during refrigerated storage of orange–carrot juice treated by high pulsed electric field and comparison with pasteurized juice. *Journal of Food Engineering* 73: 245–339.

Tsong, T.Y. 1990. Review: on electroporation of cell membranes and some related phenomena. *Bioelectrochemistry and Bioenergetic* 24: 271–295.

Tsong, T.Y. 1991. Electroporation of cell membranes. *Biophysics Journal* 60: 297–306.

Van Loey, A., Verachtert, B., and Hendrickx, M. 2002. Effects of high electric field pulses on enzymes. *Trends in Food Science and Technology* 60: 1143–1146.

Varnam, A.H. and Sutherland, J.P. 1999. Fruit juices. In *Beverages: Technology, Chemistry and Microbiology*, eds. A.H. Varnam, J.P. Sutherland, A. Varnam, and J. Sutherland, pp. 26–72. New York: Aspen Publications.

Vega-Mercado, H. 1996. Inactivation of *Escherichia coli* by combining pH, ionic strength and pulsed electric fields. *Food Research International* 29: 117.

Vega-Mercado, H., Powers, J.R., Barbosa-Cánovas, G.V., and Swanson, B.G. 1995. Plasmin inactivation with pulsed electric fields. *Journal of Food Science* 60: 1143–1146.

Vega-Mercado, H., Martín-Belloso, O., Qin, B.L., Chang, F.J., Góngora-Nieto, M.M., Barbosa-Cánovas, G.V., and Swanson, B.G. 1997. Non-thermal food preservation: pulsed electric fields. *Trends in Food Science and Technology* 8: 151–157.

Weaver, J.C. and Chizmadzhev, Y.A. 1996. Theory of electroporation: a review. *Bioelectrochemistry and Bioenergetics* 41: 135–160.

Wu, H. and Pitts, M.J. 1999. Development and validation of a finite element model of an apple fruit cell. *Postharvest Biology and Technology* 16: 1–8.

Yeom, H.W., Streaker, C.B., Zhang, Q.H., and Min, D.B. 2000. Effects of pulsed electric fields on the quality of orange juice and comparison with heat pasteurization. *Journal of Agricultural and Food Chemistry* 48: 4597–4605.

Yeom, H.W., Evrendilek, G.A., Jin, Z.T., and Zhang, Q.H. 2007. Processing of yogurt-based products with pulsed electric fields: microbial, sensory and physical evaluations. *Journal of Food Processing and Preservation* 28: 161–178.

Yu, L.J., Ngadi, M.O., and Raghavan, G.S.V. 2008. Effect of temperature and pulsed electric field treatment on rennet coagulation properties of milk. *Journal of Food Engineering* Submitted, in review.

Zárate-Rodríguez, E., Ortega-Rivas, E., and Barbosa-Cánovas, G.V. 2000. Quality changes in apple juice as related to nonthermal processing. *Journal of Food Quality* 23: 337–349.

Zhang, Q.H., Monsalve-Gonzalez, A., Barbosa-Cánovas, G.V., and Swanson, B.G. 1994a. Inactivation of *E. coli* and *S. cerevisiae* by pulsed electric fields under controlled temperature conditions. *Transactions of the ASAE* 37: 581–587.

Zhang, Q., Monsalve-Gonzalez, A., Qin, B., Barbosa-Cánovas, G.V., and Swamson, B.G. 1994b. Inactivation of *Saccharomyces cerevisiae* in apple juice by square-wave and exponential-decay pulsed electric fields. *Journal of Food Process Engineering* 17: 469–478.

Zhang, Q.H., Barbosa-Cánovas, G.V., and Swanson, B.G. 1995a. Engineering aspects of pulsed electric field pasteurization. *Journal of Food Engineering* 25: 261–281.

Zhang, Q., Qin, B., Barbosa-Cánovas, G.V., and Swanson, B.G. 1995b. Inactivation of *E. coli* for food pasteurization by high-strength pulsed electric fields. *Journal of Food Processing and Preservation* 19: 103–118.

Zhang, Q.H., Qiu, X., and Sharma, S.K. 1997. Recent development in pulsed processing. *New Technologies Yearbook*. Washington, DC: National Food Processors Association.

Zimmermann, U. 1986. Electrical breakdown, electropermeabilization and electrofusion. *Reviews of Physiology, Biochemistry and Pharmacology* 105: 176–257.

Zimmermann, U., Pilwat, G., Beckers, F., and Rieman, F. 1976. Effects of external electric fields on cell membranes. *Bioelectrochemistry and Bioenergetics* 3: 58–83.

17 Solid–Liquid Separations in Food Processing: Relevant Aspects in Safety and Quality

Enrique Ortega-Rivas, Jocelyn Sagarnaga-Lopez, and Hugo O. Suarez-Martinez

CONTENTS

17.1 INTRODUCTION

Some separation techniques, such as membrane separations, can be used to remove microorganisms and pasteurize, or even sterilize fluid foods, without an increase

TABLE 17.1

Some Solid–Liquid Separation Techniques Relevant in Food Processing

Types of Separation	Techniques	Examples
Insoluble solids from liquids	Sedimentation	Wastewater treatment, separation of sugar crystals from mother liquor, cleaning of incoming raw materials
	Centrifugation	Clarification of juices and beverages, dewatering of vegetable oils, desludging of animal oils
	Hydrocyclone separation	Corn and potato starch refining, cottonseed oil processing, extraction of soluble coffee, clarification of fruit juices
	Filtration	Sugar extraction, dewatering of starch, separation of gluten suspensions, refining of edible oils, clarification of juices and beverages
Soluble solids from liquids	Membrane separations	Desalting of brine, recovery of lactose from whey, sugar recovery in confectionary, clarification and sterilization of fruit juices, brewing products, etc.
	Crystallization	Refining of sugar from sugarcane and sugar beet, salt manufacture, winterization of oil
Liquids from solids	Pressing	Extraction of oil from seeds, removal of crude juice from fruits, expressing juices from sugarcane
	Drying	Heated-air dehydration of fruits and vegetables, drying of meat and meat products, fluidized- bed drying of grains

in temperature. By using physical removal of microorganisms, safety is virtually guaranteed, while sensory attributes are efficiently preserved. This chapter will review and discuss the effects, mainly quality effects, resulting from using some separation techniques, totally or partially, as food preservation methods.

Solid–liquid separation is an important industrial process used for recovery of solids from suspensions and/or purification of liquids. Most of the process industries in which particulate slurries are handled use some form of solid–liquid separation and yet the subject is not adequately covered in most higher education courses (Svarovsky 2000). In the food industry, solid–liquid separations are widely used in a number of tasks such as concentration and clarification of fruit juices, reduction of microorganisms in fermentation products, separation of coffee and tea slurries, desludging of fish oils, recovering of sugar crystals, treatment of wastewater, etc. A basic classification of solid–liquid separations in the food industry is given in Table 17.1. Many slurries and suspensions treated in the food industry behave as non-Newtonian liquids but the current theory describing the solid–liquid separation process normally applies to Newtonian suspensions only. Practically, there is little theory developed for solid–liquid separations dealing with non-Newtonian fluids and it is believed that more studies on the subject are needed.

17.2 THEORETICAL BACKGROUND

Important applications of solid–liquid separations in food processing generally involve handling of suspensions. Theoretical aspects of fluid mechanics dealing with characteristics of suspensions will, therefore, be reviewed. Since concentration can vary widely in solid–liquid separations, particle–fluid interactions are of the utmost importance. Another relevant aspect has to do with efficiency of separation, and the different ways of expressing it according to the particular solid–liquid separation technique used.

17.2.1 DYNAMICS OF PARTICLES SUBMERGED IN FLUIDS

If a particle moves relative to the fluid in which it is suspended, the force opposing the motion is known as the drag force. Knowledge of the magnitude of this force is essential if the particle motion is to be studied. Conventionally, the drag force F_D is expressed as

$$F_D = C_D A \frac{\rho u^2}{2} \tag{17.1}$$

where
 u is the fluid–particle relative velocity
 ρ is the fluid density
 A is the area of the particle projected in the direction of the motion
 C_D is the coefficient of proportionality known as the drag coefficient

Assuming that the drag force is due to the inertia of the fluid C_D will be constant and dimensional analysis shows that C_D is generally a function of the particle Reynolds number, i.e.,

$$Re_p = \frac{u x \rho}{\mu} \tag{17.2}$$

where
 x is the particle size
 μ is the medium viscosity

The form of the function depends on the flow regime. This relationship for rigid spherical particles is shown in Figure 17.1. At low Reynolds numbers under laminar flow conditions when viscous forces prevail, C_D can be determined theoretically from Navier–Stokes equations and the solution is known as Stokes' law and represented by

$$F_D = 3\pi\mu u x \tag{17.3}$$

This is an approximation that gives the best results for $Re_p \rightarrow 0$; the upper limit of its validity depends on the error that can be accepted. The usually quoted limit for the Stokes region of $Re_p = 0.2$ is based on an error of about 2% in the terminal settling velocity. Equations 17.1 through 17.3 combined give another form of Stokes' law as follows:

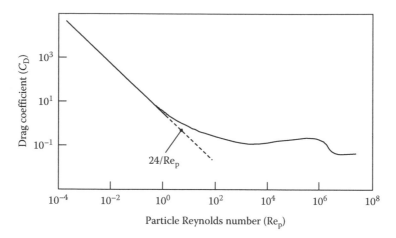

FIGURE 17.1 Common models for non-Newtonian fluids (time independent).

$$C_{\mathrm{D}} = \frac{24}{\mathrm{Re}_{\mathrm{p}}} \quad \left(\mathrm{Re}_{\mathrm{p}} < 0.2\right) \tag{17.4}$$

Equation 17.4 is represented as a straight line in Figure 17.1. For $\mathrm{Re}_{\mathrm{p}} > 1000$, the flow is fully turbulent with inertial forces prevailing, and C_{D} becomes constant and equal to 0.44 (the Newton region). The region in between $\mathrm{Re}_{\mathrm{p}} = 0.2$ and 1000 is known as the transition region, and C_{D} is either described in a graph or by one or more empirical relations.

For a particle of mass m under the influence of a field of acceleration a, the equation of motion is

$$m\left(\frac{\mathrm{d}u}{\mathrm{d}t}\right) = ma - ma\left(\frac{\rho}{\rho_s}\right) - F_{\mathrm{D}} \tag{17.5}$$

where
ρ_s is the density of the solids
t is the time

Many solid–liquid separations of industrial interest are concerned with fine particles, which are the most difficult to separate; so the Reynolds numbers are low, often less than 0.2, due to the low values of x and u. Therefore, it is reasonable to consider only the Stokes region. If this condition is applied, considering also that the times necessary for the particle velocity to reach very close to the terminal settling velocity are very short and being the field of acceleration gravity, Equation 17.5 can be solved to give

$$u_t = \frac{x^2\left(\rho_s - \rho\right)g}{18\mu} \tag{17.6}$$

in which g has replaced a and u_g is known as the terminal settling velocity under gravity.

As can be gathered from Equation 17.6, for a specific case of solid–liquid separation all the parameters, but one, are fixed. The particle size, densities, and viscosity cannot be easily modified, so the only variable that may be adjusted to increase the fluid–particle relative velocity is the gravity force. One alternative to the gravity field is a centrifugal field. If the suspension is subjected to a centrifugal force, the inclusion of the centrifugal term in Equation 17.6 leads simply to

$$u_r = \frac{x^2 (\rho_s - \rho) R\omega^2}{18\mu} \tag{17.7}$$

where
   u_r is the radial settling velocity
   R is the radius of the particle position
   ω is the angular velocity

As the concentration of the suspension increases, particles get closer together and interfere with each other. If the particles are not disturbed uniformly, the overall effect is a net increase in settling velocity since the return flow caused by volume displacement predominates in particle-sparse regions. This is the well-known effect of cluster formation, which is significant only in nearly monosized dispersions. With most practical widely dispersed suspensions, clusters do not survive long enough to affect the settling behavior and, as the return flow is more uniformly distributed, the settling rate steadily declines with increasing concentration. This phenomenon is referred to as hindered settling and can be theoretically approached in three different manners: as a Stokes' law correction by introduction of a multiplying factor; by adopting effective fluid properties for the suspension different from those of the pure fluid; and by determination of bed expansion with a modified version of the well-known Carman–Kozeny equation. All the approaches can be shown to yield essentially identical results.

Svarovsky (1984) reviewed some important correlations accounting for the hindered settling effect and demonstrated that their differences are minimal. According to this, the simple Richardson and Zaki equation is an obvious choice in practice. Such relation can be expressed as

$$\frac{u}{u_s} = (1 - C)^{4.65} \tag{17.8}$$

where
   u is the settling velocity at concentration C
   u_s is the settling velocity of a single particle

The above relationship applies only to free, particulate separation unaffected by coagulation or flocculation and where all the particles are of uniform density.

17.2.2 FLOW OF SUSPENSIONS

According to rheology, which embraces the study of flow behavior in a very general way, there are two main types of flow: viscous and elastic. Whereas the first occurs in fluids, the second is common in solids. An intermediate behavior of flow in-between was also found. Newtonian theory provides for the simplest cases of viscous behavior, in which the stresses are related to the velocity gradients existing at time of the observation. Flow of suspensions is normally studied in rheology and, in the most general sense, any fluid response not explainable by Newtonian theory may be termed non-Newtonian. In practice, however, non-Newtonian fluids as opposed to Newtonian are those in which the viscosity is not a constant value, but is a function of the imposed shear rate. The suspension resultant of dissolving solid in liquids is normally of non-Newtonian characteristics. A wide variety of nonlinear relationships between stress and shear rate have been used. Possibly, the most common one is known as the power-law, graphically represented in Figure 17.2. Fluids described by the power-law are generally known as power-law fluids. If the slope n is less than 1, the material flows more easily the faster it is sheared, and the apparent viscosity decreases with increasing shear rate. These fluids are known as shear-thinning or pseudoplastic fluids. If, on the other hand, the slope n is greater than 1, the apparent viscosity increases with the increasing shear rate. Such fluids are termed shear-thickening or dilatant fluids.

The rheological properties of suspension have been studied since the beginning of last century. The first research was done by Einstein (1906, 1911), in his classical study of the viscosity of dilute suspension of rigid spheres. His approach is purely hydrodynamic, and his model consists of an isolated sphere situated in a simple shear

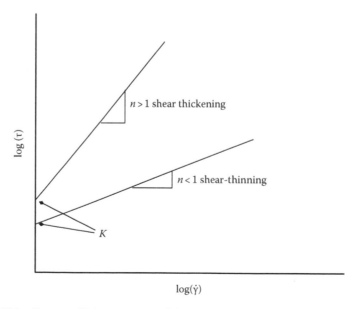

FIGURE 17.2   Drag coefficient versus particle Reynolds number for spherical particles.

flow filed in an infinite fluid. Under the assumptions that purely viscous behavior prevails and that no slip occurs at the tube wall, the power-law equation for laminar flow is in the form (Ortega-Rivas 2007):

$$\tau_w = \frac{D\Delta P}{4L} = K'\left(\frac{8v}{D}\right)^{n'} \tag{17.9}$$

where
  τ_w is the shearing stress at the wall of the tube
  D is the tube diameter
  ΔP is the pressure drop
  L is the length of the tube
  v is the mean velocity
  n' is the slope of the line when the data are plotted on logarithmic coordinates

According to concepts previously discussed, for $n' = 1$, the fluid is Newtonian, for $n' < 1$, pseudoplastic, and for $n' > 1$, dilatant. The term K' is the consistency index; as the name suggests, the larger its value the thicker or more viscous the fluid.

It may be shown (Ortega-Rivas and Svarovsky 1998) that the constant K' may be related to the analogous power-law constant K as follows

$$K' = K\left(\frac{3n'+1}{4n'}\right)^{n}\left(\frac{8v}{D}\right)^{n-n'} \tag{17.10a}$$

if the fluid obeys the power-law, $n = n'$, and

$$K' = K\left(\frac{3n+1}{4n}\right)^{n} \tag{17.10b}$$

Since Equation 17.9 rigorously portrays the laminar flow behavior of the fluid (provided n' and K' are evaluated at the correct shear stress), one may use it to define a Reynolds number applicable to all purely viscous fluids under laminar flow conditions. This dimensionless group can be derived simply by the substitution of $(D\Delta P/4L)$ from Equation 17.9 into the usual definition of the fanning friction factor, i.e.,

$$f = \frac{D\Delta P/4L}{\rho v^2/2} \tag{17.11}$$

Such a substitution leads to

$$f = \frac{16\gamma}{D^{n'}v^{2-n'}\rho} \tag{17.12}$$

where $\gamma = K'8n'-1$ and all the remaining components as already defined.

By letting $f = 16/\text{Re}$ as for Newtonian fluids in laminar flow, the above-mentioned generalized Reynolds number can be obtained as

$$\text{Re}* = \frac{D^{n'}v^{2-n'}\rho}{\gamma} \tag{17.13}$$

If the equation is desired in terms of K instead of K', Equation 17.10b may be substituted into Equation 17.13 and

$$\text{Re}* = \frac{D^n v^{2-n}\rho}{K8^{n-1}\left(\dfrac{3n+1}{4n}\right)^n} \tag{17.14}$$

For Newtonian fluids, $n' = 1$ and $K' = \mu$, so γ reduces to μ, and $\text{Re}*$ in Equation 17.13 transforms to the familiar $Dv\rho/\mu$ showing that this traditional dimensionless group is merely a special restricted form of the more general described here.

17.2.3 EFFICIENCY OF SEPARATION

The solid phase of suspensions to be treated in a solid–liquid separation equipment generally consists of an immense number of particles of diverse sizes and shapes. All this population of particles needs to be identified or characterized. The frequency of occurrence of particles of every size present, arranged, and presented in a statistical manner is known as the particle size distribution. The most common way of presenting such distributions as well as the types of distributions important to particle technology can be found in the specialized literature (Barbosa-Cánovas et al. 2005). In solid–liquid separation, it is important to know the particle size distribution in order to identify which part of it will be separated, and transform this quantity in a measure of efficiency. The common way of presenting particle size data for solid–liquid separation purposes is in the form of a plot. Normally the particle size axis is the X or horizontal and the particle amount axis is the Y or vertical. What is plotted on the particle size axis is a matter of which quantity used to represent size of individual particles suits better to a specific problem. The most logic parameter plotted for spherical particles is the diameter. For irregularly shaped particles, however, the choice of a quantity to represent individual particle size can be a real problem. A variety of quantities has been used for representing particle size, and is described in the literature (Ortega-Rivas 2005). The most common ones are equivalent sphere diameters, equivalent circle diameter, and statistical diameters. The equivalent sphere is the diameter of a sphere, which has the same property as the particle itself. Such a property could be the settling velocity. A diameter derived from the settling velocity is known as Stokes diameter and is a very useful quantity for solid–liquid separations, especially to those techniques in which the particle motion relative to the fluid is the governing mechanism. The amount of particle matter which belongs to specified size classes on the particle axis may be also represented in several ways.

Number of particles and mass of particles are the most commons ones, but surface area and volume are used as well. For solid–liquid separation techniques, the most convenient form of expressing the amount of particle matter is by mass, because the balances necessary to define performance are normally mass balances. In general, particle size distribution can be presented as frequencies, $f(x)$, or cumulative frequencies, $F(x)$, which are related by

$$f(x) = \frac{dF(x)}{dx} \tag{17.15}$$

The graphical representation of a particle size distribution is usually plotted in a cumulative form. In a typical cumulative plot, points are entered showing the amount of particulate material contributed by particles below or above a specified size. Hence, the curve presents a continuously rising or deceasing character. These oversize and undersize distributions, as illustrated in Figure 17.3, are simply related by

$$F(x)_{\text{oversize}} = 1 - F(x) \tag{17.16}$$

where $F(x)$ is the cumulative fraction undersize.

A cumulative plot will, therefore, include a broad range of particle sizes. It is often convenient, however, to refer to a single characteristic size for a system. Many characteristic sizes have been proposed, most of them involving a mathematical formula. An important one, which can be read off any cumulative plot of the particle size data, is the median particle size. It is defined as the particle size for which the particle amount equals 50% of the total. If the particle size is represented by a number, such a point is called the number median size. If mass is used as the measure of particle amount, this parameter is known as the mass median size. The distribution between

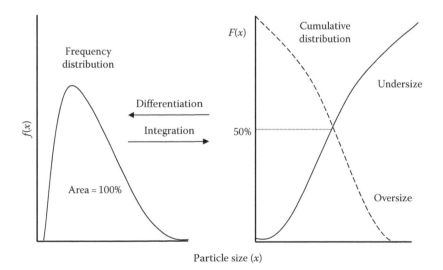

FIGURE 17.3 Relationship between frequency and cumulative distributions.

number and mass median is very important, since they differ generally by a considerable amount. Such a difference means that number and mass cumulative plots do not agree for the same system of particles. The weight of a particle, which varies as the cube of its diameter, accounts for this mentioned disagreement.

For practical purposes, it is reasonable to fit an analytical function to experimental particle size distribution data, and then handle this function mathematically in further treatment. It is, for example, very much easier to evaluate mean sizes from analytical functions from experimental data. Several different distribution functions can be found in the literature. All of them should be treated as empirical equations as they very rarely have any theoretical relation to the process in which the particles were produced.

To evaluate the efficiency of separation, it is necessary to take into account that solid–liquid separation is basically an imperfect process. Whereas the underflow is always wet slurry, the overflow can be considered as a turbid liquid. A single efficiency number can never be capable of fully describing the result of separation, except when it is ideal. The imperfection of solid–liquid separation has caused the need to express efficiency by different means. The first and most obvious definition of separation efficiency is simply the overall mass recovery as a fraction of the feed flow rate. According to Figure 17.4

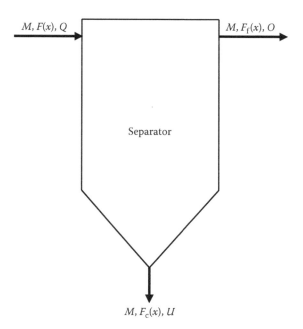

FIGURE 17.4 Schematic diagram of a separator. M, mass flow rate of solids in the feed; M_c, mass flow rate of separated solids; M_f, mass flow rate of nonseparated solids; $F(x)$, cumulative percentage oversize of feed solids; $F_c(x)$, cumulative percentage oversize of separated solids; $F_f(x)$, cumulative percentage oversize of nonseparated solids; Q, volumetric flow rate of feed suspension; U, volumetric flow rate of underflow suspension; and O, volumetric flow rate of overflow suspension.

$$E_t = \frac{M_c}{M} \tag{17.17}$$

where all the components are as defined in Figure 17.4.

If there is no accumulation of solids in the separator then

$$M = M_c + M_f \tag{17.18}$$

and there is a choice of three possible combinations of the material streams for the total efficiency testing. It can be shown (Trawinski 1977) that if all the operating conditions are equal, the most accurate estimation of the local efficiency comes from the two leaving streams.

The total efficiency defined by Equation 17.17 includes all particle sizes present in the feed solids. If only a narrow range of particle sizes is of interest, another efficiency of separation particular to that range can be defined. A mathematical expression of such partial efficiency is

$$E_p = \left(\frac{M_c}{M} \right)_{x_1/x_2} \tag{17.19}$$

where x_1 and x_2 represent the particle size limits of a definite range.

If the particle size range in Equation 17.19 becomes infinitesimal, the obtained efficiency corresponds to a single particle size x and it is known as the grade efficiency, defined by

$$G(x) = \left(\frac{M_c}{M} \right)_x \tag{17.20}$$

The grade efficiency has become a very useful definition, since most industrial powders consist of an infinite number of differently sized particles. Thus, a single particle size really corresponds to a range of particles having almost similar sizes. Therefore, the grade efficiency of most separation equipments is a continuous function of x. This function is seldom expressed analytically but graphically. An S-shaped curve is usually obtained for separators in which inertial or gravity body forces perform the separation.

As the value of the grade efficiency has the character of probability plotting the probability for any given size, fraction against particle size would give a curve as shown in Figure 17.5. This sort of curve is normally called a grade efficiency curve.

17.3   SOLID–LIQUID SEPARATIONS USED IN FOOD PROCESSES

Separation techniques in the food industry have been employed for long time and examples of some applications have been mentioned earlier. While solid–liquid separations in food processing have been typically focused on quality aspects, such as clarification, some technologies have had the capability of separating or retaining microorganisms, such as ultracentrifugation or ultrafiltration (UF), so they have also been useful in pasteurizing and sterilizing.

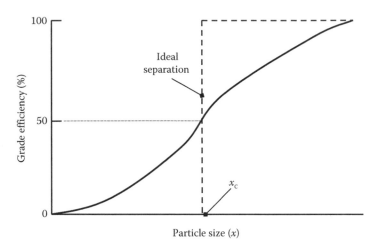

FIGURE 17.5   Grade efficiency curve for dynamic separators.

17.3.1 Centrifugal Techniques

As previously discussed, apart from gravity, centrifugal force can also be used to separate solids in suspension. Centrifugal sedimentation is a technique in which the suspension is subjected to a centrifugal field, to cause the solid particles to move radially apart from the center of rotation and lead to separation. A significant density difference between solid and liquid is a necessary prerequisite in sedimentation of any kind. Centrifugal sedimentation can be carried out in two types of equipment: rotating wall devices and fixed wall devices. Rotating wall devices are also known as centrifuges. The most popular fixed wall device is the hydrocyclone. The term "centrifugation" has been traditionally used to imply the operation of centrifugal sedimentation in rotating wall devices.

17.3.1.1 Centrifugation

As stated above, centrifugation may be defined as a unit operation involving the separation of materials by the application of centrifugal force. In solid–liquid separations, typical applications of centrifugation are centrifugal clarification, desludging or slurry centrifugation, and centrifugal filtration. The separation of immiscible liquids (liquid–liquid centrifugation) is also widely used in important applications of food industry.

Centrifugal clarification is the term used to describe the removal of small quantities of insoluble solids from a fluid by centrifugal means. If a dilute suspension containing solids with a greater density than the liquid is fed to a rotating cylindrical bowl, the solids will move toward the bowl wall. If an outlet is provided for the liquid near the center of rotation, then those particles of solid which reach the bowl wall will remain in the bowl. Those particles which do not reach the bowl wall will be carried out in the liquid. The fraction remaining in the bowl and the fraction passing out in the liquid will be controlled by the feed rate, i.e., the dwell time, in the bowl.

If a solid particle of diameter x moves radially in a liquid within a rotating bowl, at its terminal velocity under laminar flow conditions, the radial velocity of the particle will be represented by Equation 17.7. The time required for a particle to travel an elemental radial distance, dR, is

$$dt = \frac{dR}{u_r} = \frac{18\mu}{\omega^2(\rho_s - \rho)x^2} \cdot \frac{dR}{R} \tag{17.21}$$

Assuming that half of all those particles present in the feed with a particular diameter, x_c, are removed during their transit through the bowl, those particles with diameters greater than x_c will be mostly removed from the liquid, whereas those particles with diameters lesser than x_c will be likely to remain in the liquid. In this context, x_c, as here defined, is known as the "cut point" or "critical" diameter.

If clarification is taking place in a simple cylindrical centrifuge with cross section as shown in Figure 17.6, all particles of diameter x_c contained in the outer half of the cross-sectional area of the ring of liquid will reach the bowl wall and will be removed from the liquid. The maximum distance that a particle in this zone has to travel to reach the bowl wall is $R_2 - [(R_1^2 + R_2^2)/2]^{1/2}$, as indicated in Figure 17.6. The time required for a particle of diameter, x_c, to travel this distance is

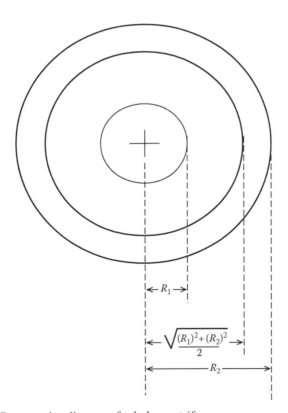

FIGURE 17.6 Cross-section diagram of tubular centrifuge.

$$t = \frac{18\mu}{\omega^2 (\rho_s - \rho) x_{50}^2} \int_{[(R_1^2 + R_2^2)/2]^{1/2}}^{R_2} \frac{dR}{R} \quad (17.22)$$

Integrating Equation 17.22 and substituting limits, the following relation is obtained:

$$t = \frac{18\mu \ln\left(\dfrac{R_2}{\left[(R_1^2 + R_2^2)/2\right]^{1/2}}\right)}{\omega^2 (\rho_s - \rho) x_{50}^2} \quad (17.23)$$

The minimum residence time for a particle in the bowl is V/Q, where V is the volume of liquid held in the bowl at any time and Q is the volumetric flow rate of liquid through the bowl. Thus, for a particle of diameter x_c to be separated out

$$\frac{V}{Q} = \frac{18\mu \ln\left(\dfrac{R_2}{\left[(R_1^2 + R_2^2)/2\right]^{1/2}}\right)}{\omega^2 (\rho_s - \rho) x_{50}^2} \quad (17.24)$$

Equation 17.24 may be written in the form

$$Q = 2\left[\frac{g(\rho_s - \rho) x_{50}^2}{18\mu}\right]\left[\frac{\omega^2 V}{2g \ln\left(\dfrac{R_2}{\left[(R_1^2 + R_2^2)/2\right]^{1/2}}\right)}\right] \quad (17.25)$$

The first term of Equation 17.25 includes the previously discussed expression for Stokes' law, i.e., Equation 17.6. Thus, another way of expressing Equation 17.25 is as follows:

$$Q = 2u_g \Sigma \quad (17.26)$$

where
u_g is the terminal settling velocity of a particle of diameter x_c in a gravitational field
Σ is the characteristic parameter of any given centrifuge; equivalent to the area of a gravity settling tank with similar settling characteristics to the centrifuge, i.e., one which will remove half of all particles of diameter x_c

Different values of Σ are given in the literature (McCabe et al. 2001). For a simple cylindrical bowl centrifuge

$$\Sigma \approx \frac{\pi \omega^2 b (3R_2^2 + R_1^2)}{2g} \tag{17.27}$$

where b is the height of the bowl. Also, for a disc-bowl centrifuge

$$\Sigma = \frac{2\pi \omega^2 (S - 1)(R_x^3 - R_y^3)}{3g \tan \Omega} \tag{17.28}$$

where
   S is the number of discs in stack
   R_x and R_y are the outer and inner radii of stack, respectively
   Ω is the conical half angle of discs (Trowbridge 1962)

17.3.1.2 Hydrocyclone Separation

The use of hydrocyclones is another possibility to separate suspended solids (SS) from a liquid taking advantage of the centrifugal force. As shown in Figure 17.7, a hydrocyclone consists of a cono-cylindrical body, which promotes vortex formation when a suspension is pumped through it. The vortex originates a centrifugal force, which causes coarse particles to migrate against the cyclone wall and be

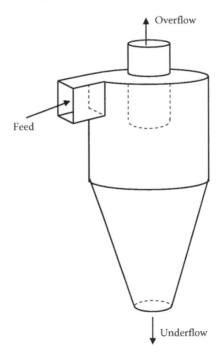

FIGURE 17.7 Diagram of a hydrocyclone.

discharged by the underflow orifice. Fine particles remain around the central axis of the cyclone and are carried out by the overflow stream. Hydrocyclones are easily manufactured and modified and have been well tested in thickening, clarification, classification, and other operations in many industries Svarovsky (1984).

Similar to centrifuges, hydrocyclones may be evaluated in terms of separation efficiency by means of the *cut size* or *cut point* (x_c or x_{50}). The cut size, which is the only single number that in some way represents the separating capability of a hydrocyclone, is the particle size at which the grade efficiency $G(x)$ curve shows a value of 50%. A grade efficiency curve for a hydrocyclone is derived from screen analysis data on the feed, overflow, and underflow streams, and is a continuous representation of the overall mass recovery as a fraction of the mass flow rate. Since most suspended powders and fine particulate systems can be represented by a continuous size distribution, the grade efficiency curve is really derived from a stepwise calculation, drawing a line through the midpoints of size intervals (Trawinski 1977). Consequently, for hydrocyclones and dynamic separators, the grade efficiency curve is an S-shaped cumulative plot in which the 50% point represents a limit value. Thus, particles with this limit size have a 50% probability of being separated. In other words, all particles above the cut size are generally discharged in the underflow, whereas those below the cut size are normally carried away in the overflow. Ortega-Rivas (1989) reviewed some of the numerous expressions utilized for determining cut size, which have been reported in the literature.

The grade efficiency curve, as described above, gives a plot which does not pass through the origin. This can be explained bearing in mind that a hydrocyclone is a flow divider, so the underflow always contains a certain quantity of very fine particles which simply follow the flow, and are split in the same ratio as the liquid. The apparent finite efficiency for fine particles is therefore equal to the underflow-to-throughput ratio R_f, and a "corrected" or "reduced" cut size x'_{50} will practically assess the performance of hydrocyclones when derived from the reduced grade efficiency $G'(x)$. All these definitions are given in Figure 17.8.

Since hydrocyclones do not have any rotating parts and the vortex action to produce centrifugal force is obtained by pumping the feed suspension tangentially into

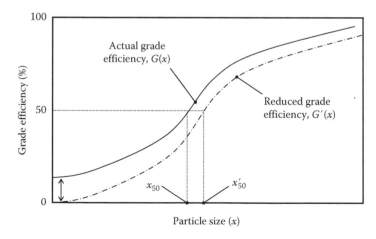

FIGURE 17.8   Grade efficiency and reduced grade efficiency curve for hydrocyclone.

the conocylindrical body, the literature is full of studies of the effects of the relative geometric proportions on pressure drop or capacity and separation efficiency. Using this information, a hydrocyclone geometry could be selected to obtain an optimum performance in terms of cut size. In this sense, possibly the best opportunity to predict hydrocyclone performance is the use of a dimensionless scale-up model well described elsewhere (Ortega-Rivas and Svarovsky 1993). Three dimensionless groups can be used to describe hydrocyclone operation and performance: the Euler number, Eu, the Reynolds number, Re, and the Stokes number, Stk_{50}. For best application of the relationships among dimensionless groups, certain proportions must be unchanged. Such proportions are generally reported as a function of the diameter of the hydrocyclone. There are several different standard hydrocyclone designs in which proportions remain the same regardless of size. One of the most efficient designs for separation is called the Rietema cyclone (Rietema 1961), whose proportions are illustrated in Figure 17.9.

The Euler number, which is a pressure loss factor, is defined as the limit of the maximum characteristic velocity v obtained by a certain pressure drop ΔP across the cyclone. It can be expressed as

$$Eu = \frac{2\Delta P}{\rho v^2} \tag{17.29}$$

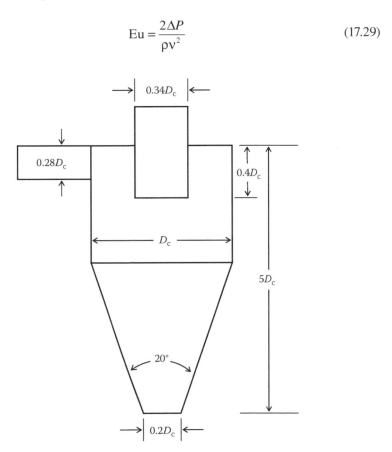

FIGURE 17.9 Dimensions of Rietema's standard hydrocyclone.

where
ρ is the liquid density
v is the superficial velocity in the cyclone body

The Reynolds number defines flow features of the system and, in the case of hydrocyclones, the characteristic dimension may be taken as the cyclone body diameter D_c:

$$\text{Re} = \frac{D_c v \rho}{\mu} \tag{17.30}$$

where
ρ is the liquid density
μ is the average viscosity

The Stokes number may be derived from basic fluid mechanics theory and is defined as follows:

$$\text{Stk}_{50} = \frac{x_{50}^2 (\rho_s - \rho) v}{18 \mu D_c} \tag{17.31}$$

where
x_{50} is the cut size
ρ_s is the solid density
D_c is the hydrocyclone diameter

All the above equations use the superficial velocity in the cyclone body as the characteristic one, i.e.,

$$v = \frac{4Q}{\pi D_c^2} \tag{17.32}$$

where Q is the feed volumetric flow rate.

The dimensional analysis gives two basic relationships between the above-mentioned dimensionless groups:

$$\text{Stk}_{50} \text{Eu} = \text{constant} \tag{17.33}$$

$$\text{Eu} = k_p (\text{Re})^{n_p} \tag{17.34}$$

where k_p and n_p are constants derived for a family of geometrically similar hydrocyclones. These relationships have been tested over a range of conditions by different authors (Nezhati and Thew 1988, Antunes and Medronho 1992, Chen et al. 2000). Their plots are typical, as shown in Figure 17.10 for Equation 17.34.

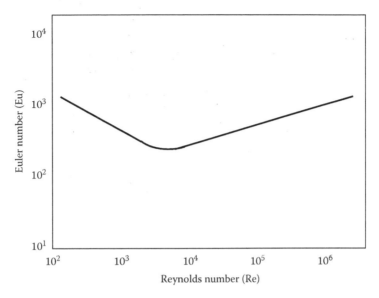

FIGURE 17.10   Typical plot of Eu versus Re for hydrocyclones.

At higher concentrations, the feed concentration as a fraction of volume C has to be included as an additional dimensionless group. Svarovsky and Marasinghe (1980) reported the following expression for the effect of high solid concentrations in the feed:

$$\text{Stk}_{50}(r) = k_1(1 - R_f)e^{k_2 c} \tag{17.35}$$

where $\text{Stk}_{50}(r)$ includes the previously described reduced cut size, which takes into account the "dead flux" effect of very fine particles simply following the flow and split in the same ratio as the liquid. As stated earlier, R_f is the underflow-to-through-put ratio. The correlation has proved to hold well for concentrations above 8% by volume, and the values of the constants k_1 and k_2 were found to be 9.05×10^{-5} and 6.461, respectively, for limestone and an AKW® (Amber Kaolinwerke GmbH, Hirschau, Germany) hydrocyclone of 125 mm in diameter.

An exhaustive study for concentrations up to 10% by volume was carried out by Medronho and Svarovsky (1984) in order to verify the applicability of Equations 17.33 through 17.35. They employed three geometrically similar hydrocyclones of Rietema's optimum geometry and obtained the following relations:

$$\text{Stk}_{50}(r)\text{Eu} = 0.047\left[\ln(1/R_f)\right]^{0.74} e^{(8.96C)} \tag{17.36}$$

$$\text{Eu} = 71(\text{Re})^{-0.116}(D_i/D_c)^{-1.3} e^{(2.12C)} \tag{17.37}$$

$$R_f = 1218(D_u/D_c)^{4.75}(\text{Eu})^{-0.30} \tag{17.38}$$

where D_i, D_c, and D_u are the inlet, body, and underflow diameters of the hydrocyclone, respectively.

For concentrations higher than 10% by volume, many practical slurries show non-Newtonian behavior and it can be shown (Ortega-Rivas and Svarovsky 1993) that Reynolds and Stokes numbers can be re-expressed to consider such behavior. The correlations derived under this mentioned consideration are the following:

$$\text{Stk}^*_{50}(r)\text{Eu} = 0.006\left[\ln(1/R_f)\right]^{2.37} e^{(6.84C)} \tag{17.39}$$

$$\text{Eu} = 1686(\text{Re}^*)^{-0.035} e^{(-3.39C)} \tag{17.40}$$

$$R_f = 32.8(D_u/D_c)^{1.53}(\text{Re}^*)^{-0.34} e^{(3.70C)} \tag{17.41}$$

where $\text{Stk}^*_{50}(r)$ and Re^* are the "generalized" Stokes and Reynolds numbers, respectively, meaning that they include the parameters of characterization of non-Newtonian suspensions, i.e., the fluid consistency index K' and the flow behavior index n, instead of the medium viscosity (Ortega-Rivas and Svarovsky 1993). The term generalized is used to imply that, for Newtonian suspensions, Stokes and Reynolds numbers above would reduce to the common forms normally found in the literature.

17.3.2 FILTRATION

Filtration may be defined as the unit operation in which the insoluble solid component of a solid–liquid suspension is separated from the liquid component by passing the suspension through a porous barrier which retains the solid particles on its upstream surface, or within its structure, or both. The solid–liquid suspension is known as the feed slurry or prefilt, the liquid component that passes through the membrane is called the filtrate, and the barrier itself is referred to as the filter medium. The separated solids are known as the filter cake, once they form a detectable layer covering the upstream surface of the medium. The flow of filtrate may be caused by several means. Pressure and vacuum are two conventional ways of driving the suspension across the medium. Gravity and centrifugal forces may also be used for suspension medium crossing. Gravity filtration has limited use in food processes but is applied to water and sewage treatment. In general terms, filtration theory applies to cases where cake buildup occurs. Some of the more fundamental treatments of filtration theory are reviewed by Dickey (1961). In the initial stages of filtration, the first particles of solid to encounter the filter medium become enmeshed in it, reducing its open surface area and increasing the resistance it offers to the flow of filtrate. As filtration proceeds, a layer of solids builds up on the upstream face of the medium and this layer, or cake, increases in thickness with time. Once formed, this cake in fact becomes the primary filtering medium. Filtrate passing through the filter encounters three types of resistance: a first resistance offered by channels of the filter itself, a second one because of the filter medium presence, and a third one due to the filter cake. The total pressure drop across the filter is equivalent to the sum of the pressure drops resulting from these three resistances. Usually the pressure drop due to the channels of the filter is neglected in calculations. If $-\Delta P$ is the total pressure drop across the filter and $-\Delta P_c$ and $-\Delta P_m$ are the pressure drops across the cake and medium, respectively, then

$$-\Delta P = -\Delta P_c - \Delta P_m \qquad (17.42)$$

The pressure drop across the filter cake may be related to the filtrate flow by the expression (McCabe et al. 2001):

$$-\Delta P_c = \frac{\alpha \mu w V}{A^2}\left[\frac{dV}{dt}\right] \qquad (17.43)$$

where
  α is the specific resistance of the cake
  μ is the viscosity of filtrate
  w is the mass of solids deposited on the medium per unit volume of filtrate
  V is the volume of filtrate
  A is the filter area normal to the direction of filtrate flow

"α" physically represents the pressure drop necessary to give unit superficial velocity of filtrate of unit viscosity through a cake containing unit mass of solid per unit filter area. It is related to the properties of the cake by

$$\alpha = \frac{k(1-\varepsilon)S_0^2}{\varepsilon^3 \rho_s} \qquad (17.44)$$

where
  k is the constant
  ε is the porosity of the cake
  S_0 is the specific surface area of the solid particles in the cake
  ρ_s is the solid density

If a cake is composed of rigid nondeformable solid particles, α is independent of $-\Delta P_c$ and does not vary throughout the depth of the cake, and is known as incompressible cake. However, if the cake contains nonrigid, deformable solid particles or agglomerates of particles, the resistance to flow will depend on the pressure drop and will vary throughout the depth of the cake. In this case, the cake is called compressible and an average value of the specific resistance for the entire cake must be used in Equation 17.43. This average specific resistance must be measured experimentally for any particular slurry.

By analogy with Equation 17.43, the filter medium resistance may be defined by the following relation:

$$-\Delta P_m = \frac{R_m \mu}{A}\left[\frac{dV}{dt}\right] \qquad (17.45)$$

where R_m is the pressure drop across the medium.

It is reasonable to assume that R_m is constant during any filtration cycle and that it includes the resistance to filtrate flow offered by the filter channels. For this case, Equations 17.42, 17.43, and 17.45 can be combined to give

$$\frac{dV}{dt} = \frac{A(-\Delta P)}{\mu\left[\dfrac{\alpha w V}{A} + R_m\right]} \tag{17.46}$$

Equation 17.46 is a general expression for the filtrate flow rate.

When the pressure drop is maintained constant, Equation 17.46 may be integrated thus

$$\int_0^t dt = \frac{\mu}{A(-\Delta P)}\left[\frac{\alpha w}{A}\int_0^V V\,dV + R_m\int_0^V dV\right] \tag{17.47}$$

or, substituting limits and transposing for time t:

$$t = \frac{\mu}{(-\Delta P)}\left[\frac{\alpha w}{2}\left(\frac{V}{A}\right)^2 + R_m\left(\frac{V}{A}\right)\right] \tag{17.48}$$

Equation 17.48 is a general expression for the filtration time during constant pressure filtration. In order to use it, values of α and R_m must be determined experimentally. This can be done by rewriting Equation 17.48 in the following form:

$$\frac{dt}{dV} = KV + B \tag{17.49}$$

where

$$K = \left[\frac{\alpha w \mu}{A^2(-\Delta P)}\right] \tag{17.50}$$

and

$$B = \left[\frac{R_m \mu}{A(-\Delta P)}\right] \tag{17.51}$$

As can be gathered, Equation 17.49 represents a straight line if dt/dV is plotted against V. Therefore, if a constant pressure filtration is carried out and values of V for different values of t are recorded, a graph of dt/dV versus V can be constructed as shown in Figure 17.11. The slope of this line is K and the intercept on the ordinate when $V = 0$ is B. Thus, by using such a graph, values of α and R_m can be directly determined from Equations 17.50 and 17.51.

For incompressible cakes, Equation 17.48 can be used directly at different pressures. However, for compressible cakes, the relationship between α and $-\Delta P$ needs to be determined experimentally by performing filtration runs at different constant pressures. Empirical equations may be fitted to the results obtained. Two such equations have been suggested (McCabe et al. 2001):

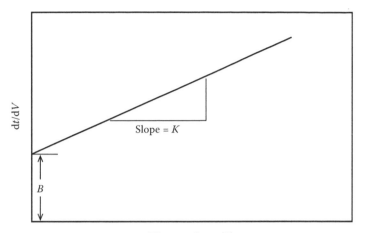

FIGURE 17.11 Plot of results from a constant pressure filtration run.

$$\alpha = \alpha_0 (-\Delta P)^s \tag{17.52}$$

$$\alpha = \alpha'_0 [1 - \beta(-\Delta P)^{s'}] \tag{17.53}$$

where α_0, α'_0, s, s', and β are empirical constants. Having determined values for α_0 and s, Equations 17.52 and 17.48 can be used for constant pressure filtration calculations at different pressures.

If filtration is carried out at constant rate then

$$\frac{dV}{dt} = \text{constant} \quad e = \frac{V}{t} \tag{17.54}$$

Equation 17.54 may be rewritten as follows:

$$-\Delta P = \left[\frac{\mu \alpha w V}{A^2 t}\right] V + \left[\frac{\mu V R_m}{A t}\right] \tag{17.55}$$

or

$$-\Delta P = K'V + B' \tag{17.56}$$

Once again it can be seen that Equation 17.56 represents a straight line if $-\Delta P$ is plotted against V. The slope of the line is K' and the intercept to the $-\Delta P$ axis when $V = 0$ is B'. Thus, for incompressible cakes α and R_m can again be determined by experimental means. Equation 17.55 can then be used for cycle calculations (Foust et al. 1960).

For compressible cakes, the relationship between α and $-\Delta P_c$ must, again, be experimentally determined. If a relationship of the form shown in Equation 17.52 is assumed to apply, then Equation 17.43 may be modified to

$$(-\Delta P_c)^{1-s} = [-(\Delta P - \Delta P_m)]^{1-s} = \frac{\mu \alpha_0 w V}{A^2} \frac{V}{t} \qquad (17.57)$$

which in turn may be written as

$$[-(\Delta P - \Delta P_m)]^{1-s} = K''t \qquad (17.58)$$

where

$$K'' = \frac{\mu \alpha_0 w}{A^2} \left[\frac{V}{t}\right]^2 \qquad (17.59)$$

If it is assumed that $-\Delta P_m$ is constant throughout the constant-rate filtration then by plotting t versus ΔP and passing a smooth curve through the points and extrapolating the curve to the $-\Delta P$ axis, an approximate value for $-\Delta P_m$ can be obtained. If $-(\Delta P - \Delta P_m)$ is then plotted against t on log–log paper and a straight line is obtained, the slope of this line is $(1 - s)$. Thus, s can be calculated and K'' and α_0 can be derived from Equations 17.58 and 17.59, respectively. If the first log–log plot of $-(\Delta P - \Delta P_m)$ is not a straight line, further approximations for $-\Delta P_m$ need to be made (McCabe et al. 2001).

17.3.3 Membrane Separations

Membrane separations are techniques used industrially for removal of solutes and emulsified substances from solutions by application of pressure onto a very thin layer of a substance with microscopic pores, known as a membrane. Membrane separation processes include reverse osmosis (RO), UF, microfiltration (MF), dialysis, electrodialysis, gas separation, and pervaporation. Of the membrane separation techniques, RO, UF, MF, and electrodialysis have been widely used commercially (Girard and Fukumoto 2000). UF has found many applications in food processing and has been successfully employed in a number of liquid "cold sterilization" and clarification applications. In UF, membrane pore size ratings are generally in the range of 0.001–0.020 µm (Chen et al. 2004).

UF is often compared with RO although the mechanism of separation is quite different. This difference is illustrated in Figure 17.12. As can be seen, in RO a rejection is based on electrostatic repulsion due to formation of a pure layer of water over the membrane, and the virtual charges on this layer reject charges of ionic-free species of salt solutions. Simultaneously, by a complex mechanism of sorption, diffusion, and desorption, pure water passes through the membrane performing the separation process. UF membranes, on the other hand, are porous in nature, with a rigid and highly voided structure, and function in a manner analogous to a screen or sieve (Figure 17.12b). The pore network is randomly distributed,

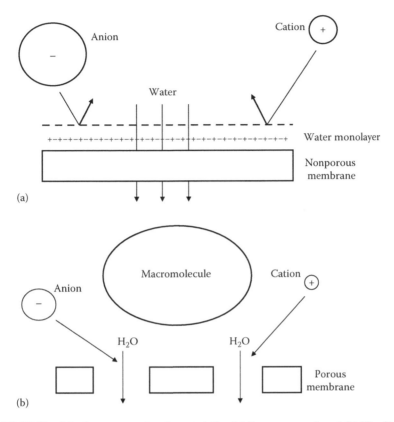

FIGURE 17.12 Membrane separation characteristics. (a) Reverse osmosis and (b) Ultrafiltration.

with pores passing directly through the membrane. The separating ability is based primarily on particle size, wherein particles and molecules larger than the largest pore are completely retained, whereas species smaller than the smallest pore are totally permeated. Therefore UF is an extension of conventional filtration, with its separating ability extending to the molecular level. In general terms, the mechanism of separation in RO is known as salt rejection, whereas that in UF is called organic rejection.

The molecular weight rating of a UF membrane is expressed in terms of a rejection coefficient against a species of specific molecular weight (Chen et al. 2004). Ideally, the membranes will have a sharply defined molecular weight cutoff (MWCO), as illustrated in Figure 17.13. Such an ideal membrane will retain all species greater than the MWCO but will allow all smaller ones to pass. Membranes are available in a number of increments in MWCO, which ranges from 1,000 up to 100,000 Da. The importance to food applications lies in the fact that these membrane characteristics give specificity in terms of permeating soluble sugar and flavor components and retaining SS, microorganisms, spores, and other particulates responsible for spoilage. In such a way, a number of food liquids are "cold sterilized" and largely retain their characteristic flavor and aroma.

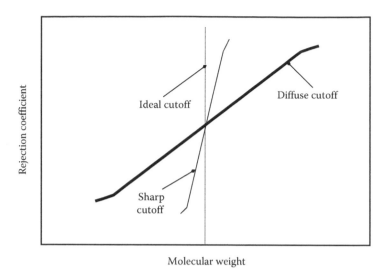

FIGURE 17.13   Representation of MWCO.

A general solute balance at the membrane surface in an unstirred cell is given by (Wijmans et al. 1984)

$$\frac{\partial C}{\partial t} - J\left(\frac{\partial C}{\partial y}\right) = D\left(\frac{\partial^2 C}{\partial y^2}\right) \qquad (17.60)$$

where
    C is the concentration
    t is the time
    J is the volumetric permeate flux
    D is the diffusion coefficient of solute in the solvent

In a selectively water permeable membrane, as permeation proceeds, the solid buildup on the membrane upstream surface exceeds the bulk concentration of the solution. This results in an increase in the osmotic pressure of the solution thereby, decreasing the driving force for permeation, or in formation of a gel layer on the surface of the membrane. A decrease in driving force for an increase in resistance for permeation results in a decrease in the volumetric permeation flux. Two mathematical models were developed to predict the volumetric flux of solute through the membrane: the osmotic model and the gel layer model. The former assumes that the decline in volumetric permeate flux is due to a reduction in driving force, whereas in the latter, the decline in volumetric permeate flux is assumed due to the formation of a gel layer on the membrane upstream surface. The volumetric flux J through the membrane is given by

$$J = \frac{\Delta P - \Delta \pi}{R_{\mathrm{m}}} \qquad (17.61)$$

The pure water permeability $J*$ is given by

$$J* = \frac{\Delta P}{R_m} \tag{17.62}$$

From Equations 17.61 and 17.62, the following relation is obtained:

$$J = \frac{J*(1-\Delta\pi)}{\Delta P} \tag{17.63}$$

where
  ΔP is the hydraulic pressure drop
  $\Delta\pi$ is the difference between the osmotic pressure on the feed and permeate sides
    of the membrane
  R_m is the membrane resistance

The permeation flux at any time is given by (Trettin and Doshi 1981)

$$J = \frac{\Delta P - \Delta\pi_g}{R_m - R_g} \tag{17.64}$$

where
  $\Delta\pi_g$ is the osmotic pressure difference between the gel layer and the permeate
  R_g is the gel layer resistance

The concentrations of solute on the feed side of the membrane C_m and the limiting flux J_α are given by (Wijmans et al. 1984)

$$C_m = C_b \exp\left(\frac{j}{k}\right) \tag{17.65}$$

$$J_\alpha = k \ln\left(\frac{C_g}{C_b}\right) \tag{17.66}$$

where
  C_b is the bulk concentration on the feed side of the membrane
  C_g is the gel concentration of the solution
  k is the mass transfer coefficient

Although UF may be considered to be an extension of conventional filtration, the operating mode of the equipment is different. As shown in Figure 17.14, in UF, the fluid moves continuously across the membrane surface. The arrangement is known as cross-flow filtration and is used as means of sweeping the membrane surface to control the buildup of foulants and particulate matter. UF may be integrated in processing lines in more than one way. The simplest technique is the batch configuration illustrated in Figure 17.15a. In this mode, an initial volume of liquid is circulated through the UF system, and permeate is continuously removed until a final volume is achieved. On the

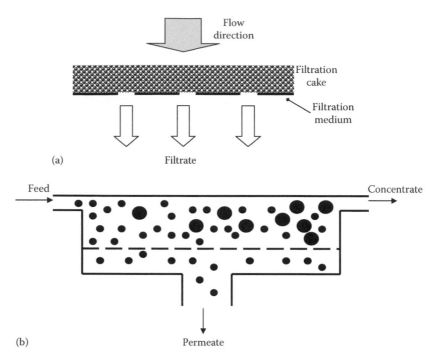

FIGURE 17.14 Mode of operation of UF membrane: (a) dead-end filtration and (b) cross-flow filtration.

other hand, recirculating topped-off batch is a variation shown in Figure 17.15b. This arrangement is more practical because it does not require the use of a large feed tank and allows the specification of a smaller prefilter prior to the system.

In the food industry and related fields, membranes are mainly used for concentration, purification, fractionation, and recovery of solvents and some by-products. In the past, most commercialization was limited to dairy and fruit processing, but, more recently, membrane technology has been extended to other industries. The increase in use of membrane separation techniques in the food and related industries is due to their many advantages such as energy conservation, elimination of denaturing of food valuable components, by-products recovery, possibility of cold pasteurizing, etc. Membrane separations are applied in the meat, dairy, and oil industries, and in processing of fruit juices, vegetable juices and pulps, sugar solutions, and beverages.

17.4 PROCESSING EFFECTS OF SPECIFIC SEPARATION TECHNIQUES

17.4.1 EFFECTS ON PROCESSING BY CENTRIFUGATION

Tubular-bowl machines and disc-bowl separators have found diverse applications in the food processing industry. Some important examples include dewatering of vegetable and fish oils, clarification of sweet juices and fermenting products, separation

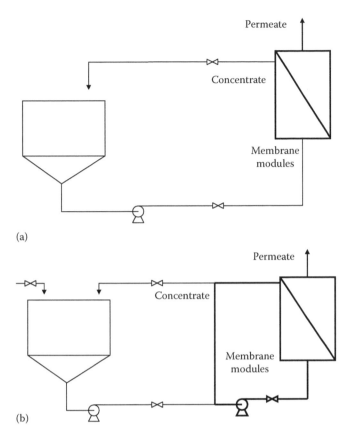

FIGURE 17.15 UF systems: (a) standard batch configuration and (b) topped-off batch system with recirculation loop.

of cream from milk, recovery of yeasts, dewatering of different starches, etc. In treatment of fruit juices, beer, and wines particularly, the quality of the clarified products may be directly related to the centrifugation efficiency in terms of removing proteic fractions as to avoid undesirable reactions such as browning. Sapers (1991) investigated filtration and centrifugation as a means of preventing enzymatic browning in minimally processed fruit juices. Centrifugation prevented browning in pear juice and in Granny Smith, Golden Delicious, and Red Delicious but not McIntosh apple juice, provided that foam was excluded and the relative centrifugal force was sufficient.

Centrifugation has also been used solely as the clarification step in fruit juice processing. A type of apple juice, for example, slightly clearer than the unclarified juice but considerably more opaque than the filtered juice is obtained by treating the pressed screened juice in a centrifuge only (Ortega-Rivas et al. 1997).

In some other applications, for example, in starch refining, the use of "sour liquid" (liquid fermentate from traditional fermentation processes) aids sedimentation and competes with centrifugation in the quality of the obtained product. For instance, mung bean starch produced by sour liquid processing had better quality than that from centrifugation, which is the technique commonly used in China to produce starch noodle (Liu and Shen 2007). The structures of mung bean starch from the two different processing methods were studied and it was shown that light transmissivity of mung bean starch from sour liquid processing was higher than that from centrifugation. The size of mung bean starch from sour liquid processing was as large as that from centrifugation. In another study, Li et al. (2008) reported on properties and qualities of vermicelli made from the sour liquid processing. Vermicelli made from sour liquid starch had more significant mesh structure than vermicelli made from centrifugation. Total cooking loss value of vermicelli made from sour liquid starch was significantly lower than that from centrifugation starch.

17.4.2 QUALITY EFFECTS BY HYDROCYCLONE TREATMENT

In terms of biological fluids and suspensions, the standard application for small hydrocyclones is for starch refining and they have been extensively used in corn and potato starch refining giving good results (Bednarski 1992). Hydrocyclones have been used to separate gossypol from cottonseed protein in cottonseed oil processing. They have also been used as separators in multistage mixer–separator extraction systems for soluble coffee. Hydrocyclones have been employed in thickening of wastewater sludge with promising results (Bednarski 1996, Ortega-Rivas and Medina-Caballero 1996). Sephadex, yeasts, and blood cells have been separated using hydrocyclones (Rickwood et al. 1992). Yeast recovery in different fermentation processes has also been well performed by hydrocyclones (Cilliers and Harrison 1996, Yuan et al. 1996, Martins da Mata and Medronho 2000). Another application reported is that of kieselguhr recycling for filters in the brewing industry (Rickwood et al. 1996).

The dimensionless scale-up model previously described has been tested in applications of apple juice clarification (Ortega-Rivas et al. 1997) and wastewater treatment (Ortega-Rivas and Medina-Caballero 1996). For the juice clarification, low values of Reynolds numbers were obtained due to the high viscosity of the feed, so the flow was not as turbulent as it was commonly found in small diameter hydrocyclone operation. The Euler number values were practically constant, which may also be attributed to the high consistency of the feed suspensions. For these reasons, relationships of Eu against Re were not properly derived. It is believed that the behavior of the feed suspension was not adequately approached in terms of using an appropriate pump. A higher capacity slurry pump for viscous fluids could have provided better pressure drops with the consequent effect in having more defined cut sizes according to the experienced pressures and obtaining more variability in Euler and Reynolds numbers. Despite all these mentioned difficulties, from experimental data a relationship similar to Equation 17.35 was obtained (Ortega-Rivas et al. 1997) as follows:

$$\mathrm{Stk}_{50}(r) = 7 \times 10^{-5}(1 - R_{\mathrm{f}})e^{(-6.47C)} \tag{17.67}$$

As can be seen, the constant values in Equations 17.35 and 17.67 show remarkable similitude. Apart from the opposite trend in the concentration correction factor, such values are, practically, similar. Taking into account that the conditions of derivation of both equations were totally different, i.e., inert against biological material and different hydrocyclone diameter and operating variables, the good comparison gives reasons to believe that the dimensionless approach may be extended to separation of biological materials. With regard to the wastewater case (Ortega-Rivas and Medina-Caballero 1996), practically all the feed suspensions were pseudoplastic. Therefore, a series of dimensionless relationships like those described by Ortega-Rivas and Svarovsky (1993) were attempted. The developed correlations are presented below:

$$Eu = 3.388 \times 10^{9} (Re*)^{-1.542} e^{(-1.325C)} \qquad (17.68)$$

$$R_{f} = 59.326 (D_{u}/D_{c})^{4.182} (Re*)^{-0.269} \qquad (17.69)$$

Apart from the fact that in Equation 17.68 there was not a significant effect of the concentration, both expressions showed adequate similarity to those suggested elsewhere (Svarovsky and Marasinghe 1980, Ortega-Rivas and Svarovsky 1993). This may be considered an important verification of the applicability of the dimensionless scale-up model of hydrocyclone operation, due to the use of real food systems used in the study.

17.4.3 Processing Effects due to Filtration

Prior to the wide commercialization of membrane separations, conventional filtration was the final clarification step in many food applications, particularly in fruit juices and fermented products, such as beer and wines. Similar to the case of centrifugation, filtration effects have focused on quality aspects, but not on microbial safety. An extensive study of control of enzymatic browning in fruit juices by filtration and centrifugation was reported (Sapers 1991). The capacity of raw apple, grape, and pear juices to undergo browning was associated with particular fractions capable to be removed by conventional filtration, using bentonite or diatomaceous earth as filter aids. Bayindirli et al. (1989) modeled apple juice filtrations and found that De La Garza and Boulton's exponential filtration model (De La Garza and Boulton 1984) successfully described the apple juice filtration, while Sperry's most widely accepted equation (Sperry 1917) did not. De La Garza and Boulton's model includes the cake resistance as an exponential function of the filtrate volume. The model parameters were easily obtained with linear regression. Values of these parameters were used to interpret the effects of various operation variables on filtration rates. Sahin and Bayindirli (1991) presented a study using De La Garza and Boulton's exponential model of wine filtrations to simulate data satisfactorily in the cake filtration of sour cherry juice. There was an inverse relation between the filtration rate and the resistance. Increase in the amount of precoating decreased the filtration rate. However, increase in filter aid dose increased the filtration rate. Further increase in filter aid dose and pressure over some limiting value gave no further increases in infiltration rate.

Maximea and Lameloise (2003) presented results of investigations on filterability of viscous complex food salting solutions. The main focus of such investigations was to find out the suitable coupling for the regeneration of complex concentrated ternary brine (salt/corn syrup/water) after fish fillet salting using filtration. Experiments were designed for analytical purposes with three selected waste brines originating from different clean solutions and salting process conditions, and processing. Processing plant considerations (e.g., economics, environment, technical availabilities) were also taken into account. In this regard, several chemical and physical pretreatments were first conducted to increase the size of SS and to reduce the proteinaceous matters in the suspension. Treated suspensions and their separated phases after settling were subjected to various filtration experiments. Laboratory and industrial scale cake filtration experiments using a rotary vacuum precoat filter (RVPF) were also conducted to find the optimal filtration conditions with the help of a unifying filterability parameter. The best coupling appeared to be pH pretreatment, followed by a settling time adjusted to the salting process conditions. Using body feed for the SS-rich phase, a differential RVPF filtration of the settled phases allowed the regeneration of solution with minimal filter aid consumption. The investigations also included filtering processes using membranes.

17.4.4 Safety and Quality on Separations by Membranes

Membrane separations have found many applications in microbial control and preservation of sensory attributes of various food systems. Dairy products, fruit juices, and different fermentation liquors may be the typical examples of the commercial success of the technology of membrane separations applied to foods.

17.4.4.1 Effects on Milk and Dairy Products

Concentration of whole milk by membranes has been a common application of this technology. In a particular study, whole milk was concentrated by 50 kDa pore membranes and the effects on retention of total solids, proteins, and fat content were investigated. The membrane used retained, satisfactorily, the main nutrients in the concentrate (Garoutte and Amundson 1982). Whey processing represented one of the first fields of application of membrane processes in the dairy industry (Zadow 1987, Maubois 1991). MF has been used to reduce the total number of lactic acid bacteria (Pafylias et al. 1996) and other microorganisms in the permeate compared to the microbiological condition of the whey (Gesan et al. 1993, Al-Akoum et al. 2002). The process also resulted in defatting of the whey and was considered as a gentle sterilization method (Eckner and Zottola 1991, Pearce et al. 1991).

Permeate flux and chemical oxygen demand (COD) reduction were investigated in dairy processing waters using different nanofiltration (NF) and RO membranes (Al-Akoum et al. 2004). Dairy process waters were simulated by ultrahigh temperature processed skim milk diluted 1:3 to obtain an initial COD of 36,000 mg O_2/L. In NF the highest permeate flux was obtained with a particular membrane type, which yielded also the highest permeate COD. In concentration tests, the permeate flux decreased with increasing volume reduction ratio (VRR) to reach 25 L/h·m^2, while the permeate COD soared to 1050 mg O_2/L. The performance of UF and NF membranes was investigated for utilization in concentration of whey protein and lactose (Atra et al. 2005). Such performances could be characterized in terms of permeate flux, membrane retention, and yield, where their parameters are determined by pressure, recycle flow rate, and temperature. The influence of these parameters on milk- and whey protein and lactose concentration was measured. The experiments were carried out using laboratory scale UF and NF units. The permeate flux, protein, and lactose content in permeate and concentrate fractions were measured during the experimental runs. Comparing the separation behavior of the membranes, it was found that the investigated membranes are suitable for concentration of the milk- and whey proteins and lactose with high flux and retention. The filtration characteristics were obviously influenced by the process parameters.

Balannec et al. (2002) studied a new combination of membrane-based cheese production procedure in order to propose a significant increase in the cheese yield by incorporating the whey proteins. The performances of dairy process water treatment with membranes for recovery of milk constituents and water were analyzed in terms of COD and ion rejection. The dead-end filtration experiments permitted to compare several NF and RO membranes. It was found that single membrane operation is insufficient for producing water of composition complying with the requirements for drinking water. Because of the high COD level of the dairy process waters and despite high rejection of lactose, COD, and milk ions, concentration in permeate remained too high even with RO membranes. To reach the goal for target reuse of the purified water in the dairy plant, a finishing step must be added.

A vibratory shear-enhanced process (VSEP) was used for the separation of casein micelles from whey proteins of skim milk reconstituted from low-heat milk powder, which has a similar protein content as fresh milk (Al-Akoum et al. 2002). This paper compared the performances of MF with a $0.1 \mu m$ pore Teflon membrane and of a UF polyethersulfone (PES)-one with 150 kD cutoff membrane. The critical flux for stable operation was investigated in MF by increasing the permeate flux in steps while monitoring the transmembrane pressure. It was found to be $50 L/h \cdot m^2$ at a VRR of 2 and the maximum frequency of 60.75 Hz (τ_ω 34 Pa). The UF membrane minimizes casein loss in the permeate with a little reduction in permeate flux. Whey protein transmission in UF was found to be 65%–70% for α-lactalbumin and 25%–30% for β-lactoglobulin. For casein micelle separation from whey proteins with the VSEP, the use of a UF membrane (with 150 kDa cutoff) permitted to minimize micelle loss while adequately transmitting acceptable whey proteins.

17.4.4.2 Properties of Membrane-Treated Fruit Juices

UF can be used as a unique operation for the clarification and pasteurization of fruit juices, such as apple juice due to its operating principle (Ortega-Rivas 1995). The separating capability of UF can be identified in terms of the rejection coefficient of a membrane against a specific molecular weight. As previously stated, the rated MWCO defines the smallest particle or molecule which can be retained by a membrane. The processing of some liquid foods is important because these membrane characteristics will give specificity in terms of permeating soluble sugars and flavor components while retaining SS and large molecules. Since microorganisms, spores, and other particulates responsible for spoilage are retained, diverse liquid foods treated by UF are said to be cold sterilized (Ortega-Rivas 1995).

Some investigations have been reported regarding sensory quality of apple juice treated by membrane filtration. Heatherbell et al. (1977) clarified apple juice by UF and obtained a stable clear product. Rao et al. (1987) studied retention of odor-active volatiles using different UF membrane materials. Padilla and McLellan (1989) investigated the effect of MWCO of UF membranes on quality and stability of apple juice. Comparison of MF and UF has also been done (Wu et al. 1990), and it has been found that the microfiltered juice contained more soluble solids and it was more turbid compared with the ultrafiltered juice. The use of mineral membranes for apple juice clarification has also been reported (Ben Amar et al. 1990). In general terms, the use of UF to clarify apple juice results in fresh appearance because the product is cold sterilized, avoiding undesirable reactions triggered by the conventional thermal process. The color of juices is an important quality aspect, and it has been argued that the advantage of thermal processing is the inactivation of enzymes which cause browning of different fruit juices (Sapers 1991). However, since it has also been suggested (Vamos-Vigyazo 1981) that enzyme activity is associated with particulate fractions, UF might also be used to separate such particulates from raw apple juice in order to prevent, or greatly reduce, enzymatic browning.

Membrane separations have been compared with conventional separation methods in terms of quality aspects, such as color. Juices produced by membrane filtration and traditional methods have mostly been shown to have similar properties. Heatherbell et al. (1977) compared apple juice clarified by UF and gelatin fining. The composition of both juices was similar except the UF juice showed a more intense color. The gelatin-fined juice had been pasteurized before fining, which reduced oxidative reactions that may have caused color development. Bottled nonpasteurized UF juice was informally judged to have superior flavor to canned UF or canned gelatin-fined juice. However, the bottled juice did develop a sediment after storage. Slight differences have also been noted between membrane and traditionally clarified juices. Rao et al. (1987) found the retention of odor-active volatiles in UF apple juice was intermediate to traditional plate and frame filtration and vacuum drum filtration. Plate and frame filtration gave the highest retention. Rwabahizi and Wrolstad (1988) found that strawberry juice clarified through a 10 kDa hollow fiber membrane had an average of 55% anthocyanin loss compared with 17% loss by conventional filtration. Drake and Nelson (1986) compared commercial apple juice made by 50 kDa MWCO at 75°C and pasteurized at 85°C with commercial juices made using a conventional plate and frame filtration system. The UF juice had lower turbidity, 5% higher soluble solids, and less color than the other juice. It was also sensory rated as lighter in color with more of an initial watery mouthfeel.

Investigations have been also focused on membrane structure and configuration. Some studies have found that membrane material can have some influence on UF and MF processing. Braddock (1982) used UF and RO to recover limonene from citrus processing waste streams. Membrane flux rates declined after contact with limonene (around 0.11%). Polysulfone membranes had the most severe decline followed by cellulose acetate and Teflon. Rao et al. (1987) studied the retention of eight odor-active volatiles in apple juice filtered through a 50 kDa polysulfone and a 30 kDa polyamide hollow fiber membrane. The permeate from the polyamide membrane contained more volatiles than the polysulfone membrane. Ben Amar et al. (1990) found that a 0.2 μm ceramic and a 0.2 μm carbon gave similar fluxes for apple juice under similar conditions. Capannelli et al. (1992) tested polysulfone, polyvinylidene fluoride (PVDF), and ceramic membranes of various MWCOs with orange and lemon juices. With a given set of working conditions, the flux was largely independent of membrane material and MWCO. This was attributed to fibrous deposits that developed at the membrane surfaces and acted as a dynamic semipermeable barrier. Riedl et al. (1988) examined how membrane structure influenced the aggregation of apple juice colloidal particles on porous MF membrane surfaces. Using atomic force microscopy, PES and PVDF membranes were found to have rough surface structures that produced looser surface fouling layers than the dense fouling layers observed on smooth surfaced membranes such as nylon and polysulfone. As a result, the fouling layers of PES and PVDF membranes demonstrated less flux resistance per unit thickness and higher overall flux performance.

Microbial inactivation capability of membrane separations has been also tested. Heatherbell et al. (1977) filtered apple juice (3.8×10^4 cfu/mL) with a 50 kDa MWCO polysulfone membrane to a twofold concentration. The permeate had low counts (<1 cfu/mL), while the concentrate reached 1×10^5 cfu/mL. No evidence of spoilage was detected in aseptically bottled product after storage at room temperature for 6 months. Grohmann and Feuerpeil (1987) mentioned that mineral membranes (e.g., ceramic) that can be thermally sterilized provide maximum operating safety in the production of a cold sterile juice using tangential flow systems. They were able to produce a cold sterile apple juice with a 0.2 μm ceramic membrane on a large scale. With polymeric membranes, minimal operating pressures should also be used to prevent damage and thus contamination from occurring. Cross-flow filtration can be combined with a final dead-end filtration to commercially produce cold-sterilized juice products and ensure microbial stability. Reid et al. (1990) described the setup for aseptic filtration and bottling of beer into polyethylene terephthalate containers. The bottle rinsing, filling, and capping operations were maintained in a filtered air environment to minimize contamination. Sterile water was also used for the bottle cleaning operation.

In terms of some physicochemical properties and their relation to quality aspects of the sensory type, Tronc et al. (1998) used bipolar membranes to lower the pH of cloudy apple juice. By circulating apple juice on the cationic side of the bipolar membrane where the H^+ ions are generated, the pH was reduced from 3.5 to 2.0. This pH reduction completely inhibited polyphenol oxidase (PPO) activity compared with the control. Following this temporary acidification, the pH was returned to its initial value by recirculating the juice on the anionic of the bipolar membrane where the OH^- were produced. Although the pH readjustment partially reactivated PPO, the color of cloudy apple juice remained stable during storage. In acidic media, the free carboxyl groups of amino acids are protonated and the negative charges are neutralized. Inhibition of PPO likely resulted from the change in tertiary structure of the protein caused by electrostatic repulsion between acids and positively charged amino groups (McCord and Kilara 1984, Zemel et al. 1990).

Combined quality effects and operating procedures of membrane separation processing in apple juice can be exemplified with the work of Zárate-Rodríguez et al. (2001), who presented a study on the effects of pore size of UF membranes on the quality of treated apple juice. Fresh apple juice was processed using a UF unit, with polysulphone membranes of 10 and 50 kDa pore size. Transmembrane pressures of 103, 120.5, 138, and 155 kPa were studied. Recovery percentages of 0, 25, 50, and 75

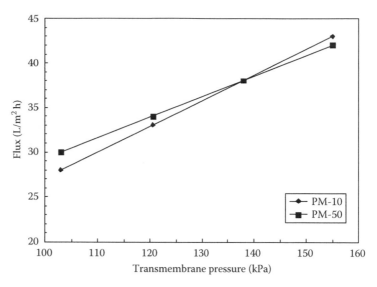

FIGURE 17.16 Transmembrane pressure versus flux of apple juice for PM-10 and PM-50 UF membranes. (Adapted from Zárate-Rodríguez, E. et al., *Int. J. Food Sci. Technol.*, 36, 663, 2001.)

were tested for the smaller pore membrane, as well as 0, 10, 20, 30, 40, 50, and 60 for the larger pore membrane. The responses to these factors were evaluated for the quality attributes pH, acid content, soluble solids, and color. In terms of efficiency, Figure 17.16 shows a plot of flux in liters per square meter per hour against transmembrane pressure for the 50% recovery percentage of both membranes. It can be observed that the flow rate was very similar for any of the employed membranes, but the trend reversed for the last reading (155 kPa) showing a larger transmembrane pressure for the smaller pore size. This was possibly due to a different effect of fouling buildup, where SS responsible for it could approach better the size of the larger pores causing blockage and obstructing the flow. However, for conventional operating transmembrane pressures (below 150 kPa), the expected trend of the higher yield as a function of pore size is observed. In terms of physicochemical properties, pH, acid content, and soluble solids, did not change significantly, but presented less variability for the smaller pore membrane treatment. For example, Figures 17.17 and 17.18 present the relationship between pore size and contents of soluble solids and acidity, respectively, as a function of transmembrane pressure. As can be observed, all values of these variables were higher for the large pore membrane. Although both membranes were supposed to allow chemical compounds responsible for color and flavor to permeate, possibly the smaller pore, coupled with SS fouling, could have had a retentive effect of these substances. The registered significant differences in soluble solids and acid content may be attributed to the possibility of more retention of organic molecules in the smaller pore membrane. Finally, relative color changes were observed for both membranes, with an evident browning trend, which was more detectable for the larger pore membrane treatment.

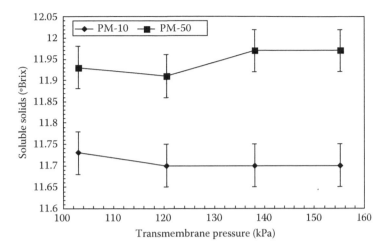

FIGURE 17.17 Transmembrane pressure versus soluble solids (°Brix) of apple juice for PM-10 and PM-50 UF membranes. (Adapted from Zárate-Rodríguez, E. et al., *Int. J. Food Sci. Technol.*, 36, 663, 2001.)

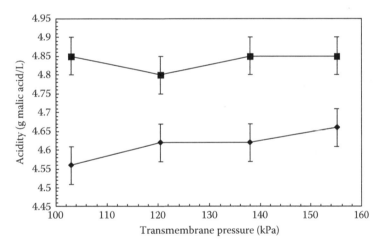

FIGURE 17.18 Transmembrane pressure versus acidity (malic acid, g/L) of apple juice for PM-10 and PM-50 UF membranes. (Adapted from Zárate-Rodríguez, E. et al., *Int. J. Food Sci. Technol.*, 36, 663, 2001.)

Membrane separations have also been compared with alternative nonthermal methods of food preservation. Zárate-Rodríguez et al. (2000) reported that UF and high-voltage pulsed-electric fields (PEF), used as nonthermal preservation techniques for pasteurization of apple juice, were very efficient in microbial inactivation as well as in quality preservation. However, both treatments resulted in juices with significant color changes. A summary of their main comparing results is given in Table 17.2. These authors suggested that more research was needed in order to understand color changes, which are very important in the overall quality of fruit juices.

TABLE 17.2
Effect of Nonthermal Processing Technique on Some Quality Attributes of Pasteurized Apple Juice

Attribute	Untreated[a]	UF-Treated[a]	PEF-Treated[a]
Soluble solids	12.42 ± 0.09^b	11.85 ± 0.16^c	12.39 ± 0.19^c
pH	3.75 ± 0.10^b	3.68 ± 0.04^c	3.73 ± 0.03^b
Acidity	5.46 ± 0.07^b	4.76 ± 1.29^c	5.53 ± 0.87^b
Color ratio	0^b	0.21 ± 0.11^c	-0.54 ± 0.23^d

Source: Adapted from Zárate-Rodríguez, E. et al., *J. Food Qual.*, 23, 337, 2000.

Note: Mean ± standard deviation ($n = 4$).

[a] Any two means in the same row followed by the same letter are not significantly ($p > 0.05$) by Student's *t*-test.

17.5 CONCLUDING REMARKS

Solid–liquid separations have been extensively used in the food industry, mainly as a final stage for purification of liquids or for recovering of solids. These separation technologies had been considered as mechanical operations and had been focused only on quality aspects. The development and scaling of membrane separations to allow their wide commercialization transformed the perception of solid–liquid separations as transformation technologies exclusively. Nonthermal pasteurization of different fluid food systems is now a reality using different alternatives of solid–liquid separations. The microbial safety, nutritive quality, and sensory value of different liquid foods can be guaranteed by means of processing operations based, totally or partially, on solid–liquid separation technologies.

REFERENCES

Al-Akoum, O., Ding, L.H., and Jaffrin, M.Y. 2002. Microfiltration and ultrafiltration of UHT skim milk with a vibrating membrane module. *Separation and Purification Technology* 28: 219–234.

Al-Akoum, O., Jaffrin, M.Y., Ding, H.L., and Frappart, M. 2004. Treatment of dairy process waters using a vibrating filtration, system and NF and RO membranes. *Journal of Membrane Science* 235: 111–122.

Antunes, M. and Medronho, R.A. 1992. Bradley hydrocyclones: Design and performance analysis. In *Hydrocyclones-Analysis and Applications*, eds. L. Svarovsky and M.T. Thew, pp. 3–13. Dordretch, the Netherlands: Kluwer Academic Publishers.

Atra, R., Vatai, G., Bekassy-Molnar, E., and Balint, A. 2005. Investigation of ultra- and nanofiltration for utilization of whey protein and lactose. *Journal of Food Engineering* 67: 325–332.

Balannec, B., Gésan-Guiziou, G., Chaufer, B., Rabiller-Baudry, M., and Daufin, G. 2002. Treatment of dairy process waters by membrane operations for water reuse milk constituents concentration. *Desalination* 147: 89–94.

Barbosa-Cánovas, G.V., Ortega-Rivas, E., Juliano, P., and Yan, H. 2005. *Food Powders: Physical Properties, Processing, and Functionality*. New York: Kluwer Academic/ Plenum Publishers.

Bayindirli, L., Özilgen, M., and Ungan, S. 1989. Modeling apple juice filtrations. *Journal of Food Science* 54: 1003–1006.

Bednarski, S. 1992. A new method of starch production form potatoes. In *Hydrocyclones-Analysis and Applications*, eds. L. Svarovsky and M.T. Thew, pp. 309–326. Dordretch, the Netherlands: Kluwer Academic Publishers.

Bednarski, S. 1996. The thickening of biological solids in hydrocyclones. In *Hydrocyclones 96*, eds. D. Claxton, L. Svarovsky, and M.T. Thew, pp. 151–159. London: Mechanical Engineering Publications.

Ben Amar, R., Gupta, B.B., and Jaffrin, M.Y. 1990. Apple juice clarification using mineral membranes: Fouling control by backwashing and pulsating flow. *Journal of Food Science* 55: 1620–1625.

Braddock, R.J. 1982. Ultrafiltration and reverse osmosis recovery of limonene from citrus processing waste streams. *Journal of Food Science* 47: 946–948.

Capannelli, G., Bottino, A., Munari, S. et al. 1992. Ultrafiltration of fresh orange and lemon juices. *Lebensmittel-Wissenschaft und-Technologie—Food Science and Technology* 25: 518–522.

Chen, W., Zydek, N., and Parma, F. 2000. Evaluation of hydrocyclone models for practical applications. *Chemical Engineering Journal* 80: 295–303.

Chen, W., Parma, F., Parkat, A., Elkin, A., and Sen, S. 2004. Selecting membrane filtration systems. *Chemical Engineering Progress* 100: 22–25.

Cilliers, J.J. and Harrison, S.T.L. 1996. The effect of viscosity on the recovery and concentration of micro-organisms using mini-hydrocyclones. In *Hydrocyclones 96*, eds. D. Claxton, L. Svarovsky, and M.T. Thew, 123–133. London: Mechanical Engineering Publications.

De La Garza, F. and Boulton, R. 1984. The modeling of wine filtrations. *American Journal of Enology and Viticulture* 35: 189–195.

Dickey, G.D. 1961. *Filtration*. New York: Van Nostrand Reinhold.

Drake, S.R. and Nelson, J.W. 1986. Apple juice quality as influenced by ultrafiltration. *Journal of Food Quality* 9: 399–406.

Eckner, K.F. and Zottola, A. 1991. Potential for the low temperature pasteurization of dairy fluids using membrane processing. *Journal of Food Protection* 54: 793–797.

Einstein, A. 1906. Eine neue bestimmung der moleküldimensionen (a new determination of the molecular dimensions). *Annalen der Physik* 324: 289–306.

Einstein, A. 1911. Berichtigung zu meiner arbeit: eine neue bestimmung der moleküldimensionen (correction of my work: a new determination of the molecular dimensions). *Annalen der Physik* 339: 591–592.

Foust, A.S., Wenzel, L.A., Clump, C.W., Maus, L., and Bryce Anderson, L. 1960. *Principles of Unit Operations*. New York: John Wiley & Sons.

Garoutte, C. and Amundson, C. 1982. Ultrafiltration of milk with hollow fiber membranes. *Journal of Food Process Engineering* 5: 191–202.

Gesan, G., Merin, U., Daufin G., and Maugas, J.J. 1993. Performance of an industrial cross-flow microfiltration plant for clarifying rennet whey. *Netherlands Milk Dairy Journal* 47: 121–124.

Girard, B. and Fukumoto, L.R. 2000. Membrane processing of fruit juices: A review. *Critical Reviews in Food Science and Nutrition* 40: 91–157.

Grohmann, M. and Feuerpeil, H.P. 1987. Cold sterile apple juice-cold sterile filtrates with tangential flow filtration. *Confructa Studien* 31: 171–174.

Heatherbell, D.A., Short, J.L., and Strubi, P. 1977. Apple juice clarification by ultrafiltration. *Confructa* 22: 157–169.

Li, Z., Liu, W., Shen, Q., Zhenga, W., and Tan, B. 2008. Properties and qualities of vermicelli made from sour liquid processing and centrifugation starch. *Journal of Food Engineering* 86: 162–166.

Liu, W. and Shen, Q. 2007. Structure analysis of mung bean starch from sour liquid processing and centrifugation. *Journal of Food Engineering* 79: 1310–1314.

Martins da Mata, V. and Medronho, R.A. 2000. A new method for yeast recovery in batch ethanol fermentations: Filter aid filtration followed by separation from filer aid using hydrocyclones. *Bioseparation* 9: 43–53.

Maubois, J.L. 1991. New applications of membrane technology in the dairy industry. *Australian Journal of Dairy Technology* 46: 91–95.

Maximea, D. and Lameloise, M.-L. 2003. Filterability and filtration experiences of viscous complex food salting solutions. *Chemical Engineering Research and Design* 81: 1150–1157.

McCabe, W.L., Smith, J.C., and Harriot, P. 2001. *Unit Operations in Chemical Engineering.* New York: McGraw-Hill.

McCord, J.D. and Kilara, A. 1984. Control of enzymatic browning in processed mushrooms (*Agaricus bisporus*). *Journal of Food Science* 48: 1479–1484.

Medronho, R.A. and Svarovsky, L. 1984. Test to verify hydrocyclone scale-up procedure. In *Proceedings of the 2nd International Conference on Hydrocyclones*, pp. 1–14, Cranfield, U.K.: BHRA The Fluid Engineering Centre.

Nezhati, K. and Thew, M.T. 1988. Aspects of the performance and scaling of hydrocyclones for use with light dispersions. In *3rd International Conference on Hydrocyclones*, ed. P. Wood, 167–180. London: Elsevier Applied Science Publishers.

Ortega-Rivas, E. 1989. *Dimensionless Scale-up of Hydrociclones for Separation of Concentrated Suspensions*. PhD thesis. Bradford, U.K.: University of Bradford.

Ortega-Rivas, E. 1995. Review and advances in apple juice processing. In *Food Process Design and Evaluation*, ed. R.K. Singh, pp. 27–46. Lancaster, PA: Technomic Publishing Co. Inc.

Ortega-Rivas, E. 2005. Handling and processing of food powders and particulates. In *Encapsulated and Powdered Foods*, ed. C. Onwulata, pp. 75–144. Boca Raton, FL: CRC/Taylor & Francis.

Ortega-Rivas, E. 2007. Hydrocyclones. In *Ullmann's Encyclopedia of Industrial Chemistry*, pp. 1–27. Weinheim, Germany: Wiley-VCH.

Ortega-Rivas, E. and Medina-Caballero, S.P. 1996. Wastewater sludge treatment by hydrocyclones. *Powder Handling* and *Processing* 8: 355–359.

Ortega-Rivas, E. and Svarovsky, L. 1993. On the completion of a dimensinless scale-up model for hydrocyclone separation. *Fluid/Particle Separation Journal* 6: 104–109.

Ortega-Rivas, E. and Svarovsky, L. 1998. Generalized Stokes number for modeling settling of non-Newtonian slurries in dynamic separators. *Advanced Powder Technology* 9: 1–16.

Ortega-Rivas, E., Meza-Velásquez, F., and Olivas-Vargas, R. 1997. Reduction of solids by liquid cyclones as an aid to clarification in apple juice processing. *Food Science and Technology International* 3: 325–331.

Padilla, O.I. and McLellan, M.R. 1989. Molecular weight cut-off of ultrafiltration membranes and the quality and stability of apple Juice. *Journal of Food Science* 54: 1250–1254.

Pafylias, I., Chelyan, M., Mehaia, M.A., and Saglam, N. 1996. Microfiltration of milk with ceramic membranes. *Food Research International* 29: 141–146.

Pearce, R.J., Marshall, S.C., and Dunkerley, J.A. 1991. Reduction of lipids in whey protein concentrates by microfiltration: Effect of functional properties. *IDF Special Issue* 9201: 118–129.

Rao, M.A., Acree, T.E., Cooley, H.J., and Ennis, R.W. 1987. Clarification of apple juice by hollow fiber ultrafiltration: Fluxes and retention of odor-active volatiles *Journal of Food Science* 52: 375–377.

Reid, G.C., Hwang, A., Meisel, R.H., and Allcock, E.R. 1990. The sterile filtration and packaging of beer into polyethylene terephthalate containers. *Journal of the American Society of Brewing Chemists* 48: 85–91.

Rickwood, D., Onions, J., Bendixen, B., and Smyth, I. 1992. Prospects for the use of hydrocyclones for biological separations. In *Hydrocyclones—Analysis and Applications*, eds. L. Svarovsky and M.T. Thew, pp. 109–119. Dordretch, the Netherlands: Kluwer Academic Publishers.

Rickwood, D., Freeman, G.J., and McKechnie, M. 1996. An assessment of hydrocyclones for recycling kieselguhr used for filter in the brewing industry. In *Hydrocyclones 96*, eds. D. Claxton, L. Svarovsky, and M.T. Thew, pp. 161–172. London: Mechanical Engineering Publications.

Riedl, K., Girard, B., and Lencki, R.W. 1988. Influence of membrane structure on fouling layer morphology during apple juice clarification. *Journal of Membrane Science* 139: 155–166.

Rietema, K. 1961. Performance and design of hydrocyclones. *Chemical Engineering Science* 15: 198–325.

Rwabahizi, S. and Wrolstad, R.E. 1988. Effects of mold contamination and ultrafiltration on the color stability of strawberry juice and concentrate. *Journal of Food Science* 53: 857–861.

Sahin, S. and Bayindirli, L. 1991. Assessment of the exponential model to sour cherry juice filtrations. *Journal of Food Processing and Preservation* 15: 403–411.

Sapers, G.M. 1991. Control of enzymatic browning in raw fruit juice by filtration and centrifugation. *Journal of Food Processing and Preservation* 15: 443–456.

Sperry, D.R. 1917. The principles of filtration. *Chemical and Metallurgical Engineering* 17: 161–165.

Svarovsky, L. 1984. *Hydrocyclones*. London: Holt Rinehart & Winston.

Svarovsky, L. 2000. *Solid–Liquid Separation*. London: Butterworths-Heinemann.

Svarovsky, L. and Marasinghe, B.S. 1980. Performance of hydrocyclones at high feed solid concentrations. In *Proceedings of International Conference on Hydrocyclones*, pp. 127–142. Cranfield, U.K.: BHRA The Fluid Engineering Centre.

Trawinski, H.F. 1977. Hydrocyclones. In *Solid/Liquid Separation Equipment Scale-up*, ed. D.B. Purchas, pp. 241–287. Croydon, U.K.: Uplands Press.

Trettin, D.R. and Doshi, M.R. 1981. Pressure independent ultrafiltration—Is it gel limited or osmotic pressure limited? *ACS Symposium Series* 2: 154–173.

Tronc, J.-S., Lamarche, F., and Makhlouf, J. 1998. Effect of pH variation by electrodialysis on the inhibition of enzymatic browning in cloudy apple juice. *Journal of Agricultural and Food Chemistry* 46: 829–833.

Trowbridge, M.E.O'K. 1962. Problems in the scaling-up of centrifugal separation equipment. *Chemical Engineering* 40: 73–86.

Vamos-Vigyazo, L. 1981. Polyphenol oxidase and peroxidase in fruits and vegetables. *CRC Critical Reviews in Food Science and Nutrition* 15: 49–127.

Wijmans, J.G., Nakao, S., and Smolders, C.A. 1984. Flux limitation in ultrafiltration: Osmotic pressure model and gel-layer model. *Journal of Membrane Science* 20: 115–124.

Wu, M.L., Zall, R., and Tzeng, W.C. 1990. Microfiltration and ultrafiltration comparison for apple juice clarification. *Journal of Food Science* 55: 1162–1163.

Yuan, H., Thew, M.T., and Rickwood, D. 1996. Separation of yeast with hydrocyclones. In *Hydrocyclones 96*, eds. D. Claxton, L. Svarovsky, and M.T. Thew, pp. 135–149. London: Mechanical Engineering Publications.

Zadow, J.G. 1987. Whey production and utilization in Oceania in trends in whey utilization. *IDF Bulletin* 212, pp. 12–16. Brussels: International Dairy Federation.

Zárate-Rodríguez, E., Ortega-Rivas, E., and Barbosa-Cánovas, G.V. 2000. Quality changes in apple juice as related to nonthermal processing. *Journal of Food Quality* 23: 337–349.

Zárate-Rodríguez, E., Ortega-Rivas, E., and Barbosa-Cánovas, G.V. 2001. Effect of membrane pore size on quality of ultrafiltered apple juice. *International Journal of Food Science and Technology* 36: 663–667.

Zemel, G.P., Sims, C.A., Marshall, M.R., and Balaban, M. 1990. Low pH inactivation of polyphenol oxidase in apple juice. *Journal of Food Science* 55: 562–563.

Part IV

Integrative Approach: Safety and Quality in Food Plant Design

18 The SAFES Methodology: A New Approach for Real-Food Modeling, Optimization, and Process Design

Pedro Fito, Marc Le Maguer, José Miguel Aguilera,
Luis Mayor, Lucía Seguí, Noelia Betoret,
and Pedro José Fito

CONTENTS

18.1 INTRODUCTION

The new conception of food process engineering provides new opportunities for product design, since it aims to develop and control processes in order to produce food products with specific properties of quality, safety, and nutrition as demanded by consumers. Several authors have recently referred to this concept as "food engineering for product design" (Aguilera 2006, Fito et al. 2007a), "food process engineering for product quality, targeted food design, reverse engineering, or from the fork to the farm" (European Technology Platform "Food for Life" 2005).

When processing complex food (polymer, colloidal, or cellular) the product changes may be observed as differences in quality factors such as food composition, nutritional facts, taste and flavor, aspect, shape and size, color, texture, safety, etc. These changes in food properties may be explained by the physical and chemical phenomena produced on the line with the process progression as structure changes, chemical or enzymatic reactions, phase transitions, etc. take place. Thus the final properties of food products as appreciated by consumers are the result of the changes in the food structure and composition as a consequence of the working conditions on the processing line (Figure 18.1).

The concern of consumers on safety, health, and well being offers also new opportunities in products designed for the brain–gut axis (Figure 18.1). Although the connection between the gut function and the brain has been known to the medical profession for more than a century, the food industry has only recently begun to focus on using the positive role of nonnutritive factors in oxidative-stress reduction, anti-carcinogenicity, blood glucose regulation, body-weight reduction, gastrointestinal

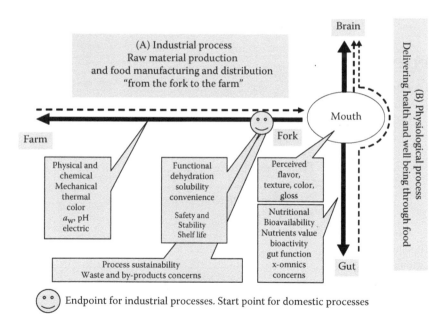

FIGURE 18.1   Two axis of the "complete" food process and the food product properties as related with them. The first step (A) is the "regular" industrial process, whereas the second one (B) represents the interactions between food and consumer. Food structure changes through the whole process: In part A, food structure is preserved or partially destroyed, or even the food is conveniently restructured. Also some components may change or move into (or out of) the food system phases. In part B, the structure is partially destroyed in the mouth (mastication) and in the stomach (digestion), hydrolyzed, and some components absorbed. In both cases the changes in food structure are related with the changes in food properties, establishing the "structure–properties–process relationships" ensemble.

improvement, gastric emptying, etc. Beneficial effects of these compounds are largely dependent on their interactions with the encapsulating matrix or microstructure, whether natural or man-made (Norton et al. 2007, Parada and Aguilera 2007).

The scientific basis needed to understand the changes that occur in foods during processing has been enormously expanded over the past half-century. A large body of research in food chemistry, food microbiology, and food processing now exists. Great progress has also been made during the past two decades in the study of food material properties (Rao et al. 2005, Aguilera and Lillford 2007) and their relation with structure (Aguilera 2005). In spite of such efforts, very few models to date have attempted to include the food structure and the full complex composition of the food material. Furthermore, most of these attempts were directed at the modeling of heat and mass transfer phenomena with some rare incursion in the deformation of the material from a structural point of view (Fito 1994, Yao and Le Maguer 1996, 1997a,b, Barat et al. 1999, Andres et al. 2004, Tremeac et al. 2007). Finally all these were specific to a given product and process, i.e., fruits and osmotic dehydration and drying, dough and bread baking. But these were rare instances in which a deliberate effort had been made to incorporate in some detail the internal structure of the material as well as the changes that took place during the manufacturing process to this same structure and its components.

It is deduced from above that there is a need for a new way of food model development, for an approach that takes into account the complexity of the food material, which considers "real foods" as multiphasic and multicomponent systems in which the structure plays a key role in product properties (physical, nutritional, sensory, safety). This new model might be shared among scientists, industry R&D personnel, and regulatory or policy-making bodies. Such an approach would have to be able to integrate a large field of knowledge beyond the usual realm of what has been included under the meaning of modeling until today: it has to be able to represent the micro- and macrostructure of the food material, provide sufficient information to assess the safety, quality, and nutritional attributes of the food as well as taking into account the conditions of the manufacturing process.

In the last few years, the SAFES (systematic approach to food engineering systems) approach (Fito et al. 2007a) evolved and proved to be a good methodology for the analysis of real food systems by bringing together a variety of information pieces and organizing it in a way that reflected the structure of the food material. This methodology may be also suitable for developing complex models that integrate the existing knowledge in the different related scientific areas. The formal framework was developed; a certain number of essential concepts emerged and were applied to a number of case studies (see the works appearing in the special issue of the *Journal of Food Engineering*, 83(2), 2007; Argüelles et al. 2007, Barrera et al. 2007a,b, Betoret et al. 2007, Campos-Mendiola et al. 2007, Castelló et al. 2007, Chenoll et al. 2007a,b, Elías-Serrano et al. 2007, Fito et al. 2007b, Heredia et al. 2007, Seguí et al. 2007). This approach recognizes the complexity of the food system and organizes around a simplified version of it through the "structured phases" concept, a large amount of information pertaining to composition, thermodynamic properties, reaction kinetics, and quality attributes among others. Furthermore, this information is presented in a matrix form which mathematically lends itself to easy performing the necessary calculations.

In the following sections, the SAFES methodology will be comprehensively described.

18.2 DESCRIPTION OF THE FOOD SYSTEM THROUGH THE SAFES APPROACH

Food systems have been traditionally simplified to monophasic systems in which only two or three components are considered relevant for the analysis, thus making possible the use of models based on the fundamentals of chemical process engineering. Real foods, however, are multicomponent and multiphasic systems which, as all biological materials, have a significant degree of complexity (chemical, physical, structural, etc.) which allow them to be functional. Although food complexity is considered a fact (Aguilera 2005, Mebatsion et al. 2008), the traditional approach has been largely used by food engineers, thus not recognizing the food system complexity neither being able to describe the food functionality, quality, or safety. Through the SAFES methodology, the food system is systematically analyzed and all relevant information (micro- and macrostructural, physical, chemical, nutritional, sensory, etc.) pertaining not only to the food system but also to the process (process parameters, standard operating procedures, environmental constraints, etc.) is compiled and organized.

18.2.1 CONCEPT OF STRUCTURED PHASES

A real food at any stage of its transformation process can be described as made up of various phases in which the food components are present in different physical or aggregation states, from gas to liquid to solid. Furthermore, from a practical and quantitative point of view, it is important to consider the functionality and properties of the food, which are directly related to food structure. Here raises the definition of structured phases, which are defined as the phases that constitute the real food, that are able to describe its structure and, as a direct consequence, to represent its properties and its functionality. Therefore, for a real food the structured phases constitute a simplified but rigorous representation of the food system, among which the components are distributed, and which are responsible for food properties and functionality.

To illustrate the concept of structured phases, two examples of very common food systems are presented in Figures 18.2 and 18.3: a whipped cream and the apple parenchyma. Figure 18.2 shows the microstructure of an industrial whipped cream; in this food system, a foam structure with a phase of gas bubbles dispersed in an emulsion (oil–water) containing fat granules may be observed, as well as an aqueous solution as continuous phase. The description of such a complex product as well as the analysis of its properties and changes throughout any process implies the recognition of the phases that play some role in the product behavior. As a first approach, it would be convenient to include the following structured phases: gas, insoluble dispersed solids, liquid fat, solid fat, and aqueous solution or liquid phase. All food components that are considered of interest (e.g., water, fat, sugar, gas, etc.) would then be distributed among the phases, and the corresponding state of aggregation noted (solid, adsorbed, vitreous, rubbery, liquid, and gas), depending on the state variables.

Figure 18.3 shows a magnification of parenchyma tissue of apple "Granny Smith" observed by Cryo-SEM. Several cells can be observed (diameter ranging from 200 to 300 μm) and their cell wall, cell membrane, and middle lamella are identified. All these elements, which are water insoluble, can be simplified as a food insoluble

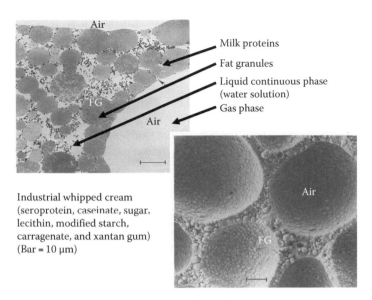

FIGURE 18.2 Sample of an industrial whipped cream observed by Cryo-SEM (right) and TEM (left). (Adapted from Fito, P. et al., *J. Food Eng.*, 83, 173–185, 2007. With permission from Elsevier.)

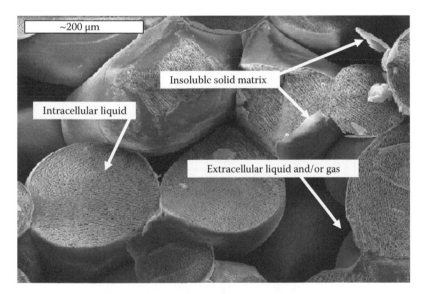

FIGURE 18.3 Parenchyma tissue of apple Granny Smith observed by Cryo-SEM. (Adapted from Fito, P. et al., *J. Food Eng.*, 83, 173–185, 2007. With permission from Elsevier.)

matrix. On the other hand, the liquid phase is identified inside the cells, this being another structured phase. In addition, several extra/intercellular spaces exist, which may be both filled with liquid or gas; hence, an extracellular liquid phase (ELP) and a gas phase might also be defined.

The decision about whether a component or a structured phase must be considered in the SAFES analysis or it should be included together with others in a simplified component or phase depends on its significance in order to explain the food properties and functionality, or on the modifications that it undergoes during processing.

Since structured phases have well-defined components, they are amenable to standard thermodynamic and engineering calculations involving the usual variables such as composition, temperature, pressure, volume, or free energy. It thus becomes possible to perform calculations for each of these elements based on the most current state of knowledge in the area of science that best represents them.

18.2.2 SAFES DESCRIPTIVE MATRIX

In the SAFES methodology, the descriptive matrix (M) represents the fingerprint of the food system (Figure 18.4). It consists of a set of matrices M_i ($i = 1, 2..., n$), which gather all the essential information concerning structure as well as the property–structure relationships of the food system that are considered of interest. According to this, the SAFES matrices may be divided into two subgroups: those that depict the structure (set of structure matrices) and those that describe the material properties as related to the structure (set of properties matrices).

18.2.2.1 Structure Matrices

The structural description of the system may be outlined by three structural matrices (M_1, M_2, and M_3) that rigorously organize the structural information of the system in

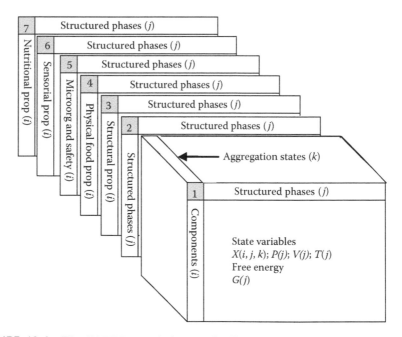

FIGURE 18.4 The SAFES descriptive matrix (M). Mathematical arrangement of the structure–properties ensemble for a food system.

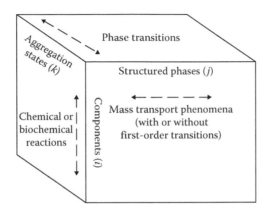

FIGURE 18.5 3-D mass distribution matrix for a food system (M_1).

a schematic configuration. The first one (M_1) arranges the simplified "space of phases and components" in a three-dimensional (3-D) matrix (Figure 18.5) that includes the distribution of mass between phases (x_{ij}^k). In this matrix, the concentration of a given component (i) in a given phase (j) is positioned as a function of its physical state (k). This matrix may be represented as shown in Figure 18.5 where the components, phases, and aggregation states are located on the three axes (i, j, k). In this arrangement, the discontinuous arrows stand for the possible changes in the local values of x_{ij}^k that may occur as a consequence of mass transfer, phase transitions, and/or chemical reactions in the system. In addition, it is also possible to convert this 3-D representation into a 2-D one in order to be able to draw it in a plane. In this case, the mass distribution matrix is converted into a 2-D structure where the aggregation states are included in the (i, j) plane as subrows (Figure 18.6). The latter arrangement also permits the insertion of the vectors that represent the state variables in each of the phases (P_j, V_j, T_j). The main properties of M_1 are

1. The value $\sum_i \sum_j \sum_k x_{ij}^k = m$, where m is the overall mass of the system
   a. If $m = 1$, x_{ij}^k are mass fractions (concentrations).
   b. If $m \neq 1$, x_{ij}^k represent the mass value of component i, in the j phase, in an aggregation state k.
2. The ratio $V_j / \sum_i \sum_k x_{i,j}^k$ is the specific volume of phase j.
3. The ratio $\sum_i \sum_j \sum_k x_{i,j}^k / \sum_j V_j = \rho$ is the food system average density.
4. $\sum_j V_j$ is the volume of the food system when $m \neq 1$.
5. $\sum_j V_j$ is the average specific volume of the food system when $m = 1$.

Let us consider again the apple tissue. First of all, the thermodynamically possible combinations of the three axes in Figure 18.6 "component–phase–aggregation state" have to be studied and decided. These combinations will appear as white cells in M_1, whereas nonthermodynamically viable combinations will be represented as dotted gray cells. Next, each white cell has to be fulfilled with the particular value of one component, in its corresponding state of aggregation, in the corresponding phase.

Phase		M_1				
Component	State	Phase$_1$	Phase$_2$	...	Phase m	Whole food
Component$_1$	G					$x_{10}^0 = \sum_1^m x_{1j}^0$
	L		x_{12}^1	$x_{1..}^1$		$x_{10}^1 = \sum_1^m x_{1j}^1$
	A	x_{11}^2			x_{1m}^2	$x_{10}^2 = \sum_1^m x_{1j}^2$
	R					
	V					
	K					
	Total	$x_{11} = \sum_{k=0}^5 x_{11}^k$	$x_{12} = \sum_{k=0}^5 x_{12}^k$	$x_{1..} = \sum_{k=0}^5 x_{1..}^k$	$x_{1m} = \sum_{k=0}^5 x_{1m}^k$	$x_{1o} = \sum_{k=0}^5 x_{1o}^k = \sum_{k=0}^5 \sum_{j=1}^m x_{1j}^k$
⋮	G					
	L					
	A					
	R	$x_{..1}^3$				$x_{..0}^3 = \sum_1^m x_{..j}^3$
	V	$x_{..1}^4$				$x_{..0}^4 = \sum_1^m x_{..j}^4$
	K					
	Total	$x_{..1} = \sum_{k=0}^5 x_{..1}^k$	$x_{..1} = \sum_{k=0}^5 x_{..1}^k$	$x_{..} = \sum_{k=0}^5 x_{..}^k$	$x_{..n} = \sum_{k=0}^5 x_{..n}^k$	$x_{..0} = \sum_{k=0}^5 x_{..0}^k = \sum_{k=0}^5 \sum_{j=1}^m x_{..j}^k$
Component$_n$	G					$x_{n0}^0 = \sum_1^m x_{nj}^0$
	L		x_{n2}^1	$x_{n..}^1$		$x_{n0}^1 = \sum_1^m x_{nj}^1$
	A	x_{n1}^2				$x_{n0}^2 = \sum_1^m x_{nj}^2$
	R					
	V					
	K					
	Total	$x_{n1} = \sum_{k=0}^5 x_{n1}^k$	$x_{n1} = \sum_{k=0}^5 x_{n2}^k$	$x_{n1} = \sum_{k=0}^5 x_{n..}^k$	$x_{nm} = \sum_{k=0}^5 x_{nm}^k$	$x_{n0} = \sum_{k=0}^5 x_{n0}^k = \sum_{k=0}^5 \sum_{j=1}^m x_{nj}^k$
Whole food		$x_{o1} = \sum_{i=1}^n \sum_{k=0}^5 x_{i1}^k$	$x_{o2} = \sum_{i=1}^n \sum_{k=0}^5 x_{i2}^k$	$x_{o..} = \sum_{i=1}^n \sum_{k=0}^5 x_{o..}^k$	$x_{om} = \sum_{i=1}^n \sum_{k=0}^5 x_{om}^k$	$x = \sum_{k=0}^5 \sum_{i=1}^n \sum_{j=1}^m x_{ij}^k = 1$

FIGURE 18.6　2-D mass distribution matrix for a food system (M_1) including the state variable vectors (P_j, V_j, T_j).

In this way, the mass fractions are distributed in matrix cells of a 2-D matrix as observed in Figure 18.7.

The descriptive matrix M_1 provides a global sight of the food structure by representing the simplified space of phases and components; however, this does not represent the structure of the food material accurately enough. In order to describe the food system in a proper manner, it is also essential to describe how these phases interact within the system, i.e., the interfaces and their behavior. According to this, the matrix of interfaces (M_2) is built; this being the second structural matrix of the SAFES food fingerprint.

In the matrix of interfaces, the structured phases are placed both in columns and rows; the resulting matrix (M_2) is symmetric in the (i, j) plane and has a null diagonal. Each of the interfaces may be characterized by different properties

Phase		Apple Parenchyma M_1				
Component	State	Solid Matrix	Intracellular Liquid	Extracellular Liquid	Gas	Whole Food
Water	G				0.000	0.000
	L		0.711	0.046		0.757
	A	0.097				0.097
	R					
	V					
	K					
	Total	0.097	0.711	0.046	0.000	0.854
Soluble solids	G					
	L		0.123	0.008		0.131
	A	0.000				0.000
	R					
	V					
	K					
	Total	0.000	0.123	0.008	0.000	0.131
Insoluble solids	G					
	L					
	A					
	R	0.015				0.015
	V	0.000				0.000
	K					
	Total	0.015	0.000	0.000	0.000	0.015
Gas	G				0.000	0.000
	L					
	A					
	R					
	V					
	K					
	Total	0.000	0.000	0.000	0.000	0.000
Whole food/mass		0.112	0.834	0.054	0.000	1.000
Whole food/volume (cm³)		0.085	0.789	0.515	0.282	1.230
Temperature (K)		298	298	298	298	298
Pressure (Pa)		101,325	101,325	101,325	101,325	101,325

FIGURE 18.7 Mass distribution matrix (M_1) of a Granny Smith apple parenchyma in its natural state.

depending on the product/process studied; hence, like the mass distribution matrix, the interface matrix is a 3-D structure (cube) that may be represented in a 2-D plane by transforming the third dimension in subrows. In this case, the subrows represent the interfaces properties (IP_k), which are chosen for convenience, e.g., interfaces area or thickness.

Coming back to the Granny Smith apple, the matrix of interfaces for the apple parenchyma would be like as shown in Figure 18.8. White cells represent the existing interfaces, whereas null combinations (diagonal) appear dark gray. So far two are the matrices defined. The two of them are 2-D representations of 3-D structures with the same j axis (structured phases). The remaining structural matrix needed to describe the product structure accurately is the structural properties (SP) matrix (M_3). Once the phases and interfaces have been well defined, it is necessary to collect any structural features of the phases that are crucial to depict the structure and, consequently, to define the product properties and be able to assess its quality and safety attributes. In this case, the matrix consists of a 2-D plane in which columns are the structured phases (j) and rows (i) are the required structural features.

Let us continue with the example above. For a plant tissue, as well as for any food material, the precise description of the elements that constitute the tissue is essential in order to model mass transfer and assess quality attributes. Some of these

Phase		Apple Parenchyma M_2 (Interfaces)				
Phase	Property	Solid Matrix	Intracellular Liquid	Extracellular Liquid	Gas	External Phase
Solid matrix	IP_1					
	IP_2					
	...					
	IP_n					
Intracellular liquid	IP_1					
	IP_2					
	...					
	IP_n					
Extracellular liquid	IP_1					
	IP_2					
	...					
	IP_n					
Gas	IP_1					
	IP_2					
	...					
	IP_n					
External phase	IP_1					
	IP_2					
	...					
	IP_n					

FIGURE 18.8 2-D representation of the matrix of interfaces (M_2) of apple parenchyma. IP stands for the interfaces properties, ranging from 1 to n.

Phases	Apple Parenchyma M_3 (Structural Properties)				
Structural Properties	Solid Matrix	Intracellular Liquid	Extracellular Liquid	Gas	Whole Food
SP_1: Mean cell volume					
SP_2: Mean ratio major to minor axes					
SP_2: Porosity					
SP_3: Mean diameter of gas voids					
⋮					
SP_n					

FIGURE 18.9 Matrix of SP (M_3) of apple parenchyma. SP stands for some structural properties. The number of rows (n) will be as long as needed for describing the structure.

structural features would be the cellular size and shapes, which can be defined in terms of cellular equivalent diameter or the ratio major to minor axes of the equivalent ellipses. An example of this structural matrix is shown in Figure 18.9. Up to now, the matrices that describe the food composition and structure have been defined. It is important to underline that there is not only one proper way of building these matrices: first, it is the user who must decide which structural phases are considered of interest for the application being tested; second, any property included in the interfaces or SP matrix is introduced in the analysis under the user's decision.

18.2.2.2 Properties Matrices

One of the most important potential applications of the SAFES approach relies on the development of the properties matrices. The approach suggests that any property of the food (P) can be obtained as a function of the value of the property on each of the phases that constitute it (P_j), taking into account the specific weight of each phase over the total (x_j). Depending on the nature of the property itself, some may be additive or averageable; however, many others can be calculated only by means of a specific function. Whichever case, the challenge consists in obtaining the functions that, based on structure–properties relationships, allow estimating the value for the whole food.

The properties matrices are essentially similar to the matrix of SP (M_3). In order to organize the information, the properties are gathered by their nature, as shown in Figure 18.4 (physical, nutritional, sensorial, microbiological, and safety properties), each representing a different property matrix. Within any matrix, each row represents a different property; each column, as in any SAFES matrix, represents a structured phase.

Let us continue with the apple parenchyma. The physical properties (PP) matrix will have the structure shown in Figure 18.10. The number of rows, or properties listed, is as long as the number of properties considered of interest; therefore, for n properties, the PP matrix is represented as follows (some specific properties have been included for illustration).

An example of the development of a row of the PP matrix is shown next. The effective thermal conductivity of a porous product was modeled by Keey (1972) and has been proved to be useful for predicting the thermal conductivity of apple tissue (Martínez-Monzó et al. 2000):

$$\frac{1}{K} = \frac{1-a}{(1-\varepsilon) \cdot K_{\mathrm{M}} - \varepsilon \cdot K_{\mathrm{G}}} + a \left(\frac{1-\varepsilon}{K_{\mathrm{M}}} - \frac{\varepsilon}{K_{\mathrm{G}}} \right) \tag{18.1}$$

where

K is the effective thermal conductivity of the porous product

K_{M} is the thermal conductivity of the matrix (including the solid matrix [SM] and the liquid phase)

K_{G} is the thermal conductivity of gas filling the voids

ε is the product porosity

a is the distribution factor that quantifies perpendicular pore volume fraction of the sample

Phases	Apple Parenchyma M_4 (Physical Properties)				
Physical Properties	Solid Matrix	Intracellular Liquid	Extracellular Liquid	Gas	Whole Food
PP$_1$: Density					
PP$_2$: Thermal conductivity					
PP$_3$: Thermal diffusivity					
PP$_4$: Specific heat					
$\vdots$					
PP$_n$					

FIGURE 18.10 Matrix of PP (M_4) of apple parenchyma. PP$_i$ stands for the physical properties considered of interest for the analysis/process.

On the other hand, the equations proposed by Choi and Okos (1986) estimate the thermal conductivity of the food as an additive property, taking into account the content in water and solids. According to this, K_M may be estimated from the following equations, considering apple solids as carbohydrates:

$$K_M = x_w K_w + (1 - x_w) K_s \tag{18.2}$$

$$K_w (\text{W/m K}) = 0.59075 - 9.8601 \times 10^{-4} T \,(^\circ\text{C}) \tag{18.3}$$

$$K_s (\text{W/m K}) = 0.19306 - 8.4997 \times 10^{-4} T \,(^\circ\text{C}) \tag{18.4}$$

Since terms of Equation 18.2 are additive and function of mass fractions, the previous equations can be applied individually to each of the phases, excluding the gas one, which are the SM, the intracellular liquid phase (ILP), and the ELP. Therefore, the conductivity of these phases can be directly calculated by introducing on the properties matrix Equations 18.3 and 18.4 and multiplying by the corresponding mass fractions in the mass distribution matrix (M_1). Considering the apple tissue's composition distributed in Figure 18.7, and assuming a temperature of 25°C

$$K_{SM} = x_w^{SM} K_w + x_s^{SM} K_s = 0.097 K_w + 0.015 K_s = 0.061 \text{ W/m K} \tag{18.5}$$

$$K_{ILP} = x_w^{ILP} K_w + x_s^{ILP} K_s = 0.711 K_w + 0.123 K_s = 0.452 \text{ W/m K} \tag{18.6}$$

$$K_{ELP} = x_w^{ELP} K_w + x_s^{ELP} K_s = 0.046 K_w + 0.1008 K_s = 0.029 \text{ W/m K} \tag{18.7}$$

K_w and K_s are calculated from Equations 18.3 and 18.4.

The thermal conductivity of the matrix (K_M) needed to apply Equation 18.1 corresponds to the value obtained after adding the results of Equations 18.5 through 18.7. The value of the gas conductivity can be obtained from the literature $(K_G = 0.0295 \text{ W/m K})$ (Rha 1975, Martínez-Monzó et al. 2000). The porosity of the sample ($\varepsilon = 0.229$) was obtained from Betoret et al. (2007), as did for the data used for building the mass distribution matrix (M_1). Finally, considering that apples behave as a structure with pores predominantly distributed in parallel sense ($a = 0$) (Martínez-Monzó et al. 2000), Equation 18.1 can be written as a function of the thermal conductivity of the phases that constitute the food and a structural property ε, i.e., Equation 18.8. At the same time, the thermal conductivity of phases SM, ILP, and ELP is a function of mass fractions:

$$K = (1 - \varepsilon) \cdot (K_{SM} + K_{ILP} + K_{ELP}) - \varepsilon \cdot K_G \tag{18.8}$$

In this way, the thermal conductivity of apple tissue can be calculated as a function of the same property in each of the phases considered for the SAFES analysis (Figure 18.11). Once the full descriptive matrix is built, cells corresponding to the thermal conductivity of the SM, ILP, and ELP will call directly to mass fractions in M_1.

Likewise, the cell that yields the apple tissue thermal conductivity (K) will call for ε in the SP matrix, and for K_{SM}, K_{ILP}, and K_{ELP} in the thermal conductivity row. Applying the equations given above, the row corresponding to the thermal conductivity in the PP matrix (M_4) turns out as shown in Figure 18.12.

Another example of food property matrix is the nutritional properties (NP) matrix (M_5). Figure 18.13 shows the NP matrix for the apple parenchyma. As in the case of the PP matrix, some NP have been included for illustration, and two of them

Phases	Apple Parenchyma M_4 (Physical Properties)				
Physical Properties	Solid Matrix	Intracellular Liquid	Extracellular Liquid	Gas	Whole Food
Thermal conductivity	$=x_w^{SM}K_w+x_s^{SM}K_s$	$=x_w^{ILP}K_w+x_s^{ILP}K_s$	$=x_w^{ELP}K_w+x_s^{ELP}K_s$	$K_G=0.229$	$=(1-\varepsilon)\cdot(K_{SM}+K_{ILP}+K_{ELP})-\varepsilon\cdot K_G$

FIGURE 18.11 Matrix of PP (M_4) row that corresponds to the thermal conductivity. The thermal conductivity of the food is calculated as a function of the thermal conductivity of the four structural phases previously defined. Cells call for values in the mass distribution (M_1) and structural matrices (M_3) to calculate the whole food property.

Phases	Apple Parenchyma M_4 (Physical Properties)				
Physical Properties	Solid Matrix	Intracelluar Liquid	Extracelluar Liquid	Gas	Whole Food
Thermal conductivity (W/mK)	0.061	0.452	0.030	0.0295	0.412

FIGURE 18.12 Row of the physical matrix properties which corresponds to the thermal conductivity, showing the values obtained by means of the equations introduced in the cells (5–8).

Phases	Apple Parenchyma M_5 (Nutritional Properties)				
Nutritional Properties	Solid Matrix	Intracellular Liquid	Extracellular Liquid	Gas	Whole Food
NP_1: Energy (kcal/g)	0.0140	0.4760	0.0306	0	0.5206
NP_2: Nutritional composition	0.1120	0.8340	0.0540	0	1.0000
NP_{21}: Water	0.0970	0.7110	0.0460	0	0.8540
NP_{22}: Protien	0	0.0030	0.0002	0	0.0032
NP_{23}: Fat	0	0.0040	0.0002	0	0.0042
NP_{24}: Carbohydrates	0.0035	0.1070	0.0070	0	0.1175
NP_{25}: Organic acids	0	0.0060	0.0004	0	0.0064
NP_{26}: Fiber	0.0115	0	0	0	0.0115
NP_{27}: Minerals	0	0.0030	0.0002	0	0.0032
NP_3: Bioavailability	(*)	(*)	(*)	(*)	(*)
NP_4: Glycemic index	(*)	(*)	(*)	(*)	(*)
$\vdots$					
NP_n					

FIGURE 18.13 Nutritional properties matrix (M_5) for apple parenchyma, where NP_i stands for the nutritional properties considered of interest for the food/process description and (*) for not available.

(energy and nutritional composition) have been developed. The nutritional components considered are those of a typical nutrition database, i.e., USDA (2008), but in this case and for a better understanding a simplified version of the nutritional components is shown, grouping all the minerals in the same row and including vitamins in the organic acids and fat rows. Values of the different nutrients for the apple tissue have been obtained from Senser and Scherz (1991).

As observed in Figure 18.13, the nutritional components are distributed in the different structured phases considered in the apple tissue. Since data obtained from literature consider the apple tissue as a whole, and there are some differences on chemical composition between literature data (Senser and Scherz 1991) and Betoret et al. (2007) data due to natural variability of food composition, some hypothesis were taken into account to distribute the nutritional components in the structured phases:

1. Compositional ratios of the apple parenchyma can be considered equal for both apple varieties.
2. SM of fresh apple is made up of fiber, adsorbed water, and starch.
3. Proteins, fats, carbohydrates, organic acids, and minerals are dissolved or suspended in the intracellular and extracellular liquids at the same concentration.

Energy values were obtained from the well-known energy conversion factors of 4.0 kcal/g for protein, 9.0 kcal/g for fat, and 4.0 kcal/g for carbohydrates. These values are related conceptually to the metabolizable energy, since they are based on the heats of combustion of protein, fat, and carbohydrate with corrections for losses in digestion, absorption, and urinary excretion of urea (FAO 2003).

Although these energy conversion factors are a good approximation of the actual energy contribution to energy intake, the high diversity in composition, digestibility, and dietary proportion of each nutrient in each food product results in a persistent source of imprecision in the calculation of dietary energy that should be corrected (Ferrer-Lorente et al. 2007).

As suggested for PP, it is expected that the prediction of the energy values by means of the structured phases, taking also into account the food system structure, will lead to accurate models applicable to a wide range of food products. This may be also the case of other NP such as glycemic index, which shows significant variability among food products of similar composition but different structure (Riccardi et al. 2003). As observed in Figure 18.13, there are some rows where the information is not currently available; these could open future research areas that will contribute to fully develop the food descriptive matrix, and lead to a significant progress in the prediction of food properties.

18.2.3 Process Matrix

The descriptive matrix has been previously introduced as a tool that permits both describing the food system by recognizing its structural features, and relating food properties to food structure by identifying structure–property relationships. During processing, the food system undergoes changes that may affect any of its properties, since both its composition and its structure may be modified.

18.2.3.1 Definition of Stage of Changes

Processes are traditionally divided into unit operations (UOs) that usually adopt the name of the main macroscopic change undergone by the product (e.g., drying, evaporation, cooling, extraction, crystallization, freezing, etc.). However, this macroscopic change is in many cases the result of several important changes that take place in the food system, which may occur simultaneously or not. The SAFES methodology introduces here the term "stage of changes," which corresponds to the period of time between two "critical points," these referring to any specific moment at which a significant change is produced in the food. According to this definition, a UO might be divided into two or more stages of changes as long as the changes in the food system occur consecutively so that one ore more critical points are identified in the UO.

Examples of critical points that would lead to the division of a UO in stages of changes are

- Appearing/disappearing of a phase in the system; for instance, when in drying operations the loss of water promotes the appearance of a new solid phase as a consequence of soluble solid precipitation.
- Phase transitions such as starch gelatinization in baking or ice formation in freezing.
- When a change in mass and/or energy transport properties is produced, e.g., in a plant tissue when there is loss of cell compartmentation.
- When the pattern followed by chemical or microbiological reactions changes as a consequence of changes in temperature, pH, activation, or inactivation of enzymes, etc.

18.2.3.2 Matrix of Changes

In process engineering, mass and energy balances are a set of equations that allow calculating the changes undergone by the product, during a UO or a process, in terms of mass and energy. These relationships may be used for process control, design, and optimization. In the SAFES approach, it is the matrix of changes (MC) that collects this information; it is the mathematical tool that allows gathering in a single representation the changes that are traditionally calculated by means of mass and energy balances.

The MC has the task of representing the changes undergone by the food system during a stage of changes. It may be calculated by subtracting two consecutive mass distribution matrices (M_1), i.e., the matrices that represent the same food system at the beginning and end of the stage of changes that is being analyzed. However, this determination requires the basis of calculus of one of the matrices involved to be transformed, which leads to the definition of the "*transformed matrix*," i.e., a mass distribution matrix whose basis of calculus has been changed in order to calculate a MC.

In terms of notation, a transformed matrix can be distinguished from a mass distribution matrix by adding two more subindexes into brackets: the first one referring to the basis of calculus, and the second one to the stream in which the food product is being described. According to this, the mass distribution matrix for the stream that enters stage of changes n would be $M_{1(n-1, n-1)}$, the mass distribution matrix for the stream that exists stage of changes n would be $M_{1(n, n)}$, and the transformed matrix

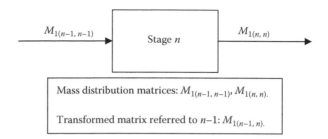

FIGURE 18.14 SAFES diagram representing stage n and streams that enter and leave the stage. Mass distribution descriptive matrices and MC are also represented.

that describes the stream that leaves n referred to the stream that enters n would be notated as $M_{1(n-1, n)}$. The MC that represents the changes undergone by the food during stage n would be $MC_{1(n-1, n)}$ and would be calculated as $MC_{1(n-1, n)} = M_{1(n-1, n)} - M_{1(n-1, n-1)}$ (Figure 18.14).

In a MC, the resulting values in the matrix cells may be major, minor, or equal to zero and indicate the changes that have occurred during the stage. Each cell differing from zero indicates a change in some component, phase, or state of aggregation during the stage of changes analyzed; a positive value indicates a gain and a negative a loss. Accordingly, a zero value indicates that no changes have occurred in this component, in this phase, at this specific aggregation state. Having defined the transformed matrices, it must be deduced that it is possible to refer the food system belonging in one point of the process to any other point of the process and not necessarily to the immediately previous. Accordingly, the MC may be calculated not only for one stage of changes, but also for two or more, or even for the whole process. Obviously, values of change resulting when comparing not consecutive points of the process lose a lot of significant information about what has really happened in each one of the stages. The arrangement of all the possible matrices (descriptive, transformed matrices, and matrices of changes) belonging to a transformation process configures the process matrix (PM).

The matrix of one process with N stages of change would be made up from $(N + 1)^2$ matrices distributed as follows:

$N + 1$	Descriptive matrices
$0.5[(N + 1)^2 - (N + 1)]$	Transformed matrices
$0.5[(N + 1)^2 - (N + 1)]$	Matrices of change

For example, in a process with four stages of change there are five streams of matter (m_0, m_1, ..., m_4). The description of the product of each stream may be represented by the corresponding descriptive matrices ($M_{0,0}$, $M_{1,1}$, ..., $M_{4,4}$). The PM, for this example, has been represented in Figure 18.15. The cells in the diagonal of this PM are occupied by the five descriptive matrices corresponding to the five process streams. The cells corresponding to the right-up side of the diagonal would be occupied by the transformed matrices and the left-down side of the diagonal by

$M_{0,0}$	$M_{0,1}$	$M_{0,2}$	$M_{0,3}$	$M_{0,4}$
$MC_{1,0}$	$M_{1,1}$	$M_{1,2}$	$M_{1,3}$	$M_{1,4}$
$MC_{2,0}$	$MC_{2,1}$	$M_{2,2}$	$M_{2,3}$	$M_{2,4}$
$MC_{3,0}$	$MC_{3,1}$	$MC_{3,2}$	$M_{3,3}$	$M_{3,4}$
$MC_{4,0}$	$MC_{4,1}$	$MC_{4,2}$	$MC_{4,3}$	$M_{4,4}$

FIGURE 18.15 PM of a process with four stages of change.

the matrices of change. Descriptive matrices and PM may be used both to analyze data obtained in experiments performed in laboratory or pilot plant and develop specific models or to predict them by using previously developed models. They are the major advantages provided by the use of this approach. The real innovation is that the structured phases are being modeled individually, and are linking them through their interfaces to predict the behavior of the overall material. This permits the use of advance knowledge, which is usually developed at the level of the structured phases. For example, it is possible to model thermodynamically a complex aqueous liquid phase with very powerful mathematical models as a function of temperature, pressure, and composition and deal with its interaction with a surrounding macromolecular matrix in which it is embedded.

18.3 CONCLUSIONS

The SAFES methodology has been presented here as a new tool for real food modeling and product/process design. Although the methodology has been tested previously in different works and modeling real food is now partially achievable thanks to the recent development of the related scientific areas, future developments which provide data required by the approach and open new research lines are also needed.

The current state of the art and its further development should be included in a friendly and powerful computer aided procedure. It will allow for the full implementation of the SAFES methodology, the integration and harmonization of data and models organized in a nexus to represent in a flexible but rigorous manner real complex foods.

The sophisticated description of the food material now achievable, in particular the structure modeling and the detailed spatial distribution of key components, will permit meaningful correlation work to be done on the safety, nutrition, and sensory attributes. The organization and integration of that knowledge through the SAFES methodology will allow for a significant step forward in the modeling of food products, processes, and process control.

ACKNOWLEDGEMENTS

The author Luis Mayor wishes to acknowledge his post doctoral grant from Fundación Cajamurcia, Spain.

NOTATIONS

a	distribution factor (Equation 18.1)
A	adsorbed
ELP	extracellular liquid phase
G	gaseous
ILP	intracellular liquid phase
IP	interfaces properties
K	crystalline
K	effective thermal conductivity (Equation 18.1; W/m K)
L	liquid
m	mass (kg)
M	descriptive matrix
MC	matrix of changes
NP	nutritional properties
P	pressure (N/m^2)
P	property
PP	physical properties
PM	process matrix
R	rubbery
SM	solid matrix
SP	structural properties
T	temperature (K)
UO	unit operation
V	volume (m^3)
V	vitreous
x	mass fraction (kg/kg)

Greek Symbols

ρ	density (kg/m^3)
ε	porosity (m^3/m^3)

Subscripts

ELP	extracellular liquid phase
G	gas
ILP	intracellular liquid phase
M	matrix
S	solids
SM	solid matrix
w	water

Superscripts

ELP   extracellular liquid phase
ILP   intracellular liquid phase
SM   solid matrix

REFERENCES

Aguilera, J.M. 2005. Why food microstructure? *Journal of Food Engineering* 67: 3–11.
Aguilera, J.M. 2006. Seligman lecture 2005—Food product engineering: Building the right structures. *Journal of the Science of Food and Agriculture* 86: 1147–1155.
Aguilera, J.M. and Lillford, P.J. 2007. *Food Materials Science: Principles and Practice*. New York: Springer.
Andres, A., Bilbao, C., and Fito, P. 2004. Drying kinetics of apple cylinders under combined hot air-microwave dehydration. *Journal of Food Engineering* 63: 71–78.
Argüelles, A., Castelló, M., Sanz, P., and Fito, P. 2007. Application of SAFES methodology in Manchego-type cheese manufacture. *Journal of Food Engineering* 83: 229–237.
Barat, J.M., Albors, A., Chiralt, A., and Fito, P. 1999. Equilibration of apple tissue in osmotic dehydration: Microstructural changes. *Drying Technology* 17: 1375–1386.
Barrera, C., Betoret, N., Heredia A., and Fito, P. 2007a. Application of SAFES (systematic approach to food engineering systems) methodology to apple candying. *Journal of Food Engineering* 83: 193–200.
Barrera, C., Chenoll, C., Andrés, A., and Fito, P. 2007b. Application of SAFES (systematic approach to food engineering systems) methodology to French fries manufacture. *Journal of Food Engineering* 83: 201–210.
Betoret, N., Andrés, A., Seguí, L., and Fito, P. 2007. Application of safes (systematic approach to food engineering systems) methodology to dehydration of apple by combined methods. *Journal of Food Engineering* 83: 186–192.
Campos-Mendiola, R., Hernández-Sánchez, H., Chanona-Pérez, J.J. et al. 2007. Non-isotropic shrinkage and interfaces during convective drying of potato slabs within the frame of the systematic approach to food engineering systems (SAFES) methodology. *Journal of Food Engineering* 83: 285–292.
Castelló, M.L., Fito, P.J., Argüelles, A., and Fito, P. 2007. Application of the SAFES (systematic approach to food engineering systems) methodology to strawberry freezing process. *Journal of Food Engineering* 83: 238–249.
Chenoll, C., Betoret, N., Fito, P.J., and Fito, P. 2007a. Application of the SAFES (systematic approach to food engineering systems) methodology to the sorption of water by salted proteins. *Journal of Food Engineering* 83: 250–257.
Chenoll, C., Heredia, A., Seguí, L., and Fito, P. 2007b. Application of the systematic approach to food engineering systems (SAFES) methodology to the salting and drying of a meat product: Tasajo. *Journal of Food Engineering* 83: 258–266.
Choi, Y. and Okos, M.R. 1986. Thermal properties of liquid foods. In *Physical and Chemical Properties of Food*, ed. M.R. Okos, pp. 35–77. St. Joseph, MI: American Society of Agricultural Engineers.
Dickinson, E. and Stainsby, G. 1982. *Colloids in Food*. London: Applied Science.
Elías-Serrano, R., Tellez Medina, D.I., Dorantes Álvarez, L. et al. 2007. Acid-salt conversion by means of electrodialysis: Application of the systematic approach to food engineering systems (SAFES) methodology. *Journal of Food Engineering* 83: 277–284.
European Technology Platform "Food for Life" 2005. *European Technology Platform "Food for Life": The Vision for 2020 and Beyond*. Available at http://etp.ciaa.be
Ferrer-Lorente, R., Fernandez-Lopez, J.A., and Alemany, M. 2007. Estimation of the metabolizable energy equivalence of dietary proteins. *European Journal of Nutrition* 46: 1–11.

Fito, P. 1994. Modelling of vacuum osmotic dehydration of food. *Journal of Food Engineering* 22: 313–328.

Fito, P., Chiralt, A., Betoret, N. et al. 2001. Vacuum impregnation and osmotic dehydration in matrix engineering: Application in functional fresh food development. *Journal of Food Engineering* 49: 175–183.

Fito, P., Le Maguer, M., Betoret, N., and Fito, P.J. 2007a. Advanced food process engineering to model real foods and processes: The "SAFES" methodology. *Journal of Food Engineering* 83: 173–185.

Fito, P.J., Castelló, M.L., Argüelles, A., and Fito, P. 2007b. Application of the SAFES (systematic approach to food engineering systems) methodology to roasted coffee process. *Journal of Food Engineering* 83: 211–218.

Food and Agriculture Organization of the United Nations (FAO). 2003. Food energy—methods of analysis and conversion factors, FAO Food and Nutrition Paper No. 77 (ISSN 0254–4725). Rome: FAO.

Heredia, A., Andrés, A., Betoret, N., and Fito, P. 2007. Application of the SAFES (systematic approach of food engineering systems) methodology to salting, drying and desalting of cod. *Journal of Food Engineering* 83: 267–276.

Keey, R.B. 1972. *Drying: Principles and Practice*. Oxford, U.K.: Pergamon Press.

Martínez-Monzó, J., Barat, J.M., González-Martínez, C., Chiralt, A., and Fito, P. 2000. Changes in thermal properties of apple due to vacuum impregnation. *Journal of Food Engineering* 43: 213–218.

Mebatsion, H.K., Verboven, P., Ho, A.T., Verlinden, B.E., and Nicolaï, B.M. 2008. Modelling fruit (micro) structures, why and how. *Trends in Food Science and Technology* 19: 59–66.

Norton, I., Moore, S., and Fryer, P. 2007. Understanding food structuring and breakdown: Engineering approaches to obesity. *Obesity Reviews* 8(Suppl. 1): 83–88.

Parada, J. and Aguilera, J.M. 2007. Food microstructure affects the bioavailability of several nutrients. *Journal of Food Science* 72: R21–R32.

Rao, M.A., Rizvi, S.S.H., and Datta, A.K. 2005. *Engineering Properties of Foods*. Boca Raton, FL: CRC Press.

Rha, C.K. 1975. Thermal properties of food materials. In *Theory, Determination and Control of Physical Properties of Food Materials*, ed. C.K. Rha, pp. 311–355. Dordrecht, the Netherlands: Riedel.

Riccardi, G., Clemente, G., and Giacco, R. 2003. Glycemic index of local foods and diets: The Mediterranean experience. *Nutrition Reviews* 61: 56–60.

Seguí, L., Barrera, C., Oliver, L., and Fito, P. 2007. Practical application of the SAFES (systematic approach to food engineering systems) methodology to the breadmaking process. *Journal of Food Engineering* 83: 219–228.

Senser, F. and Scherz, H. 1991. Der Kleine "Souci-Fachmann-Kraut", *Lebensmitteltabelle für die Praxis, 2*. Stuttgart, Germany: Auflage Wissenschaftliche Verlagsgesellschaft mbH.

Tremeac, B., Datta, A.K., Hayert, M., and Le-Bail, A. 2007. Thermal stresses during freezing of a two-layer food. *International Journal of Refrigeration* 30: 958–969.

United States Department of Agriculture (USDA). 2008. National Nutrient Database for Standard Reference. Available at http://www.nal.usda.gov/fnic/foodcomp/search

Yao, Z. and Le Maguer, M. 1996. Mathematical modeling and simulation of mass transfer in osmotic dehydration processes. Part I: Conceptual and mathematical model. *Journal of Food Engineering* 29: 349–360.

Yao, Z. and Le Maguer, M. 1997a. Mathematical modelling and simulation of mass transfer in osmotic dehydration processes. Part II: Simulation and verification. *Journal of Food Engineering* 32: 21–32.

Yao, Z. and Le Maguer, M. 1997b. Mathematical modelling and simulation of mass transfer in osmotic dehydration process. Part III: Parametric and sensitivity analysis. *Journal of Food Engineering* 32: 33–46.

19 Development and Implementation of Food Safety Programs in the Food Industry

Sergio Nieto-Montenegro and *Katherine L. Cason*

CONTENTS

19.1 INTRODUCTION

The distance from farm to table is currently longer and more complicated than ever. It includes many steps that were not present 50 years ago. This can introduce new and

---

* Written while working at Department of Food Science and Human Nutrition, Clemson University, Clemson, South Carolina.

more resistant foodborne pathogens into the food chain due to changing production methods, processes, practices, and habits (Kaferstein and Abdussalam 1999). Much of our food is not locally grown; it is now produced and shipped across the country by large corporations. This has changed the face of food safety, resulting in the need for adaptive changes to the current system to keep the food supply safe (Shank and Carson 1997). The previous chapters in this book have presented up-to-date information regarding science and product/process technologies that, when applied alone or in combination, help food companies to produce safe foods while maintaining desirable levels of quality. In addition to these technologies, food companies also are required to have programs in place to achieve the safety and quality of foods. Food quality and safety are issues that are closely related; and, in most instances, food safety management within a food company lies under the jurisdiction of the quality control team.

It is important to differentiate between food quality and food safety. A succinct definition of *quality* has been given by Juran: quality means "fitness for use," where fitness is defined by the customer (Wikipedia 2007). Meanwhile, food safety refers to the inability of a food to cause harm or illness when it is eaten. Quality programs are good for business and some companies receive premium prices for products with outstanding quality. Furthermore, food products that comply with specific quality parameters may experience increased sales, with quality even serving as an effective marketing tool. Food safety, as opposed to quality, is an attribute that must be ingrained within the food. Food companies should not use it as a marketing tool. It is the obligation of food companies to produce foods that do not cause harm to consumers. In most cases, failing to comply with a quality point does not harm the consumer. However, failing to comply with food safety standards can bring devastating consequences to consumers and the food company. Food safety parameters are nonnegotiable; and unfortunately, in most instances, there is no increase in sales or any premium price paid because a food is safe for consumption, or because a company has a food safety program in place.

Having a food safety program in place can help food companies to minimize the risk of product contamination. Developing and implementing a food safety program in the food industry can be demanding, since there are many challenges and barriers to overcome during this process. The challenges related to this task generally are closely related to the size and type of operation within the food company where these actions will take place. Other external factors such as regulatory environment and customer requirements could add to the complexity of a food safety program. There is no "one size fits all" food safety program. In order to succeed, each program needs to be custom designed for the operation where it will be implemented and is adequate to manage the risk that is being handled. To put this in perspective, the food safety program in a meat processing plant will differ greatly from the program designed for mushroom growers and packers, which will also differ from a program designed for the cheese industry or the food service industry. Each company must develop its own food safety program, according to its specific needs, while taking into consideration the specific challenges of its own operation and always following any governing laws that exist for that specific product or region. Some companies might be required to have a full Hazard Analysis and Critical Control Points (HACCP) program in place, whereas others only will be required to have a comprehensive program to cover good agricultural practices (GAPs) or good manufacturing practices (GMPs). There is a wealth of information about food safety programs, guidelines, and/or management systems, which could overwhelm the people responsible for developing and implementing a food safety program. It is not the intention of this chapter to detail the different food safety programs that are available. These programs have been widely discussed elsewhere (FDA 1998, NACMCF 1998, Brown 2000, Mortimore and Wallace 2001, Bihn and Gravani 2006, FDA 2006). However, not much has been published about the challenges and implications that exist when developing and implementing a food safety program. The objectives of this chapter are (1) to present general challenges and (2) to discuss practical solutions and generalities for developing and implementing general food safety programs within a food production facility.

19.2 WHAT IS A FOOD SAFETY PROGRAM?

A food safety program is a set of actions and systems designed to minimize the occurrence of food safety hazards within a food production, processing, and/or preparation facility. The program is documented through a written food safety plan that addresses all possible hazards, establishes control measures, and specifies corrective actions. Food safety standards may be established by governments, industry groups, and/or individual companies. A single activity or an isolated event, such as training employees in hygiene practices and hand washing, or keeping records of production logs, is not considered a food safety program. All the activities in a food safety program have the ultimate goal of preventing the physical, chemical, and/or biological contamination of a product, so the food will not cause harm or sickness to the consumers who eat it. It is important that the methodologies and procedures that are included in the development and implementation of the program, and that are aimed at preventing or minimizing the product contamination, are based on science and research.

19.3 WHAT TRIGGERS THE IMPLEMENTATION OF A FOOD SAFETY PROGRAM?

As stated above, the ultimate goal of developing and implementing a food safety program within a company is to prevent food from becoming contaminated or adulterated with physical, chemical, or biological contaminants. In reality, several forces individually or in combination drive the implementation of a food safety program within a company.

19.3.1 CUSTOMER REQUIREMENTS

This might be the driving force that motivates and even makes companies start thinking about having or keeping their food safety program in good shape. It is common got food companies to be asked or forced by their customers to have a full food safety program in place, if they want to keep selling products to them. Based upon customer requirements, companies might need to set up a supplier or vendor food safety program to protect their food products along the entire food production chain. In other words, some customers not only require food safety programs in the companies from which they buy directly, but also from these companies' suppliers. Some customers might even take things a step further, by auditing the supplier of a food company vendor. For instance, more often we are seeing that some produce growers must have a GAP's program in place, in order to sell their product to processors or packers. Some produce buyers such as retailers, grocery stores, or restaurant chains are asking their suppliers to insist that their vendors and suppliers have multiple certifications and complete extensive paperwork; and, in some instances, companies are required to go through all these steps just to be considered a potential supplier. Customers make sure that their supplier's food safety program is in place and is working properly through the use of their own auditors or by using third party auditing companies.

19.3.2 COMPETITION AND/OR INDUSTRY GROUP AGREEMENTS

Some other food companies decide to take a stance on food safety program development because their competitors are taking action related to food safety and they do not want to put themselves at a disadvantage. Recently, and in an effort to produce safer foods, industry trade groups—as for the green leafy vegetable, tomato, strawberry, and mushroom industries, among many others—are in the process of developing or have developed their own food safety guidelines, which would work as recommendations for food production. In addition, some of them have drafted these guidelines into marketing agreements that must be adopted by companies across the industry. Some of these guidelines have been developed by groups of people with very diverse backgrounds, including consultants, people in the industry, academia, and government. Among the advantages of these efforts are that these guidelines (1) are tailored to the industry, (2) take into consideration common practices within the specific industry, and (3) comply with food safety regulations already in place.

Small companies are the ones that most often struggle to implement a food safety program. Industry trade groups have played a key role in food safety program development and implementation, since these trade groups often have acted as a facilitator during the development process and have offered technical assistance to their affiliates. Small companies can benefit greatly from these guidelines and the services that are offered by trade groups.

19.3.3 COMPLIANCE WITH LOCAL, STATE, FEDERAL, OR INTERNATIONAL LAWS OR CODES

Foods that are introduced into the marketplace always must meet regulatory requirements; therefore, companies must adapt themselves to the regulatory environment in which they exist. Some regulations are tougher than others. There are examples of regulations that have mandated companies to adopt food safety programs. For instance, the meat and poultry industries were forced to adopt food safety programs in 1996, when the U.S. Food Safety Inspection Service (FSIS) established that food companies within this category must comply with certain food safety requirements that included but were not limited to the development and implementation of Sanitation Standard Operating

Procedures (SSOP), microbial testing programs during slaughtering, establishment of pathogen reduction standards, and the implementation of the HACCP program. Implementation of this rule was done gradually, based upon the size of the company, and it definitely forced everyone within that industry to become serious about food safety programs (Federal Register 1995, Federal Register 1996). Another example of companies being forced to develop and implement food safety programs was the 1995 U.S. Food and Drug Administration (FDA) rule that mandated the application of HACCP principles in the fish and seafood processing industry (Federal Register 1995). In 2001, another group of food companies was forced, by an FDA rule, to adopt a food safety program. This time, the government wanted to ensure the safe and sanitary processing of fruit and vegetable juices, and it mandated that companies adopt an HACCP plan at all their production facilities (Federal Register 2001).

Additionally, the global economy has compelled certain food companies to adopt or develop food safety programs to enter into new markets, so that they can expand and become more profitable than their current market will allow. If a food company wishes to sell its products within the international marketplace, it needs to comply with the local food safety requirements of each country to which its food products are being shipped.

These examples illustrate how the intervention of the government can force food companies to adopt food safety programs. It might appear that the hand of the government can push companies toward producing safe food for consumption; but this approach is subject to failure, since there is a need for these laws to be enforced. Having a food safety law in place does not guarantee that all food production companies will comply with it. Government agencies, even in the wealthiest countries, are underfunded and lack of human resources necessary to perform policing tasks. It is impossible to have inspectors in every place ensuring that all companies are doing what they ought to do. There will be instances in which companies will perform well when the inspector walks through the door, but return to old, less-safe habits as soon as the inspector leaves. Regulatory laws can make companies adopt a food safety program, but they cannot make sure that all companies run their programs adequately. Food companies should implement and develop food safety programs for their operations, because they are fully convinced that these are necessary to minimize the risk of product contamination and protect their customers, not because they fear the government or legal enforcement.

19.3.4 SELF-MOTIVATION BY PERCEIVED BENEFITS AND RISKS

Some companies become self-motivated to implement a food safety program, because they see that food safety programs can bring many benefits, such as improvements in organizational culture, worker morale and customer confidence, higher efficiency, less waste, and more control of their operation. In an ideal world, all food safety programs should have been initiated because of this way of thinking; but, unfortunately, this has not been the case. Perceived risk that the food produced by a company can be the source of an illness outbreak also can trigger the implementation of a food safety program at a given food production facility. This perceived risk could be generated from product microbial testing, problems with competitors' products, or even information from the media. Information flows extremely quick; news can be generated and spread within minutes, so that word of a tainted food product can spread immediately around the world. Food companies do not want to harm their consumers and become the next headline story in the media. In several instances news stories have forced companies to seriously start thinking about having a fully functional food safety program in place.

19.4   FINANCIAL IMPLICATIONS OF FOOD SAFETY PROGRAMS

Three words are often heard when food safety comes to the discussion table: time, human resources, and money. Ultimately, the word that creates the greatest reaction is money, and this can be translated into a barrier against the implementation of safer practices. Several authors have found that cost is a common barrier for program development across different industries (WHO 1999, Panisello and Quantick 2001, Bas et al. 2007). It is always possible to acquire additional human resources to run a food safety program, but this is going to cost the company money. Similarly, time spent working to develop or maintain a food safety program costs money. As mentioned at the beginning of this chapter, in most cases, having a food safety program in place does not allow companies to obtain a premium price for their products. Some food companies have tried to apply a food safety surcharge, but these attempts have failed.

Based upon the size and nature of a company, starting a program may require allocating resources to totally new areas within the company or to areas that have been forgotten. An example of this is formal training of human resources in food safety, an area that will be covered later in this chapter. Also, the allocation of human resources to run a food safety program can add to the list of costs related to food safety programs. However, it is advisable for companies to hire one or more people to run their food safety program and to enforce food safety rules. Marler (2007) recommends that food company CEO's invest money to hire qualified people to run their food safety programs; and also to listen carefully and try to understand what those they have hired for this purpose have to say. Ultimately, they are the experts and have a clearer idea of what is happening within the company, with respect to food safety. Celaya et al. (2007) also outlined the need for qualified technical employees to adequately implement HACCP programs. Large companies might have the means to allocate resources for program development and implementation; but resources can be a significant challenge to overcome for small companies. In some instances, these companies might be more interested in the overall efficiency of their operation than in making sure their food safety program is working. Large companies might encounter other problems not related to money, such as poor communication between their different departments. Protecting customers must be the number one priority when implementing food safety programs; but, unfortunately, some companies appear to believe that they are exempt from any illness outbreak. Regardless of the size of the company, it is important for food companies to think about the implications and costs associated with an outbreak or a product recall, and weigh them against the costs of implementing a food safety program. Being the cause of a food-borne outbreak can ruin not only the company that caused it, but it can also cause consumers to shun all similar products, even if they have been produced by other companies. Consequently, a single outbreak can affect entire clusters of industries. Several past examples illustrate how entire industries have been affected by outbreaks. For example, the spinach industry was severely affected by the 2006 outbreak, with estimates of losses as high as 100 million dollars; even by the end of 2007, spinach consumption had not reached preoutbreak levels. The green onion outbreak in 2003 helped to put the restaurant chain that served the green onions out of business, even though the green onions were not contaminated at the restaurant. To summarize, if a company wants to be competitive in the marketplace, it needs to invest in a food safety program and keep allocating resources to that program to make sure that it is functioning properly over the years.

19.5 PLANNING STEPS FOR THE DEVELOPMENT AND IMPLEMENTATION OF A FOOD SAFETY PROGRAM

Planning is very important for the development and implementation of a food safety program. The following seven planning steps lay out and condense the basic activities that are required to develop and implement a food safety program within a food production operation (Figure 19.1). These are general steps that can be further broken down into many different areas; nonetheless, they can serve as general guidelines for companies that are starting to develop and implement their food safety programs. Following these planning steps can help food companies to develop necessary concepts, think, and plan systematically. This can be a very effective and efficient tool for developing a successful food safety program, since the work will be easier to do, resources can be allocated more efficiently, and food safety-related work will

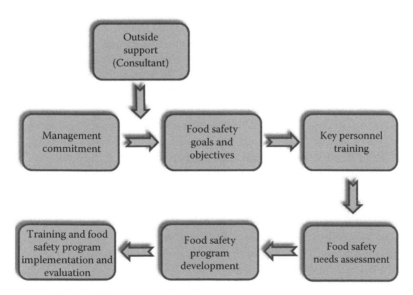

FIGURE 19.1 Food safety program planning steps.

be performed faster, since every step will be controlled. These steps also can help companies that already have a program in place to rethink and redesign their current food safety programs. The seven steps are

- Management commitment
- Outside support
- Food safety goals and objectives
- Key personnel training
- Food safety needs assessment
- Food safety program development
- Training and food safety program implementation and evaluation

19.5.1   Obtaining Management Commitment

Much has been said, both formally and informally, about the role that top management plays in food safety; and it could be said that management commitment is the single most important factor when implementing a food safety program. Regardless of what triggers the development and implementation of a food safety program, management's commitment to food safety is the first step to and the driving force behind developing and maintaining such a program (NACMCF 1998, WHO 1999, Panisello and Quantick 2001). This commitment is not only necessary at the top of the organization, but also at all management levels, from top to bottom. Management support increases the likelihood of success in any food safety program within an industry. Management commitment goes far beyond providing resources only to develop and implement the food safety program. Top management should work closely with the quality control team and listen to it during all phases of program development and implementation. Top management also should commit to aiding the quality control team in overcoming the day-to-day nuisances involved in program implementation. It should be aware of the role that it should play as leader in this endeavor; and, in some instances, such as in small companies, managers may become role models. Simple things, such as following the posted rules at all times, can make a big difference with respect to program success or failure. It is almost impossible to create an adequate organizational culture focused on food safety, if there is no true management support. Implementing a food safety program requires not only the support of top management, but also that all the people within the company understand and really make a serious long-term commitment to it. Food safety programs cannot become fully operational overnight. Getting everyone within the company to understand the implications of implementing a food safety program is not an easy task and takes time. A program cannot be created by a single person, a food safety team must be assembled, and this team must own the program. Management also will be in charge of helping to form this food safety team that will be responsible for undertaking the responsibility of development of the company's food safety program. Ideally, someone within the food safety team will be assigned the leadership role, in terms of program development and implementation. The leader must not only be accountable, but also must delegate roles to others and ensure that everyone who has been delegated responsibilities does what they ought to. Teams in small companies might consist of just two or three people. Large companies must consider carefully the size of the food safety team they want, in order to make it as efficient as possible. Preferably, the team will include an outsider (consultant), a quality control person, plus other qualified personnel. It is advisable that company management and the team meet on a regular basis to evaluate the program's progress.

19.5.2   OUTSIDE SUPPORT

Food safety programs must be built from within; no one knows the food operation of the company better than the people who work there. However, outside support can be of great benefit, especially to small operations that generally do not have adequate human resources or are very new in the food safety arena (WHO 1999, Mortimore 2001). Outside support, such as consultants, also can be of great help in obtaining management commitment. Small and large companies can benefit from outside support, since admitting an outsider with "new eyes" can help to provide a fresh perspective on operations. Additionally, the outside support can help to develop the plan and perform periodic examinations of the food safety program, so that emerging food safety problems can be addressed. It is strongly advised that companies obtain outside support. On the other hand, it is crucial that companies are very careful choosing which outside support they select. Trade groups, other food companies with which the company does business, and people in academia are great sources for finding qualified consultants. Food safety programs must be customized to address the specific risks and operation at which they will be implemented. A quick fix, such as the "one size fits all" programs offered by some consultants, will not work in the long run. It is important to be clear regarding the services that will be rendered and the deliverables that the food company will receive. Finally, it is important to note that some food companies might find easier and desirable to commission the full development of their food safety program to an outsider. This could be a recipe for failure, since the development and implementation of the food safety program must be a shared task and an effort between the outsider and the company's team. The food safety team and outside consultant(s) must work very closely together to make decisions. As noted above, outside support can be of great help and something of which companies can take advantage; but, ultimately, it is the responsibility of each company to make sure that its own food safety program is implemented properly and followed on a daily basis.

19.5.3   FOOD SAFETY GOALS AND OBJECTIVES

Companies must decide which program they need and be willing to develop and implement internally. Management must provide the necessary resources and work with the food safety team to identify overall short- and long-term food safety goals and objectives. These often serve as the guiding principles for everyone working toward food safety in the organization. Everyone on the food safety team needs to understand what these goals and objectives are. They also need to know what their role is in meeting these. In addition, it is important to define a realistic time line for accomplishing these goals. For instance, Panisello and Quantick (2001) have mentioned that HACPP has become the "exchange currency" of food safety programs. The full development and implementation of an HACCP program can take anywhere from 6 months to a full year, though this depends upon several environmental and production factors of that specific operation (Georgakopoulos 2006). One also must not forget that, if a company has or is in the process of developing and implementing an HACCP plan, there are other food safety programs, called "prerequisite program" that must be fully operational beforehand. Additionally, a company must define and document its own food safety policies and Standard Operation Procedures (SOP) which are specific to the company and that describe the activities necessary to complete food safety tasks. Prerequisite programs have been known for several years. They are also known as GAPs and/or current good manufacturing practices (cGMPs). The prerequisite programs include, but are not limited to: employee personal hygiene practices, employee training, plant premises, a pest control program, crisis management, equipment, cleaning and sanitation procedures, sanitary plant design, waste management, product recall, and trace back plans. These programs are the foundation of more complicated programs, like HACCP. Some companies might not want or be able to develop a full HACCP program; implementing a GAPs and/or a GMPs program then would be the furthest they might be able to go to minimize the risk of product contamination. Within the fresh-cut produce industry, some instances exist in which it is still not possible to develop and implement a full HACCP program, since data are insufficient to back up the program (De Roever 1999). Some companies have made the mistake of calling their food safety program an HACCP plan, where it clearly is nothing more than a collection of procedures related to food safety, without any unifying plan.

Some companies, especially large companies, want to take it a step further and manage their food safety programs as a single package. To achieve this, their objectives must be set to do so, something which can be accomplished by using a more comprehensive food safety management system. Such a system can be a valuable resource. Several examples exist, but two comprehensive programs that have been used are (1) The Safe Quality Food (SQF) program which combines three quality management systems, the CODEX, ISO, and HACCP systems and (2) ISO 22000, which harmonizes quality and safety issues within a food company (ISO 2005, FMI 2007). A number of organizations support these two management systems; but food companies first must understand that the key to the successful implementation of a food safety program is to develop a food safety foundation first, utilizing a more simplistic approach, before trying to adopt these innovative and effective management systems. Not doing so can result in the application of an incomplete and, therefore, ineffective food safety management system. Within this step, research must be done by the food safety team to find out which program suits the needs of the company, prior to establishing any goals or objectives or to making the decision to focus on the developmental and implementation steps of the program.

19.5.4 KEY PERSONNEL TRAINING

Although technology is needed to keep food safe and its quality desirable, the implementation of food safety programs does not rely on sophisticated technology, but on human resources. The World Health Organization (2000) has suggested that the education and training of food handlers is a good approach for preventing food-borne illnesses, since mishandling of food by workers has caused most outbreaks. Even within the food safety team, some employees might not be fully familiar with food safety terminology and the concepts related to food safety programs; and few, if any, would know how to start a food safety program from scratch. Consequently, it is vital that they be provided with the necessary knowledge and skills. It is advisable to start food safety training with the members of the team. Each member of the food safety team must be trained at least to the level of responsibility that they hold. Proper employee training is paramount if steps are to be accomplished correctly the first time. A study conducted by Sertkaya et al. (2006) explored the largest food safety problems within the industry in the United States. Using the Delphi technique and a panel of experts, the authors found that deficient training of employees was at the top of the list of 10 food safety problems faced by food manufacturers. The panel of experts in this study also concluded that training could help to solve some of the other problems included on this list. A lack of food safety training can lead to significant risks anywhere when food handlers lack adequate food safety skills (Seaman and Eves 2005). Implementing a food safety program in a company means that new behaviors and habits must be adopted. Food safety programs sometimes can be confusing. A given food safety program can easily overwhelm people within a company and, as a consequence, be implemented erroneously. Training key people in food safety and teaching them the new practices, operational procedures, and behaviors that are needed on a daily basis, is a necessary first step to prevent this. The nature of training differs relative to the responsibilities of the employee. It is important to take food safety training seriously, and to devote enough time and allocate enough resources to it. This will give the company a head start in food safety, after which, further food safety training can help to make the process of developing and implementing a food safety program that much smoother a process. Georgakopoulos (2006) mentioned that companies that invest in training their human resources can have a competitive advantage over companies that are less committed to this.

19.5.5 Food Safety Needs Assessment

A need can be defined as the gap between what is and what ought to be. A food safety needs assessment is a systematic approach to establishing a company's needs. A food safety needs assessment in a food company that is trying to develop and implement its own food safety plan might help it to identify where it realistically stands with respect to food safety, as well as the steps that must be taken to accomplish its ultimate food safety objectives. Developing a needs assessment is not as complicated as one might think. Surprisingly, some companies might find, through a needs assessment, that they already have several of the requirements of a food safety program in place. However, some of these activities that they have been doing over time might be scattered around the company, never having been written down formally or in a single place. A food safety program must have a clear, formal written plan.

The food safety team needs to know and understand specifically what needs to be accomplished. Furthermore, members need to know who is responsible for completing each task. Then, they must understand what everyone needs to know in order to successfully complete these food safety tasks. Finally, they need to compare where the company stands with respect to food safety, relative to the goals that have been set. The needs assessment is developed directly from this process, with the help of people who work at the company and perhaps as an outside consultant. As stated above, outside consultants can be an excellent resource for identifying the food safety programming needs of a company. Sometimes, a food safety team may think that it knows every single detail of its company's operation, regardless of its size. But, surprisingly, assessments often reveal specific issues and patterns that have been overlooked. The team must look to answer specific questions such as: What are the company's food safety priorities? What are the food safety requirements for the product that is being handled? Where does the company stand with regard to food safety? What already is available; for example, is a GAP or GMP program already in place? What are the resources available for developing and implementing the food safety program? What previously has been done? What food safety issues need to be addressed in the program? Introducing an outsider with fresh eyes to an operation also may help a company to discover all of the issues that can impact upon food safety issues, and help it to better understand what needs to be accomplished. A mock or internal food safety audit can be a powerful tool to help determining food safety needs. Methods for conducting a needs assessment include, but are not limited to: direct observations, a short survey, interviewing personnel in key positions, review the operation's records and logs, and a literature review. Combining methods is recommended. The data obtained will help the team to prioritize food safety needs and identify the costs of the food safety program. This information can also be used to develop a checklist of those steps that must be undertaken first. After developing a prioritized list of food safety needs and clearly outlining one's objectives, it is time to start developing the program.

19.5.6 FOOD SAFETY PROGRAM DEVELOPMENT

Within this step, the food safety team actually develops the plan. The format of a food safety program can vary greatly, depending upon diverse factors. However, it is important not to deviate too far from the initial objectives. Work closely with the chosen outside help, but remember that food safety programs must be developed from within.

The food safety team must write or redesign and test SOP, assign roles, and set up rules that need to be followed by all employees. Answering the following questions can help to develop the food safety SOP and some of the routines that are needed in the program:

Who does what, by when, and how?
   Who is the person responsible for doing a specific task (what)?
   The answer to the "when" question must clearly state when the task must be completed.
   "How" should provide details on how to perform each specific activity.

The food safety team also must develop an internal system to document the food safety program. Checklists to monitor the food safety training tasks must be developed and tested. If the chosen program calls for testing, the team must develop sampling plans and/or work with external laboratories to do so. This is a crucial step where the team and leaderships shows up. The success of any food safety program relies upon a coordinated effort by all members within the organization where it is being implemented (Bas et al. 2007).

19.5.7   TRAINING AND PROGRAM IMPLEMENTATION AND EVALUATION

Before starting the actual implementation of the food safety program, it is important to create the appropriate working environment and conditions in the operation where the program is going to be implemented. It is essential to provide the equipment and supplies that employees will need to complete the requested tasks. At this point the food safety team will be eager to start running the program; however, it is advisable to not start until everything within the company is ready and in place. Otherwise, the food safety program likely will fail. In order to realistically minimize the risk of product contamination through a food safety program, the program must be implemented properly and on a daily basis. The program must be sustainable over time. The food safety team also must consider how to cope with production constraints that can affect the program.

Within this step, training plays another important role, it is important to train all the employees within the company on the basics of the program, plus give additional training to employees who work directly with the product and the ones who would be directly responsible for documenting the program. Procedures that might seem familiar, logical, and easy to do for the food safety team may be very difficult for someone who is not familiar with the tasks. The food safety team must not take things for granted, and assume that people who will be running the food safety program on a daily basis will do everything they are told to do. Basic food safety training is needed to provide the specific skills to deal with the daily tasks of the program. Additionally, all food safety programs and third party audits call for line employee training in personal hygiene, plus other practices. This is another level of training, and it is different from the training received by the food safety team. This training might include teaching all the employees what they must do on a regular basis. The type of training must be chosen according to the program objectives set in the previous planning step, and the training must be based on skill provision and demonstration. Employees must be provided with the "why's" of these new practices. Food safety training programs can provide a wealth of knowledge and skills; but even the best training program cannot ensure that necessary behavioral changes will occur just because employees have had training. This is important to keep in mind when planning company training. Very little research has monitored food safety programs specifically to see if there have been significant changes in the long-term food safety behaviors of employees. Behavioral change has become one of the major challenges in the implementation of food safety programs. The team could have developed an excellent program but the people at the production floor can make it fail if there are no changes in behaviors and implementation of new practices. There is no evidence that food safety practices improve in direct relation to the acquisition of knowledge (Rennie 1994, Ehiri et al. 1997, Clayton et al. 2002). In order for food safety behaviors and practices to become habitual, enforcement and monitoring are essential and team must get involved in this task. It has been mentioned that management commitment is the key to the success of any food safety program. But it is difficult for managers and owners to make sure that the food safety practices learned during training sessions are followed on a daily basis. Research has shown that supervisors and middle management are critical to maintaining appropriate food safety behaviors at work, because they deal directly with the daily routines that occur on the farms or in the packing houses, and have experience in the complicated associated tasks. Therefore, training must be supplemented by supervisory enforcement of food safety rules. Management support of supervisors in this role will contribute to the success of the food safety program (Nieto-Montenegro et al. 2008). Companies must make sure that the program is working on a daily basis; otherwise, all the effort put into the previous steps will be a waste of time and resources. A system to constantly monitor the program is needed. Do not confuse this monitoring with the critical control points monitoring step; this step refers to monitoring the entire food safety program. Comprehensive and continuous internal and external audits can be the easiest ways to evaluate the effectiveness of a program, since they can verify that everything is working in the food safety program and can identify areas wherein modifications need to be made or corrective action taken. New employees must be trained in food safety as soon as they start working for the company. After the program has been put in place, the food safety team must be maintained, and must continue to work closely with management to analyze and make decisions about any changes to the program that might be needed over time.

19.6 CONCLUSIONS

Regardless of the food safety program that is being implemented, it is important that it is applied properly. Instances exist in which companies believe they have a food safety program in place, such as HACCP, when all they truly have is a filing cabinet filled with virtually meaningless paperwork. A false sense of security can be very dangerous. Food safety programs must be taken seriously at the food company level. Perceptions that such programs are nothing more than a waste of time and a pile of paperwork must be avoided. Having an incomplete or improperly working food safety program could result in your company being the center of the next food-borne outbreak on the news. Properly trained employees are crucial to succeed in program development and implementation of a food safety program. Applying food safety practices within a company requires hard work, from its planning, development, and implementation to its daily maintenance and monitoring. Ultimately, a food safety program's success is dependent upon management support, relevant employees having the technical knowledge to do things properly, and the commitment of all involved to keep it running properly on a daily basis.

REFERENCES

Bas, M., Yuksel, M., and Cavusoglu, T. 2007. Difficulties and barriers for the implementing of HACCP and food safety systems in food businesses in Turkey. *Food Control* 18: 124–130.

Bihn, E.A. and Gravani, R.B. 2006. Role of good agricultural practices in fruit and vegetable safety. In *Microbiology of Fresh Produce*, ed. K.R. Matthews, pp. 21–54. Washington, DC: ASM Press.

Brown, M.H. 2000. HACCP in the meat industry. In *Implementing HACCP in a Meat Plant*, ed. M. Brown, pp. 177–201. Boca Raton, FL, CRC Press.

Celaya, C., Zabala, S.M., Pérez, P. et al. 2007. The HACCP system implementation in small businesses of Madrid's community. *Food Control* 18: 1314–1321.

Clayton, D.A., Griffith, C.J., Price, P., and Peters, A.C. 2002. Food handlers' beliefs and self-reported practices. *International Journal of Environmental Health Research* 12: 25–29.

De Roever, C. 1999. Microbiological safety evaluations and recommendations on fresh produce. *Food Control* 9: 321–347.

Ehiri, J.E., Morris, G.P., and McEwen, J. 1997. Evaluation of a food hygiene training course in Scotland. *Food Control* 8: 137–147.

Federal Register. 1995. Procedures for the safe and sanitary processing and importing fish and fishery products. Final Rule. *Federal Register* 60: 65095–65202.

Federal Register. 1996. Pathogen reduction: Hazard analysis and critical control point systems (HACCP). Final Rule. *Federal Register* 61: 38805–38855.

Federal Register. 2001. Hazard analysis and critical control point (HACCP): Procedures for the safe and sanitary processing and importing of juice. Final Rule. *Federal Register* 66: 6137–6202.

Food and Drug Administration. 1998. Guide to minimize microbial food safety hazards for fresh fruits and vegetables. FDA/CFSAN web site: http://www.cfsan.fda.gov/~dms/prodglan.html

Food and Drug Administration. 2006. Current good manufacturing practice in manufacturing, packing, or holding human food. Web site: http://www.access.gpo.gov/nara/cfr/waisidx_06/21cfr110_06.html

Food Marketing Institute. 2007. The SQF program: A brief guide. The SQF Institute, Arlington, VA. Website: http://www.sqfi.com/

Georgakopoulos, V. 2006. Application of HACCP in small food businesses. In *Food Safety. A Practical and Case Study Approach*, eds. A. McElhatton and R.J. Marshall, pp. 239–252. New York: Springer.

International Standard Organization (ISO). 2005. ISO 22000: Food safety management systems—requirements for any organization in the food chain. ISO 22000:2005(E). Geneva, Switzerland: International Standard Organization.

Kaferstein, F. and Abdussalam, M. (1999). Food safety in the 21st century. *Bulletin of the World Health Organization* 77: 347–351.

Marler, W. 2007. Food safety and the CEO: Keys to bottom line success. *Food Safety Magazine* 13: 32–39.

Mortimore, S. 2001. How to make HACCP really work in practice. *Food Control* 12: 209–215.

Mortimore, S. and Wallace, C. 2001. *HACCP*. Oxford, U.K.: Blackwell Science.

National Advisory Committee on Microbiological Criteria for Foods (NACMCF). 1998. Hazard analysis and critical control point principles and application guidelines. *Journal of Food Protection* 61: 1246–1259.

Nieto-Montenegro, S., Brown, J.L., and LaBorde, L.F. 2008. Development and assessment of pilot food safety educational materials and training strategies for Hispanic workers in the mushroom industry using the Health Action Model. *Food Control* 19: 616–633,

Panisello, P.J. and Quantick, P.C. 2001. Technical barriers to hazard analysis critical control point (HACCP). *Food Control* 12: 165–173.

Rennie, M.D. 1994. Evaluation of food hygiene education. *British Food Journal* 96: 20–25.

Seaman, P. and Eves, A. 2005. The management of food safety—the role of food hygiene training in the U.K. service sector. *Hospitality Management* 25: 278–296.

Sertkaya, A., Berlind, A., Lange, R., and Zink, D.L. 2006. Top ten food safety problems in the United States food processing industry. *Food Protection Trends* 26: 310–315.

Shank, F.R. and Carson, K. (1997). Food safety: A U.S. perspective. *Food Science and Technology Today* 11: 218–224.

Wikipedia Contributors. 2007. *Quality* Wikipedia, The Free Encyclopedia, Web site: http://en.wikipedia.org/wiki/Quality

World Health Organization. 1999. *Strategies for Implementing HACCP in Small and/or Less Developed Businesses*. WHO/SDE/PHE/FOS/99.7. Geneva, Switzerland: World Health Organization.

World Health Organization. 2000. *Foodborne Disease: A Focus for Health Education*. Geneva, Switzerland: World Health Organization.

Index

T - #0593 - 071024 - C0 - 234/156/26 - PB - 9780367385118 - Gloss Lamination